Microbes and Non-Flowering Plants

Impact and Applications

Microbes and Non-Flowering Plants

Impact and Applications

Mamatha Rao

Department of Botany

Bangalore University, Bangalore,

INDIA

Ane Books Pvt. Ltd.

New Delhi ♦ Chennai

Microbes and Non-Flowering Plants : Impact and Applications

Mamatha Rao

Price: Rs. 1195/-
Pages : 578 (HB)
Size: Royal (6.25" x 9.5"
Paper Quality (GSM) : 70

Published by

Ane Books Pvt. Ltd.
4821, Parwana Bhawan, 1st Floor, 24 Ansari Road,
Darya Ganj, **New Delhi** - 110 002, India
Tel.: +91(011) 23276843-44, Fax: +91(011) 23276863
e-mail: kapoor@anebooks.com, Website: www.anebooks.com

Branch Office:

- Avantika Niwas, 1st Floor, 19 Doraiswamy Road, T. Nagar,
  Chennai - 600 017, Tel.: +91(044) 28141554, 28141209
  e-mail: anebooks_tn@airtelmail.in

ISBN : 978-81-8052-173-7

Printed at : R.K. Digital Process, New Delhi.

To

Prakruthi

With the able assistance of

Savitha Murthy

FOREWORD

Plants are essential for the survival of human beings and the planet Earth. While flowering plants are widely studied and their contribution to human welfare is well recognised, the contributions of non-flowering plants are relatively unknown.

The non-flowering plants are a heterogeneous assembly of the micro organisms (microbes) that may be viruses, bacteria, fungi, and the micro-algae on one hand and the macro-fungi, macro-algae, lichens, bryophytes, pteridophytes and the gymnosperms, on the other. The non-flowering plant groups inhabit very diverse habitats and exhibit an enormous diversity of physical form and structure. They exhibit both positive and negative impact on human welfare, and find applications in food, medicine and industry.

The microbes play amazing role in human welfare. Infectious diseases are caused by viruses, bacteria and fungi. Microbes provide us antibiotics and vaccines to protect ourselves from pathogens and diseases. Microbes are the tools and raw materials of biotechnology and of innumerable industries that provide us a wide array of bio-products including fermented foods and beverages. Our agriculture and the health of our environment are dependent upon the roles microbes play in biogeochemical cycles. Lichens, bryophytes and pteridophytes also have different uses. Gymnosperms largely cater to our timber needs.

The diversity and applications of microorganisms and non-flowering plants are largely unknown outside specialist groups of biologists. In the context of conserving our plant diversity and planning for a more effective and sustainable utilisation of our plant resources, a greater understanding of non-flowering plants and their current uses is very much needed. The present volume authored by Dr Mamtha Rao is a laudable effort in this direction. The book extensively covers very wide and diverse aspects of microorganisms and non-flowering plants, their impact and applications. Such a stock taking forms an essential basis for a better understanding of the significance of these life forms.

I congratulate Dr Mamatha Rao for her novel idea to synthesise existing information and knowledge of microorganisms and non-flowering plants for the benefit of the scientific community. I am optimistic that this book will stimulate interest in this field and pave the way for the compilation of utilitarian databases for future use. This book should find a place in all libraries to serve the needs of every category of stakeholders.

S Shantharam
President
Biologistics International, LLC
Ellicott City, MI 21042
USA

PREFACE

Scientists have always looked and searched for novel products from plants and their utilisation for human health and environment. The focus of search has always been the flowering plants, the non-flowering plants remained unexplored and unexploited. These plants with a tremendous potential as therapeutic agents should be tapped by future researchers in biology, chemistry and medicine.

This book is a modest effort to consolidate information that is wide spread in many diverse publications and to create awareness among the students, teachers, researchers, medical fraternity and the common man about the importance of non-flowering plants as potential source of medicine and research.

Non-flowering plant groups are a part of postgraduate curricula of Botany in many Indian Universities. The present book is an outcome of my postgraduate teaching experience of all the 10 groups of non flowering plants and my personal interest in plant based medicines for the health of common man.

The contents of the book have been divided into 24 chapters. Each chapter opens with a brief introduction. Diversity of non-flowering plants and their use as bioresources, biotechnological potential, biomolecules from these plants and natural products from endophytes used in medicine have been dealt in separate chapters. The microbes which include viruses, mycoplasmas, bacteria, yeasts and fungi and non-flowering groups such as algae, bryophytes, pteridophytes and gymnosperms were dealt separately, giving a concised account of their use as food supplements, medicine and in industries. The microbial products such as antibiotics, vaccines, toxins are added to give an insight into the application of microbial metabolites. Impact of viruses, mycoplasmas, bacteria, yeasts and fungi causing various diseases in plants, insects, birds, animals and humans have been discussed in detail. A short account of role of microbes in biogeochemical cycles and biodegradation has been added to emphasise the role of microbes in environmental pollution control.

In the present-day life of stress and emergence of new diseases, non-flowering plants might pave the way for finding an answer for simple, inexpensive and an alternative solution.

Compilation and writing of a book with diagrams and tables require technical expertise. I thankfully acknowledge the competence and ever-willing help of Mr Guruprasad V R in every stage of preparation of the text. He has been my strength in every step.

My special thanks to my teacher Prof C Kameswara Rao, former Chairperson, Department of Botany, Bangalore University, for teaching me the skill of scientific writing, for his support, guidance and critical analysis. I thank you Sir for everything.

The backbone of this book is Mr Sunil Saxena, Ane Books Pvt. Ltd. My heartfelt thanks for his encouragement and trust in my ability, for making my vision a reality. His support is the driving force behind this book.

The major part of a scientific book is its intricately drawn and correctly represented diagrams that give the right picture of the contents. I thank Mr. Sunil for lending his creativity to the botanical diagrams.

My heart felt thanks to Smt. Usha and Leelavathi, from Bangalore University library and Mr. Patan Shetty, Regional Institute of Ayurveda for their ever willing help.

Mr Hemanth for his encouragement and interest in the book which made me sustain the difficulties.

My thanks to Dr S Shantharam, Biologistics International, Ellicott City, USA for his interest in this work.

Mr Anil Kumar, Spandana Publishing House, Bangalore for initiating the conception of this book and Ms Shalini, B.U for the help rendered.

I thank all my teachers who taught me Botany.

Though I have tried my best to present the information pedantically, errors and omissions might have crept in. I accept them with humility. Comments, suggestions and healthy criticism are the stepping-stones for future improvement of a book and I welcome them.

Mamatha Rao

CONTENTS

CHAPTER 1

INTRODUCTION

1.1 BIODIVERSITY

The term 'biodiversity' is the short form of biological diversity, used to convey the total number, variety and variability of living organisms in a given area, whether on land, in fresh water bodies and the seas or the whole surface of the earth. Biodiversity embraces the whole of life and includes all the micro-organisms, plants and animals, on earth. Biodiversity also refers to the genes of organisms, the environment they live in and the products and services they offer to keep man well provided for all his needs of food, medicine, clothing and shelter and to keep the planet healthy to sustain life.

Biodiversity and related issues are now a serious global concern. An aware-ness of the issues involved is not really new, though the urgency for action was driven home in stages in the past three decades, the most important force being the Earth Summit (or UN Conference on Environment and Development) at Rio de Janeiro, June 1992, where the document 'Agenda 21', defined the programme for biodiversity estimation, conservation and sustainable use.

1.2 BIOLOGICAL RESOURCES AND BIOPROSPECTING

The term 'biological resources' denotes those components of biodiversity, which maintain the current or potential human uses.

'Bioprospecting' is the systematic scientific search for new chemical compounds in nature, for potential economic exploition. Searching for the organisms with easily visible form is not difficult, but the microscopic forms of bacteria, fungi and algae are often difficult and so various techniques have been developed. Bioprospecting in micro-organisms is important because they are the largest future source of various products and processes and are of critical importance to the sustainability of life in nature because they are involved in the recycling of elements.

The techniques of exploration for new organisms remained conventional but the isolation and characterization of chemical compounds have become increasingly sophisticated. Even environmental DNA is now being harnessed for commercial production of certain substances without the need to grow the actual micro-organism.

1.3 DUAL IMPACT OF ORGANISMS AND ENVIRONMENT

All organisms on earth live in a delicate, intricate and interactive balance among themselves and with the environment. All organisms in a given environment are interdependent and have a mutual impact among themselves and on the environment. Some of this impact from the organisms or the environment is beneficial and sustains healthy life, while some of the impact is detrimental to the organisms and/or the environment. Man has not really mastered handling the negative impact of other organisms, such as those that cause disease. He has also not learnt much about handling the impact of the environment on him and/or the impact of his actions on other organisms and the environment. He has not fully harnessed or even realized the potential of bioresources and the serious risk of indiscriminately exploiting them.

1.4 BIOPRODUCTS IN THE SERVICE OF MAN

Starting some 10,000 years ago, in the form of agriculture and animal husbandry, man has domesticated a number of species of plants and animals for his benefit. Man has also used plants and animals to resource a diverse variety of products needed for use in food, medicine, clothing, shelter and recreation. Modern science and technology have enabled finding newer applications of organisms and their products in the service of man. Such efforts have also led to huge business enterprises providing biosourced products and livelihood.

1.5 GROUPS OF LIVING ORGANISMS

Biologists have devised systems of classification for an orderly handling of millions of organisms that inhabit the earth. While even a common man can distinguish between a plant and an animal in most situations, and the several different kinds under each group, in certain other situations even biologists are at a loss in precisely ascribing a particular organism to the group of plants or animals.

Viruses, that cause several plant, animal and human diseases, are really not organisms. They become active in the environment of a host cell and outside this biological system they are mere chemical entities. For this reason their nomenclature does not involve the species concept, applied to the other groups of organisms.

Bacteria, fungi, algae, lichens, bryophytes, pteridophytes, and gymnosperms are different groups of plants. These groups are defined in the respective chapters and contexts in this book.

Often arguments are raised on whether bacteria and fungi are plants at all. Since certainly they are not animals, they tend to be regarded as plants for all general purposes. The terms 'micro-organisms' or 'microbes' generally refer to mycoplasms bacteria, fungi and some algae, that are very small in size, so require a microscope to study them, though some of them produce large visible structures.

The algae are green plants, some of which are microscopic, while some are large enough for the naked eye to locate them. The lichens are the first to colonize in any non-aquatic environment. They are not microscopic and are made of an algal component and a fungal component. Bryophytes are not microscopic, but are rather insignificant plants of moist habitats. The pteridophytes include the ferns, which are common in gardens. None of these produce seeds. All these plant groups together are often referred as the 'cryptogams' or 'lower plants'.

Gymnosperms produce seeds but without fruits. The group of plants that is most involved in our day-to-day experience is the angiosperms, which produce seeds enclosed in fruits. They produce flowers, small or big. While some botanists refer only the angiosperms as the flowering plants, some include the gymnosperms also in the ambit of this term. The term 'higher plants' refers to gymnosperms and angiosperms together.

In this book, the term microbes refers to the lower groups such as mycoplasma, viruses, bacteria, fungi and algae where as lichens, bryophytes, pteridophytes and gymnosperms are considered as non flowering plants.

1.6 THE NEED FOR A DATABASE OF PLANTS

There is a vast amount of diverse kinds of information on plants, spread over thousands of different kinds of publications. There is an urgent need to bring this massive information together in one place, in order to help the interest groups to evaluate and utilize the existing knowledge and use this as a basis for new knowledge and applications.

An effective utilization and conservation of natural resources requires systematically evaluated information on all aspects of the resources. This can be achieved only when all available information on the subject is organized in the form of 'Databases', which are an orderly and structured packets of information, facilitating storage, modification and retrieval of information, in any of several forms or combinations. Without the aid of computers, databases are difficult to both organize and operate.

The objectives behind the compilation of plant databases are:

a) It is of utmost importance to provide information, to a variety of users, on the botany, taxonomy and chemical constitution of economically useful species. Correct information on the identity of such species is of great significance in several issues such as conservation and Intellectual Property Rights.

b) The prime objective of database of plants is to provide all available information on them, in one place. This will bring to light the little known uses of plants. Such information will help to introduce new plant resources into other areas and also to bring to light new uses of existing species. These measures will help in reducing the pressure on the over stretched conventional plant resources.

c) Information on the chemistry, nutritional and therapeutic properties of plants helps us to choose foods rich in a particular compounds of nutritional value or those with an appropriate therapeutic use. Large sections of urban population spend millions of rupees on nutriceuticals, the dietary supplements, like vitamins, minerals (particularly iron and zinc), antioxidants, etc., most of which can be obtained from edible parts of plants easily and inexpensively. The public should be advised on food items that are rich in nutritional supplements.

d) Almost every plant has its own therapeutic properties, in addition to its other uses. A number of minor ailments have been managed for millennia, through using plants in folk and home medicine, in most parts of the world, more particularly in the developing countries. With acculturation and consequent neglect, this fund of knowledge is fast disappearing. Plant databases help to keep this information alive.

e) The industry is forever looking for new plant based chemicals, as structural modification of such compounds is far easier and cheaper than *de novo* design and total synthesis of compounds. Proteins, amino acids, oils and a variety of other chemical compounds are used in various industries. A database will highlight the sources of phytochemical compounds with economic potential and provides an easier access to information and source material.

1.7 THE NEED FOR A DATABASE OF NON-FLOWERING PLANTS

The microbes have attracted the attention of scientists from various disciplines all over the world all the time. A voluminous amount of information is available on their taxonomy, biology, chemistry and their application in medicine, food and industry, but this is spread in several thousand diverse kinds of publications. There is hardly any compiled information on the economic potential of viruses, mycoplasms, bacteria, algae and fungi.

There are about 30,000 species of non-flowering plants recorded from India. In addition to native species, several economically useful species have been introduced.

The enormous economic potential of a very large number of species of the non-flowering plants such as lichens, bryophytes, pteridophytes and gymnosperns has not been fully realized and much less so in India. This hampers a complete or sustainable utilization of these species. It is of utmost importance that all the available information is compiled and put in one place to facilitate an easy access to it. This will help in a periodical evaluation of the potential of the species involved. A database as the only means of achievng this objective.

In addition to published information, there is a lot more valuable unrecorded information on the uses of plants with people in the remote and tribal areas. This culture based information, constitutes Ethnobotany, Ethnomedicine and Ethnopharmacology (Kameswara Rao, 2000), will be lost forever, if not recorded the soonest.

There is an unlimited scope for bioprospecting diverse chemical compounds in non-flowering plants. For such efforts it is necessary to know what is already known, in order to save time and effort and to prevent duplication of efforts.

1.8 OBJECTIVE AND SCOPE OF THE PRESENT VOLUME

The major effort of this book is to:

a) Highlight information on i) the diversity of microbes and non-flowering plants ii) nutritive and therapeutic value of these plants in human health and welfare.
b) To provide a bird's eye view of the impact and economic potential of the microbes and non-flowering plants in various industries
c) To create awareness among different communities on the importance of non-flowering plant biorsources.

There is enormous amount of literature and innumerable number of diverse applications of the flowering plants. The non-flowering plants are no less important. They are the focus of this book as they are little known and much less exploited, compared to the angiosperms, the true flowering plants. The shift of focus to the non-flowering plants for their enormous potential as bioresources is long over due.

The present volume's an attempt to highlight the importance of the microbes and non flowering plants as food, medicine and industry.

The material in this book may not be everything that is available on the non-flowering plants. It is a significant attempt as an accessible nucleus of information that provides the foundation for further development of a utility centered database on this important group of plants.

CHAPTER 2

PLANT DIVERSITY, CONSERVATION AND SUSTAINABLE USE OF PLANT BIORESOURCES

2.1 INTRODUCTION

The concepts of biodiversity, bioresources and bioprospecting and their importance in harnessing the biological wealth for greater human welfare were introduced in Chapter 1. In order to be able to appreciate the significance of these concepts, it is necessary to understand them in a scientific perspective and in relation to the group of organisms in question.

Studies of the diversity of both microbes and higher plants are focused on one or more aspects of species diversity, genetic diversity and ecosystem diversity. For reasons inherent in their physical and chemical organisation, which determines their physiological role in the environment, microbes and higher plants vastly differ from each other in almost every respect. Consequently, studies of their diversity though based on similar criteria, require different methods.

2.1.1 Species Diversity

Species diversity represents the numbers and the degree of variability of families, genera and species.

The exact number of species of all living organisms is not known, even approximately. About 2,75,000 species of flowering plants are scientifically known and some estimates indicate that an equal number of species of flowering plants awaits scientific discovery. Wilson (1988) estimated that the number of species of all organisms falls between 5 and 30 million. At present, probably about 1.5 million species of all organisms have been scientifically described and named. A lot more species are yet to be known to science. Ignorance is thus the basic cause for the enormity and insolvability of several aspects of the species problem, much more so in the developing countries.

2.1.2 Genetic Diversity

Genetic diversity is the amount of genetic variability occurring within a species, which is the sum total of all genetic information contained in the genes of all individual organisms of that species. In nature, variability within a species is expressed in physical and visible characteristics (for example, several hundred varieties of rose) or in not so readily discernible features such as yield or chemical constitution. Some of this variability is due to the influence of environmental factors and some is based in the genetic constitution (genotype). The gene-based variability is very important in issues of biodiversity.

One complete set of genes of an organism is its 'genome'. In prokaryotic organisms which have simple cell structures without nuclei (such as bacteria) there is only one set of genes while in most others with complex cell differentiation (eukaryotes) there are two (in some more than two) sets of genes. The number of genes in the genome of an organism is widely variable. Some species of bacteria have about 1,000 genes, the yeast (*Saccharomyces cerevisiae*) has 6,100 genes and up to 40,000 genes are known in some flowering plants like rice. The number of genes in man was expected to be about over 1,30,000, involving about three billion nucleotides in 46 chromosomes. But now it is known that the number of genes in man is about 30,000 and would not be more than 40,000, even if future discoveries substantially add to the known number of genes.

The total genetic component of a species, represented by all of the individuals of that species in the entire world, constitutes its 'gene pool'. Virtually no two individuals of the same species are 100 per cent genetically identical. For this reason, it is possible to identify individuals by DNA finger printing, as for example, in paternity disputes or forensic issues.

The genetic variability of economically useful species/varieties and their wild relatives, together forms a continued basic supply of traits for breeding new and improved varieties and for this reason these are called the 'genetic resources' of a country/region.

In general, a large number of species of non-flowering plants in India require a thorough taxonomic handling a) for their accurate botanical description and taxonomic determination, b) to understand the magnitude of their diversity and usefulness, and c) to determine the need for and the extent of conservation.

2.2 MICROBIAL DIVERSITY

No life form exists in isolation from that of the microbial world. Plants, animals and human lives have microflora that form commensal, symbiotic, parasitic and pathogenic relationships. The beneficial or symbiotic relationships are ignored by us while pathogenic harmful relationships are highlighted. Micro-organisms associated with life forms are highly diverse and include Chlamydia, Archaea, Bacteria, Viruses, Mycoplasma, Algae and lower Fungi.

Microbial diversity is far greater by ecological, phenotypic and genetic measure than any other taxonomic group. Soils are the most complex of habitats within the biosphere containing estimated 10^{30} microbial cells. Understanding microbial diversity in terms of their abundance, number of species and community difference between different ecozones gives an insight to its depth as the ecosystem, and the value of microbial biotechnology.

The solar system is estimated to be 4.6 billion years old. The age of the earth is approximately 4 billion years estimated from the oldest known sedimentary rock (The Isua) at the west coast of Greenland. Evidence point to the occurrence of abundant microbial life at least 3.5 billion years ago (Lazcano and Miller, 1994; Ogunseitan 1994).

Nearly 80 percent of natural history of life on earth is exclusively comprised of microbial communities. Flint like siliceous rocks (Cherts) contain fossils of bacteria and cyanobacterial cells embedded in silica (Kasting, 2001). Stromatolites are fossilized microbial communities that form laminated structures and found in precambian carbonate sediments (600 million years) made of lime stone, dolomite or silicate.

The antiquity of microbial life involves the presence of biochemical compounds in the sedimentary rocks such as lipids, porphyrins and kerogens considered as reliable fingerprints of microbial groups.

2.2.1 Microbial Diversity and Environmental Evolution

Understanding the history of microbial life on earth can provide clues for the future of environmental evolutions. There are number of evidences that earth is a "microbial planet" in the sense that microorganisms predate other life forms both in terms of number and distribution. Microbial life has profound influence on the integrity and functioning of global ecosystems.

The global biodiversity assessment program suspects that more than one million species of prokaryotes exist in nature and only 5,000 have been described. Species and speciation are the fundamental units and the determinants of microbial diversity. Microbial speciation and diversity are directly relevant to finding out of origin of life and evolution of environment.

In the history and evolution of earth microorganism have been responsible for the global environmental change. Deeper insight into nature of speciation that generates molecular and physiological diversity among microorganisms is needed to facilitate the establishment of comprehensive inventories of microbial diversity.

2.2.2 Microbial Conservation

Human health related concern, expansion of biotechnology as an industry, and the production of new, novel therapeutics from microbes have provided the rationale for establishment of Microbial resource centre (MIRCEN) network by United Nations Educational, Social and Cultural Organization (UNESCO) which manages distribution, utilization, management of microbial gene pools, conservation of microorganisms in developing countries for promoting economic and environmental application of microbes.

The largest and best known microbial conservation centre is American Type Culture Collection (ATCC) with 18,000 strains of 750 genera of bacteria, 500 bacteriophages, and 3600 type cultures. The fungal collection includes 27,000 strains of fungi and yeasts distributed among 1,500 genera and 7,000 species, 2,000 genetic strains of *Saccharomyces cerevisiae* and other yeasts. The other collections include Chlamydias, Rickettsia and Viruses.

In addition microbial database (http://www.tigr.org/tdb/mdb/mdbcomplete.html) and genomes online Database (Gold) provides a World Wide Web resource.

2.3 HIGHER PLANT DIVERSITY

There is vast information on the diversity of higher plants. Pooled up global data have resulted in certain important concepts related to qualification, quantification and conservation of the diversity of the higher plants. Many of these concepts cannot be applied to the microbes. The more important issues, related to the diversity of higher plants on which there is a considerable international agreement, are discussed here.

2.3.1 Mega Biodiversity Centers

Mega biodiversity centers constitute amazing dimensions of diversity and are also areas that require a close scientific attention to find ways and means of sustainable utilization through conservation in the future interests of humanity. India is one among the 12 mega biodiversity countries recognized in the world. With unique biogeographic location, diversified climatic conditions and colossal ecodiversity and geodiversity, India harbours over six per cent of the world's known flora and fauna, in spite of having only two per cent of the total landmass of the world. India is very rich in terms of both species and genetic diversity. So far more than 1,27,000 species of plants, animals and micro-organisms have been reported from India.

2.3.2 Hotspots of Biodiversity

Hotspots are areas featuring exceptional concentrations of endemic species and experiencing exceptional loss of habitat (Myers, 1988; Myers *et al.*, 2000).

Endemic species (genera, families) are restricted in their geographical distribution to defined areas and do not occur outside these areas, in their native state. The endemic status of species of a particular area should be periodically reviewed, in order to be certain.

Now 25 hot spots are recognized in the world (Myers *et al.*, 2000), which harbour 44 per cent of all endemic species of plants (300,000) and 35 per cent of all endemic vertebrates. The Western Ghats along with Sri Lanka and the Eastern Himalayas and the Andaman-Nicobar islands, along with the Indo-Burma region, are recognized, as the two megadiversity hotspots. These two areas also figure in the list of eight hottest hotspots (Myers *et al.*, 2000) in the world.

The following 24 areas, often termed as the micro-hotspots, have also been identified in India: Andamans, Nicobars, Agasthyar hills, Annamalai hills, Nilgiri-Silent valley, Palni hills, Shimoga-Kanara, Mahabaleshwar, Konkan, Satpura ranges, Tirupathi hills, Visakhapatnam hills, Deccan hills, Chotanagpur plateau, Kutch, Aravalli hills, Khasia-Jainitia hills, Patkoi-Lushai hills, Cachar-Mikir hills, Arunachal Pradesh, Sikkim Himalayas, Garhwal Himalayas, Lahul Himalayas and Kashmir Himalayas

The endemism of Indian biodiversity is primarily confined in the North-East, Western Ghats, North-West Himalaya and the Andaman and Nicobar islands. About 30 per cent of the world's recorded flora is endemic to India, including about 200 species of Pteridphytes and 4,950 species of Angiosperms. About 3,500 endemic species are found in the Himalayas and adjoining regions and 1600 in the Western Ghats alone.

2.4 LOSS OF BIODIVERSITY

There has been a rapid decline in the biodiversity of the world, more particularly during the past two decades or so. Biodiversity losses have been alarming in the developing countries in the tropics. The underdeveloped countries are generally less aware of the degree of biodiversity loss in their countries and its consequences.

Biodiversity losses occur due to habitat destruction, over harvesting, pollution, inappropriate and often accidental introduction of exotic plants and animals, etc. Habitat destruction is often related to development projects like land conversion for agriculture, construction of dams and the like. Biodiversity is also lost due to sudden natural calamities like floods, cyclones, hurricanes and earthquakes. Conservation of biodiversity is one of the paramount concerns the world over. Governments, Non-Governmental Organizations (NGOs), scientists, are all preoccupied with the problem of devising ways and means of conserving biodiversity, or at least retarding the rapid rate of its loss.

2.5 CONSERVATION OF BIODIVERSITY

Conservation is the planned management of natural resources, to retain the natural balance, diversity and evolutionary change in the environment (Lincoln *et al.*, 1982). It is a protective measure taken a) to prevent the loss of genetic diversity of a species, b) to save a species from becoming extinct and c) to protect an ecosystem from damage so as to promote its sustained utilization.

The aims and objectives of biodiversity/species conservation programmes are:

a) to promote conservation of biodiversity in defined habitats and geographical areas;
b) to identify components of biodiversity in a region for conservation and sustainable use;
c) to promote *in situ* conservation and sustainable management of ecosystems, natural habitats and species populations;
d) to establish and formulate strategies for *ex situ* conservation for the components of biodiversity through an integrated scientific research;
e) to study the intrinsic and extrinsic factors causing the depletion of species/populations and to formulate and adopt scientifically oriented conservation strategies;
f) to identify biodiversity conservation regions and 'hot spot' areas (that require urgent measures) in the world for concentrated research and protection;
g) to promote research in traditional and ethnic knowledge areas on and utilization of, natural resources and to encourage equitable sharing of benefits arising out of utilization of such knowledge systems;
h) to promote research on wild relatives, land races and cultivars of cultivated species;
i) to maintain database systems on economically important groups of plant and animal species as a source information on them;
j) to evolve guidelines for the management of protected areas so as to enable the policy makers and managers to take effective measures to conserve biodiversity and species; and
k) to disseminate information on the importance of conservation and sustainable utilization of biodiversity, through popular and scientific publications and other educational programmes.

Since the criteria and action plans differ very widely, often within a particular country, depending upon the needs of a specific situation, no uniform approach is practicable. Biological conservation has its own impacts on environment and the society, which need to be addressed carefully before any programme is implemented (Balakrishna, 1999).

2.6 OPTIONS FOR BIODIVERSITY CONSERVATION

Options and priorities for biodiversity conservation have been discussed at length in different contexts in flora at different levels (Chauhan, 1993; Khoshoo, 1996).

The following options exist for the conservation of live material:

2.6.1 *Ex situ* Conservation

Complete organisms are conserved in botanical gardens, arboreta, zoological parks, zoos, etc. It is impossible to conserve this consciously or unconsciously selected diversity permanently, in this 'outside the site' conservation practice. Biobanks provide for the conservation of pollen, seeds, sperms, eggs, embryos, somatic cells, tissues, organs, etc., by various methods. *Ex situ* conservation results in the cessation of biological evolution and so generally provides for the conservation of only the selected genotypes of a given time.

2.6.2 *In situ* Conservation

Multi-use wild lands to conserve biodiversity in the form of biosphere reserves, national parks, sanctuaries, watersheds, protected landscape and ethnobio-logical reserves. *In situ* conservation, the 'on the site' practice, allows continued biological evolution and may influence changes in the component of biodiversity from time to time and hence it is conservation of an uncertain status.

2.7 SUSTAINABLE USE OF BIODIVERSITY

An often suggested alternative to development or conservation options, is sustainable use of biodiversity, which is hoped to provide for a balanced utilization and conservation of biodiversity. Sustainable utilization has not fully satisfied either the developmentalists or the conservationists.

2.8 INTERNATIONAL ORGANIZATIONS FOR BIODIVERSITY CONSERVATION

The following international organizations advise the Governments on different aspects of biodiversity and its conservation:

2.8.1 International Union for Conservation of Nature (IUCN) The World Conservation Union

The International Union for Conservation of Nature and Natural Resources (IUCN) is the most important world body that is concerned with the conservation of nature world wide.

2.8.2 Species Survival Commission (SSC)

The Species Survival Commission (SSC) is one of the six volunteer Commissions within the IUCN, with the mission to conserve biological diversity by developing and executing programmes to study, save, restore and manage wisely, the species and their habitats. SSC is the source of information for IUCN on the conservation of species.

2.8.3 Convention on Biological Diversity (CBD)

The Convention on Biological Diversity (CBD), aims to protect the world's biological resources from further erosion or at least to slow rate of erosion down. Till CBD came into force, living organisms were considered a common heritage of the humankind, but CBD accepts them as a sovereign property of the nation states. CBD is to promote conservation of biological diversity, a sustainable use of its components and equitable sharing of the resultant benefits. Thus there is a difference in the objectives of the IUCN and the CBD, though basically both strive to conserve the world's biological resources.

2.8.4 World Wildlife Fund (WWF)

The World Wildlife Fund (WWF) is the most publicized conservation organization with centres in almost every country. There is one in India, WWF-India. For most part, the activities of WWF were concerned with the megafauna, but in recent times WWF-India has now launched programmes for conservation of medicinal plants.

2.9 REGULATION OF BIODIVERSITY, ITS CONSERVATION AND INTERNATIONAL TRADE IN ENDANGERED SPECIES

2.9.1 Convention on International Trade in Endangered Species of Wild Flora and Fauna (CITES)

CITES has established the international legal framework for the prevention of trade in endangered species and for an effective regulation of trade in others. Member states respect the recommendations of CITES presented in CITES Appendices and implement restriction on the trade of the listed species.

CITES periodically reviews the situation of trade in threatened plants and animals and publishes lists of trade restricted species in three Appendices, the degree of restriction being most severe for those in the Appendix I.

Appendix I: *Cycas beddomei*

Appendix II: *Cyathea* species (tree ferns), *Taxus wallichiana*

Appendix III: *Cycas pectinata*

2.9.2 Trade Records Analysis of Fauna and Flora in Commerce (TRAFFIC)

TRAFFIC is the body that monitors the volume of trade in endangered species and works in co-ordination with CITES and SSC, to assess the impact of trade, the objective being to manage trade sustainably. CITES and TRAFFIC provide the means for an international control of trade in the listed organisms but trade in them within a country is the responsibility of the respective governments.

2.9.3 Acts of Governments

Every country, based on the recommendations of the international organizations concerned with biodiversity and conservation and its own scientific community, has enacted laws to protect their biological resources. The Government of India also has enacted a number of different laws to protect and conserve India's biodiversity and regulate the use of natural resources. The Forest Acts, Wildlife Protection Acts, Environment Protection Acts and the Biodiversity Bill, are some of the legal instruments in this regard.

2.10 AUTHENTIC REFERENCE DOCUMENTS ON BIODIVERSITY AND ITS CONSERVATION

2.10.1 Red List of Threatened Species

The Red Lists are the most comprehensive and authoritative global surveys of threatened species. They identify species that are most threatened, thereby serving as tools to help to set priorities for conservation action and providing baseline information for monitoring status of species.

The Red Lists provide a framework for more specific information about the species under threat, such as national or regional lists, conservation strategies, alert regions, nations and communities to species of international conservation concern. The Red Lists are the technical resource to international legal instruments and are frequently used to create and strengthen national species protection Laws. The IUCN separately publishes Red Lists of Threatened Animals and Plants to publicise information on the conservation status of species. Periodical revisions of these lists are issued from time to time. The first ever IUCN Red List of Threatened Plants of 1997, containing 33,798 species names in the threatened category, was released in April 1998. (Sowdowski, 1998). The listed species belong to 369 plant families and 200 countries. They constitute 12.5 per cent of known species of vascular plants. The publication of this list is a turning point in plant conservation.

A Red List of Threatened Vascular Plant Species in India has been compiled from the IUCN Red List of plants (Kameswara Rao *et al.*, 2003) and this should be the starting point for updating the threat status of species in India. This list includes the pteridophytes and gymnosperms (Tables 2.1 and 2.2) but not the lower groups of plants, for which no such lists are readily available.

Table 2.1 Family wise Numerical Distribution of Threatened Genera and Species of Pteridophytes and Gymnosperms in India.

Family	Genera	Species
PTERIDOPHYTES		
Adiantaceae	1	1
Cyatheaceae	1	2
Dennstaedtiaceae	1	2
Dryopteridaceae	3	3
Gleichiniaceae	1	1
Isoetaceae	1	3
Lomariopsidaceae	1	3
Marratiaceae	1	1
Polypodiaceae	3	3
Selaginellaceae	1	2
Thelypteridaceae	7	9
Woodsiaceae	2	3
GYMNOSPERMS		
Cephalotaxaceae	1	2
Cyacadaceae	1	1
Pinaceae	2	3
Taxaceae	1	1

Source: Kameswara Rao *et al.*, 2003.

Table 2.2 Indian Pteridophytes and Gymnosperms in the IUCN Red List of Threatened Species

PTERIDOPHYTES	
Adiantaceae	
Coniogramme indica	R
Cyatheaceae	
Cyathea albosetacia	V
Cyathea nilgirensis	E

Dennstaedtiaceae	
Lindsaea malabarica	R
Dryopteridaceae	
Dryopteris gamblei	V
Lastreopsis wattii	EX
Polystichum wattii	EX/E
Gleichiniaceae	
Dicranopteris linearis	V
Isoetaceae	
Isoetes bilaspurensis	R
Isoetes dixitii	EX
Isoetes sampathkumarnii	EX
Lomariopsidaceae	
Elaphoglossum beddomei	R
Elaphoglossum nilgiricum	E
Elaphoglossum stigmatolepis	V
Marratiaceae	
Christensenia assamica	V
Polypodiaceae	
Dendroglossa minutula	E
Drynaria meeboldii	V
Thelypteridaceae	
Cristella clarkei	V
Cristella kaumaunica	V
Cyclograma squamaestipes	R
Oriopteris elwisii	R
Pseudocyclosorus gamblei	E
Pseudocyclosorus griseus	E
Stenogramma himalaica	V
Woodsiaceae	
Athyrium attratum	E
Athyrium duthei	R
Woodsia andersonii	EX/E
GYMNOSPERMS	
Cycadacese	
Cycas bedomeii	E
Cephalotaxaceae	
Cephalotaxus mannii	V
Cephalotaxus oliveri	V

Pinaceae	
Picea brachytila	V
Picea spinulosa	R
Pinus gerardiana	R
Taxaceae	
Amentotaxus assamica	V

Source: Kameswara Rao *et al.*, 2003.

Abbreviations: E: Endangered, **EX:** Extinct, **R:** Rare, **V:** Vulnerable

2.10.2 Red Data Books

Red Data Book of the IUCN provide in-depth analyses of the conservation status of particular groups of species. Like several other countries in the world, Red Data Books have been published from India. Unfortunately, for plants the lists of threatened species have been based on unscientific criteria.

Consequently we need to revise the Red Data Books of Plants in India on scientific methods. This work is now being carried out at the Indira Gandhi Conservation Monitoring Centre associated to the WWF-India, at New Delhi.

2.11 SYSTEM OF CATEGORIES OF THREATENED SPECIES

Species are assigned to different categories of threat basing on the degree of threat of extinction they face. Over a period of time a certain amount of confusion has crept in, in the use of the terms. The terms 'Endangered' and 'Threatened' have come to be used interchangeably and along with the term 'Extinct' have been used unscientifically, creating a lot of confusion. Aware of this situation, the SSC has reviewed the terminology and published the IUCN Red List Categories, and guidelines to determine them, which is mandatory (Anonymous, 1994). Species (or subspecific taxa) facing any threat of extinction should be assigned to one of the several categories on the basis of detailed and specific criteria, given by the IUCN (Anonymous, 1994). The 1994 version of the structure of categories of threat to plants was reviewed by the SSC Criteria Review Working Group (Anonymous, 1999).

Assigning threat status to species or their sub-categories should be based on quantitative scientific data. For the most part, such a procedure was not adopted in the developing countries. While the pteridophytes and gymnosperms are considered in the Red Lists and Red Data Books, as they are vascular plants, the groups of plants down from bryophytes have no such detailed scientific evaluation and this is the most important laccuna that has to be filled urgently.

CHAPTER 3

BIOMOLECULES FROM NON-FLOWERING PLANTS

3.1 INTRODUCTION

Organisms should be viewed as composed of systematically ordained packets of thousands of chemical compounds, in a complex, interrelated and interdependent milieu. How the organisms acquire their physical form and how they react and respond to beneficial and antagonistic internal and external factors are the consequence of chemical interactions. The physiological functions, such as nutrition, growth, reproduction and survival, are all manifestations of chemical interactions. So are the physiological disorders and diseases caused by pests and pathogens, as well as the cures and recovery. All these responses and interactions depend upon the genotype (genetic constitution) of the organisms, the environment in which they live and the interaction of the genotype and the environment.

An understanding of the chemistry of organisms is required in order to appreciate the potential of harm we suffer and the benefits we derive from organisms.

3.2 BIOMOLECULES

Organisms contain several thousands of organic and inorganic compounds belonging to almost every chemical group, now generally referred as biomolecules. All organisms contain certain compounds that are beneficial or anti-nutritional or even toxic, depending upon several factors. With time, man has learnt to put many of these compounds to several uses in nutrition, medicine and industry. In spite of two to three centuries of such effort, we have not yet realized the full economic potential of plant chemicals, more particularly those in non-flowering plants. We need to study and understand the detailed chemistry of our biological resources in order to derive the full benefit from them.

The species we use as food are the sources of major nutrients such as carbohydrates, proteins and fats, but such compounds are present even in organisms not consumed by us. In addition, we need several hundreds of other compounds present in plants in smaller quantities and these chemicals play very crucial roles in our metabolism and health. During the process of domestication of plants and animals, our ancestors have painstakingly and cleverly chosen species and their varieties that contained more of beneficial chemical compounds and less of hazardous compounds, in an effort spanning over 10,000 years of the history of agriculture and plant use. Our understanding of the chemistry of edible plant varieties will help us in choosing the right food and feed more judiciously, avoiding the hazardous chemicals. We can use the toxic chemicals to develop drugs to cure our disorders and diseases. And we need raw material from plants for our industries.

3.3 BIOINORGANIC COMPOUNDS

In addition to thousands of organic compounds that basically contain carbon and hydrogen, organisms also contain several elements and a large number of inorganic compounds. This area of biological chemistry is now rapidly developing into 'bioinorganic chemistry'. Organisms absorb and accumulate a number of elements and inorganic compounds, from water and soil.

Elements and inorganic compounds may also influence biological processes either in a beneficial or a hazardous manner (bioactivity). Mineral deficiencies cause disorders in the plants themselves and also in our body system. Arsenic and mercury are cumulative poisons, used as medicine in small doses. Selenium absorbed by plants from the soil is a toxicant.

3.4 NATURAL PRODUCTS

Metabolism is the sum total of all chemical interactions in an organism. The products of metabolism are metabolites, which in turn may be the players in other metabolic processes. Certain compounds with high molecular weight, such as the deoxyribonucleic acid (DNA), ribonucleic acid (RNA) and proteins (that include thousands of enzymes that are essential to every other metabolic function and lectins with diverse properties and roles), are considered as the primary metabolites. Other compounds, most of which are of low molecular weights, are secondary metabolites.

The secondary metabolites do not usually re-enter the metabolic processes. These are called the Natural Products that either accumulate in certain tissues/organs or excreted (in animals) which are million in number. Organisms are virtually highly systematically organized chemical factories precisely turning out an enormous number of chemical compounds in varying quantities, in a short time and at ambient temperatures.

The majority of compounds occurring in the biological systems are endlessly in a metabolic flux, due to continuous interactions, fresh synthesis, utilization or inter-conversion of compounds or some of them going into the storage sites. The nucleotides, amino acids, carbohydrates, carotenoids and organic acids occur in pools with continuous additions and depletions. Against this dynamic state, it is an intellectual and technical feat to extract, isolate, characterize and identify such a large number of chemical compounds from organisms.

While it is not always possible to explain the exact individual role of each of the large number of secondary metabolites in plants, many of them have been shown to be the chemical defense against pests, pathogens and grazers, the common sources of biotic stress. Some compounds accumulate in response to abiotic stress, such as high temperature, drought or salinity.

ADVANTAGES OF NATURAL PRODUCTS

- Natural products offer unmatched chemical diversity with structural complexity and biological potency (Verdine, 1996).
- Natural products have been selected by nature for specific biological interactions. They have evolved to bind to proteins and have drug-like properties.
- Natural products are a main source of pharmacophores. Drugs such as cyclosporin A and FK-506 are not only active as immunosuppressants but also as antiviral, antifungal and antiparasitic agents.
- Natural product resources are largely unexplored. Novel discovery strategies will lead to novel bioactive compounds. Natural product extracts are complementary to synthetic and combinatorial libraries. About 40 percent of natural-product diversity is not represented in synthetic compounds libraries.
- Natural products can guide the design of synthetic compounds (Breinbaver *et al.*, 2002).
- Research on natural products has led to the discovery of novel mechanisms of action – for example, those of immunosuppressants and guggulsterone.
- Natural products are powerful biochemical tools; they serve as "pathfinders" for molecular biology and chemistry, and in the investigations of cellular functions.

3.5 BIOSYNTHESIS, SEMISYNTHESIS, STRUCTURAL MODIFICATION AND ARTIFICIAL SYNTHESIS

The natural synthesis of different chemical compounds in biological systems utilizing the elements and/or other simpler chemical compounds in the environment of the cell is biosynthesis, also called *in vivo* (in the living organism) synthesis. The complicated routes of biosynthesis of different compounds are called biosynthetic pathways, which virtually constitute a maze. Semi-synthesis is synthesizing a biological compound in the laboratory, using naturally occurring precursor compounds. In certain situations biochemists modify a part of the structure of a naturally occurring compound in order to produce a more effective or more useful or less expensive product. This is structural modification. When compounds are synthesized totally artificially in the laboratory, starting from elements and/or simple artificially synthesized compounds, it is artificial synthesis or *in vitro* (literally, in glass the test tube) synthesis.

With the availability of extensive computer databases on the physico-chemical characteristics of chemical compounds and their functional properties, it is now possible to design a compound with desired functions, using procedures of bioinformatics, before one attempts to actually synthesize it in the laboratory. This bypasses the conventional route that involves a lot of time and money with the results being uncertain. This computer based procedure, described as *in silico* (because computer functions are based on silicon chips), is the modern tool for the synthesis of new compounds, particularly drugs, usually based on natural models, in a very short time and is far less expensive . The *in vitro* synthesis follows the *in silico* procedure. In actual practice, all these procedures come in handy, in one or the other situation, since there is no single universally convenient or economical method of designing and/or synthesizing biomolecules.

3.6 BIOACTIVITY OF METABOLITES

Many of the primary metabolites like proteins and a very large number of secondary metabolites, either produced within an organism (endogenous) or introduced from outside sources (exogenous), induce responses in biological systems, called bioactivity. Bioactivity can be either beneficial or hazardous to organisms. We take advantage of the bioactivity of such compounds and use them in various ways.

The beneficial compounds function as precursors of other metabolically significant compounds, micronutrients, growth promoters, antioxidants, pharmaceuticals, nutriceuticals, and agriceuticals and/or participate in biosynthetic processes. In recent times, antioxidants have gained a lot of importance as they scavenge free radicals in our system. Free radicals form routinely during metabolic processes and are implicated in several diseases such as heart attack and cancer.

The hazardous compounds may function as anti-nutritional factors limiting the benefits we derive from food. Some others, called toxicants or toxins, interfere with metabolic processes by inhibiting enzymes or act in some other way to disrupt metabolic processes, causing minor to major disruption of physiological processes, or organic damage, leading to several disorders, or even death.

A very large number of bioactive compounds are beneficial at lower doses or concentrations, and toxic at higher levels. This factor is in fact crucial in the development of drugs. Many of the toxicants are useful as drugs, because usually the pathogens and the diseased cells/tissues are more sensitive to them than the healthy ones. Depending upon their effects, toxicants are also used to control pathogens, deranging their metabolism in some way, without affecting the host's metabolic processes.

3.7 CLASSIFICATION OF CHEMICAL COMPOUNDS IN ORGANISMS

In order to clearly understand different chemical compounds and their relationships, it is necessary to classify them systematically. A classification facilitates an orderly handling, understanding and conveying information. Classification is particularly helpful in dealing with the enormous number of secondary metabolites. Classification of the multitude of biomolecules is a hard task.

There are several different ways of classifying biomolecular compounds, but none is totally satisfactory or useful in all contexts. The complexity of the synthesis of thousands of chemical compounds via inter-dependent and inter-connected routes and the overlap of their physico-chemical properties make it difficult to arrive at a universally acceptable classification. With the result, the classification of biological chemicals is need based and depends upon the context.

The nucleic acids are polymers of nucleotides. Similarly, the proteins are polymers of amino acids, carbohydrates are polymers of monosaccharides (simple sugars) and lipids are polymers of fatty acids. The secondary metabolites are more difficult to handle.

The non-flowering plants contain an enormous number of chemical compounds belonging to almost every chemical group. In this book, a general-purpose classification of secondary metabolites used by plant chemists, is adopted as outlined in Box 3.1.

Box 3.1 Outline of the Classification of Secondary Metabolites in Non-flowering Plants

The following outline of classification of chemical compounds excludes the nucleic acids, proteins, carbohydrates and lipids. It mainly deals with the secondary metabolites.

A. Nitrogenous compounds (amino acids, amines, alkaloids and glycosides).
B. Phenolic compounds (phenolic glycosides, flavonoids, tannins, quinines, chromones, phenyl propanoids and anthrones)
C. Terpenoids (mono, di and sesquiterpenoids).
D. Organic acids
E. Elements and inorganic compounds

CHAPTER 4

VIRUSES

4.1 INTRODUCTION

Viruses are tiny particles, which cannot be seen under the optical microscopes. They cause several plant, animal and human diseases. They are strictly parasitic and cannot be cultivated away from the living cells. Outside a biological system, viruses are mere chemical entities. They are host specific and contagious, and transmitted by air, water, and soil or even food. Antibiotics cannot cure a viral infection.

Biochemical, electron microscopic and X-ray crystallographic studies are required to understand the structure of virus. The Tobacco mosaic virus (TMV) was the first to be studied in detail.

Viruses are either rod shaped or polyhedral or variants of these two basic structures, such as cubical or tadpole shaped. Usually plant and animal viruses are non motile while the bacterial viruses are motile. A virus consists of a coat of protein called capsid composed of capsomeres, enclosing the core of nucleic acid, which may be DNA or RNA. On the basis of hosts and the nucleic acid component, viruses have been divided into plant viruses (RNA only), vertebrate viruses (DNA or RNA), insect viruses (DNA or RNA), mycoplasma viruses, bacterial viruses or bacteriophages (DNA or RNA), cyanophages infecting cyanobacteria (DNA only) and fungal viruses (RNA). The nucleic acid in the virus may be either linear or circular. DNA of most animal viruses is linear double stranded DNA (dsDNA) or single stranded DNA (ssDNA), except papova virus where the DNA is super coiled circular dsDNA. The RNA in the animal viruses is a linear double or single stranded molecule.

Illustrations of representative species of viruses are given in Figures 4.1 to 4.6 to provide an overview of their appearance.

4.2 VIRAL INFECTIONS OF MICROBES

Viruses infect different micro-organisms such as mycoplasmas, bacteria, fungi and algae. The following is the list of various viruses infecting microbes.

Mycoplasma Viruses

The viruses that infect *Mycoplasma* are grouped into filamentous, isometric , enveloped and short tailed viruses.

Acholeplasma Viruses

L1, L2, L3 and L172 were isolated from *Acholeplasma laidlawii* strains. M1 virus was isolated from *Acholeplamsa modicum*. O1 virus was isolated from *Acholeplasma oculii*.

Spiroplasma Viruses

Virus like particles observed during characterization of *Spiroplasma*, were designated as C1, C2, C3 and SPV4.

Bacterial Viruses (Bacteriophages)

Bacterial viruses are bacteriophages which can have single stranded DNA or RNA. Bacteriophages have head and a tail with a tadpole like appearance. They can be complex phages, lambda phages, polyhedral phages, helical or fliamentous phages.

Fungal Viruses (Myco viruses)

Fungal viruses are precisely virus like particles with dsRNA observed in higher fungi like Ascomycetes, Basidiomycetes and Deuteromycetes. These are latent virus particles that do not produce any pathogenicity, but cause damage to the host, as for example the virus-like particles in cultivated mushrooms results in poor yield.

Mycoviruses are being exploited as interferon inducers to combat viral replication. However, the interferons produced by them are unsuitable for human application due to adverse side effects.

Blue Green Algal Viruses (Cyanophages)

The cyanophages (blue green algal viruses) are named according to the host they infect, as for example, *Lyngbya Plectonema Phormidium* virus (LPP-1). They are dsDNA viruses with a long contractile tail and a lipid containing head surrounded by the capsid. They are being effectively used to control excessive algal growth in water reservoirs.

4.3 PLANT VIRUSES

Plant viruses infect plants and cause variety of systemic diseases. They are transmitted by physical contact, mechanical wounds, by grafting and vegetative propagation or through insect vectors.

The viral surface is made up of definite number of protein subunits (capsomeres), which are spirally arranged in the elongated viruses and packed subunits on the sides of spherical/polyhedral viruses.

All most all plant viruses are RNA viruses, containing a single or double stranded RNA molecule, with a few exceptions such as the Cauliflower Mosaic Virus (CMV) and the Dahlia Mosaic Virus (DMV) which contain DNA. Plant viruses can cause serious crop losses. Though plant viruses are normally harmful, some can be used as gene vectors in genetic engineering.

4.4 DISEASES CAUSED BY VIRUSES

4.4.1 Plant Diseases

4.4.1.1 Plant diseases caused by rod shaped viruses

Tobamo Viruses

Tobacco Mosaic Virus named after the nature of symptom-a mosaic pattern of diseased spots on the tobacco leaves. The Tobamo viruses cause tobacco mosaic disease. The symptoms consists of mottling, chlorosis, curling, distortion and dwarfing of leaves and plants. The Tomato Mosaic Virus infects tomato, Pepper Green Mottle Virus infects pepper, and Odontoglossum Ring spot Virus infecting orchids are commercially important Tobamo Viruses.

Furo Viruses

Furo Viruses are fungus-transmitted rod shaped viruses. They include soil borne Wheat Mosaic Virus and Peanut Clump Virus, both transmitted by *Polymyxa graminis.* Potato Mop-top Virus is transmitted by *Spongospora subterranea.* Beet Necrotic Yellow-vein Virus causes rhizomania disease of sugar beets. Furo viruses cause symptoms that vary with the host. Infected plants show mottling or rings on the leaves and roots may be excessively branched (rhizomania).

Hordei Viruses

Named after barley (*Hordeum*). Barley stripe Mosaic Virus (BsMV) infects grain crops and grasses. Infected plants show mosaics, chlorotic spots, yellow brown stripes, dwarfing, rosetting and necrosis.

Tobra Viruses

The name Tobra Virus is derived from Tobacco Rattle Virus, which infects tobacco. Two other viruses Pea Early browning Virus and Pepper ring-spot

Virus are also Tobra viruses. The virus causes necrotic areas on stems and leaf veins and crumpling of the leaves of the infected host (tobacco) and necrotic areas on the tuber (in potato).

4.4.1.2 Plant diseases caused by filamentous viruses

Carla Viruses

Named after carnation latent virus, the genus Carla Virus has more than 50 types, out of which Pea Streak Virus and Poplar Mosaic Virus cause serious diseases. The disease is completely symptomless. Some interact synergistically with other viruses when symptoms appear. For example, Lily Symptomless Virus interacts with Cucumber Mosaic Virus to cause Fleck disease.

Capillo Viruses and Tricho Viruses

Capillo viruses include Apple stem grooving virus infecting apple and Citrus Tatter leaf Virus affecting citrus plants. Tricho viruses include Apple Chlorotic Leaf spot Virus of apple. The infected plants exhibit, tattered and torn leaves, grooves on the stem, chlorosis and necrosis.

Clostero Viruses

Clostero viruses means, "thread like viruses". They include Beet yellows Virus and Citrus Tristeza Virus and Lettuce yellows Virus. The symptoms of the disease include yellowing of the leaves, necrosis and stem pitting.

Poty Viruses

Named after Potato Virus Y (PVY), Poty viruses are the largest group of plant viruses. They cause some of the most serious plant diseases. They include Bean common Mosaic Virus (BcMV), Bean yellow Mosaic Virus (ByMV), Beet Mosaic Virus (BtMV), Celery Mosaic Virus (CeMV), Lettuce Mosaic Virus (LMV), Papaya Ringspot Virus (PRSV), Soybean Mosaic Virus (SoyMV), Sugarcane Mosaic Virus (ScMV), Turnip Mosaic Virus (TuMV), Water melon Mosaic Virus (WmMV) and others. They also infect ornamental plants, such as Tulip breaking virus affecting tulip flower, but interestingly this adds to the value of tulips by the appearance of streaks and shades.

The disease appears primarily as mosaics, mottling, chlorotic rings and include streaking on the leaves, necrosis and yellowing of leaves.

Rymo Viruses

Named after Rye Grass Mosaic Virus they cause disease of grain crops such as barley, wheat, rice and oats. Rymo viruses include Agropyron Mosaic Virus, Oat Necrotic Mottle Virus, Barley yellow Mosaic Virus, Rice Necrosis Mosaic Virus and Wheat Spindle streak Mosaic Virus. The infection symptoms streak in foliage, flowers, fruits and stems along with stunting, malformation of fruits and necrosis of tissue.

4.4.1.3 Plant diseases caused by single stranded RNA viruses

Bromo Viruses

Bromo viruses are important pathogens of agronomical and horticultural crops. They infect graminaceous and legume plants.

Cucumo Viruses

Named after Cucumber Mosaic Virus (CMV), these are a group of viruses that include Tomato Aspermy Virus (TAV), Peanut Stunt Virus (PSV) and Cucumber Mosaic Virus (CMV) causing mosaics, stunting of plants, leaf and fruit malformation and discoloration. Cucumber Mosaic Virus infects gladioli, melons, pepper, spinach, tomatoes, celery, beet, beans, bananas and crucifers.

Como Viruses

Named after Cowpea Mosaic Virus (CpMV). Como viruses infect legumes (bean, cowpea, soybean and clover) and squash (Squash Mosaic Virus) and radish (Radish Mosaic Virus). They cause stunting, mosaics and malformations.

Ila Viruses

Named after their description as "isometric labile ringspot viruses" , the group includes Tobacco streak Virus, Apple Mosaic Virus, Rose Mosaic Virus, Prunus Necrotic Ring spot Virus, Prunus Dwarf Virus, Plum line Pattern Virus, Citrus leaf Rugose Virus and Citrus variegation Virus. They cause line patterns, ring spots, mosaics and distortion on foliage and blossoms.

Luteo Viruses

Named after the latin word *luteus* which means yellow. Luteo Viruses are a large group of 30 viruses infecting plants causing yellowing symptoms. They include Barley yellow dwarf Virus, Potato Leaf fall Virus, Beet Western yellows Virus and Potato leaf roll Virus (upward rolling of leaves).

Nepo Viruses

Nepo viruses stand for nematode transmitted viruses. Some of the important Nepoviruses are Tobacco Ring spot Virus, Cherry leaf Roll Virus, Grape Vine fan leaf Virus, Arabis Mosaic Virus, Raspberry Ring spot Virus and Tomato Ring spot Virus, Raspberry Leaf curl Virus and Raspberry yellow dwarf virus.

The symptoms include green or yellow mosaic rings on the leaves, smaller asymmetric leaves, fan like leaves in (Grape vine fan leaf) rolling of leaves, pale rings and spots, exudation of gum (Cherry leaf fall), stem pitting and necrosis (Tomato ring spot).

Waika Viruses

Named after Rice Waika (stunting) Viruses, Waika Viruses include Rice Tungro spherical Virus and Maize Chlorotic Dwarf Virus. Tungro infected rice plants are stunted and show mottling and yellowish orange discoloration of the leaves, where as plants infected by maize chlorotic dwarfing, shows stunted growth and tearing of leaves.

4.4.1.4 Plant diseases caused by isometric double stranded RNA viruses

Reo Viruses

Reo Viruses includes Rice gall dwarf Virus causing galls and tumours, Wound tumour Virus infect dicotyledons, inducing vein enlargement, tumours, stunting, darkening of leaf color, leaf distortion and galls on veins, Rice Black streaked dwarf Virus, Maize Rough dwarf Virus, Oat sterile dwarf Virus, Oryza Virus and Rice ragged stunt Virus cause dwarfing, leaf distortion, galls on veins and suppression of flowering.

4.4.1.5 Plant diseases caused by negative RNA (ss RNA) viruses

Rhabdo Viruses

Named after the Greek word *Rhabdos* (rod). Rhabdo Viruses infect vegetables and graminaceous plants. They include Lettuce Necrotic yellows Virus, Potato yellow Dwarf Virus, Rice Transitory yellowing Virus and Wheat striate Mosaic

Virus. The symptoms of the disease include, yellowing of the fruits and leaves, necrosis and mosaics.

Tenui Viruses

Tenui viruses meaning thin, filamentous viruses cause diseases of rice and corn, which includes Rice grassy stunt Virus, Rice stripe Virus and Maize stripe Virus. The leaves of infected plants show chlorotic white or yellow colour, stunted plant growth and proliferation of tillers.

Tospo Viruses

Named after Tomato spotted Wilt Virus, they have extremely wide host range of ornamental, vegetables and fruits such as papaya, lettuce, dahlia, peanut and pineapple. They include Tomato spotted wilt Virus, Impatiens necrotic spot Virus and Water Melon silver mottle Virus. Infection symptoms appear as chlorotic and necrotic rings, lines or spots on the leaves, stems and fruits, wilting of leaves, malformed fruits, stunting and necrosis of the host.

4.4.1.6 Plant diseases caused by double stranded DNA viruses

Badna Viruses

Named after the bacilliform DNA viruses, Badna Viruses include Tungro Bacilliform Virus infecting rice plants, Banana streak Virus (BsV) infecting banana and Cocoa swollen shoot Virus infecting cocoa. The infected plants develop necrosis of roots, chlorosis of leaves and pods.

Caulimo Viruses

Named after Cauliflower Mosaic Virus, they cause mottling and mosaics on vegetables and ornamental plants. Important Caulimo Viruses include Cauliflower Mosaic Viruses, Dahlia Mosaic Virus and Carnation etched ring Virus.

4.4.1.7 Plant diseases caused by single stranded DNA viruses

Gemini Viruses

Named because of geminate (paired) particles, some of the important Gemini Viruses are Bean Golden Mosaic Virus, Squash Leaf curl Virus, Tobacco Leaf curl Virus, Tomato golden Mosaic Virus, Tomato Mottle Virus, Tomato yellow leaf curl Virus, Maize streak Virus and Wheat dwarf Virus. Infected plants exhibit bright mottles, yellow mosaics, curling of leaves and distortion.

4.4.1.8 Plant diseases caused by isometric single stranded DNA viruses

The isometric single stranded DNA viruses include Banana bunchy top Virus (BBTV), Subterranean Clover stunt Virus, Coconut Foliar decay Virus and Faba bean Necrotic yellows Virus. The infected plants appear stunted with decay of foliage, necrosis and dark streaks on the petiole and bunchy top (Agrios, 1997).

4.4.2 Insect Diseases

Three families of viruses, Baculoviridae, Iridoviridae and Reoviridae infect insects. Baculoviridae are dsDNA viruses with enveloped cylindrical virons. They mainly infect Lepidoptera, Hymenoptera, Coleoptera and Diptera. Iridoviridae are non-eveloped, icosahedral linear dsDNA viruses that infect arthropods like beetles. Reoviridae are dsRNA viruses with non-enveloped icosahedral virons infecting mainly Lepidoptera.

More than 250 viruses have been found to be pathogenic to insects. Insect viruses cause latent infections like Densonucleosis infecting wax moth (*Galleria mellonella*). Sigma Virus of fruit flies, Bee Paralysis Virus, Sac Brood Virus of honeybee and Flacherie Virus of silkworm larvae are few of the insect viruses. Silk worms are also infected by nuclear polyhedrosis virus and cytoplasmic virus resulting in severe economic losses.

The insect infecting viruses are being studied to develop eco-friendly bioinsecticides. The nuclear polyhedrosis virus of the family Baculoviridae is effectively used against boll worm of cotton. A commercial product 'Elcar' is available in the market. Cytoplasmic granulosis virus of the same family is used in controlling *Dendrolimus spectabilis*, a caterpillar infection of pine trees. (Pelczar *et al.*, 1993).

4.4.3 Bird and Animal Diseases

Alpha Viruses

Alpha viruses are a group of Toga viruses which mainly infect birds and animals and are transmitted to humans by mosquitoes. The symptoms include a characterized viremia which results in fever, joint pains and skin rashes in humans. Eastern, Western and Venezuelan strains of Alpha viruses cause Equine encephalitis (Pelczar *et al.*, 1993).

Flavi Viruses

Flavi viruses infect birds (swine) and animals (monkey) and are transmitted to humans by mosquitoes and ticks. Most of the flaviviral infection results in fever and encephalitis.

Influenza A Virus

Avian influenza or the bird flu, is caused by H5N1, a subtype of Influenza A virus. The disease affects mainly birds, is contagious and causes epidemics, resulting in the death of a very large number of birds. The disease may occasionally occur in humans who are in direct contact with infected birds, particularly poultry. Several other strains also cause bird flu and vaccines are available against each of them.

Arena Viruses

Arena viruses cause asymptomatic infections in rodents.

Bunya Viruses

Bunya viruses cause hepatitis in domestic animals. They are transmitted by mosquitoes, biting flies, or ticks.

Foot and Mouth Disease Virus

Foot and Mouth Disease Virus is a causative agent of foot and mouth disease in cattle, sheep, goats and swine. Humans are succeptible to infection. The transmission of the disease is through cuts and abrasions on the skin. The symptoms in animals include lesions on the tongue, teats and foot.

Bovine Parvo Virus

Bovine parvo virus cause diarrhoea and abortion in cattle.

Rhabdo Viruses

Rhabdo viruses cause diseases in humans, animals (vertebrates and invertebrates) and plants. In animals, Vesicular Stomatitis Virus (VSV) (a type of Rhabdo virus), causes vesicular lesions in the mouth and other areas of the body of the cattle, pigs and horses, which can be aquired in a benign form.

Retro Viruses

Retro viruses cause feline leukaemia, a blood cancer in cats.

4.4.4 Human Diseases

4.4.4.1 Viral diseases of the skin

Small Pox

Small pox is caused by Variola Virus. The disease is characterized by rashes on the skin consisting of pustules which appear first along the hair line and later on other parts of the body followed by fever. Transmission is through droplets or contact with the infected patients.

Chicken Pox

Chicken pox is caused by Varicella-Zoster Virus in children, which is characterized by vesicular skin rash. Activation of latent virus fom previous Varicella infection causes Herpes Zoster (Shingles) in adults, which is characterized by vesicular eruptions and painful inflammation of the sensory nerves.

Herpatic keratitis

Herpatic keratitis is a localized corneal infection caused by Herpes Simplex Virus type I. The infection results in corneal ulcers. Trifluoridine is the drug of choice.

Herpes

In animals and humans, herpes is caused by Herpes virus. Herpes is of two types, Herpes Simplex type 1 causes acute gingivostomatitis, pharyngitis, keratoconjunctivitis and cold sores. Herpes Simplex Virus type 2 causes genital herpes transmitted by sexual contact. Infection is associated with painful blisters on the cervix, vagina, urethra and anus in women and in men the lesions appear on the penis, urethra and anus, followed by fever and inflammation of the lymph glands.

Rubella (German measles or Three day measles)

Rubella Virus occurs only in humans with air borne transmission via respiratory secretions. The disease is characterized by fever, lassitude, skin rash and swelling of lymph glands below the ear and the neck. Pregnant women with this infection give birth to children with vision and hearing defects and mental retardation.

Measles (Rubeola)

Measles is caused by Morbilli Virus. It occurs in the respiratory tract and is transmitted by droplet infection and mainly affects children. The symptoms include cold, cough, fever and occurrence of Koplik's spots (bluish white spots surrounded by reddish area on the mucous membrane of the cheek and lips) and red blotchy skin rash. (Table. 4.1)

4.4.4.2 Viral diseases of the nervous system

Polio myelitis

Polio is caused by Polio Virus. It affects the nervous tissue in humans. The diseases is subclinical (no symptoms). It causes non-paralytic poliomyelitis characterized by fever and stiffness or pain in the neck and back muscles and paralytic poliomyelitis characterized by inflammation of the spinal cord, damaging of the motor neurons resulting in crippling of the limbs. Polio virus is transmitted from the throat and the nasal discharges and the faeces of the infected individuals. Sabin vaccine is administered for immunization against the disease.

Rabies

Rabies is caused by Rabies Virus in humans, characterized by severe headache, high fever with alternating stages of excitement and depression, difficulty in swallowing, hydrophobia and muscular spasms in the throat and chest. Trans-mission to humans occurs through the bite of dogs, cats, bats and monkeys.

Arboviral Encephalitis

Encephalitis is caused by mosquito born Arbovirus. The disease is characterized by chills, headache, fever, mental confusion and coma. (Table. 4.2)

4.4.4.3 Viral diseases of the respiratory tract

Influenza

Influenza is commonly called as viral "flu" and is caused by Influenza Virus. The disease is characterized by headache, chills, muscular aches and fever.

Para Influenza

Paramyxo Virus causes acute respiratory infections in infants and children. The infection ranges from mild cold, cough, leading to bronchiolitis and pneumonia (Tortora *et al.*, 1998).

Viral Pneumonia

Pneumonia is caused by Enteric virus, which develops due to complication of influenza, measles or chickenpox. (Table 4.3)

4.4.4.4 Viral diseases of the digestive system

Hepatitis

Hepatitis is caused by Hepatitis Virus, which is characterized by systemic infections primarily affecting the liver. Hepatitis is of two types. Hepatitis type A virus causes Viral Hepatitis type A which can cause acute epidemic hepatitis and infective hepatitis. The disease is characterized by loss of appetite, fatigue, abdominal discomfort and fever, followed by onset of jaundice. no vaccine is available for active immunization but passive immunization is protective. Mode of transmission is by faecal-oral route.

Viral Hepatitis type B is caused by Hepatitis B Virus. It is also known as serum hepatitis or homologous serum jaundice. The virus is transmitted by transfusion of blood, contaminated hypodermic syringe and sexual contact. (Table. 4.4)

Hepatitis Delta Virus (HDV) is unique among human pathogens as it shares several characteristics with both plant viroids and viroid like plant satellite RNAs. It is a blood borne pathogen which replicates in the liver and causes fulninant hepatitis. It is detected only in the presence of hepatitis B virus as it needs a hepatitis B virus protein for packaging, it is considered as an RNA satellite of hepatitis B virus (Acheson, 2007)

4.4.4.5 Viral diseases of the reproductive system

Genital Herpes

Genital herpes is caused by Herpes Simplex Virus type 2. The disease is characterized by lesions on the genital organs and painful urination.

Genital Warts

Genital warts are caused by over 60 serotypes of Papilloma Virus, characterized by finger like projections resembling cauliflower or smooth and flat.

Table: 4.1 Viral diseases of the skin

Disease	Pathogen	Symptoms
Small pox (Variola)	Variola virus	Pustules that may be nearly confluent on skin; systemic viral infection affects may internal organs
Chicken pox (Varicella)	Varicella-Zoster Virus	Vesicles in most cases confined to face, throat, and lower back.
Shingles (Herpes Zoster)	Herpes-Zoster Virus	Vesicles similar to chickenpox; typically on one side of waist, face and scalp, or upper chest
Herpes Simplex	Herpes Simplex Virus type 1	Most commonly as cold sores or vesicles around mouth; can also affect other areas of skin and mucous membranes
Measles (Rubeola)	Measles Virus	Skin rash of reddish macules first appearing on face and spreading to trunk and extremities
Rubella (German measles)	Rubella Virus	Mild disease with a rash resembling measles, but less extensive and disappears in 3 days or less

Table: 4.2 Viral diseases of the nervous system

Disease	Pathogen	Symptoms
Poliomyelitis	Polio Virus	Transmitted mainly with ingested water. Results in partial paralysis
Rabies	Rabies Virus	Usually transmitted by animal bites; characterized by severe headache, high fever, hydrophobia and difficulty in swallowing
Arboviral encephalitis	Arbo Virus	Mosquito-borne; disease is characterized by chills, headache, fever and mental confusion

Table: 4.3 Viral diseases of the respiratory system

Disease	Pathogen	Symptoms
Viral pneumonia	Enteric Virus	Symptoms of influenza and measles.
Para Influenza	Para Influenza Virus	Cold, cough, leading to bronchiolitis.
Respiratory syncytial viral (RSV) disease	Respiratory syncytial Virus	A serious respiratory disease of infants.
Influenza	Influenza Virus	An illness characterized by chills, fever, head ache and muscular aches. Virus changes antigenic character rapidly.

Table: 4.4 Viral diseases of the digestive system

Disease	Pathogen	Symptoms
Hepatitis A	Hepatitis A Virus (HAV)	Nausea, vomiting, jaundice, pain in the abdomen, dark coloured urine, loss of appetite.
Hepatitis B	Hepatitis B Virus (HBV)	Loss of appetite, nausea, vomiting, itching all over the body, pain in the liver, pale coloured stools
Hepatitis C	Hepatitis C Virus (HCV)	Nausea, vomiting, diarrhoea, pain in theliver, dark coloured urine.
Hepatitis D	Hepatitis D Virus (HDV)	Joint pains, nausea, loss of appetite enlarged liver
Hepatitis E	Hepatitis E	Same as Hepatitis D Virus (HEV)

Source: Tortora *et al.*, 1998

4.4.4.6 Fevers caused by viruses

Chikungunya

An urban disease resembling Dengue, occurring mainly in Africa, the Indian subcontinent and South east Africa. The disease caused by Arbovirus is transmitted by *Aedes* mosquitos and is characterized by fever associated with severe joint pain.

Dengue Fever

Dengue fever is caused by Dengue fever virus and transmitted by mosquitoes. Symptoms include fever, muscle and joint pain with skin rash. Reinfection causes Dengue haemorrhagic fever.

Lassa Fever

Lassa fever is caused by Arena virus. It causes infections in rodents but are transmitted to humans by contact with rodent urine and faeces. Human infection includes meningoencephalitis and haemorrhagic fevers. it is highly contagious.

Yellow Fever

Yellow fever is caused by Yellow Fever virus. It affects mainly monkeys and transmitted to humans through mosquitoes causing a disease in which liver, kidney and other organs are infected. The symptoms include fever, chill, head ache, nausea, jaundice, albuminuria (albumin in the urine), kidney damage, hematemesis and hepatomegaly.

4.4.4.7 Other viral diseases in humans

Acquired Immune Deficiency Syndrome (AIDS)

Immune deficiency is normally acquired by aging where in the immune system becomes less responsive to foreign antigens and T-cell function deteriorates. This is due to slow inactivation of thymus gland. Humoral immunity deficiency, on the other hand is due to inactivity of B-cell production in the bone marrow resulting in Acquired Immuno Deficiency Syndrome (AIDS). Immunodeficiency can also result due to viral infection. The main causal agent of AIDS is Human Immuno deficiency Virus (HIV-I and HIV-2). The location of origin of HIV is thought to be Central Africa.

The most common symptoms of fully developed AIDS are various infections caused by opportunistic organisms. the virus is transmitted by sexual contact, contaminated hypodermic syringes and blood transfusion.

Aplastic crisis

Aplastic crisis is caused by Parvo Virus. This is a disease which infects persons with sickle cell anaemia. it is also implicated in Erythema Infectiosum, a contagious disease in children.

Mononucleosis

Epstein-Barr virus is the causative agent of infectious mononucleosis, which affects humans between the age group of 15 to 20. The disease is characterized by fatigue, sore throat, fever, swollen cervical lymph nodes and increased level of monocytes in the blood (mononucleosis).

Common cold

Common cold is caused by Rhino virus and is spread by nasal and throat discharges. no vaccine exists for prevention of common cold.

Mumps

Mumps virus is spread by droplets of infected saliva. Although it is a childhood disease, it also attacks adults. It is characterized by painful swelling of the parotid glands. In adults, inflammation of the testes in males and ovaries in females also occur.

4.4.4.8 Viruses and cancer

Several types of cancer are known to be caused by oncogenic viruses. The connection between virus and cancer was shown for the first time by Peyton Rous in 1911 when he proved that chicken sarcoma was transmitted to healthy chicken by a virus and this is referred to as Rous Sarcoma Virus (RSV), a retrovirus with a single stranded RNA.

Virus induced adenocarcinomas in mice were discovered in 1936. These viruses include both DNA and RNA animal viruses. The alteration of genetic material in a normal cell to make it cancerous is due to the alteration of cellular DNA through oncogenes. Michael and Varmus in 1989 proved that these cancer inducing genes are carried by viruses.

All oncogenic viruses so far known have one common feature, that is the genetic material of the virus is integrated into the host cell DNA and is replicated along with the host cell chromosomes. Integration of viruses next to cellular oncogenes stimulates transcription leading to cancer.

Oncogenic viruses that are DNA viruses are distributed in the families of Herpesviridae, Poxviridae, Papovaviridae and Hepadnaviridae. Oncogenic viruses which are RNA viruses belong to the family of Retroviridae.

The following are some common examples of cancer causing viruses:

1. Epstein – Barr Virus (EBV): It is one of the best studied human cancer virus. It is the cause of Burkitt's lymphoma, nasopharyngeal carcinoma and Hodgkin's disease. The EBV particles have been found in the cells of these cancerous tissues. The patients affected by EBV also have high level of antibodies against EBV.
2. Hepatitis B Virus (Hepadnaviridae) is associated with hepatocellular carcinoma.
3. Human papilloma Virus (Papovaviridae) is linked to cervical cancer.
4. Human T-cell Leukaemia Virus causes T-cell Leukaemia and hairy cell leukaemia.
5. Herpes Virus Type I and II are implicated in lip, mouth and cervical cancer.

4.5 ANTIVIRAL DRUGS

Antibiotics are ineffective against viruses. Vaccines and antibodies are the only certain means of combating viral infections. However, the following drugs are in use to control some of the viral infections. Amantadine for Influenza A prevention; Nucleoside analogs - Acyclovir, Ribavirin, Ganciclovir and Trifluoridine for Herpes virus and Zidovudine, Didanosine, Zalcitabine for HIV.

4.6 MEDICINAL USES OF VIRUSES

Medicinal uses in Homoeopathy

Small Pox Virus

Lymph from small pox pustule called as variolinum, is used for pain in the occiput, deafness, inflamed eyelid, oppressed breathing, fever with radiating heat, excruciating backache, wrist pain and shingles. It is a chief remedy for small pox as an internal vaccination (Boericke, 1991).

4.7 OTHER USES OF VIRUSES

Bacteriophages in Disease Control

Bacteriophages are specific type of viruses that infect bacteria. There are several bacteriophages each specific to a certain type of bacteria. Though the research on the use of bacteriophages in the control of bacterial diseases is still in the early stages, they promise to provide a specific solution to a particular pathogen. There does not seem to be any possibility of pathogenic bacteria acquiring resistance to the bacteriophages, as they do for antibiotics.

Bacteriophages in Research

Bacteriophages have been very important tools in several areas of biological research. Phages are used as an important tool for molecular cloning in biotechnology. They are useful vectors for genetic engineering experiments. The lambda being a temperate phage it injects its DNA to the host cell as a natural process and DNA gets integrated with the host DNA. (Sullia and Shantharam, 2005).

Phages have been useful in identification and classification of bacteria. Bacterial species and strains are usually heterogenous which include several biotypes and serotypes. Phages are useful in sub dividing them into phage-type or group with similar or dissimilar phage sensitivity. Phage typing is the most important application in bacterial diagnosis involving clinically important bacteria.

4.8 PRIONS

Prions (Proteinaceous infectious particles) are slow viruses with a slow progressive course. These transmiscible agents have unconventional or atypical behaviour.

They do not contain nucleic acid and have only protein as the infective component. (Fig. 4.5). Prions cause diseases in animals and humans.

Animal Diseases Caused by Prions

Prions cause various animal diseases such as Scrapie, Mink Encephalopathy and Bovine Spongiform Encephalopathy (BSE) locally known as mad cow disease.

Human Diseases Caused by Prions

A number of fatal diseases of central nervous system in humans are caused by Prions. They have long incubation time and progress slowly. Humans suffer from Creutzfeldt-Jakob disease (CJD) which is aquired by consumption of beef. They also cause Gerstmann-Straussler Scheinker Syndrome and Fatal Familial Insomnia. Several years after causing measles the same virus (Morbilli virus) causes encephalitis called Subacute Sclerosing Panencephalitis (SSPE) in humans.

4.9 VIROIDS

Viroids are nucleic acid entities that cause diseases of cultivated plants. Unlike viruses, viroids do not possess a protein coat and exist only as short infectious circular single stranded RNA molecule. Though it is not known how viroids can cause a disease, it is thought that their location in the nucleus and the inability to act as mRNA suggest their interference with gene regulation in the infected host cell. Diagramatic Illustration of Viroids is given in Fig. 4.6.

Viroids so far have been identified as pathogens of only plants. No animal or human disease has been shown to be caused by viroids. Viroids are spread from diseased to healthy plants primarily by mechanical means through tools used for cutting. No specific vectors of viroids are known. Viroids survive in nature outside the host in dead plant matter and are resistant to high temperature. Potato spindle tuber, Citrus exocortis, Chrysanthemum stunt, Tomato bunchy top, Apple scar skin, Coconut Cadang-Cadang and Cucumber pale fruit are caused by viroids.

4.9.1 Plant satellite RNA virus

Plant satellite RNA viruses are small single stranded RNAs that depend on a helper virus for replication, but show little or no sequence similarity with their helper virus. Like viroids, they have a circular, highly base-paired structure which does not encode any protein. Unlike viroids, they are encapsulated and can be transmitted more efficiently from plant to plant. Furthermore, like group A viroids, all viroid-like plant satellite RNAs studied so far have a hairpin or hammerhead structure. A number of different RNA containing plant viruses support the replication of satellite RNAs, by using their own RNA-dependent RNA polymerases (Acheson, 2007).

4.10 VIRUSOIDS

Virusoids resemble plant satellite RNA viruses in many respects. Virusoids are also small circular single stranded RNA particles. They differ from viroids in their inability to replicate themselves and in their requirements of a helper virus. Thus unlike viroids which occur independently, virusoids need helper viruses for successful infection. They may occur within the capsid of helper virus. The helper viruses associated with virusoids are Tobacco mottle virus, Solanum nodiflorum virus, and Lucerne Transient Streak Virus. All virusoids are pathogenic to only plants.

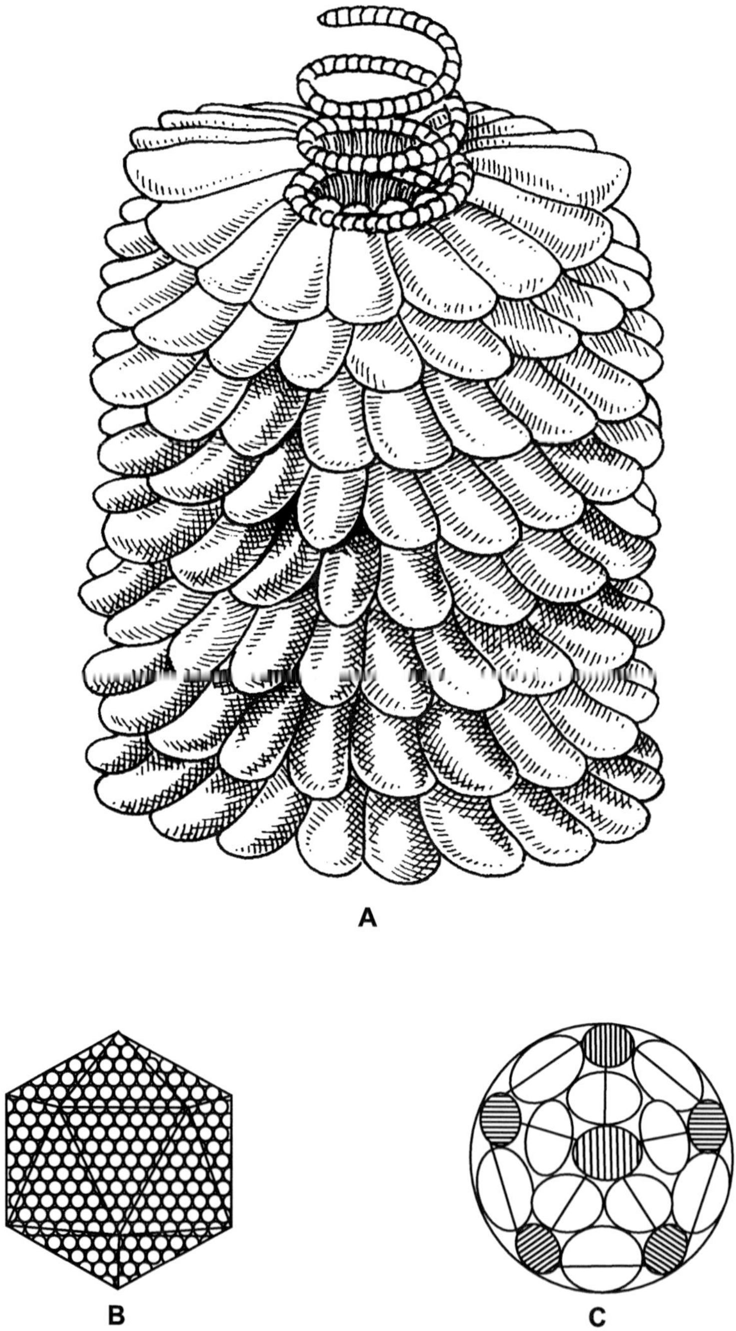

A B C

Fig. 4.1 Shapes of viruses. A : Helical; **B** : Cubical; **C** : Icosahedral
Fig. 4.1 to 4.4 Redrawn from Pelczar *et.al.*, 1993 and Purohit, S.S (1999)

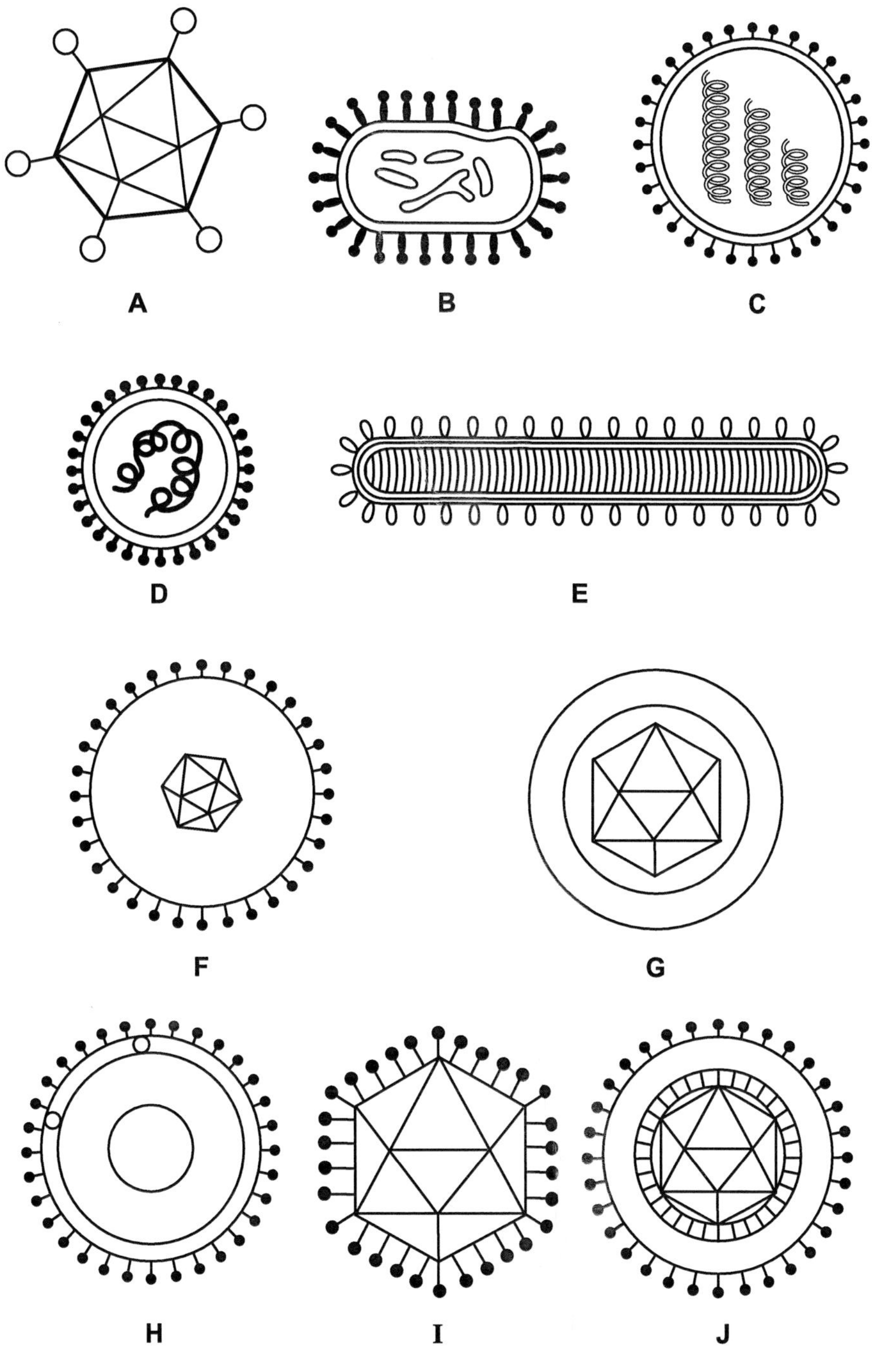

Fig. 4.2 Disease causing viruses. A : Adena virus; **B** : Arena virus; **C** : Bunya virus; **D** : Corona virus; **E** : Ebola virus; **F** : Flavi virus; **G** : Hepatitis B virus; **H** : Hepatitis D virus (Delta virus); **I** : Hepatitis E virus; **J** : Herpes virus

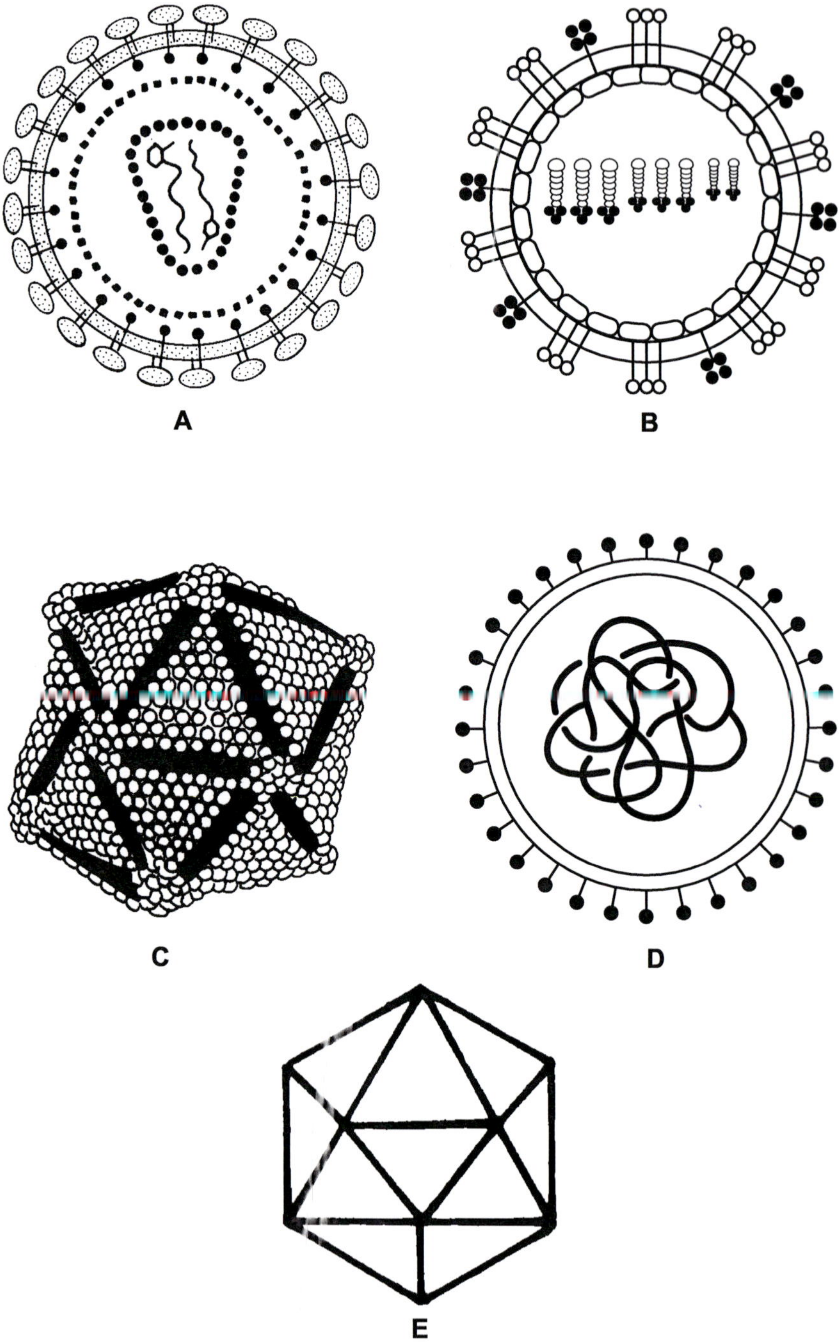

Fig. 4.3 Disease causing viruses. A : HIV virus; **B** : Influenza virus; **C** : Irido virus; **D** : Measles virus; **E** : Arbo virus (causing Chickungunya)

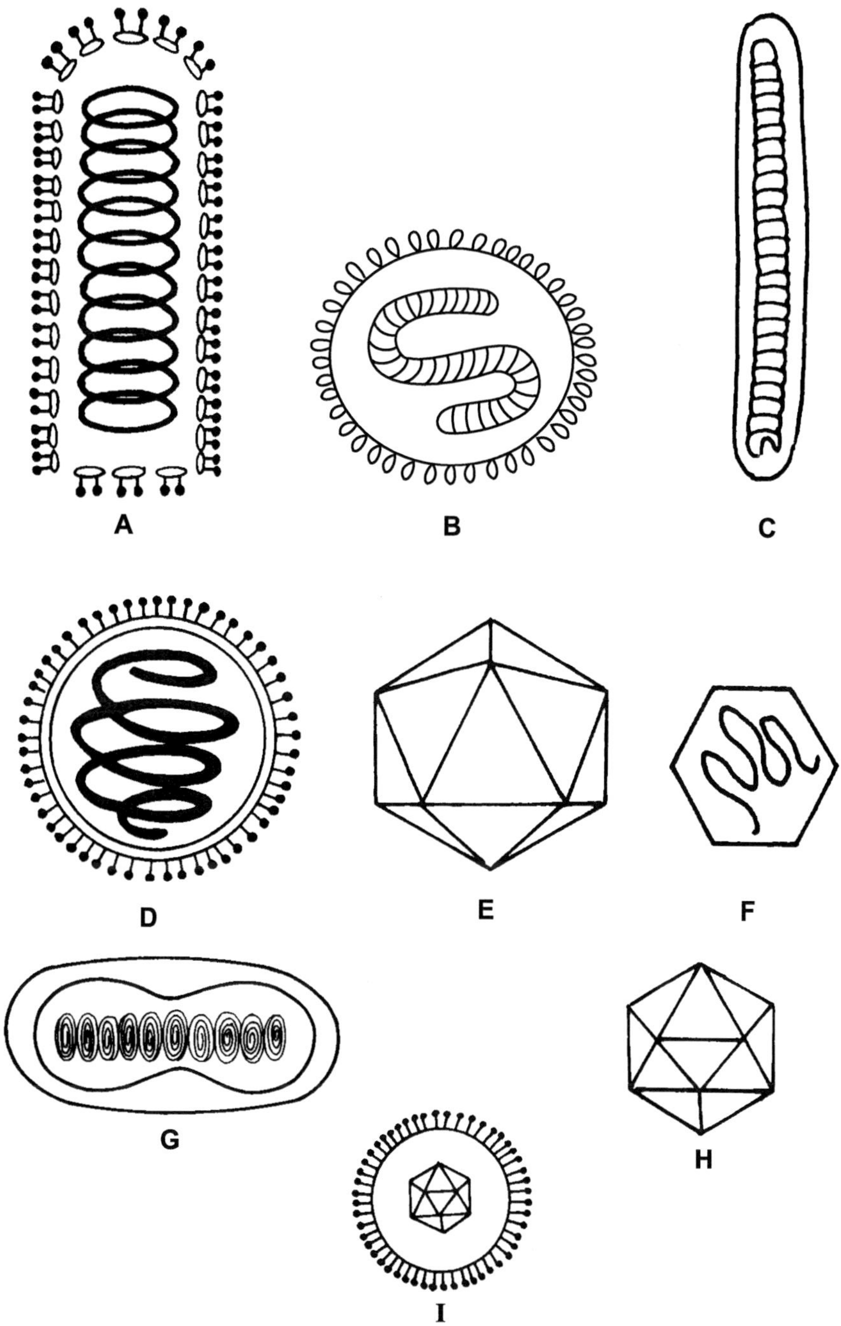

Fig. 4.4 **Disease causing viruses. A** : Lassa fever virus (Rhabdo virus); **B** : Mumps virus; **C** : Nuclear polyhedrosis virus; **D** : Orthomyxo virus; **E** : Papilloma virus; **F** : Polio virus; **G** : Pox virus; **H** : Reo virus (Toga virus); **I** : Rubi virus

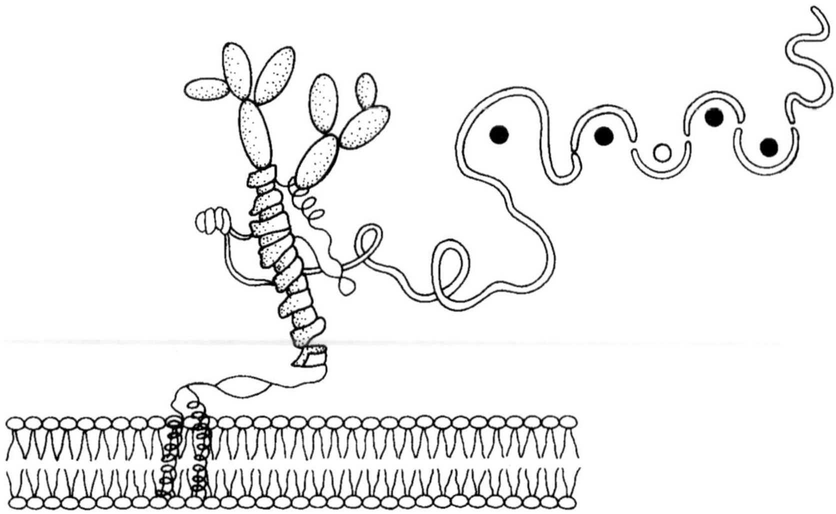

Fig. 4.5 Prion protein shown in different orientation inserted into the plasma membrane.
Courtesy: Acheson, 2007

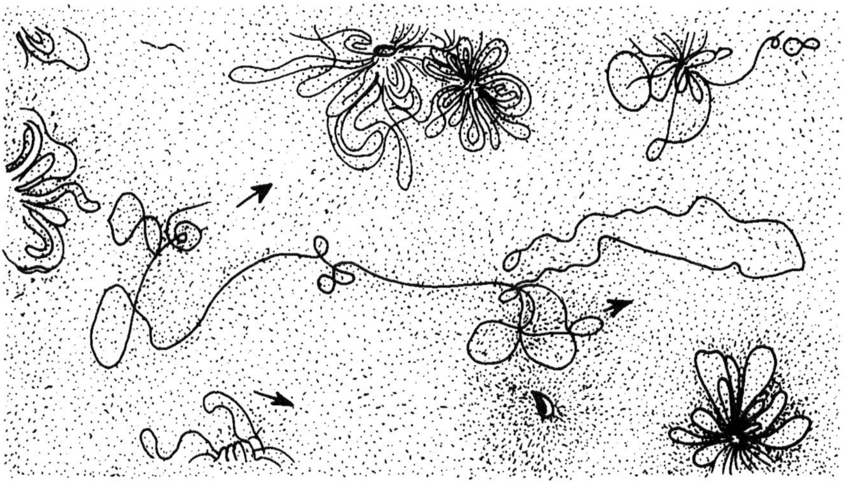

A

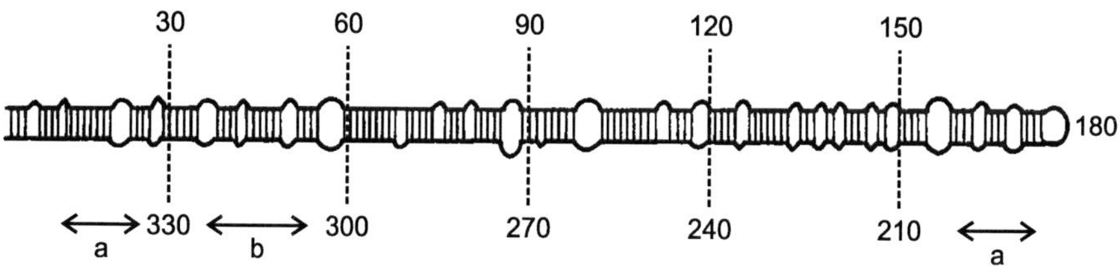

B

Fig. 4.6 **Viroid. A** : Electron micrograph of potato spindle tuber viroid particles (arrows) mixed with double stranded DNA of bacteriophage. **B** : Secondary structure of potato spindle tuber viroid.

Courtesy : Koller, T. and Sago, J.H. (**a** : Region involved in replication; **b** : Region of pathogenecity)

CHAPTER 5

MYCOPLASMAS

5.1 INTRODUCTION

Mycoplasmas are prokaryotes and most are aerobes or facultative anaerobes. Earlier they were called Bovine Pleuro Pneumonia Like Organisms (PPLO). They are regarded as intermediate between viruses and bacteria. Due to the absence of cell walls they are highly pleomorphic and are resistant to antibiotics like penicillin. They produce filaments that resemble fungi, hence the prefix 'myco'.

The cells range in size from 0.1 to 0.25mm and are bound only by cytoplasmic membrane. In the late growth phases they may be in the form of chains of coccoid bodies. The branched filaments appear like mycelia (pseudomycelia). On a solid medium they develop hemispherical flattened colonies with a fried egg appearance. They are non motile with DNA as the genetic material. Mycoplasmas may be saprophytic, parasitic or pathogenic. They are known to cause variety of plant and mammalian diseases.

Thermoplasmas are heat loving *Mycoplasma*. The colonies are flat, granulated and dark brown in colour. *Thermoplasma acidophylum* is an acid loving *Thermoplasma*.

5.2 DISEASES CAUSED BY MYCOPLASMA

Mycoplasma cause diseases in plants, animals and humans; plant diseases caused by *Mycoplasma* infection results in loss in agricultural economy.

5.2.1 Plant Diseases

Mycoplasma cause number of plant diseases that are transmitted by insect vectors through the salivary glands. Characteristic symptoms of infection by *Mycoplasma* include phyllody (conversion of floral parts into leaf like structure), witches broom (increased number of side branches with small leaves appearing like broom), yellowing and reduction in leaf size. Following are the most important plant diseases caused by *Mycoplasma*.

Sandal Spike

The symptoms of Sandal spike mainly include severe reduction in the leaf size and shortening of internodes. The affected plants do not produce any flower or fruit. Phyllody is a common phenomenon. Transmission of the disease is by the insect *Jassus indicus*. Sandal spike disease affects large tracts of natural stands of sandal wood trees and causes extensive financial loss, as no cure has yet been found.

Yellow Disease of Peanut

Symptoms of Yellow disease infection include proliferation of axillary buds, reduction in the size of the leaves, phyllody and yellowing. Infected plants do not produce nuts.

Grassy Shoot Disease of Sugarcane

Grassy shoot disease is the most serious disease of sugar cane. The disease is characterized by the production of thin lanky tillers from the base of the affected plants. Tillers bear pale yellow, thin narrow leaves giving a crowded grass like appearance. Diseased plants exhibit excessive proliferation of axillary buds.

Katte Disease of Cardamum

The symptoms of Katte disease includes the discolouration of the young leaves, leaving pale green discontinuous stripes starting from the midrib towards the periphery of leaves. The infected young plants show reduction in the leaf size and stunted growth. Fruiting is completely suppressed. The disease is transmitted by banana aphid. Katte disease causes extensive financial loss to cardamum plantations and no cure is known.

Citrus Green Disease

Infected leaves of the citrus plant become extremely small, veins and midrib turn yellow and later decolorize totally. Fruits develop yellow blotches and fall early.

Little Leaf of Tomato and Brinjal

The infected plant becomes short and bushy with a cluster or bunch of very small leaves and shortened internodes. The axillary buds exhibit excessive proliferation.

Little Leaf of Vinca

The symptoms includes reduction in the size of leaves and nodes giving a bushy appearance.

Coconut Root Wilt

The disease is of national importance. The diseased plant exhibits wilting and drooping of the leaves, yellowing and necrosis. Characteristic bending and ribbing of the leaflets of the leaves in the central whorl is the earliest symptom followed by necrosis of the inflorescence.

Suspected *Mycoplasma* Diseases of Plants

Alfalfa witche's broom	Papaya bunchy top
Apple proliferation	Green pea yellow dwarf
Areca nut yellow leaf	Peach yellows
Aster yellows	Peanut witche's broom
Bottle gourd phyllody	Pear decline
Blueberry stunt	Phloem necrosis of elm
Cassava witche's virescence	Pigeon pea sterility
Cherry buckskin	Potato purple top
Citrus greening disease	Potato stolbur
Citrus stubborn	Potato witche's broom
Clover dwarf	Rice stripe
Clover phyllody	Rice yellow dwarf
Clover stolbur	*Rubus* stunt
Clover virescence	Rubbery wood of apple and pear
Corn stunt	Safflower phyllody
Cowpea witche's broom	Sesame phyllody
Cranberry false blossom	Strawberry green petal
Cucumber phyllody	Sugarcane grassy shoot
Grassy stunt of rice	Sugarcane white leaf
Legume witche's broom	Sweet potato little leaf
Little cherry	Sweet potato witche's broom
Little peach	Tobacco yellow dwarf.
Lucerne witche's broom	
Mulberry dwarf	

Source: Mehrotra and Agarwal, 2003.

5.2.2 Animal Disease

Bovine pleuropneumonia in cattle is caused by *Mycoplasma.*

5.2.3 Human Diseases

Mycoplasma hominis and *Mycoplasma fermentans* cause infertility, arthritis, otitis and nongonococcal urethritis. *Mycoplasma pneumoniae* causes primary atypical pneumonia (walking pneumonia) affecting the upper respiratory tract and has air borne mode of transmission and *Mycoplasma genitalium* is believed to be involved in pelvic inflammatory diseases.

5.3 UREAPLASMA

Ureaplasmas were earlier designated as T-strain of *Mycoplasma*. They are distinguishable from *Mycoplasma* in their potential to hydrolyze urea by ureases and they require urea for growth. They are non-motile and similar in morphology to *Mycoplasmas*. In humans, *Ureaplasma urealiticum* occur in the mouth, respiratory and urinogenital tract, and cause diseases such as infertility in both men and women, urethritis, meningitis, pneumonia, premature birth and still birth and in cattle it causes pneumonia. The drug of choice for treatment is tetracyclines.

5.4 MOLLICUTES

Mollicutes are prokaryotic cells without cell wall and are not true mycoplasms. These wall less organisms were detected in the phloem of plants and in insect vectors and were subsequently called Mycoplasma Like Organisms (MLO). Those with a helical structure are called Spiroplasmas and those which are round to elongate but not spiral are called Phytoplasmas.

5.4.1 Spiroplasma

Spiroplasmas are helical mollicutes with cells that vary in shape from helical or slightly ovoid, to helical and branched filaments. They are 100 to 240nm in diameter. They can be cultured on nutrient media. They lack a true cell wall and are bound by a unit membrane. Though they lack flagella they are motile, moving by slow undulation of the helix.

Spiroplasmas cause plant diseases such as Stubborn disease of citrus (Citrus stubborn), Brittle root disease of horse radish and Stunt disease of corn (Corn stunt).

Symptoms of diseases caused by Spiroplasma vary a great deal and few are expressed at the same time on a tree. In general, affected trees show bunchy, growth of twigs and branches, with short internodes, stunting, and yellowing of the leaves.

5.4.2 Phytoplasma

Phytoplasmas are organisms that resemble *Mycoplasma* in all morphological aspects. They lack cell walls and are bound by unit membrane and have cytoplasm, ribosomes and strands of nuclear material. They are spheroid or ovoid or irregularly tubular to filamentous. They cannot be cultured on artificial nutrient media. They cause plant diseases such as Aster yellows of Aster, Grape yellows, X-disease of Peach, Pear decline, Coconut lethal yellowing ,Apple proliferation and Stolbur of tomato.

The symptoms of infection are uniform yellowing of the leaves, smaller leaves, shortening of the internodes, excessive proliferation of shoots, greening of flowers, followed by rapid decline and death of the plant.

5.5 L-FORMS OF BACTERIA

Bacteria often spontaneously produce variants that fail to produce cell walls. The progeny of such variants comprise populations of wall less bacteria called the L-forms which resemble mycoplasma. The L-form of bacteria are indistinguishable from mycoplasmas and phytoplasmas. They develop in living organisms during or after treatment with antibiotics. Because of the absence of cell wall they form "fried egg" type of colonies like Mycoplasmas.

The only pathogenic bacteria reported to produce L-forms are *Agrobacterium tumefaciens* the causal agent of crown gall and *Erwinia caratovora* Pv *atroseptica* causal agent of black leg disease of potato. They produced the same disease as the parent bacteria and retained the pathogenecity.

5.6 RICKETTSIAS

Rickettisias are rod shaped intracellular obligate parasites exhibiting a high degree of pleomorphism. The shape of the cell varies from rod shaped and round to oval. They are non motile and gram negative and contain double stranded DNA and RNA or both as the genetic material. They multiply by binary fission within the host cell and obtain their ATP from the host cell.

They are associated with arthropods, which serve as major host or as vectors for transmissions to vertebrates. They are also transmitted to humans by insects and ticks and cause diseases.

The family Rickettsiaceae is subdivided into three tribes and all the three tribes are pathogenic. The members of the tribe Rickettsieae are pathogenic to humans (Table 5.1), while those of the tribe Ehrlichieae are pathogenic to domestic animals. They are gram negative and live within white blood cells and maybe transmitted to humans by ticks and cause Ehrlichiosis-a fatal blood disease. Tribe Wolbachieae is pathogenic to arthropods.

Rickettsial infection in man begins in the vascular system following the bite of an infected arthropod. They proliferate in the endothelial cells and establish focal areas of obstruction in the small blood vessels due to hyperplasia of infected endothelial cells and RBC's. The clinical symptoms include characteristic spotted skin rash, headache, stupor and fever (Pelczar *et al.*, 1993).

Table 5.1 Human diseases caused by Rickettsias

Pathogen	Type of Disease
Ehrilichia species	Ehrlichiosis
Coxiella burnetii	Q fever
Rickettsia akari	Rickettsial pox or vesicular rikettsiosis
Rickettsia cononi	Marscillus fever
Rickettsia prowazekii	Typhus fever
Rickettsia sibirica	Non-asiatic rickettsiosis
Rickettsia rickettsii	Rocky mountain fever
Rickettsia tsutsugamushi	Scrub typhus or Tsutsugamushi fever
Rickettsia mooseri, Rickettsia typhii	Murine typhus fever
Rochalimaea quintana	Trench fever (Five day fever)

Source: Tortora *et al.*, [illegible]

5.7 CHLAMYDIAS

Chlamydias are obligate parasites of birds and humans, distinguished from Rickettsias by their inability to synthesize ATP and dependency on the host for the compound. The Chlamydias belong to a single family Chlamydiaceae with a single genus *Chlamydia.* They can liberate carbon dioxide from glucose and have metabolically active enzymes. They are thus called as 'energy parasites'. Cells are more or less spherical with an outer cell well resembling that of gram negative bacteria. Cell wall contains peptidoglycans and muramic acid. They possess both DNA and RNA. Chlamydias cause diseases in birds and humans. They are not contacted by arthropod vectors but by contact or by air.

5.7.1 Diseases Caused by Chlamydias

5.7.1.1 Bird disease

Psittacosis (Ornithosis)

Psittacosis is primarily an infectious disease of pigeons, chicken, turkey and ducks, caused by *Chlamydia psittaci* which can be transmitted to humans by inhalation of dust of dry avian droppings. It affects poultry workers, causing respiratory infections, headache, chills followed by convulsions, coma and death. The disease causes economical loss as domestic fowls and birds are affected (Tortora *et al.*, 1998).

5.7.1.2 Human diseases

Chlamydias cause Inclusion conjunctivitis (Chlamydial conjunctivitis) and Trachoma which is a disease of the eye characterized by vascularization and scarring of the cornea leading to blindness in humans and other diseases such as Pneumonia, Lymphogranuloma venereum, which infects males causing swelling of lymph nodes around the groin, Non-gonococcal urethritis leading to testicular swelling and inflammation of the prostate in males and fallopian tube damage in females, followed by painful urination and watery discharge (Table 5.2 and 5.3).

Table 5.2 Chlamydial diseases of the lower respiratory system

Disease	Pathogen	Symptoms
Psittacosis (ornithosis)	*Chlamydia psittaci*	Symptoms are fever, headache, chills. Most commonly spread by contact with exudates of fowl.
Chlamydial pneumonia	*Chlamydia pneumoniae*	Mild respiratory illness common in young people; resembles Mycoplasmal pneumonia.

Source: Tortora *et al.*, 1998

Table 5.3 Chlamydial diseases of the reproductive system

Disease	Pathogen	Symptoms
Nongonococcal urethritis (NGU)	*Chalmydia* species	Painful urination and watery discharge in females, with possible complications such as PID (Pelvic inflammatory disease)
Lymphogranuloma venereum (LGV)	*Chlamydia trachomatis*	Swelling of lymph nodes in groin.

Source: Tortora *et al.*, 1998.

CHAPTER 6

BACTERIA

6.1 INTRODUCTION

Bacteria are cosmopolitan in distribution and are ubiquitous in natural habitats such as soil, water and air. Few species are found in the hot springs (thermophilics) and extreme cold condition (psychrophilics).

Bacteria in soil determine its physical and chemical characteristics. Most of them occur in the upper strata of the soil with approximately 1,000,000 to 10,000,000 bacteria being present per gram of soil.

Rain water is usually free of bacterial flora. The surface water contains the maximum number and species of bacteria. Bacteria such as *Alcaligenes, Flavobacterium* and several others are found in fresh water. *Aeromonas, Alteromonas, Staphylococcus, Vibrio* and others are found in sea water. In an estuary with nutriently poor water, *Caulobacter, Nitrobacter, Nitrosomonas, Pseudomonas* are normally present. Bacteria also constitute the major flora of sewage water. Coliforms, *Clostridium, Pseudomonas* and *Streptococcus* are abundant in sewage.

Number and types of bacteria present in air depends on the place, weather and air current. Dust, soot and ash also contain bacteria. Bacteria also occur in a variety of food and food products like milk, butter, cheese and on vegetables and fruits causing their spoilage.

A large number of plant and animal diseases are caused by bacteria. They cause the most dangerous and contagious diseases in humans like cholera, tuberculosis, diphtheria and pneumonia. They also cause plant diseases such as wilts and rots resulting in economic loss.

Beneficial activities of bacteria are of equal importance. They are the producers of enzymes, organic acids, antibiotics, vitamins and several other metabolites. They are the fermentors of many foods and drinks we consume (curds, cheese, wine, vinegar). They are the fixers of nitrogen in soil and degraders of dead and decomposed organic matter. They are the most important players in biogeochemical cycles involving sulphur, phosphorus, iron and other elements.

There are four basic forms of bacteria: spherical (*Coccus*), rod shaped (*Bacillus*), spiral (*Spirillum*) and comma shaped (*Vibrio*). Some like *Azotobacter* exhibit several forms or shapes (pleomorphic). The bacterial cell has a thin, delicate cell wall covered by soft slimy layer capsule which adds to their virulence. The bacteria which exhibit positive reaction to gram staining are gram positive bacteria, while those which exhibit negative reaction are gram negative bacteria. The cell walls of these two types differ in their chemical constituents.

In the cytoplasm of the bacterial cell, ribosomes and mitochondria are present along with a circular DNA molecule. Additional small circular DNA molecules called plasmids also occur in several·bacterial species. Plasmids replicate independently and are responsible for extra-chromosomal gene expression. Several bacteria possess flagella for locomotion and fimbriae (pili) helping in adhesion.

Certain genera in different bacterial groups are more significant as exemplified by the following.

Alcaligenes latus is employed in the production of bioabsorbent with a potential use in sanitary disposable products. *Agrobacterium tumefaciens* has the ability to transfer genes into plant cells. It is the only known example of natural gene transfer between a prokaryote and eukaryote.

Bacillus is the source of industrial enzymes (proteolytic enzymes) used in laundry detergents and polysaccharide hydrolyzing enzymes used in the degradation of starch. *Bacillus thuringiensis* is highly exploited as insecticide. Bacitracin from *Bacillus subtilis* and polymixin from *Bacillus polymixa* are important antibiotics.

Deinococcus isolated from soil, meat and dust is highly resistant to ionizing radiation. The cells are bright red or pink due to high carotenoid content. They are the indicators of ionizing radiation pollution.

Eubacteria include those which degrade a number of unusual compounds, fix atmospheric nitrogen and are important in biotechnology.

Escherichia coli is the most extensively studied living organism and a favourite choice for producing foreign proteins via recombinant DNA technology. Immobilized cells are used in the synthesis of glutothion and ascorbic acid.

Gluconobacter is the major producer of gluconic acid and vinegar the two important commercial products.

Bacteriorhodopsin from *Halobacterium* is being used for the development of computer chips to store more information than a silicon chip and for faster processing of information.

Clostridium is highly pathogenic causing food poisoning. It is a producer of industrial solvents like ethanol, acetyl methyl carbinol, butanol and acetone. In World War I it was used in the manufacture of cordite. *Clostridium acetobutylicum* is the major source of acetone.

Lactobacillus, Leuconostoc, Pediococcus and Streptococcus commonly called as lactic acid bacteria are the common microbiota in human stomach. They are also the producers of butter, butter milk, yoghurt and cheese.

Rhizobium is nitrogen fixing bacteria which fixes biological nitrogen.

The sulphur bacteria has the classic fluorescent Pseudomonads—*Pseudomonas aeruginosa, Pseudomonas fluorescens, Pseudomonas putida, Pseudomonas syringae.* They are degraders of camphor, toluene and octane.

Streptococcus species are pathogenic, but are the main producers of dairy products. Species of *Streptococcus* is used in the production of hyaluronic acid, a polysaccharide that has applications in the treatment of skin disorders, arthiritis and faster wound healing after surgery .

Staphylococcus inhabiting the skin of humans and animals are responsible for spoilage of food, meat, poultry and cause skin infection (*Staphylococcus aureus*).

Thermus aquaticus is used as a source of thermo stable DNA polymerase (Taq polymerase enzyme) used in the amplification of DNA.

Thiobacilli are important organisms in the industrial extraction of copper and uranium.

Xanthomonas, related to fluorescent Pseudomonads, is the major producer of xanthan gum used in food and cosmetic industry such as salad dressing,shampoos and cold creams (German *et al.*, 1999) and in oil recovery.

Zygomonas found in palm wine, sugar cane juice and apple ciders is the major producer of ethanol.

Illustrations of representative species of bacteria are given in Fig. 6.1 to 6.3 to provide an over view of their appearance.

Box 6.1 Snow Max

Snow max, a protein produced by genetically engineered bacteria, causes water to freeze at higher than normal temperature. This genetically engineered bacteria is employed in making artificial snow in Lake Tahoe, California.

6.2 ACTINOMYCETES

Actinomycetes are saprophytic soil organisms, with characters similar to those in bacteria and fungi. Unlike the slimy bacterial colonies, actinomycete colonies are powdery. Actinomycetes have unicellular branched mycelium like bodies like several fungi but differ from them in the composition of their cell wall.

The common genera of Actinomycetes are *Actinoplanetes*, *Corynebacterium, Mycobacterium, Nocardia* and *Streptomyces.*Actinomycetes cause diseases such as actionmycosis in immunocompromised individuals and transplant recipients. Actinomycetes cause allergy in humans, the most common being ventilation pneumonitis caused by *Thermoactinomyces vulgaris,* frequently found in air-conditioned rooms. They also cause plant diseases such as potato scab and rot of sweet potato.The aerobic soil inhabiting Actinomycetes are a major interest to the pharmaceutical industry as a source of novel antimicrobial agents.

6.2.1 Economic Importance of Actinomycetes

Actinomycetes are a source of a very large number of antibiotics (See Chapter 15). They are used as biological agents to control plant pathogens. They can convert agricultural wastes into alcohol.

Following are the major actinomycetes which produce industrially important products.

Actinoplanetes

Actinoplanetes grow in soil, forest litter and beach sand. *Micromonospora* degrades chitin and cellulose and produces the antibiotic Gentamycin.

Cellulomonas

Species of *Cellulomonas* are used in the production of alcohol and proteins from cellulose rich plant matter.

Corynebacterium

Corynebacterium species cause plant diseases like ring rot of potatoes and vascular diseases in alfalfa. *Corynebacterium glutamicum* produces L-glutamic acid, which is used in the preparation of monosodium glutamate, a food flavour and converts feed stock to glutamic acid and many other amino acids.

Frankia

Frankia, a genus of Actinomycetes, is an efficient fixer of atmospheric nitrogen in non-leguminous plants.

Mycobacterium

Mycobacterium intracellulare and *Mycobacterium kansasii cause* noncontagious tuberculosis-like infections. *Mycobacterium scrofulaceum* causes lymphodentitis in children. *Mycobacterium tuberculosis* causes tuberculosis in humans. *Mycobacterium ulcerans* causes skin ulcers. Heat killed *Mycobacterium* is used as a component of immunological adjuvants to enhance mammalian immunological response.

Nocardia

Nocardia is distributed world wide in soil and water. It degrades hydrocarbons and wax, and causes deterioration of rubber joints in water and sewage pipes. *Nocardia asteroides* is an opportunistic pathogen which causes nocardiosis in humans which affects the lungs and the central nervous system.

Streptomyces

Many species of *Streptomyces* are saprophytes in soil while some are pathogenic to animals and plants. A large number of *Streptomyces* are antibiotic producers. Streptomycin, tetracycline, chloramphenicol and erythromycin are the major antibiotics produced by the strains of different species of *Streptomyces* (See Chapter 15).

Box 6.2 Finger Printing of Sediments

Zymomonas nobilis contains hopanoids in the cell membrane. Hopanoids are triterpenes discovered as geohopanoids (geochemical transformation products) in geochemical sediments. The chemical diversity of geohopanoids in unrelated sediments has made them the tools for "finger printing" of sediments. The geohopanoid patterns are indicators of oil.

6.3 SPIROCHAETES

One of the first micro-organisms described by Leewenhoek in 1600 was a spirochaete taken from saliva and tooth scrapings. Spirochaetes are helical motile single cell organisms, with some exceptions such as *Spirochaeta plicatillis.* The helical cell is covered by a multi layered membrane referred as the outer sheath, which surrounds a protoplasmic cylinder. Around the protoplasmic cylinder are wound periplasmic flagellae and endoflagellae. Spirochaetes are classified into Spirochaetaceae and Leptospiraceae. Spirochaetes cause several human diseases (Table 6.1).

Table 6.1 Human diseases caused by Spirochaetes

Pathogen	Disease
Borrelia recurrentis	Relapsing fever
Borrelia burgdorferi	Lyme disease
Leptospira interrogans	Leptospirosis
Treponema caroteum	Skin disease
Treponema pallidum sub sp., *endemicum*	Venereal syphilis
Treponema pallidum sub sp., *pertenue*	Yews

Source: Tortora *et al.*, 1998

6.4 DISEASES CAUSED BY BACTERIA

6.4.1 Plant Diseases

Bacteria cause several diseases such as spots, blights, wilts, cankers and rots on economically important cereals, millets, fruits, and vegetables.

1. Spots and Blights

Spots and blights are the most common type of bacterial disease of plants. Spots appear on the leaves, stems, flowers and fruits. In some, the spots advance rapidly and the disease is then called blight. The spots are necrotic, circular and may be surrounded by a yellowish halo. In dicots, the spots are restricted by large veins and appear round. In monocots, for the same reason they appear as streaks or stripes. All most all spots and blights are caused by *Xanthomonas* and *Pseudomonas* (Table 6.2).

Table 6.2 Bacterial Leaf spots and blights

Pathogen	Disease
	Leaf spot
Xanthomonas campestris pv. *vesicatoria*	Spot of tomato and pepper
Xanthomonas campestris pv. *malvacearum*	Angular leaf spot of cotton
Xanthomonas eleusineae	Leaf spot of ragi
Xanthomonas indica	Leaf spot of Italian millet
Xanthomonas penniseti	Leaf spot of bajra
Xanthomonas cajani	Leaf spot of *Cajanus*
Xanthomonas ricini	Leaf spot of castor
Xanthomonas alfalfae	Leaf spot alfalfa
Xanthomonas cyamopsidis	Leaf spot of *Cyamopsis*
Xanthomonas phaseoli	Leaf spot of *Lablab*
Xanthomonas stizolibicola	Leaf spot of *Mucuna* sp.
Xanthomonas vesicatoria	Leaf spot of chillies
Xanthomonas esculenti	Leaf spot of bhindi
Pseudomonas mangiferae-indicae	Leaf spot of mango (Canker)
Pseudomonas sesami	Leaf spot of *Sesamum*
Pseudomonas syringae pv. *lachrymans*	Angular leaf spot of cucumber
Pseudomonas syringae pv. *tabaci*	Wildfire of tobacco
	Blight
Xanthomonas malvacearum	Blight of cotton (angular leaf spot)
Xanthomonas oryzae	Blight of rice
Xanthomonas vignicola	Blight of cowpea
Xanthomonas sesami	Blight of *Sesamum*
Xanthomonas phaseoli, *Xanthomonas phaseoli* var. *fuscans* and *ariindicus*	Bean blight
Pseudomonas siccatum	Seedling blight of onion
Pseudomonas avenae	Leaf blight of cereals
Pseudomonas syringae pv. coronafaciens	Halo blight of oats and cereals
Pseudomonas mori	Blight of *Morus*

2. Wilts

Vascular wilts caused by bacteria affect mostly herbaceous plants. The bacteria enter, multiply in the xylem vessel of the host and interfere with translocation of water and nutrients resulting in drooping, wilting and death (Table 6.3).

Table 6.3 Bacterial Wilts

Pathogen	Disease
Pseudomonas solanacearum	Wilt of Solanum, tobacco, potato, jute.
Corynebacterium flaccumfaciens	Wilt of bean
Erwinia stewartii	Stewarts wilt of corn
Erwinia tracheiphila	Wilt of cucurbits

3. Rots

Bacterial soft rots results in infection of fleshy plant tissues in the field or storage resulting in rotting. Rotting tissues become soft, watery, slimy and crack open. The disease occurs in vegetables like potato, carrot, cabbage, lettuce and spinach. They cause loss of production both in quality and quantity and market value of crops.

The soft rot symptoms begin as a small water soaked lesion which later become soft, mushy and discoloured. The tissues within the affected region become cream coloured and slimy mass. This is followed by development of cracks and the slimy mass exudes out. In most cases, the infected parts emit a foul odour. When root crops are affected, the lower parts of the stem is infected which may turn black and stunted, wilt and die (Table 6.4).

Table 6.4 Bacterial Rots

Pathogen	Disease
Xanthomonas campestris pv. *campestris*	Black rot of crucifers
Corynebacterium michiganense ssp. *sepedonicum*	Ring rot of potato
Pseudomonas syringae pv. *atrofaciens*	Basal glume rot of cereals
Pseudomonas rubrilineans	Top rot of sugarcane
Corynebacterium tritici	Tundu (yellow ear rot) of wheat

4. Galls

Galls are produced on the stems and roots of plants infected primarily by *Agrobacterium*, *Pseudomonas*, *Rhizobacter* and *Rhodococcus*. The galls are overgrowths or tumours of disorganized plant tissue or proliferation of plant tissues that develop into organized teratomorphic organs (Table 6.5).

Table 6.5 Bacterial Galls

Pathogen	Disease
Agrobacterium rhizogenes	Hairy root of apple
Agrobacterium rubi	Crown gall of raspberry

5. Canker and Gummosis

Primarily affecting fruit trees, like pear and apple, canker also infects ornamentals, vegetables and grains. The disease damages branches and tree trunks, kills young trees, fruits and flowers and reduces the yield by 20 to 80 percent.

Canker development is preceeded by development of sunken and dark brown areas. The cortical tissues of the cankered area are bright orange to brown. Cankers some times produce gum, which breaks through the bark and runs down (gummosis). The cankers which do not produce gum are soft, moist, and sunken with foul smell. The leaves show curling, drooping and yellowing.

Leaf infections appear as water-soaked spots, which turn brown and brittle and eventually fall. The infected fruit is flat or depressed with dark brown or black spots and the underlying tissue is spongy or gummy (Table 6.6).

Table 6.6 Bacterial Cankers and Gummosis

Pathogen	Disease
	Canker
Xanthomonas citri	Canker of Citrus
Corynebacterium michiganense ssp. *michiganense*	Canker of tomato
	Gummosis
Xanthomonas vasculorum	Gummosis of sugarcane

6. Blotches and Stripes

Some bacterial pathogens cause irregular discoloration or a pustule (blotch) or stripes on the leaves and stems, reducing the photosynthetic area (Table 6.7)

Table 6.7 Bacterial blotches and stripes

Pathogen	Disease
	Blotch
Xanthomonas annamalaiensis	Leaf blotch of bajra
Xanthomonas rubri-sorghi	Red leaf blotch of sorghum
Xanthomonas translucens f. sp. *oryzae* (= *Xanthomonas oryzicola*)	Streak leaf of rice

Pathogen	Disease
	Stripe
Pseudomonas sorghi-cola	Leaf stripe of sorghum
Pseudomonas andropogonis	Stripe of corn
Xanthomonas	Red stripe of sugarcane
Pseudomonas rubrilineans	

Source: Vishwanath, 1961; Agrios, 1997.

7. Plant Diseases Caused by Fastidious Vascular Bacteria

The fastidious vascular bacteria (previously known as Rickettsia like organisms or RLO) cause the following diseases:

1. Clover club leaf disease of clover and periwinkle.
2. Greening disease of citrus plants.
3. Citrus variegated chlorosis.
4. Leaf scald of plum.
5. Leaf scorch of almond and elm.
6. Phoney peach disease
7. Pierce disease of grapes.
8. Ratoon stunting of sugarcane.

6.4.2 Insect Diseases

Some bacteria can cause insect diseases many of which reach epidemic proportions. If the insect species is beneficial, the diseases can cause enormous loss to ecological balance. Many a times diseases of insects appeared to be nature's control mechanism for prevention of mass destruction of plants and animals. *Enterobacter aerogenes* causes diseases in locusts and grass hoppers. *Bacillus thuringiensis* causes diseases in lepidopterous larvae.

6.4.3 Animal Diseases

Anthrax

Anthrax is caused by *Bacillus anthracis* which infects only herbivorous animals, particularly cattle and sheep. Two types of anthrax is observed in animals which is based on the mode of transmission. Intestinal anthrax which develops due to ingestion of anthrax spores while grazing and cutaneous anthrax developed through scratch or abrasions on skin.

Listeriosis

Listeriosis is caused by *Listeria monocytogenes* infecting mammals and birds. The symptoms are primarily neurological and are characterized by acute encephalitis, lesions and tuberculosis in the affected animals.

Plague

Plague caused by *Yersinia pestis* mainly infects wild rodents through rat fleas.

Tularemia

Tularemia is caused by *Francisella tularensis.* The disease affects pet animals.

6.4.4 Human Diseases

6.4.4.1 Bacterial diseases of the skin

Acne

Propionibacterium acnes causes inflammation of the skin, formation of white heads, black heads and pustules. Topical application of benzyl peroxide is recommended. Antibiotics as creams and lotions are also available.

Leprosy

Leprosy is also called as Hansen's disease (after the discoverer), and is caused by *Mycobacterium leprae.* The disease is characterized by loss of sensation in certain regions of the skin that is surrounded by a border of nodules. The fingers, toes, nose and ears become disfigured and even may disintigrate.

Pseudomonas Infections

Pseudomonas aeruginosa causes burn infections and dermatitis. A blue-green pus oozing out of the infection is a characteristic symptom. Fluroquinolones are useful in treating these infections.

Staphylococcus Skin Infections

Staphylococcus aureus and *Staphylococcus epidermidis* enter the body through the nasal passage, cuts, broken skin, follicles of the skin and hair resulting in furuncles (boils) and carbuncles (hard inflamed tissue under the skin). When the hair follicle is affected it results in folliculitis. In new borns it causes impetigo and scalded skin syndrome.

Streptococcal Skin Infections

Streptococcus pyogenes causes impetigo and erysipelas in children. The disease spreads by contact. (Table 6.8)

6.4.4.2 Bacterial diseases of the nervous system

Meningococcal Meningitis

Meningococcal meningitis is caused by *Neisseria meningitides.* The disease is characterized by throat infection, and bacteremia. It usually occurs in young children.

Streptococcal Meningitis

Streptococcal meningitis is caused by *Streptococcus pneumoniae*. The disease affects children between the ages of one month and three years. In addition to meningitis it causes pneumonia, epiglottitis and otitis media.

Botulism

Botulism is caused by *Clostridium botulinum,* a paralytic disease of humans and animals, contracted by consuming contaminated food. The neurotoxin produced by the bacterium causes lethargy, excessive sleep, sluggish reflexes, general weakness and respiratory failure. Patients are treated with polyvalent antitoxin.

Tetanus

Tetanus is caused by a powerful neurotoxin produced by *Clostridium tetani.* The bacterium is distributed in soil and the intestinal tract of herbivorous animals. The toxin affects the peripheral motor nerves and the central nervous system. Symptoms include painful and violent contractions of neck and jaw muscle followed by paralysis of the thoracic muscles leading to respiratory failure. Antitoxin is given to neutralize the toxin. Immunization against tetanus is accomplished by a vaccine containing tetanus toxoid (Table 6.9)

6.4.4.3 Bacterial disease of the eye

Neonatal Gonorrheal Ophthalmia

Neonatal gonorrheal ophthalmia is a serious form of conjunctivitis in children caused by *Neisseria gonorrheae.* Infection results in pus and ulceration of the cornea.

6.4.4.4 Bacterial diseases of the mouth

Dental Caries

Streptococcus mutans causes dental caries. It also causes plaque formation, resulting in tooth decay.

Gingivitis

Gingivitis is characterized by bleeding gums and is caused by variety of *Streptococci,* actinomycetes and gram negative anaerobic bacteria.

Periodontitis

A progression of chronic gingivitis results in periodontitis. Disease is caused by *Porphyromonas* and is characterized by inflamed gums and bleeding, followed by pus formation. As the disease progresses, the bone and the tissue supporting the teeth is destroyed leading to teeth loss.

Ulcerative Gingivitis

Ulcerative gingivitis caused by *Prevotella intermedia* and is characterized by ulcers in the mouth, bleeding, pain and foul breath. (Table 6.10).

6.4.4.5 Bacterial diseases of the digestive system

Cholera

The causative agent of cholera is *Vibrio cholerae*. The symptoms of the disease include 'rice water' stools, discharge of intestinal mucus, loss of body fluids and electrolytes and violent vomiting.

Gastroenteritis

Gastroenteritis is caused by *Campylobacter jejuni* and results in diarrhoea, acute inflammation of the small and large intestine and stomach cramps. It is treated with tetracycline.

Several other bacteria such as *Bacillus cereus,Clostridium perfringens, Escherichia coli, Vibrio para haemolyticus* and *Yersinia enterocolitica* also cause gastroenteritis.

Gastric Ulcer

Colonization of *Helicobacter pylori* in the stomach leads to chronic inflammatory response resulting in peptic ulcers, gastritis and lymphoid tissue lymphoma.

Listeriosis

Listeriosis is a disease caused by *Listeria monocytogenes,* which mainly affects mammals and birds. Humans can contact the disease from animals although the mechanism of transmission is not known. It mainly affects young children and aged persons. It may affect alcoholics, diabetics and cancer patients more than the others. Symptoms are neurological and characterized by encephalitis. Ampicillin is the drug of choice.

Pseudomembranous Colitis

Pseudomembranous colitis is a severe infection of the colon, caused by *Clostridium difficile,* characterized by watery diarrhoea, inflammation of the colon wall and grey or white patches on the intestinal wall. Treatment of the disease is by oral administration of vancomycin.

Typhoid

Typhoid is caused by *Salmonella typhi.* The disease is characterized by fever, headaches and diarrhoea (Table 6.11).

6.4.4.6 Bacterial diseases of the respiratory system

Diphtheria

Corynebacterium diphtheriae causes diptheria that mainly affects children. It is an airborne disease localizing in the tonsils, throat and nose. The toxins produced by the organism can circulate through out the body resulting in general toxemia. The toxin affects the heart muscle, nerve tissue and kidneys, and necrosis of throat cells. Inflammatory exudates develop into a tough pseudomembrane that extends into the trachea, suffocating the patient. An effective treatment is by diphtheria antitoxin. Antibiotics are not effective, but they may prevent further transmission of the disease.

Legionellosis

Legionellosis is a type of bronchopneumonia in humans caused by *Legionella pneumophila.* Aquired from environmental sources, it can be controlled by the use of erythromycin.

Pulmonary Tuberculosis

Mycobacterium tuberculosis causes pulmonary tuberculosis. The bacterium lodges in the air sac in the lungs and tubercles are formed. Within the tubercles the tissue cells and the macrophages die and fuse to form an amorphous mass which erodes the wall of the bronchioles. The patient suffers loss of appetite, fatigue, weight loss, night sweats and persistent cough with blood streaked sputum. Chaemotherapy is the most effective method of treatment along with antibacterial drugs to ensure that the infection does not progress to clinical tuberculosis.

Pneumonia

Although pneumonia is caused by many other bacteria, the major causal organism is *Streptococcus pneumoniae.* The disease is characterized by inflammation of the lungs, accumulation of fluid in the lungs, cold, fever and cough. Many a time the infection results in other disorders such as endocarditis, septic arthritis, mastioditis and otitis. Penicillin is the drug of choice.

Streptococcal Pharyngitis

Streptococcal pharyngitis is caused by *Streptococcus pyogenes*. The infection results in local inflammation of the throat, fever, and tonsillitis, with enlarged lymph nodes in the neck.

Whooping Cough

Bordetella pertussis is the causative agent for 'pertussis' (whooping cough), which also induces hypoglycaemia. Erythromycin is the drug of choice for treatment. It is characterised by a paroxysmal cough, cyanosis, vomiting and haemorrhagea of the nose, eyes adn even brain may occur.(Table 6.12)

6.4.4.7 Bacterial diseases of the cardiovascular system

Bacterial Endocarditis

Bacterial endocarditis is caused by α - haemolytic *Enterococcus, Streptococcus* and *Staphylococcus*. The infection arises from tooth extraction and the symptoms include fever, anaemia, heart murmur, and rapid destruction of heart valves.

Rheumatic Fever

Rheumatic fever is an autoimmune complication due to infection by *Streptococcus* resulting in arthritis, inflammation of the heart and fever. The drug of choice is penicillin (Table 6.13).

6.4.4.8 Bacterial diseases of the reproductive system

Chancroid

Chancroid is a sexually transmitted disease. A swollen painful ulcer that forms on the genitalia is the symptom of the disease followed by infected lymph nodes and discharge of pus and lesions. The causative agent is *Haemophilus ducreyi.* The recommended antibiotics includes erythromycin and ceftriaxone.

Gardnerella Vaginosis

The causal agent of gardnerella vaginiosis is *Gardnerella vaginalis,* which causes inflammation of the vagina, fishy odour, frothy vaginal discharge, sloughing of vaginal tissue and over growth of anaerobic bacteria.

Gonorrhea

Gonorrhea is a sexually transmitted disease caused by *Neisseria gonorrheae.* In the males, the primary site of infection is urethra and in the female it is the cervix. The symptoms include painful urination and fever in males and in females vaginal discharge and abdominal pain (Table 6.14).

6.4.4.9 Bacterial diseases of the urinary system

Cystitis

Cystitis is inflammation of the urinary bladder followed by dysuria and polyuria. The causal organism is *Staphylococcus saprophyticus.* The drug of choice is ampicillin.

Pyelo Nephritis

Pyelo nephritis is the inflammation of the kidney. The causative agent is *Escherichia coli.* The drug of choice is cephalosporin (Table 6.15).

Table 6.8 Bacterial diseases of the skin

Disease	Pathogen	Symptoms
Impetigo	*Staphylococcus aureus, Streptococcus pyogenes*	Superficial skin infection, isolated pustules.
Folliculitis	*Staphylococcus aureus*	Infection of hair follicle.
Toxic shock syndrome	*Staphylococcus aureus*	Fever, rash, shock.
Necrotizing fasciitis	*Streptococcus pyogenes*	Extensive tissue destruction.
Erysipelas	*Streptococcus pyogenes*	Reddish patches on skin; often with high fever
Pseudomonas dermatitis	*Pseudomonas aeruginosa*	Superficial rash
Acne	*Propionibacterium acnes*	Inflammatory lesions originating with accumulation of sebum that rupture a hair follicle
Leprosy	*Mycobacterium leprae*	Loss of sensation in certain regions on the skin surrounded by border of nodules.

Table 6.9 Bacterial diseases of the nervous system

Disease	Pathogen	Symptoms
Meningococcal meningitis	*Nesseria meningitidis*	Affects mostly children under the age 2. Symptoms include throat infection
Streptococcal meningitis	*Streptococcus pneumoniae*	Symptoms includes pneumonia, epiglotitis and otitis.
Tetanus	*Clostridium tetani*	Toxin formed in contaminated wound causes upper muscle contractions, and eventual respiratory failure.
Botulism	*Clostridium botulinum*	Toxin ingested in food; causes partial respiratory failure.

Table 6.10 Bacterial diseases of the mouth

Disease	Pathogen	Symptoms
Dental caries	*Streptococcus mutans*	Accumulation of plaque; localized acid production by bacteria, forming hole in tooth.
Periodontal disease	*Porphyromonas* species	Presence of bacterial plaque initiates inflammatory response that destroys bone and tissue.
Gingivitis	*Streptococcus* species	Bleeding gums.
Ulcerative gingivitis	*Prevotella intermedia*	Ulcers in the mouth, foul breath,

Table 6.11 Bacterial diseases of the digestive system

Disease	Pathogen	Symptoms
Typhoid fever	*Salmonella typhi*	Pathogen is shed in human faeces. Symptoms include high fever, disseminated infection.
Cholera	*Vibrio cholerae*	Exotoxin causes diarrhoea with loss of water and electrolytes.
Enterotoxigenic *E. coli* gastroenteritis	*Escherichia coli*	Watery diarrhoea that resembles mild form of cholera; typical traveler's diarrhoea.
Parahaemolyticus gastroenteritis	*Vibrio parahaemolyticus*	Exotoxin causes cholera like diarrhoea, but generally milder.
Enteroinvasive *E. coli* gastroenteritis	*Escherichia coli*	Enterotoxin causes Shigella-like dysentery.
Enterohemorrhagic *E. coli* gastroenteritis	*Escherichia coli* O157: H7	Causes haemorrhagic colitis (bloody stools) and haemolytic uremic syndrome (blood in urine and possible kideny failure).

Disease	Pathogen	Symptoms
Campylobacter gastroenteritis	*Campylobacter jejuni*	Diarrhoea, acute inflammation of intestine and stomach cramps.
Helicobacter peptic ulcer	*Helicobacter pylori*	Pathogen is adapted to survive in stomach; presence leads to peptic ulcers.
Yersinia gastroenteritis	*Yersinia enterocolitica*	Symptoms are abdominal pain and diarrhoea.
Clostridium perfringens gastroenteritis	*Clostridium perfringens*	Diarrhoea, nausea or vomiting; probably caused by different toxins.
Pseudomem-branous colitis	*Clostridium difficile*	Inflammation of the colon and watery diarrhoea.
Listeriosis	*Listeria monocytogenes*	Usually transmitted by contaminated food; symptoms are primarily neurological, characterised by encephalitis.

Table 6.12 Bacterial diseases of the respiratory system

Disease	Pathogen	Symptoms
Streptococcal pharyngitis	*Streptococcus pyogenes*	Inflamed mucous membrane of the throat (strepthroat).
Diphtheria	*Corynebacterium diphtheriae*	Bacterial exotoxin interferes with protein synthesis, damages the heart, kidney and other organs; causes membrane formation in the throat.
Pertussis (whooping cough)	*Bordetella pertussis*	Cilia in upper respiratory tract inactivated; mucous accumulates followed by spasms of intense cough.
Tuberculosis	*Mycobacterium tuberculosis, Mycobacterium bovis*	Chronic cough, fever, wasting of the body tissue.
Streptococcal pneumonia	*Streptococcus pneumoniae*	Infected alveoli of lung fills with fluids, interferes with oxygen uptake
Legionellosis	*Legionella pneumophila*	Potentially fatal pneumonia that tends to affect males who drink or smoke heavily. Pathogen grows in water and in air- conditioning towers.

Table 6.13 Bacterial diseases of the cardiovascular system

Disease	Pathogen	Symptoms
Bacterial endocarditis	Mostly α-haemolytic *Streptococci*	Bacteria lodges in heart valves and may cause fatal damage.
Pericarditis	*Streptococcus pyogenes*	Affects sac surrounding heart, rapidly progressive.
Rheumatic fever	*Group A α-haemolytic Streptococci*	Probably an autommune condition; repeated streptococcal infections result in antibodies that damage heart tissue.

Table 6.14 Bacterial diseases of the reproductive system

Disease	Pathogen	Symptoms
Chancroid (soft chancre)	*Haemophilus ducreyi*	Painful ulcers of genitals, swollen lymph nodes on groin.
Gardnerella vaginosis	*Gardnerella vaginalis*	Fishy odor, frothy vaginal discharge.

Table 6.15 Bacterial diseases of the urinary system

Disease	Pathogen	Symptoms
Cystitis (urinary bladder infection)	*Escherichia coli, Staphy lococcus saprophyticus*	Difficulty or pain in urination.
Pyelonephritis (kidney infection)	*Escherichia coli*	Fever, back or flank pain.

Source: Tortora *et al.*, 1998.

6.4.4.10 Other human diseases caused by bacteria

Anthrax

Anthrax is caused by *Bacillus anthracis.* The grazing animals acquire infection after ingesting the endospores in the soil. Humans contact the disease by handling the hides of infected animals and the bacterium enters through a cut in the skin. The symptoms include pustules, septicemia and pneumonia.

Brucellosis (undulant fever)

Brucellosis is caused by *Brucella abortus, Brucella melitensis* and *Brucella suis.* Domesticated animals and unpasteurized milk constitute the source of infection. The bacteria enter through the breaks or cuts in the skin. Clinical signs are fever that spikes each evening (undulant fever). Tetracycline is the drug of choice.

Gangrene

Gangrene is caused by *Clostridium perfringens* in humans. The disease results from contamination of wounds. The disease is characterized by toxemia along with the development of gas in the tissues resulting in tissue damage particularly the connective tissue. The treatment is the surgical removal of the affected part, antibiotics to stop bacterial growth and hyperbaric oxygen; (placing the patient in a chamber containing pure oxygen.) As the pathogen is anaerobic, its growth is inhibited.

Otitis Media

Otitis media is caused by *Streptococcus pneumoniae* and *Moraxella catarrhalis,* infecting the middle ear, causing inflammation, pus and pain. First choice of therapy is penicillin.

Scarlet Fever

Scarlet fever is caused by *Streptococcus pyogenes.* The toxin released by the bacterium causes pinkish red skin rash with strawberry like spots on the tongue, resulting in peeling skin.

Septicemia, Sepsis and Septic Shock

Although blood is generally sterile, a moderate number of micro-organisms can enter the blood stream without causing any harm. However, if the defences of the blood and lymphatic system fail, microbes undergo uncontrolled proliferation in the blood resulting in septicemia. The clinical symptoms include chills, fever and inflamed lymph vessels causing red streaks under the skin, running along the arms and legs. The toxic condition resulting from septicemia is sepsis. The endotoxins released by the bacteria cause a drop in blood pressure resulting in septic shock. The disease is caused by gram negative bacilli and a few gram positive rods.

Tularemia

Tularemia is caused by *Francisella tularensis,* transmitted from pet animals, especially rabbits or consumption of undercooked meat of diseased animals or bite of deer flies, with fever as the main symptom. Other symptoms include development of papules on the skin of fingers or hands which later become an open sore (Tortora *et al.,* 1998; Pelczar *et al.,*1993) (Table 6.16).

Table 6.16 Other human diseases caused by bacteria

Disease	Pathogen	Symptoms
Anthrax	*Bacillus anthracis*	Symptoms include pustules, septicemia and pneumonia.
Brucellosis	*Brucella* sp.	Mostly acquired by ingesting cow or other milk, contact with animal carcasses. Clinical sign is fever in the evening.
Gangrene	*Clostridium perfringens*	Caused by contamination of an open wound by clostridial endospores; toxins destory tissues; symptoms include toxemia, development of gas in the connective tissue
Otitis media	*Streptococcus pneumoniae, Moraxella catarrhalis*	Infecting the middle ear, causing inflammation, pus and pain.
Plague	*Yersinia pestis*	Classically transmitted by fleas on rats. Symptoms include swelling of the lymph nodes and fever
Tubremia	*Francisella tularensis*	Symptoms include fever, development of papules on the of skin fingers and hands

6.4.5 Opportunistic Bacterial Pathogens

Opportunistic pathogens are organisms that ordinarily do not cause disease in their normal habitat in a healthy person. For example the microbes that are considered as members of the normal micro biota of the mouth can cause tooth decay and gum disease or when the host is weak the microbes that are usually harmless can cause disease. Some times people carry micro-organisms that are generally pathogenic but may not cause disease in them (carriers). *Neisseria meningitidis* which resides benignly in the respiratory tract can cause meningitis. *Streptococcus pneumoniae* a normal resident of the nose and throat can cause pneumonia.

6.4.6 Nosocomial Infections

Nosocomial infections are infections contacted from hospitals as a result of weakened state of the host. The microorganisms in the hospital environment are opportunistic pathogens or those which have become drug resistant. The infections result from catheters, tracheotomy, ventilators, intravenous therapy and from hospital sinks and flower vase water (Table 6.17).

Table 6.17 Bacterial nosocomial infections

Bacteria	Infections
Proteus, Enterobacter, Enterococcus, Serratia marcescens, Staphylococcus aureus	Bacteremia, septicemia, gastrointestinal inflammation, neonatal meningitis, urinary and respiratory tract infection, endocarditis, cutaneous infections, puerperal sepsis.

6.5 MEDICINAL USES OF BACTERIA

6.5.1 Bacterial Prebiotics

Prebiotics are a category of functional foods, some of which are undigested oligosaccharides in the human body. These are non-digestible oligosaccharides (NDOs) which form the substrate for *Bifidobacterium* and *Lactobacilllus* in the colon and enhance the growth of these prebiotic bacteria. These are also used as dietary fibres, bulking agents, sugar substitutes, fat replacers and functional food ingredients. Many of the commercially available NDOs produced by microbial enzymes are given in Table 6.18. The health benefits of prebiotics are given in Table 6.19 (Ouwehand *et al.*, 2002).

6.5.2 Bacterial Probiotics

Probiotics are dietary supplements containing potentially beneficial bacteria or yeasts (See Chapter 7), but lactic acid bacteria are most commonly used. In typical cases, live active bacteria in raw milk, curds, yoghurt and similar products are used as probiotics.

The ancient Indian practice of drinking freshly drawn raw milk and urine therapy, are ethnic probiotic practices. Lactic acid bacteria have been in use for a long time for the production of lactic acid from lactose, the sugar in milk. Lactic acid gives sour taste to milk foods and acts as a preservative.

Probiotics are conceived to assist the body's naturally occurring flora within the digestive tract to re-establish themselves, after a disturbance caused by disease, use of drugs, alcohol stress, toxins and the like. In such situations, bacteria in probiotics form temporary colonies making way for the restoration of normal gut flora.

Some common probiotics of bacterial origin include *Bifidobacterium bifidum, Bifidobacterium breve, Bifidobacterium infantis, Bifidobacterium longum, Lactobacillus acidophilus, Lactobacillus casei, Lactobacillus johnsonii, Lactobacillus plantarum* and *Lactobacillus rhamnosus.*

Some like *Lactobacillus bulgaricus* and *Streptococcus thermophilus* found in yoghurt do not have a probiotic effect. There are also some cases of fraud in probiotic marketing with such species as *Lactobacillus sporogenes, Lactobacillus bifidus* and *Lactobacillus caucasicus,* which are unknown to science.

Table 6.18 Commercially availabe NDO's produced by microbial enzymes

Oligosaccharide	Substrate	Enzyme used
Fructo-oligosaccharides Fructofuranosyl (Fn) type Non-reducing glucosyl (GFn) type	Inulin	α-fructo furanosidase
α– galacto–oligosaccharides	Lactose	α–galactosidase
Genito–oligosaccharides	Starch / Maltodextrose	α–galactosidase
Glucosyl sucrose	Maltose / Sucrose	Glucano transferase
Isomalto–oligosaccharides	Starch / Maltose	α–glucosidase
Lactosucrose	Lactose/ Sucrose	Levansucrase
Palatinose (Isomaltulose)	Sucrose	Sucrose – 6- glucosyl mutase

Source: O'Toole and Lee, 2006.

Table 6.19 Health benefits from prebiotic bacteria.

Source species	Benefits
Lactobacillus casei Shirota *Lactobacillus reuteri* SD2112 *Lactobacillus rhamnosus* GG	In treating diarrhoea
Bifidobacterium lactis Bb12	Inflammatory bowel disease
Lactobacillus johnsonii	Reduction in colonization of *Helicobacter pylori*
Lactobacillus plantarum 299V	Reduction of irritable bowel syndrome; lowers LDL cholesterol
Bifidobacterium lactis Bb12	For travelers diarrhoea and antibiotic associated diarrhea; treatment of allergies

6.5.3 Use of Bacterial Toxins as Therapeutics

Most of the pathogenic bacteria produce toxins that kill host cells. Pathogenecity of the bacteria is measured by the potency of the secreted toxins. These bacterial toxins can be harnessed as molecules with beneficial effects. The following are some such examples.

Neurotoxins

Botulinal toxin is a neurotoxin secreted by *Clostridium botulinum* which is the causal agent of botulism – a fatal food poisoning. The toxin is used as a general therapy (approved by FDA of U.S.A.) for dystonia – a painful neurological disorder affecting large populations in North America and is characterized by involuntary muscle contraction that makes body movement painful.

Before the use of botulinal toxin was known, the therapy available was surgical destruction of nerve endings affected by spasms which was very expensive and risky since dystonia involves sustained contraction of muscle. Neuromuscular blocking effect of botulinal toxin alleviates the muscle spasm caused by the disease. The method of treatment is injecting the toxin to the spasm affected site. Botulinal toxin therapy is cheaper and less dangerous but the effect is temporary. Repeated treatment is administered for continuous relief. Botulin is also used in cosmetic industry as botox injections for removing ageing lines, crow's feet and laughter lines.

Chimeric toxins

Another group of cytotoxins (exotoxins) produced by *Pseudomonas aeruginosa* and Pertussis toxin produced by *Bordetella pertussis* are used therapeutically. These toxins contain the enzymatic domain that carries out the catalytic function of the toxin. Introduction of the enzymatic domain of the toxin into the cytoplasm of the tumour cells in a specific manner will be able to in killing those cells selectively. This idea led to the study of chimeric toxin. A chimeric toxin is chemically attached to a protein molecule with high specificity for certain receptors found on the surface of eukaryotic cells, so that the action can be directed towards the targeted cells. Toxin targeting is very useful in killing those cells selectively. This idea led to the study of chimeric toxin. A chimeric toxin is chemically attached to a protein molecule with high specificity for certain receptors found on the surface of eukaryotic cells, so that the action can be directed towards the targeted cells. Toxin targeting is very useful and effective for killing solid tumour cells.

6.5.4 Medicinal uses of Bacterial Toxins in Homoeopathy

Botulin toxin

Botulin is a toxin produced from *Clostridium botulinum* which is used in the homoeopathic system of medicine for curing food poisoning, double vision, blurred vision, difficulty in swallowing and breathing, choking sensation, dizziness, cramps in the stomach, weakened facial muscles and severe constipation.

Gonorrheal toxin

In the homoeopathic system of medicine, the concentrated particles from gonorrheal sore is used for treating various disorders. It is known as medorrhinum and is used for chronic ailments. It is an effective remedy for chronic rheumatism, irritability of nervous system, intolerable pain, chronic catarrh, edema of limbs, weak memory, restlessness, fear of darkness, burning pain in the occiput, pain in the eye ball, dandruff, pulsuation in the ears, pharyngeal catarrh, ravenous hunger, excessive thirst, craving for liquor, salt and sweets, violent pain in the liver and spleen, nocturnal aneuresis, offensive menses, renal colic, leucorrhoea with fishy odour, sycotic warts on the genitals, ovarian pain, metrorrhagia, menstrual colic, nocturnal emissions, enlarged and painful prostate, painful urination, dry night cough, burning of hands and feet, copper coloured spots on the skin, tumour on the skin and night sweat (Boericke, 1991).

Tuberculinum toxin

The bacterial extract (Tuberculinum) is given in renal affection, chronic cystitis, incipient tuberculosis, low receptive power, fatigue, aversion to work rapid emaciation, chronic diarrhoea, bluish pallor of the skin, mental retardation, nervous weakness, epilepsy, arthritis, mania, melancholy, insomnia, depression, neuralgia, meningits, polyuria, boils on nose, otorrhoea, benign mammary tumours, enlarged tonsils, bronchorrhoea, shortness of breath,broncho-pneumonia in children, chronic eczema, measles,psoriasis, sleepiness, remittent fever, profuse sweating, pthisis, *Staphylococcus* and *Pneumococcus* infection of the lungs. It is a leading remedy when prescribed drugs fail to act on diseases.

Box 6.3 Fluorescent Pseudomonads

Pseudomonas aeruginosa, Pseudomonas fluorescens, Pseudomonas putida and *Pseudomonas syringae*, produce a bluish green or yellowish green pigment. Pseudomonads when viewed under U.V. light exhibit fluorescence. Luminescent pseudomonads are being used as bicontrol agents against root rots, seedling blights caused by species of *Fusarium, Pythium, Phytophthora* and *Rhizoctonia.*

6.6 BACTERIAL ENZYMES IN GENETIC ENGINEERING

Enzymes such as DNA polymerase and restriction enzymes have been isolated and purified from different species of bacteria. They are used in gene manipulation such as cloning and sequencing.

DNA polymerase required for generating DNA fragments in both pro and eukaryotic cells is isolated from thermophilous bacteria inhabiting hot springs. DNA polymerases are used to make the sticky ends of DNA fragments to blunt ends. Taq polymerase is isolated from *Thermus aquaticus*. They are also used to incorporate radioactive labels in DNA via random primer labelling where nucleotides containing radioactive isotopes are built into DNA strands to determine nucleic acid sequences.

DNA ligase is another enzyme widely used in genetic engineering and isolated from *E.coli* is called as 'super glue' that joins the two DNA fragments together. Restriction endonucleases used to cut and paste DNA fragments into plasmic vectors (cloning) are used as templates in genetic manipulation, RNA synthesis and as probes. These were discovered in *E.coli* in 1970. The first enzyme to be isolated was EcoRI by Robert. N. Yoshimori in San Fransisco. A list of restriction endonucleases from various bacterial sources are given in Table 6.20

Table 6.20 List of restriction endonucleases from bacteria

Bacterial species	Restriction endonucleases
Bacillus amyloliquefaciens	*Bam* HI
Caryophanon latum	*Cla* I
Diplococcus pneumoniae	*Dpn* I
Escherichia coli RY13	*Eco* RI
Escherichia coli J62	*Eco* RV
Haemophilus influenzae Rd	*Hind* III
Haemophilus influenzae Rf	*Hind* FI
Moraxella bovis	*Mbo* I
Neisseria denitrificans	*Nde* I
Serratia marcescens	*Sma* I

Source: Chow *et al.*, 2006

Box 6.4 Bioluminiscence

Bioluminiscence is the emission of light by eukaryotic or prokaryotic organisms where chemical energy is converted into bluish green light energy. Bacteria like *Beneckea, Photobacterium* and *Vibrio* are bioluminiscent. *Photobacterium phosphoreum* is a marine bacterium used extensively as a bioluminescent indicator of pollutants in marine water to monitor its quality. Genes controlling the light emission of *Photobacterium* are located in the lux operon and is transferred to *E.coli* making it a transgenic bioluminescent bacterium. Such transgenic bacteria are used to develop biosensors to detect pollutants in industrial effluents.

6.7 OTHER USES OF BACTERIA

Production of Ethanol

Clostridium thermocellum, *Clostridium thermohydrosulfurium*, *Thermoanaerobium brockii*, *Thermobacteroides acetoethylicus* and *Zygomonas mobilis* are employed in the production of ethanol.

Bacteria in Leaching of Metals

Bacteria play an important role in the extraction of metals from their ores. Iron oxidising bacteria like *Pseudomonas aeruginosa, Ferrobacillus ferro-oxidans, Ferrobacillus sulfo-oxidans,* and *Thiobacillus thio-oxidans* oxidize suphide metals into sulphates and sulphuric acid. These bacteria are also helpful in leaching of copper, nickel, molybdenum, zinc, silver, uranium and arsenic from their sulphates. Bacterial leaching is economical for extraction from low grade minerals.

Retting of Fibers

Retting is a process in which plant fibers are separated by activity of microbes like *Clostridium butyricum* and *Clostridium felsineum.* They hydrolyse pectic substances from the middle lamella and separate the fibres.

Curing of Tobacco and Tea

Curing is a process where the leaves of tobacco and tea are allowed to ripen by fermentation to give characteristic taste and aroma. *Micrococcus candiscens* is utilized for curing.

Bacteria as Alternative Energy Source

As our supply of fossil fuels diminish, interest in the use of renewable energy source also increase. Biomass, the collective organic matter formed by living organisms can be the source of future energy. Microbes are being used for bioconversion of biomass into alternative energy source and also can decrease the amount of waste materials requiring disposal. Methane is the most convenient energy source formed from bioconversion. The aerobic methane oxidizing bacteria can be used for production of methane from the biomass.

Bacteria can be employed in converting solar energy for combating energy crisis. Autotrophic bacteria can also be utilized for producing inorganic energy by solar thermal method.

Bacteria have immense industrial applications. Many industrial products or raw materials for the products come from bacteria (See Chapter 20). They can be utilised in bioremediation (See Chapter 23). They form a major component of biogeochemical cycles (See Chapter 22) and contribute to industrial products which are of therapeutic value (See Chapter 14, 15, 16 and 20). They are the major contributors to alcoholic beverages, fermented food products.

Box 6.5 Spider Silk

Spiders secrete different kinds of silks varying in adhesiveness, flexibility and strength for spinning webs. Drag line silk, the strongest of these is made up of polypeptides with short amino acid side chains. Production of spider silk proteins using genetically engineered *E. coli* have opened the way for making new fibres of exceptional strength and for a new class of composite materials.

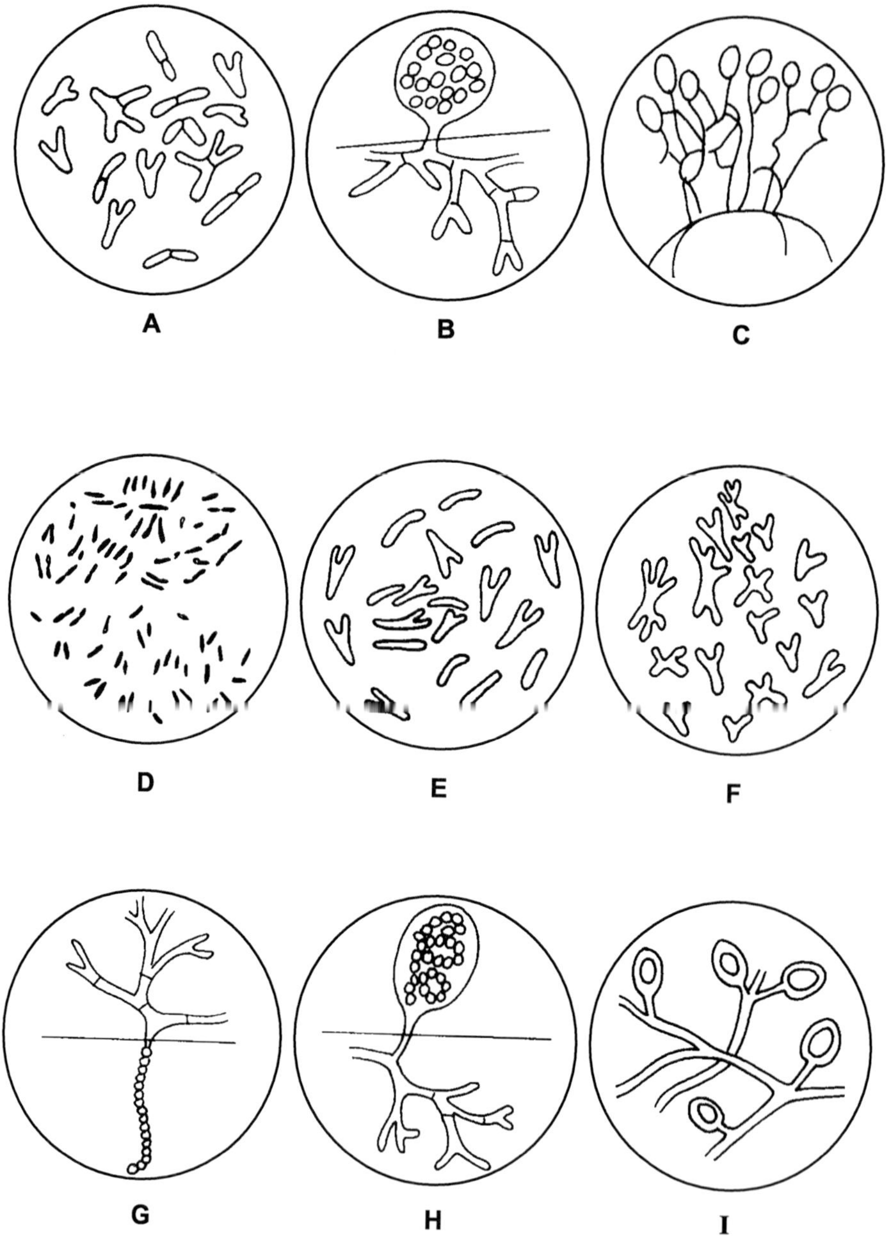

Fig. 6.1 **Bacteria.** *Actinomycetes.* **A** : *Actinomyces israelii* ; **B** : *Actinoplanes*; **C** : *Ampulariella* ; **D** : *Corynebacterium* sp. ; **E** : *Mycobacterium* sp.; **F** : *Nocardia asteroides* ; **G** : *Streptomyces* sp. ; **H** : *Streptosporangium* sp.; **I** : *Thermoactinomyces vulgaris.*

Fig. 6.1 to 6.3. Redrawn from Pelczar *et.al.,*(1993) and Purohit, S.S. (1999)

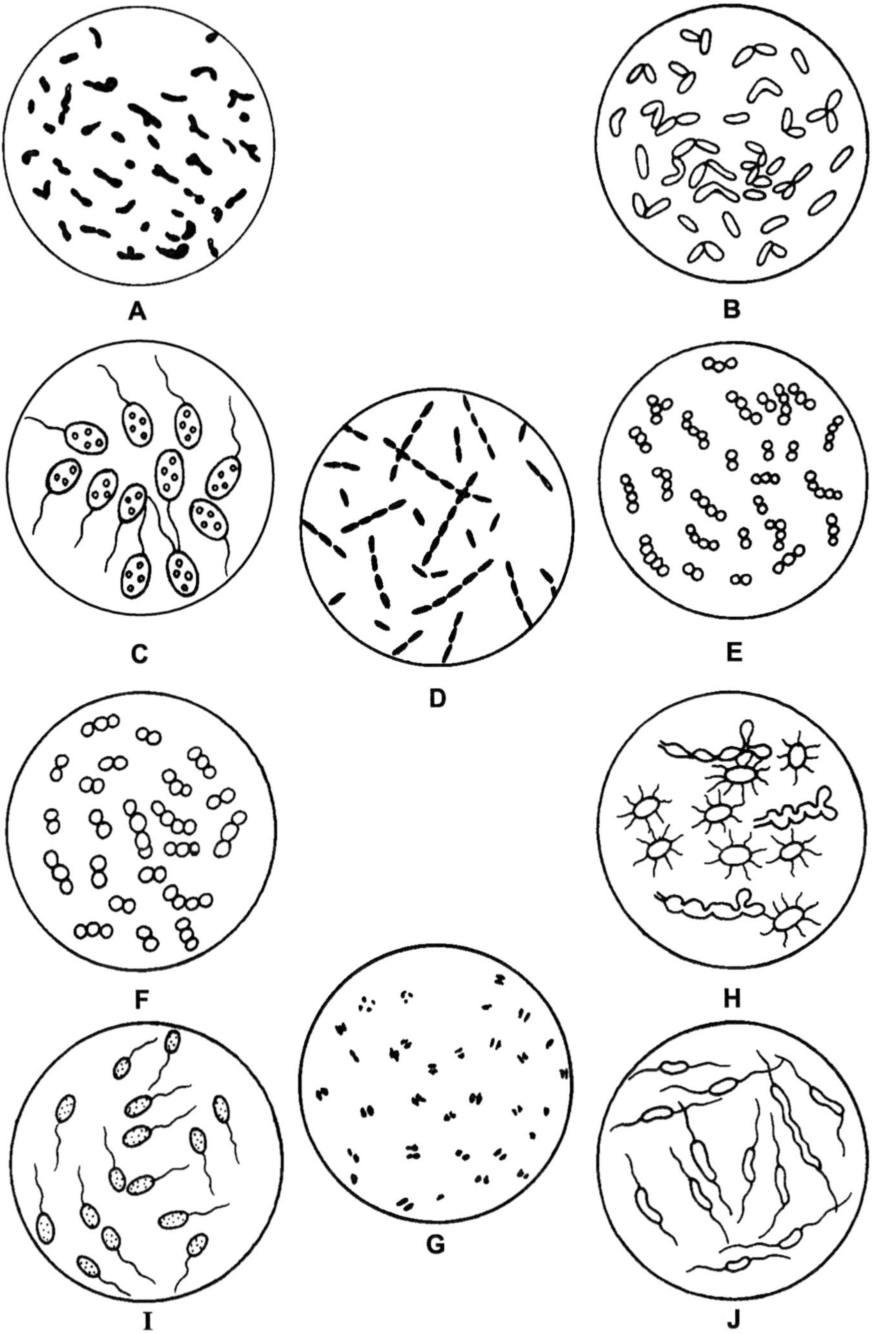

Fig. 6.2 **Bacteria. A** : *Arthobacter* sp. ; **B** : *Brevibacterium linens* ; **C** : *Chromatium* sp. ; **D** : *Lactobacillus* sp. ; **E** : *Lactococcus* sp. ; **F** : *Leuconostoc mesenteroides* ; **G** : *Pediococcus* sp. ; **H** : *Rhodomicrobium vannielii* ; **I** : *Rhodopseudomonas* sp. ; **J** : *Rhodospirillum rubrum*

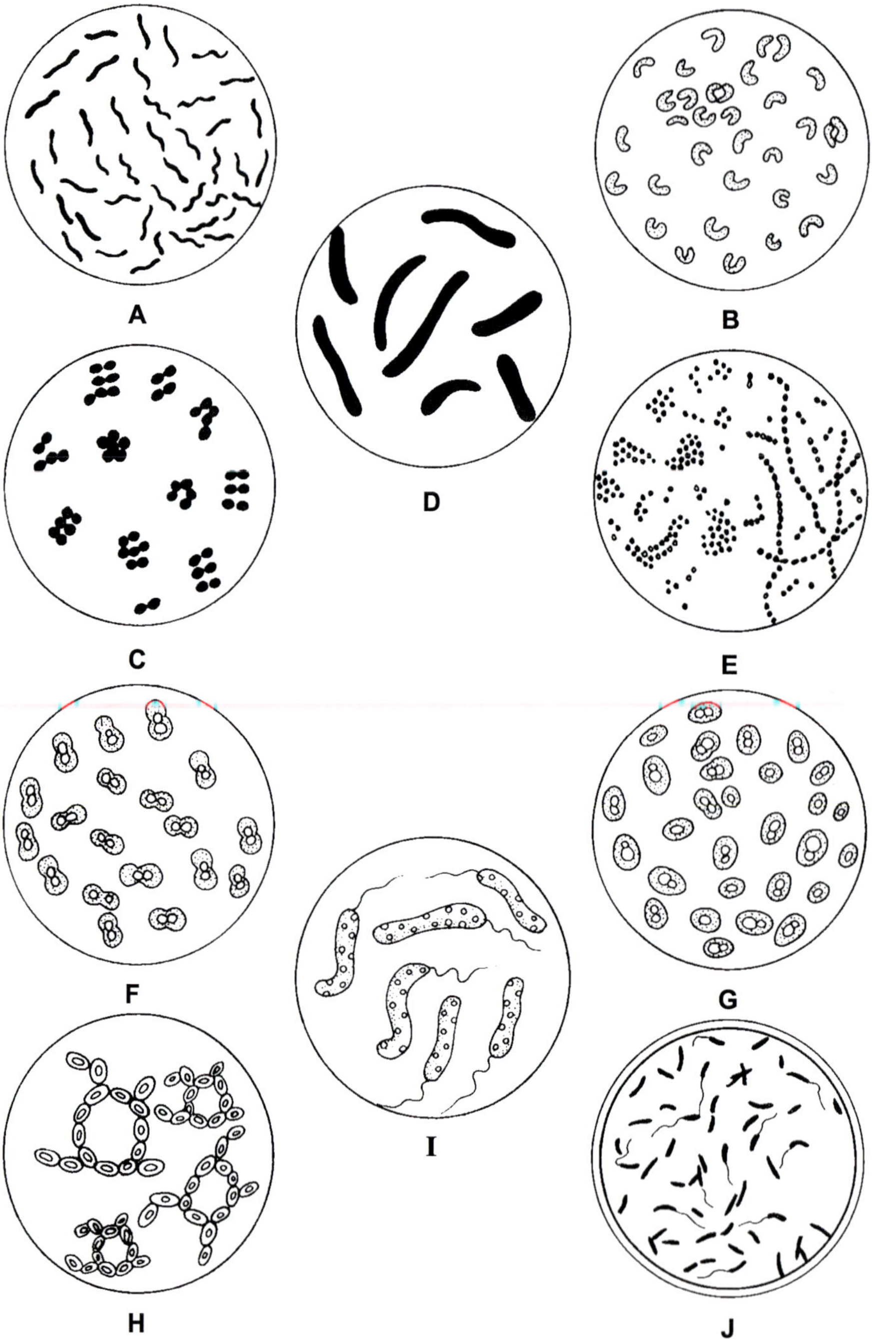

Fig. 6.3 **Bacteria. A** : *Aqua spirillum* ; **B** : *Rhodocyclus purpureus* ; **C** : *Sarcina* sp. ; **D** : *Spirillum volutans* ; **E** : *Staphylococcus* and *Streptococcus* sp. ; **F** : *Streptococcus pneumoniae* ; **G** : *Thiocapsa reseopersicina* ; **H** : *Thiodictyon elegans* ; **I** : *Thiospirillum jenense* ; **J** : *Vibrio* sp.

CHAPTER 7

YEASTS

7.1 INTRODUCTION

The use of yeasts to make bread and alcohol has been recorded for thousands of years. The Babylonians (circa 6000 B.C.) and Egyptians (circa 5000 B.C) have left written accounts of their production of beers, wines and bread which warranted the use of yeasts.

Yeasts are abundantly present in soil, decaying vegetables and fruits. Aptly called Saccharomycetes, the 'Sugar fungi', yeasts grow luxuriantly where there is plentiful supply of sugar. More than 1,000 species of yeasts are known, mainly from Saccharomycetes and non-Saccharomycetes genera such as *Candida* and *Cryptococcus.*

Yeasts are unicellular organisms, the cells being minute, spherical or oval. Individually the cells are hyaline but in colonies they appear white, cream or slightly brownish. One group of yeasts reproduce vegetatively by forming buds from mature cells (budding yeasts) while the other group forms daughter cells by cell division. Yeast cells may form long branched chains resembling filaments. Another group, the dimorphic yeasts, may form mycelia like other fungi.

The cell wall is composed of cellulose like substance. The cells contain granular cytoplasm, mitochondria, fat, glycogen and a central area traversed by chromatin threads, a nucleolus and a centrosome.

Yeasts show a great diversity in terms of colour and geometry. Following is a list of yeasts which are differently coloured (pigmented yeasts). *Saccharomyces cerevisiae* (cream), *Geotrichum candidum* (white), *Aureobasidium pullulans* (black), *Phaffia rhodozyma* (pink), *Rhodotorula rubra* (red) *and Cryptococcus laurentii* (yellow).(Walker and White, 2005) Many of the pigmented yeasts have biotechnological uses such as astaxanthin from *Phaffia rhodozyma* as aquacultural feed supplement for farmed salmon.

Yeasts are the mediators of fermentation of simple sugars. Two most important application of yeasts are to leaven bread using strains of baker's yeast and the other to produce alcoholic beverages using strain of brewer's yeast. In the absence of oxygen, fermentative yeasts produce their energy by converting carbohydratаes into carbon dioxide and ethanol (ethyl alcohol) or lactic acid. In brewing, the ethanol is bottled, while in baking the carbon dioxide raises the bread, and the ethanol evaporates.

7.2 YEASTS AS HUMAN PATHOGENS

7.2.1 Diseases of the Mucous Membrane

Species of *Candida* are the most common yeasts in the human body. The transition from a commensal to a harmful pathogenic state eventually depends on a decrease in host resistance, changes in the local ecology or changes in intrinsic fungal virulence. The histopathological changes related to the infection include inflammatory cell infiltration of the epithelium and connective tissue, intracepithelial microabscesses, epithelial intercellular edema, epithelial atrophy, hyperplasia and dysplasia.

7.2.1.1 Candidiasis

The genus *Candida* has approximately 200 species that are ubiquitous in the environment. However, a dozen of the species are associated with human commensalism. They are *Candida albicans, Candida dubliniensis, Candida famata, Candida glabrata, Candida guilliermondii, Candida inconspicua, Candida kefyr, Candida krusei, Candida lusitaniae, Candida norvegensis, Candida parapsilosis* and *Candida tropicalis.*

Candidiasis is a disesase of the mucous membrane caused by *Candida* infection which is carried innocuously by humans on the epithelial surface of the mouth, skin and gastrointestinal tract. Though they are harmless in healthy individuals when the host's immune defenses are compromised they cause several diseases.

There are different types of candidiasis. The most important ones are listed below.

Alimentary candidiasis

Alimentary candidiasis is caused by the species of *Candida.* It usually infects oesophagus and the intestine of HIV, cancer patients and diabetics. It is characterized by the development of lesions in the oral cavity, thrush, glossitis, stomatitis and angular chelitis.

Bronchial candidiasis

Bronchial candidiasis is caused by *Candida* species. The symptoms include bronchitis with cough and production of sputum.

Chronic mucocutaneous candidiasis (CMC)

Chronic mucocutaneous candidiasis is caused by *Candida albicans.* It infects the epithelial surface of the skin, oral mucosa, upper respiratory tract, gastrointestinal and urinary tract and genitals. It is an aquired or congenital form of candidiasis often associated with autoimmune diseases and is normally chronic where the patient has persistent and recurrent candidiasis of the skin, nails and mucous membranes. The infection results due to failure of T lymphocytes to produce cytokines, leucocyte dysfunction, hypoparathyroidism, diabetes and polyglandular auto immune disease.The symptoms include classic lesions which are verrucous and warty with horn like projections growing out from the skin (Sullia and Shantharam, 2005).

Mucocutaneous candidiasis can occur as a) chronic localized mucocutaneous candidiasis of the nails and skin of hands and feet, b) chronic localized mucocutaneous candidiasis with granulomatous lesions on the face and scalp and c) chronic localized mucocutaneous candidiasis with endocrine disorders. Other forms occur on groin, inter and submammary folds, gluteal folds, inter digital spaces and umbilicus. Triggering factors are obesity, friction and diapers.

It often affects people whose fingers have been continually immersed in detergents that macerate the nail folds and cuticle. The candidiasis lesions are painful erythematous swelling around the affected nail and finally result in total destruction of the nail.

Candidiasis of gallbladder, pancreas and peritoneum

The infection of *Candida* occurs due to dialysis, gastrointestinal ulcers, colitis and surgery. Symptoms of infection are fever and abdominal pain.

Gastrointestinal candidiasis

Gastrointestinal candidiasis caused by *Candida* species is often associated with haematological malignancies with ulceration of stomach, duodenum and intestine leading to peritonitis.

Hepatic and hepatospleenic candidiasis

Hepatic candidiasis caused by *Candida* species infects mainly immunocompromised patients or those with severe neutropenia and acute leukemia. Symptoms include fever, hepatospleenomegaly and abcesses containing small number of pseudohyphae.

Oesophageal candidiasis

Infection by *Candida albicans, Candida glabrata* and *Candida krusei* results in oesophageal candidiasis. It is frequently associated with AIDS. Symptoms include pain in the retroesternal region, dysphagia, nausea and vomiting with white mucosal plaques and erythema in the oesophagus. The infection is associated with Herpes simplex virus or Cytomegalo viral infection.

Oral candidiasis

Candida albicans is often isolated from warm blooded animals including humans where it exists as part of the normal microbiota of mucous membrances. It causes oral candiadiasis, a disease of the mucous membrane of the mouth and alimentary tract. More serious infections can involve endocarditis, septicemia and meningitis. Prolonged use of antibiotics, steroids, obesity and alcoholism are predisposing factors for *Candida* infection.

There are four types of oral candidiasis (a) acute pseudomembranous candidiasis (thrush), (b) acute atrophic candidiasis, (c) chronic atrophic candidiasis (denture stomatitis) and (d) chronic hyperplastic candidiasis.

Osteoarticular candidiasis

Prostatic or rheumatoid joints are susceptible for *Candida* infection. Arthritis may result in conjunction with candidiasis. The infection may be due to articular corticosteroid injection or surgery. Pain in the knee is the normal symptom.

Pulmonary candidiasis

Pulmonary candidiasis by *Candida* is a primary disease characterized by cough, low grade fever, night sweats, weight loss and production of mucoid gelationus sputum with bloody tinge. Pulmonary candidiasis may result in pneumonia (Candida pneumonia) and is very difficult to diagnose.

Urinary tract candidiasis

Risk factors of infection by *Candida* species include catheters and advanced lower urinary tract infections. It is normally transient and asymptomatic.

Vulvo vaginal candidiasis

Vulvo vaginal candiadiasis is a common mucosal infection affecting women with high vaginal pH, sexual activity, pregnancy, contraceptives may contribute to the infection. Symptoms are vulvar pruritus, burning erythema, dyspareunia and creamy white curd like discharge. The infection can occur in complicated (severe) or uncomplicated forms (mild). The complicated form is associated with candidiasis at other locations including mouth, skin or nails.

Neo-natal candidiasis

Low birth weight and prolonged intravascular chatheterization and use of broad spectrum antibiotics are main cause of neonatal candidiasis. The symptoms include generalized erythematous vesicular rash.

Candidial endocarditis, myocarditis and pericarditis

Endocarditis, myocarditis and pericarditis are the most common form of cardiac candidiasis. Intravenous drug abuse, catheters, antibiotics, heart surgery are predisposing factors. Clinical symptoms include fever, cardiac murmurs, anaemia, arterial emboli, myocardial abcess and purulent pericarditis.

Candidial suppurative phlebitis

Inflammation of the vein is known as phelibitis. *Candida* infection occurs in immunocompromised patients and whose cental vein is linked to local catheter. Symptom includes localized phlebitis.

Candidial meningitis

Candidial meningitis by *Candida* is a rare condition seen in low birth weight neonates and in patients with haematological malignancies and intra cerebral prosthetic devices. Symptoms are fever and meningeal irritation.

Candidial endophthalmitis

Endophthalmitis is internal infection of the eye. Candidial endophtalmitis by *Candida* species is associated with candidemia or drug abuse. Lesions are often localized near the macula along with chorioretinitis and patients complain of cloudy vision.

Candida septicemia

Candida septicemia is also called as candidemia. Candidial dissemination by the blood stream can lead to formation of multiple micro abcesses. The infection results from prolonged use of catheters, surgery, corticosteroid abuse, parental nutrition and burns. The symptoms include skin lesions.

Candida allergy

Candida species can cause allergic rhinitis, sinusitis, asthma, pneumonitis, urticaria and atopic eczema.

7.2.2 Diseases of the Nervous System

Cryptococcosis

Nervous affections, menengitis caused by *Cryptococcus neoformens* infection is due to contact with soil contaminated by pigeon droppings and inhalation of dried droppings. It mainly affects central nervous system.

7.2.3 Diseases of the Reproductive System

Candidial balanitis

Balanitis is the inflammation of glans penis and prupuce. Infection occurs in uncircumcised men with poor hygiene and the symptoms include erythema, pruritus and vesicular pustules on the penis.

Gardnerella vaginosis

The causal agent is *Candida albicans.* The symptoms include inflammation of the vagina resulting in reduction in the lactic acid bacteria and over growth of anaerobic bacteria, fishy odour and frothy vaginal discharge and sloughing of vaginal tissue (Tortora *et al.*, 1998).

Vaginitis

Vaginitis is caused by *Candida* species. The symptoms include the presence of a yellow milky discharge from the vagina. Patches of greyish-white pseudomembrane are seen on the vaginal mucosa.

7.2.4 Opportunistic Yeast Pathogens

Candida glabrata is an opportunistic agent in systemic and superficial infections especially in immunocompromised patients and also patients with septicemia, pulmonary infections and endocarditis (Palma Carlos and Palma Carlos, 2004)

7.3 YEASTS AS SOURCE OF FOOD

7.3.1 Enriched Yeasts

With increasing demand for organic and non-genetically modified additives in human and animal feeds, interest in production of "organic vitamins and minerals" has increased in recent years. To overcome the concern, minerals have been incorporated into *Saccharomyces cerevisiae*. The two most commonly available of these are selenium yeast (Selplex®) and chromium yeast (Cofactor III ™). Selenium yeasts have many advantages over traditional inorganic selenium. These include reduction of environmental pollution due to selenium and increased bioavailability (Murphy and Horgan 2005).

7.3.2 Yeast Extracts

Since ancient times, oriental cultures have been using microorganisms to transform or produce food. Yeasts play a crucial role in food processing by improving texture, digestibility, nutritional value, flavour and appearance. The first industrially produced yeast for nutritional purpose was *Torula* in Germany where it was incorporated into soup and sausages. Concentrates of yeast invertase and β-galactosidase and proteins are used in food and feed industry and in laboratories.

7.4 MEDICINAL USES OF YEASTS

7.4.1 Yeasts as Probiotics

Probiotics are nutritional supplements from yeasts.The role of yeast as probiotics has been much overlooked. *Saccharomyces boulardii* is one of the important probiotics, which is also employed in treating various diarrhoeal diseases. *Saccharomyces cerevisiae* is reported to bind enterotoxin from enterobacteria to the surface of the yeast. Live toxin-bound cells of the yeast are detected in the small intestine, showing the potential of yeasts as probiotics (Hansen and Jakobsen, 2004).

The major benefits from probiotic administration are: (a) management of lactose intolerance, (b) prevention of colon cancer, (c) lowering of serum cholesterol, (d) lowering of blood pressure, (e) improving immune function and prevention of infections, (f) reducing inflammation, (g) improving mineral absorption, and (h) growth of harmful bacteria .

Probiotics are commonly recommended by nutritionists but physicians also prescribe them in cases of lactose intolerance in children or after a course of antibiotics that disturb the balance of the gut flora and in certain situations like candidiasis. The probiotics are expected to strengthen the immune system. (see Chapter 6)

7.4.2 Candidal Antigens

Immunodominant antigenic moieties from *Candida* species have been targeted for the production of monoclonal antibodies. Aspartic proteinase (SAP2), mannoproteins and mannans are the leading immunodominant candidal antigens. Intra vaginal and intra nasal immunisation with SAP 2 secreted by *Candida* confers protection against vaginal candidiasis.

7.4.3 Yeasts in The Treatment of Cancer and Tumour

Yeasts are used in traditional medicine in treating various types of cancer and and tumour (Table 7.1)

Table 7.1 Yeasts in the treatment of cancer

Source species	Types of cancer/tumour
Candida albicans	Oral cancer; leukaemia
Saccharomyces cerevisiae	Indurations; tumours of foot, liver, spleen and vulva; cancer of the kidney; whitlow; fibroma.
Torulopsis glabrata	Leukaemia
Cryptococcus albidus	Myeloid and lymphatic chronic leukemia.

Source: Hartwell, 1982

7.4.4 Medicinal Uses in Homoeopathy

Saccharomyces cerevisiae, used in the homoeopathic system of medicine is given as '*Torula cerevisiae*'. It is useful in treating nausea, wheezing, irritability, nervousness, improper digestion, backache, disturbed sleep, constipation with moldy odour, boils, itching eczema around ankles head and neck pain.

7.5 OTHER USES OF YEASTS

Yeasts play a very important role in dairy, food and brewing industry. Yeasts are employed in the production of fermented beverages such as beer and whisky and fermented foods such as bread, cheese, chocolates, soy sauces and pastes. They are effective organisms in biodegradation of hydrocarbons, explosives and bioconversion of distillery wastes. Industrially very important organic acids, amino acids, vitamins, sugars, growth hormones, solvents, enzymes, alternate sweeteners and single cell proteins are sourced from yeasts.

Microbial production of ethanol has become an important source of valuable fuel. Brazil is the major producer of ethanol as an automotive fuel as ethanol combustion is less polluting than gasoline in the manufacture of dyes, soaps, polishes, plastics, synthetic rubber, acetic acid, acetaldehyde, ether and butanol. Yeasts that are commonly used in the industrial production of alcohol are *Saccharomyces cerevisiae, Saccharomyces uvarum (Saccharomyces carlsbergensis), Saccharomyces diastaticus, Kluyveromyces fragilis and Kluyveromyces lactis.* Different species of yeasts are used in the production of ethanol (Table 7.2).

In addition to their positive uses yeasts have a negative impact in causing wood decay. A few species produce toxins causing food spoilage. Raw milk is spoiled by *Candida diffluens, Candida famata* and *Candida lipolytica.* Cream and butter are spoiled by *Cryptococcus laurentii, Rhodotorula rubra* and *Trichosporon cutaneum* (See Chapter 21).

Uses of yeasts in various industries and the source species is summarised in Table 7.3.

Table 7.2 Ethanol production by yeasts

Carbohydrate source	Source species
Cellobiose	*Candida wickerhamii, Kluyveromyces cellobiovorus.*
Glucose + Sucrose	*Saccharomyces cerevisiae, Saccharomyces carlsbergensis, Saccharomyces oviformis, Schizosaccharomyces pombe.*
Inulin	*Candida pseudotropicalis, Candida macedoniensis, Candida kefyr, Candida membranaefaciens, Debaryomyces castellii, Kluyveromyces fragilis, Kluyveromyces marxianus, Kluyveromyces thermotolerans, Saccharomyces kluyveri, Torulopsis sp.*
Lactose	*Brettanomyces sp., Candida pseudotropicalis, Debaryomyces sp.*
Starch	*Candida tropicalis, Candida shehatae, Candida wickerhamii, Kluyveromyces marxianus, Saccharomyces diastaticus, Schwanniomyces castelli*
Xylose	*Candida tropicalis, Candida shehatae, Clavispora sp., Kluyveromyces marxianus, Kluyveromyces cellobiovorus, Pichia tannophilus, Pichia stipitis.*

Source: Benitez and Codon, 2004.

Table 7.3 Industrial uses of yeasts

Product	Source Species
Organic acids	
Citric acid	*Candida guilliermondii, Saccharomyces lipolytica*
Itaconic acid	*Candida sp., Rhodotorula*
Fumaric acid	*Candida sp.,*
Malic acid	*Candida sp.,*
α – Ketoglutaric acid	*Candida sp.,*
Stearic acid	*Cryptococcus sp.,*
Dicarboxylic acid	*Candida sp.,*
Amino acids	
Phenylalanine	*Rhodotorula sp.,*
Lysine	*Cryptococcus laurentii, Saccharomyces cerevisiae, Saccharomyces lipolytica*
Proteins	*Hansenula polymorpha, Pichia pastoris*

Vitamins	
Riboflavin	*Candida famata*
Vitamin D2	*Saccharomyces cerevisiae*
Biotin	*Saccharomyces cerevisiae*
Pyridoxina	*Pichia sp.*,
Sugars	
Trehalose	*Rhodococcus erythropolis*
Raffinose	*Saccharomyces cerevisiae*
Growth hormones	
Ethylene	*Candida sp.*,
Antioxidants	
Superoxide dismutase	*Saccharomyces cerevisiae* var., *ellipsoids*
Lipids	
Sophorolipids	*Candida bogoriensis, Torulopsis bombicola*
Trehalose lipids	*Rhodococcus erythropolis*
Peptidolipid	*Candida petrophilum*
Industrial solvents	
Ethyl alcohol	*Kluyveromyces fragilis, Saccharomyces cerevisiae, Saccharomyces uvarum*
Glycerol	*Candida glycerinogenes, Candida magnoliae*
Enzymes	
Invertase	*Saccharomyces cerevisiae*
Lactase	*Kluyveromyces lactis, Kluyveromyces fragilis, Saccharomyces lactis, Torula cremoris*
Lipase	*Candida antarctica, Candida cylindracea*
	Candida rugosa, Candida paralipolytica
	Saccharomyces lipolytica
Xylanase	*Humicola grisea, Humicola lanuginosa*
	Myceliophthora albomyces, Saccharomonospora viridis
Novel enzymes	
Alcohol dehydrogenase	*Rhodococcus erythropolis*
Carybonyl reductase	*Candida magnoliae, Candida parapsilosis*
Adenosyl methionine	*Saccharomyces sake*

Alternate sweeteners	
Erythritol	*Candida magnoliae*
Mannitol	*Candida magnoliae*
Xylitol	*Candida tropicalis*
Bioflavours	*Candida versatilis* *Zygosaccharomyces rouxii*
Production of Single Cell Proteins	*Candida sp., Debaryomyces sp., Hansenula sp., Kluyveromyces sp., Rhodotorula sp., Saccharomyces sp., Torulopsis sp.,*
Beverages	
Beer	*Saccharomyces cerevisiae, Saccharomyces carlsbergensis*
Sorghum beer	*Saccharomyces cerevisiae*
Whisky	*Saccharomyces cerevisiae*
Wine	*Kloekera sp.,*
Fruit wines	*Candida sp., Saccharomyces sp., Torulopsis sp.,*
Honey wine	*Saccharomyces cerevisiae*
Rice wine	*Saccharomyces uvarum*
Maize wine	*Saccharomyces apiculata, Saccharomyces pastorianus*
Sake	*Saccharomyces sake*
Palm toddy	*Kloeckera javanica, Pichia ohmeri* *Saccharomyces chevielier, Saccharomyces marxianus*
Fermented meat and fish products	*Saccharomyces sp., Torulopsis sp.,*
Fermented foods	*Candida vini, Debaryomyces sp., Hansenula sp., Paecilomyces sp., Rhodotorula sp., Saccharomyces cerevisiae, Torulopsis sp.,*
Soy sauces	*Hansenula sp., Rhodotorula sp., Saccharomyces rouxii, Torulopsis versatile*
Milk based products	
Cheese	*Candida catenulata, Candida famata, Candida lipolytica, Candida spherica, Debaryomyces hansenii, Dekkera anomala, Kluyveromyces marxianus, Pichia membranaefaciens, Saccharomyces cerevisiae, Yarrowia lipolytica*
Koumiss	*Kluyveromyces marxianus, Torula koumiss*

Biodegradation of hydrocarbons	*Candida elegans, Cryptococcus albidus, Debaryomyces vanrijae*
Biodegradation of TNT	*Candida dusenii, Candida nebularis, Candida odora, Candida stercorus, Candida versicolor, Saccharomyces commune, Saccharomyces granulatus, Saccharomyces variegatus*
Bioconversion of distillery wastes	*Candida sp., Hansenula sp., Kluyveromyces sp., Saccharomyces sp.,*
Adsorption of metals	*Kluyveromyces marxianus*

Source: Crueger and Creuger, 2005

7.6 NON SACCHAROMYCETES IN WINERY

Although Saccharomycetes are mainly involved in brewing industry, in the production of beer and wine, non Saccharomycetes also have a role to play in improving aroma, colour and quality of wine by producing certain enzymes (Table 7.4)

Table 7.4 Enzyme production by non saccharomycetes

Enzymes	Source species
Protease	Species of *Candida, Kloeckera* and *Pichia.*
Beta glucosidase	Species of *Debaryomyces, Hansenula, Kloeckera, Kluyveromyces, Metschnikowia, Saccharomycodes, Zygosacchromyces, Schizosaccharomyces, Hanseniaspora.*
Esterase	Species of *Rhodotorula, Debaryomyces* and *Brettanomyces*
Pectinase	Species of *Cryptococcus, Debaryomyces* and *Rhodotorula*

Source: Querol *et al.*, 2004

CHAPTER 8

FUNGI

8.1 INTRODUCTION

Fungi are ubiquitous, found in air, water, soil, on plant or animal bodies. The cottony growth on bread, jam, jelly and on vegetables like beans and pea, moldy appearance on leather shoes, bags and old furniture during rainy season, the bracket like growth (*Polyporus*) on tree trunk, the white mushrooms on the ground, the round ball like puff balls, black powdery mass inside wheat grains and barley, black powder with a bitter taste in groundnuts are all examples of fungi. Despite the great diversity in form and appearance they are all grouped under the group fungi, on account of great many common characters.

They live on dead organic matter (saprophytes) or within living plants and animals (parasites). Fungi exhibit different degrees of parasitism. Some fungi cannot survive outside a living host (obligate parasites) while some can also become saprophytes (facultative saprophytes). Like wise there are obligate saprophytes and facultative parasites.

They do not possess roots, stems, leaves and chlorophyll. A few fungi are unicellular but majority of them are multicellular except slime molds that are with cell walls. The vegetative plant body consists of tubular, branching, septate (with cross walls) or aseptate (with out cross walls) microscopic threads called hyphae. Its tangled mass is known as mycelium. The hyphal cells contain nuclei, cytoplasm and vacuoles. Cell wall is made up chitin which is highly resistant to degradation.

Fungi cause various types of diseases in plants such as leaf spots, anthracnoses, mildews, blasts, rusts, rots,smuts and many others. They mainly infect cereals, millets, vegetables, fruits, fibre and oil crops, etc. (Table 8.1 to 8.13). Though less known, they also cause disease in humans mainly affecting the skin and reproductive system. Fungal spores are known to cause allergy. Fungi are immensely valued as biotechnological systems for biodegradation of various industrial and agricultural effluents.

Illustration of representative species of Fungi are given in Fig 8.1 to 8.9 to provide an over view of their appearance.

8.2 DISEASES CAUSED BY FUNGI

8.2.1 Plant Diseases

Table 8.1 Fungal diseases of cereals

Pathogen	Disease
	Leaf spots
Cercosporidium oryzae	Narrow brown leaf spot of rice
Cochliobolus miyabeanus *Drechlera oryzae*	Brown spot of rice
Dilophospora rolfsii	Dilophospora leaf spot of wheat
	Anthracnose
Colletotrichum graminicola	Anthracnose of cereals
	Powdery Mildews
Blumeria graminis	Powdery mildew of cereals
Erysiphe graminis	Powdery mildew of wheat and barley
	Downey Mildews
Sclerophthora macrospora	Downey mildew of cereals
	Blights
Alternaria triticina	Leaf blight of wheat
Cochliobolus sp.	Rice blight
Cochliobolus victoriae	Victoria blight of oats
Pythium sp.	Cereal blight
Rhizoctonia sp.	Sheath and stem blight of rice
Rhizoctonia solani	Sheath blight of rice
Species of *Fusarium,Helminthosporium* and *Rhizoctonia*	Seed and seedling blight of oats
	Blasts
Magnaphorthe grisea, Pyricularia oryzae	Rice blast
	Rots
Cochliobolus sativus	Root rot of wheat and barley
Gibberella fujikuroi	Foot rot of rice(Bakanae disease)
Helminthosporium sativum	Foot rot of wheat
Helminthosporium sigmoideum var. *irregularae*	Irregular stem rot of rice
Leptosphaeria salvinii, Pellicularia oryzae	Stem rot of rice
Pythium graminicola	Pythium root rot of wheat
Rhizoctonia sp.	Root rot of rice

	Rusts
	Puccinia coronata Crown rust of oats
Puccinia graminis f.sp. *avenae*	Stem rust of oats
Puccinia graminis f.sp. *hordei*	Stripe rust of wheat, barley and rye
Puccinia graminis f.sp. *tritici*	Stem rust of wheat
Puccinia hordei	Leaf rust of barley
Puccinia recondita	Brown rust of wheat and rye
Puccinia striiformis	Yellow rust of wheat
	Smuts
Entyloma oryzae	Leaf smut of rice
Neovossia barclayana	Kernel rice smut
Tilletia foetida	Covered wheat smut
Urocystis tritici	Flag smut of wheat
Ustilaginoidea virens	False smut of rice
Ustilago hordei	Covered smut of barley
Ustilago kolleri	Covered smut of oats
	Bunts
Neovossia horrida	Bunt disease of rice
Neovossia indica	Kernel bunt of wheat
Tilletia caries,	
Tilletia foetida	Hill bunt of wheat
Ustilago avenae,	
Ustilago nuda,	
Ustilago tritici	Loose smut of cereals
	Other Diseases
Anguina tritici	Ear cockle of wheat
Cladosporium herbarum	Black mold of wheat
Ditylenchus angustus	Ufra disease of rice
Ephelis oryzae	Udbatta disease of rice
Pellicularia rolfsii	Sclerotial disease of barley
Trichoconis padwickii	Stack burn disease of rice
	Blotch
Pyrenophora avenae	Leaf blotch of oat
Pyrenophora sativum	Spot blotch of barley
Pyrenophora teres	Net blotch of barley
Septoria nodorum	Glume blotch of wheat
Septoria tritici	Leaf blotch of wheat

Table 8.2 Fungal diseases of millets

Pathogen	Disease
	Leaf spots
Ascochyta sorghina	*Ascochyta* leaf spot of sorghum
Cercosporidium sorghi	Leaf spot of maize
Colletotrichum graminicola	Bed leaf spot of sorghum
Curvularia penniseti	Leaf spot of bajra
Gloeocercospora sorghi	Zonate leaf spot of sorghum
Helminthosporium frumentacei	Leaf spot of barnyard millet
Phyllachora coicis	Tar leaf spot of job's tear millet
Physoderma maydis	Brown spot of maize
	Downey Mildews
Perenosclerospora sacchari	Downey mildew of sugarcane
Perenosclerospora sorghi	Downey mildew of sorghum
Sclerospora graminicola	Downey mildew of common millet and bajra
Sclerospora philippinensis	Philippine downey mildew of maize
Sclerophthora rayssiae var. *zeae*	Brown stripe downey mildew of maize
	Blights
Cochliobolus heterostrophus	Southern corn leaf blight
Helminthosporium turcicum	Turcicum leaf blight of maize
Fusarium moniliforme, *Pythium aphanidermatum*, *Rhizoctonia* sp.	Seedling blight of maize
	Rots
Phytophthora megasperma	Root and stem rot of soybean
Pythium aphanidermatum	Pythium stalk rot of maize
	Rusts
Puccinia operata	Rust of tear millet
Puccinia penniseti	Rust of bajra
Puccinia purpurea	Sorghum rust
Puccinia sorghi	Corn rust
Uromyces setariae	Italian millet rust
	Smuts
Melanopsichium eleusinis	Smut of ragi
Sorosporium paspali var. venucosum	Smut of kodo millet
Sphacelotheca destruens	Head smut of common millet

Sphacelotheca reiliana	Head smut of maize and sorghum
Sphacelotheca cruenta	Loose smut of Jowar
Sphacelotheca sorghi	Grain smut of Jowar
Tolyposporium ehrenbergii	Long smut of sorghum
Tolyposporium penicillariae, *Tolyposporium senegalense*	Smut of bajra
Ustilago crameri	Smut of Italian millet
Ustilago crusgalli	Shoot smut of barnyard millet
Ustilago maydis	Corn smut
Ustilago coicis, *Ustilago lachrymae-jobi*	Job's tear millet smut
	Other Diseases
Balansia oryzae	Udbatta disease of Italian millet
Cephalosporium acremonium	Black bundle of maize
Claviceps microcephala	Ergot of bajra
Helminthosporium leucostylum, *Helminthosporium nodulosum*, *Helminthosporium tetramera*	Helminthosporiose of ragi
Ramularia sorghi	Sooty stripe of sorghum
Sclerospora graminicola	Green ear disease of bajra

Table 8.3 Fungal diseases of pulses

Pathogen	Disease
	Leaf spots
Cercosporidium canescens, *Cercosporidium cruenta*, *Cercosporidium dolichi*, *Cercosporidium sesbaniae*, *Cercosporidium sojina*	Leaf spot of pulses
	Anthracnose
Colletotrichum lindemuthianum	Bed leaf spot of *Phaseolus* species
Glomerella lindemuthianum	Bed leaf spot of pulses
	Downy Mildews
Perenosclerospora manchurica	Downey mildew of soybean
	Blight
Ascochyta phaseolorum	Bean blight
Ascochyta rabiei	Blight of *Cicer*
Colletotrichum capsici	Seedling and pod blight of *Sesbania* and soybean

Macrophomina phaseolina	Ashy stem blight of *Phaseolus* species
	Cankers
Diplodia cajani	Stem canker of *Cajanus*
	Rots
Pellicularia filamentosa	Root rot of pulses
Pellicularia rolfsii	Collar rot of *Cicer*
Phytophthora megasperma	Root and stem rot of soybean
Pythium sp.	Seed, root soft rot of bean
Sclerotinia sclerotiorum	Stem rot of *Cicer*
	Vascular Wilts
Fusarium orthoceras var. *lathyri*	Wilt of *Lathyrus*
Fusarium orthoceras var. *entidis*	Wilt of lentil
Fusarium oxysporum var. *ciceri*	Wilt of *Cicer*
Fusarium udum	Wilt of *Cajanus*
	Other Diseases
Colletotrichum capsici	Die-back of cow pea and horse gram

Table 8.4 Fungal diseases of oil crops

Pathogen	Disease
	Leaf spots
Alternaria tenuis	Leaf spot of sunflower
Cercosporidium ricinella	Leaf spot of castor
Cercosporidium carthami,	Leaf spot of safflowe
Cercosporidium sesami	Leaf spot of *Sesamum*
	Powdery Mildews
Leveillula taurica	Powdery mildew of castor
	Downey Mildews
Perenosclerospora brassicae	Downey mildew of mustard
Plasmopara halstedii	Downey mildew of sunflower
	Blights
Alternaria brassicae	Alternaria blight of mustard
Alternaria lini	Alternaria blight of *Linum*
Phytophthora parasitica	Seedling blight of castor
	Rots
Macrophomina phaseoli	Dry root rot of groundnuts
Pellicularia rolfsii	Collar rot of groundnuts

	Rusts
Albugo candida	White rust of mustard
Melampsora lini	Rust of *Linum*
Melampsora ricini	Rust of castor
Puccinia arachidis	Rust of groundnut
Puccinia carthami	Rust of safflower
	Vascular Wilts
Fusarium vasinfectum	Wilt of *Sesamum*
	Other Diseases
Alternaria carthami	*Alternaria* disease of safflower

Table 8.5 Fungal diseases of fibre crops

Pathogen	Disease
	Leaf spots
Colletotrichum agaves	Red leaf spot of *Agave*
Alternaria sp., *Helminthosporium gossypii, Mycosphaerella gossypiana, Myrothecium roridum*	Leaf spot of cotton
	Anthracnose
Colletotrichum indica, Glomerella gossypii	Bed leaf spot of cotton
	Powdery Mildews
Leveillula taurica, Oidium erysiphoides	Powdery mildew of cotton and sunhemp
	Cankers
Rhizoctonia sp.	Cotton seedling stem canker
	Rots
Macrophomina phaseoli	Dry root rot of cotton and jute
Rhizoctonia bataticola	Root rot of cotton
	Rusts
Uromyces decoratus	Rust of sunhemp
	Vascular Wilts
Fusarium udum var. *crotalariae,*	Wilt of sunhemp
Verticillium albo-atrum	*Verticillium* wilt of cotton
	Damping Off
Phytophthora parasitica	Damping-off of cotton
	Other Diseases
Diplodia corchori	Black band of jute
Pellicularia rolfsii	Sclerotial disease of cotton and jute
Ramularia areola	Grey mildew of cotton

Table 8.6 Fungal diseases of fodder crops

Pathogen	Disease
	Leaf spots
Cercosporidium davisii	Leaf spot of *Melilotus*
Cercosporidium traversiana	Leaf spot of *Trigonella*
Pseudopeziza medicaginea	Leaf spot of alfalfa
	Downey Mildews
Perenosclerospora meliloti	Downey mildew of *Melilotus*
Perenosclerospora aestivalis,	
Perenosclerospora trifolium	Downey mildew of alfalfa
Perenosclerospora trigonellae	Downey mildew of *Trigonella*
	Rusts
Uromyces anthylidis	Rust of *Trigonella*
	Other Diseases
Physoderma alfalfae	Crown wart of alfalfa

Table 8.7 Fungal diseases of vegetables

Pathogen	Disease
	Leaf spots
Alternaria brassicae	Leaf spot of cabbage
Alternaria sp.,	Leaf spot of crucifers
Cercosporidium beticola	Leaf spot of *Beta*
Cercosporidium canavaliae	Leaf spot of *Canavalia*
Cercosporidium capsici	Leaf spot of chillies
Cercosporidium hibisci	Leaf spot bhindi
Cercosporidium melongenae	Leaf spot of *Solanum*
Cercosporidium batatae *Cercosporidium bataticola*	Leaf spot of sweet- potato
Didymella bryoniae	Leaf spot of cucurbits
Septoria lycopersici	Tomato leaf spot
	Anthracnoses
Colletotrichum lagenarium	Anthracnose of cucurbits
Colletotrichum phomoides	Anthracnose of egg plant (brinjal)and tomato
Colletotrichum capsici *Glomerella lindemuthianum*	Bed leaf spot of *Canavalia*
	Powdery Mildews
Erysiphe polygoni	Powdery mildew of beets, crucifers

Leveillula taurica	Powdery mildew of tomato, *Cyamopsis*
Leveillula taurica var. *macrospora*	Powdery mildew of *Dolichos lablab*
Sphaerotheca humuli	Powdery mildew of cucurbits
	Downey Mildews
Bremia lactucae	Downey mildew of lettuce
Perenosclerospora affusa	Downey mildew of spinach
Perenosclerospora destructor	Downey mildew of onion
Perenosclerospora parasitica, *Pseudoperanospora cubensis*	Downey mildew of cucurbits
Perenosclerospora pisi	Downey mildews of peas
	Blights
Alternaria dauci	Carrot blight
Alternaria solani	Blight disease of chillies
Alternaria solani *Alternaria palandui*	Early blight of tomato, *Solanum and* potato
Alternaria porri	Blight of onion
Alternaria sp., *Ascochyta pinodella* *Ascochyta pinodes*	Blight of crucifers
Ascochyta pisi	Blight of peas
Didymella bryoniae	Gummy stem blight of cucurbits
Pellicularia rolfsii	Blight of *Solanum*
Phytophthora colocasiae	Blight of Colocasia
Phytophthora infestans	Late blight of potato, tomato
Septoria apicola	Late blight of celery
	Rots
Ascochyta pinodella, *Ascochyta pinodes,* *Ascochyta pisi*	Foot rot of potato
Ceratostomella fimbriata	Black rot of sweet-potato
Colletotrichum capsici	Ripe rot of chillies
Fusarium coeruleum	Dry rot of potato
Fusarium cucurbitae	*Fusarium* root rot of cucurbits
Fusarium solanii	*Fusarium* root rot of brinjal, tomato and potato
Macrophomina phaseolina	Charcoal rot of potato and sweet-potato
Pellicularia filamentosa	Root rot of *Cyamopsis* and foot rot of tomato
Pellicularia rolfsii	Stem rot of cucurbits

Pathogen	Disease
Plasmodiophora brassicae	Club root of crucifers
Phytophthora sp., Pythium sp.,	Fruit rot of tomato
Pythium aphanidermatum *Pythium butleri*	Fruit rot of cucurbits
Rhizoctonia sp.,	Carrot rot, fruit rot of potato and brinjal
Rhizopus nigricans	Soft rot of sweet-potato
Sclerotium sepivorum	White rot of onion
Choanephora sp.,	Soft rot of sqash, pumpkin
Rhizopus and *Mucor* sp.,	Soft rot of vegetables
	Rusts
Albugo candida	White rust of radish and mustard
Uromyces appendiculatus	Legume rust
Uromyces fabae	Rust of beans
Uromyces pisi	Rust of pea
Puccinia porri	Rust of onion
	Smuts
Urocystis cepulae	Smut of onion
	Vascular Wilts
Fusarium annum	Wilt of chillies
Fusarium oxysporum 7.sp. lycopersici	*Fusarium* wilt of tomato
Ozonium texanum var. *parasiticum*	*Ozonium* wilt of *Cyamopsis*
	Damping Off
Pellicularia filamentosa	Damping-off of tomato and chillies
Pythium aphanidermatum	Damping-off of bhindi and chillies
Rhizoctonia sp.	Damping-off onion
	Other Diseases
Cladosporium fulvum	Leaf mold of tomato
Colletotrichum capsici	Die-back of chillies
Colletotrichum circinans	Onion smudge
Pellicularia filamentosa	Black scurf of potato
Pellicularia rolfsii	Sclerotial disease of *Cyamopsis*
Phomopsis vexans	Phomopsis blight and fruit rot of *Solanum*
Pythium sp.,	Seedling disease of peas
Spongospora subterranea	Powdery scab of potato
Synchytrium endobioticum	Black wart of potato

Table 8.8 Fungal diseases of fruit crops

Pathogen	Disease
	Leaf spots
Blumeriella sp.	Leaf spot of cherries and plums
Cercosporidium viticola	Brown leaf spot of grapes
Guignardia bidwelli	Leaf spot of grapes
Mycosphaerella citri	*Citrus* greasy leaf spot
Cercosporidium musae, Mycosphaerella musicola	Banana leaf spot (Sigatoka disease)
	Leaf Curl
Taphrina deformans	Leaf curl of peach and apricot
	Witches Broom
Taphrina sp.,	Witches broom of cherries
	Anthracnoses
Colletotrichum acutatum *Colletotrichum fragariae*	Strawberry anthracnose
Colletotrichum gloeosporioides	Anthracnose of mango
Elsinoe ampelina	Anthracnose of grapes
Elsinoe veneta	Anthracnose of raspberry
Gloeosporium musarum	Anthracnose of banana
	Powdery Mildews
Oidium mangiferae	Powdery mildew of mango
Oidium tingitanium	Powdery mildew of *Citrus*
Podosphaera leucotricha	Powdery mildew of apple and pear
Sphaerotheca macularis	Powdery mildew of strawberry
Sphaerotheca mors-uvae	Powdery mildew of gooseberry
Uncinula necator	Powdery mildew of grapes
	Downy Mildews
Plasmopara viticola	Downy mildew of grapes
	Blights
Pestalotia psidii	Grey blight of guava
	Cankers
Botryosphaeria dothidia	Apple and peach canker
Dibotryon morbosum	Black knot of plum and cherry
Eutypella parasitica	Trunk canker of maple
Leucostoma sp.,	*Leucostoma* canker of peach

Nectria galligena	*Nectria* canker of apple and pear
Phytophthora palmivora	*Citrus* bark canker
Sphaeropsis malorum	Canker of apple and pear
	Rots
Alternaria sp.,	Core rot of apple, lemons and oranges
Botryodiplodia, Gloeosporium sp.,	Stalk rot of banana
Ceratostomella paradoxa	Shoot rot of banana and Storage rot of pineapple
Glomerella cingulata	Bitter rot of apple and pear
Guignardia bidwellii	Black rot of grapes
Phytophthora citrophthora	Stem rot of *Citrus*
Phytophthora fragariae	Strawberry red rot stele
Phytophthora palmivora, Phytophthora parasitica	Brown rot of citrus and jack fruit
Pythium aphanidermatum	Foot rot of papaya
Rhizoctonia sp.,	Root rot of banana
Rhizopus artocarpi	Inflorescence rot of jackfruit
Rosellinia sp.,	Black root rot of jackfruit
Sphaeropsis malorum	Black rot of apple and pear
Mucor and *Rhizopus* sp.,	Soft rot of fleshy fruits
	Vascular Wilts
Fusarium oxysporum var. *cubense*	Panama wilt of banana
	Sooty Molds
Capnodium ramosum	Sooty mold of mango
Capnodium sp.,	Sooty mold of sapodilla
Meliola butleri	Sooty mold of *Citrus*
Meliola psidii	Sooty mold of guava
	Other Diseases
Botryosphareia prunispinosae	Die-back of peach and apricot
Colletotrichum gloeosporioides	Wither tip (die-back) of *Citrus*
Gloeosporium limetticola	Fruit fall of *Citrus*
Fomes noxius	Brown root disease of jack fruit
Macrophomina musae	Black finger of banana
Pellicularia salmonicolor	Pink disease of *Citrus*, custard apple, jack fruit, bread fruit and mango
Stachylidium theobromae	Cigar end disease of banana
Venturia inaequalis	Scab of apple and pear

Table 8.9 Fungal diseaes of plantation crops

Pathogen	Disease
	Leaf spots
Cercosporidium coffeicola	Brown eye spot of coffee
Cercosporidium longipes	Brown leaf spot of sugarcane
Cercosporidium kopkei	Yellow leaf spot of sugarcane
Pestalotia palmarum	Grey leaf spot of coconut
	Witches Broom
Rhizoctonia sp	Witches broom of cocoa
	Powdery Mildews
Oidium heveae	Powdery mildew of *Hevea*
Phyllactina corylea	Powdery mildew of *Morus*
	Blights
Colletotrichum camelliae	Brown blight of tea
Coniothyrium arecae	Leaf blight of arecanut and betel nut
Exobasidium vexans	Blister blight of tea
Macrophomina phaseoli *Phytophthora* sp.	Seedling blight of *Cinchona*
Guignardia camelliae, *Laestadia theae*	Copper blight of tea
	Cankers
Botryodiplodia theobromae	Stem canker of *Cinchona*
Ceratocystis fimbriata	Cocoa and coffee canker
Cytospora sacchari	Stem canker of sugarcane
Phytophthora palmivora	Canker of *Hevea*
	Rots
Botryodiplodia theobromae	Root rot of tea
Colletotrichum falcatum	Red rot of sugarcane
Corticium invisum	Black rot of tea
Ganoderma lucidum	Root rot of arecanut, betel nut and coconut
Helminthosporium halodes	Leaf rot of coconut
Irpex destruens	Stump rot of tea
Kretzschmaria deusta	Dry rot of *Hevea*
Pellicularia filamentosa	Collar rot of coffee
Phytophthora palmivora	Bud rot of palms and pod rot of *Hevea*, Fruit rot of arecanut, betel nut (koleroga) and coconut
Rhizoctonia sp.,	Root and crown rot of sugarcane
Rosellinia bunodes	Black root rot of pepper

Rosellinia sp.,	Black root rot of coffee and tea
Ustilina zonata	Charcoal rot of *Hevea* and tea
	Rusts
Aecidium mori	
Cerotelium fici	Rust of *Morus*
Hemileia vastatrix	Coffee rust
Puccinia sacchari	Sugarcane rust
	Damping Off
Pellicularia filamentosa	Damping-off of coffee
	Vascular Wilts
Cephalosporium sacchari	Wilt of sugarcane
	Other Diseases
Botryodiplodia theobromae	Die-back of *Hevea*
Ceratostomella paradoxa	Stem bleeding, stem crack, sun scratch of arecanut, betel nut, coconut
Colletotrichum coffeanum	Die-back of coffee
Fomes noxius	Brown root disease of *Hevea*, coffee, tea and *Cinchona*
Helminthosporium sacchari	Eye spot disease of sugarcane
Nectria cinnabarina	Die-back of tea
Pellicularia salmonicolor	Pink disease of *Hevea*, coffee, tea and Cinchona
Pleocyta sacchari	Rind disease of sugarcane
Rosellinia sp	Black root of *Cinchona*

Table 8.10 Fungal diseases of spices, condiments and narcotics

Pathogen	Disease
	Leaf spots
Cercosporidium nicotianae	Brown eye leaf spot of tobacco
Phyllosticta zingiberi	Leaf spot of ginger
	Anthracnose
Colletotrichum necator	Anthracnose of pepper
	Powdery Mildews
Erysiphe cichoracearum	Powdery mildew of tobacco
Erysiphe polygoni	Powdery mildew of coriander, cumin and poppy
	Downey Mildews
Perenosclerospora arborescens	Downy mildew of poppy
Perenosclerospora tabaccina	Downy mildew of tobacco

	Blights
Alternaria palandui	Blight of garlic
Pellicularia filamentosa	Thread blight of ginger
Ramularia foeniculi	Blight of *Foeniculum*
	Rots
Pellicularia rolfsii	Stem rot of tobacco
Pellicularia sepivorum	White rot of garlic
Phytophthora cinnamoni	Stem and root rot of *Cinnamomum*
Phytophthora parasitica var. *piperina*	Leaf rot and foot rot of *Piper*
Pythium aphanidermatum	Soft rot of ginger
Pythium butleri, *Pythium deliense,* *Pythium graminicola,* *Pythium myriotylum,* *Pythium pleroticum,* *Pythium vexans,*	Rhizome rot of ginger
Rosellinia bunodes	Black rot of peper
Choanephora sp.,	Soft rot of pepper
	Vascular Wilts
Fusarium cumini	Wilt of cumin
	Damping Off
Phytophthora palmivora	Damping-off of tobacco
Pythium aphanidermatum	Damping-off of *Foeniculum*
Rhizoctonia sp.,	Damping-off of garlic
	Other Diseases
Phytophthora palmivora	Black leg of tobacco
Phytophthora parasitica var. *nicotianae*	Tobacco black shank
Taphrina maculans	Leaf blotch of turmeric

Table 8.11 Fungal diseases of flowers

Pathogen	Disease
	Powdery mildews
Erysiphe cichoracearum	Powdery mildew of Dahlia
Microsphaera alni	Powdery mildew of Lily
	Rusts
Puccinia chrysathini	*Chrysanthemum* rust
	Rots
Sclerotinia sclerotiorum	Stem rot of *Hibiscus*
	Vascular Wilts
Rhizoctonia solani	Wilt of *Chrysanthemum*

	Damping Off
Pythium species	Damping-off of *Chrysanthemum*
	Other Diseases
Diplocarpon rosae	Black spot of rose
Septoria chrysanthemella	Leaf blotch of *Chrysanthemum*

Table 8.12 Fungal diseases of conifers

Pathogen	Disease
	Leaf spots
Elytroderma deformans	Leaf spot of pine
	Witches Broom
Elytroderma deformans	Witches broom of pine
	Canker
Gremmeniella abietina	Scleroderris canker of conifers
	Blight
Lophodermium pinastri	Needle blight of pine
	Rusts
Cronartium quercuum	Fusiform rust of pine
Cronartium ribicola	Pine blister rust
Peridermium harknessi	Gall rust of pine

8.2.1.1 Wood Rot and Decay Caused by Fungi

Slow white rot of wood is caused by species of *Daldinia, Ustilina and Xylaria*, while soft rots are caused by species of *Alternaria, Bisporomyces, Diplodia* and *Paecilomyces.*

Heavy loos of timber trees in the forest are caused by wood rotting basidiomycetes. In living trees, rotting is confined to older central wood. Even wood and wood products are attacked by basidiomycetes. The wood attacked by pathogen turns spongy, rot, crack and crumble (Table 8.13).

Table 8.13 Wood rots caused by Fungi

Pathogen	Type of Rots
Ganoderma	Basal rot in Conifers and Palms
Heterobasidion	Rots and butt rot
Inonotus	White heart rot
Lenzites	Brown cubical rot of logs, poles and hard wood
Laetiporus	Brown heart rot

Polyporus	Heart rot
Phellinus	Root rot of Conifers
Peniophora	Decay of Coniferous logs and pulp wood
Pleurotus, Schizophyllum and Trametes	White rot of hardwood

Source: Agrios, 1997

8.2.2 Bird and Animal Diseases

Fungi infect mainly domesticated birds and animals. *Trichophyton verrucosum* causes ringworm in cattle and *Trichophyton gallinae* causes infection resulting in feather fall in chicken.

Stachybotryotoxicosis

Stachybotryotoxicosis is a disease caused by toxins of *Stachybotrys atra* affecting horses, resulting in unexplained haemorrhage.

Turkey - X disease

Turkey – X disease is caused by aflatoxin from immature peanuts. When used as feed stock, the seeds or peanut cake causes liver enlargement, hardening of the liver, haemorrhage and jaundice in cattle and sheep. Even the milk of the cow becomes toxic.

Box 8.1 Nematophagous Fungi

Fungi that destroy nematodes are nematophagous fungi. Fungi such as *Arthrobotrys oligospora, Dactylaria candida, Pleurotus ostreatus* and *Pochonia chlamydospora* attack not only living but also the dead nematodes as saprophytes and parasites. The nematophaguous fungi are soil inhabitants. These can be used in biological control of plant parasitic nematodes (Jansson and Llorca, 2004).

8.2.3 Human Diseases

8.2.3.1 Fungal diseases of the skin

1. Dermatomycoses/Dermatophytoses

Fungi that colonize the hair, nails and the outer layer of the epidermis are called dermatophytes or ringworm fungi and the infections they cause are called dermatophytoses. It is generally reffered to as ringworm caused by species of *Trichophyton, Microsporum* and *Epidermophyton.*

Ringworm

Ringworm occurs in several forms and are named according to the organs on which the symptoms manifest. Some of these types are described below.

Tinea captis (capita-the head; scalp ringworm)

Tinea captis is caused by the species of *Microsporum canis, Piedraia hortae, Trichophyton meginini, Trichophyton rubrum, Trichophyton tonsurans* and *Trichosporon beigelli*. The symptoms include circular patterns of baldness on the scalp. *Microsporum canis* and *Microsporum gypseum* causes ring worm of *the scalp in children.*

Tinea barbae (barba-the beard; barber's itch)

Tinea barbae is caused by the species of *Trichophyton* and *Microsporum*. The lesions appears on the beard.

Tinea corporis (corpus-the body; body ringworm)

Tinea corporis is caused by *Epidermophyton floccosum, Trichophyton rubrum* and *Trichophyton tonsurans* affecting the legs, arms and groin.The symptoms includes ring like spots with a central scaly area.

Tinea cruris (crura-the leg;groin ringworm or jock itch)

Tinea cruris is caused by *Trichophyton rubrum* and *Epidermophyton floccosum.* The symptoms includes skin folds of the groin, and pubic region. It is highly contagious.

Tinea pedis (ped-the foot; Athlete's foot or infections of the feet)

Tinea pedis is caused by *Epidermophyton floccosum* and *Trichophyton mentagrophytes* var., *interdigitale*. The infections occurs in the interdigital spaces.

Tinea manuum (manus-the hand; infection of hands)

Tinea manuum is caused by *Epidermophyton flocccosum, Trichophyton rubrum* and *Trichophyton mentagrophytes*. The infection occurs in the hands.

Tinea unguium (unguis-the nail)

Tinea unguium is caused by *Trichophyton rubrum, Trichophyton mentagrophytes* and *Epidermophyton floccosum.* The infection usually affects the nails.

Tinea versicolor

Tinea versicolor is caused by *Malassezia furfur* which is a generalized infection of the skin (Sullia and Shantharam, 2005).

2. Subcutaneous Mycoses

Subcutaneous mycoses affects the subcutaneous area without becoming systemic.

Sporotrichosis

Sporotrichosis is caused by *Sporothrix schenckii* which affects mostly gardeners and those who work with soil. The infection results in the formation of ulcer on the hand.

Chromoblastomycosis

Chromoblastomycosis is caused by *Phialophora verrucosa* and *Fonsecaea pedrosoi*. The symptoms include dark brown pigmented nodules on the feet.

Madura mycosis

Madura mycosis is caused by *Madurella mycetomatis*. The symptoms include the destruction of subcutaneous tissue producing serious deformities and occurs due to walking barefoot on contaminated soils.

3. Systemic Mycosis

Systemic mycosen affects the subcutaneous tissue or lungs and from there it spreads to other organs of the body.

Aspergillosis

Aspergillosis is caused by *Aspergillus* species.The disease, usually seen in individuals with structural abnormalities within the lung or severely impaired resistance to infection, because of metastatic cancer, leukemia and lymphomatous diseases. The symptoms include cough.

Mucormycosis

Mucormycosis is caused by species of *Mucor* and is of the two types: Rhinocerebral mucormycosis is caused by the species of *Mucor*. The symptoms include a characteristic edema on the face, necrosis and bloody exudate. Thoracic mucormycosis is a fatal disease caused by the species of *Mucor*. The symptoms include pulmonary lesions.

Geotrichosis

Geotrichosis caused by *Geotrichum candidum* is of two types. Pulmonary geotrichosis, the symptoms of which includes a light grey, thick and mucoid sputum which is purulent and rarely blood tinged. Bronchial geotrichosis—the symptoms of which includes a chronic cough, gelatinous sputum and lack of fever.

Trichosporonosis

Trichosporonosis is caused by *Trichophyton beigelli* and *Trichophyton capitatum.* The symptoms are fever, pneumonia, skin lesions, infections of lung, eye and kidney (Sullia and Shantharam, 2005). (Table 8.14)

Table 8.14 Fungal diseases of the skin

Disease	Pathogen	Symptoms
Ringworm (tinea)	*Microsporum*, *Trichophyton*, *Epidermophyton* species	Skin lesions of highly varied appearance; on scalp causing local loss of hair.
Sporotrichosis	*Sporothrix schenckii*	Ulcer at site of infection spreading into near by lymphatic vessels.
Chromoblastomycosis	*Phialophora verrucosa* *Fonsecaea pedrosoi*	Dark brown nodules on the feet.
Maduramycosis	*Madurella mycetomatis*	Destruction of subcutaneous tissue of the feet.
Aspergillosis	*Aspergillus* species	Severe cough and sputum.
Mucormycosis	*Mucor* species	Characteristic edema on the face, bloody exudate, pulmonary lesions.
Geotrichosis	*Geotrichum candidum*	Light grey thick and mucoid sputum, chronic cough, without fever.
Trichosporonosis	*Trichophyton beigelli* *Trichophyton capitatum*	Fever, pneumonia, skin lesions, infections of lung, eye, kidney.

8.2.3.2 Fungal disease of the respiratory system

Aspergilloma

Aspergilloma is an infection of lung causing allergy of the upper respiratory tract in asthmatic patients by species of *Aspergillus.*

Blastomycosis

Blastomycosis is an infectious disease affecting the lungs caused by *Blastomyces dermatitidis.* It is characterized by chronic, granulomatous, suppurative lesions of any body tissue.

Coccidiomycosis

Coccidiomycosis is caused by *Coccidoides immitis.* The symptoms of the disease results in fever, cough and weight loss. The symptoms resemble that of tuberculosis.

Histoplasmosis

The causal organism of histoplasmosis is *Histoplasma capsulatum*. The infection results in lesions in lungs and calcification followed by cough, fever and joint pain.

Pneumocystis pneumonia

Pneumocystis pneumonia is caused by *Pneumocystis carinii,* which infects the lungs of immunosuppressed patients. The disease is treated by trimethoprim. This micro-organism is usually regarded as a protozoan belonging to Sporozoa group although some workers consider it as a fungus (Pelczar *et al.,* 1993)

8.2.3.3 Respiratory diseases caused by opportunistic fungi

Opportunsitic fungi are those which cause diseases in weak or compromised host. *Rhizopus* and *Mucor* cause pulmonary infections when individuals are exposed to spores. *Aspergillus fumigatus* and other species of *Aspergillus* which are wide spread in decaying vegetation like compost can cause pulmonary aspergillosis. (Table 8.15)

Table 8.15 Fungal diseases of the respiratory system

Disease	Pathogen	Symptoms
Blastomycosis	*Blastomyces dermatitidis*	Rare but extremely serious fungal infection. Symptoms include chronic, granulomatous, suppurative lesions on the body tissue.
Coccidioidomycosis	*Coccidioides immitis*	Fungal pathogen grows in desert soil occasionally fatal. Symptoms include fever, cough and weight loss.
Histoplasmosis	*Histoplasma capsulatum*	Fungal pathogen grows in soil, if contaminated with bird droppings.

		occasionally fatal. Symptoms include lesions in lungs, fever and joint pain.
Pneumocystis pneumonia	*Pneumocystis carinii*	A common, serious complication of AIDS; no symptoms in people with healthy immune system

Source: Tortora *et al.*, 1998

8.2.3.4 Fungal diseases of the digestive system

Aflatoxin poisoning

Aflatoxin, a mycotoxin produced by *Aspergillus flavus* infects ground nut crop. The consumption of infected ground nut causes aflatoxin poisoning in humans. The infection causes, nausea, headache and vomiting (see chapters 17 and 21).

Ergot poisoning

Mycotoxins produced by *Claviceps purpurea* on grain crops cause ergotism in human after ingestion of the grains corn, wheat and rye grains. The toxin restricts blood flow in the limbs causing gangrene. (Table 8.16) (see Chapter 17 and 21).

Table 8.16 Fungal diseases of the digestive system

Disease	Pathogen	Symptoms
Aflatoxin poisoning	Mycotoxin produced by *Aspergillus flavus*	Mycotoxin contributes to nausea, headache, and vomiting.
Ergot poisoning	Mycotoxin produced by *Claviceps purpurea*	Ingestion causes neurological or circulatory problems.

8.3 FUNGI AS SOURCE OF FOOD

The following fungi are used in salads and other culinary preparations.

Agaricus campestris, Armillaria mellea, Cantharellus cibarius, Clavaria botrytis, Clavaria fusiformis, Clitocybe dealbata, Clitocybe laccata, Clitocybe nebularia, Collybia albuminosa, Collybia microcarpa, Coprinus niveus, Craterellus cornucopioides, Entoloma, Favolus spathulotus, Hypoxylon vernicosum, Lactarius deliciosus, Lentinula edodes, Lentinus exilis, Lentinus praerigidus, Lentinus sajor-caju, Lentinus subnudus, Lentinus squarrosulcus, Lepiota cepaestipes, Lepiota mastoidea, Lepiota procera, Lycoperdon bovista,

Lycoperdon gemmatus, Lycoperdon pyriforme, Lycoperdon saccatum, Morchella esculenta, Panaeolus cyanascens, Pleurotus cretaceus, Pleurotus ostreatus, Podaxon calyptratus, Podaxon pistillaris, Schizophyllum commune, Scleroderma aurantium and *Scleroderma verrucosum* (Ambasta, 1986). Corn smut (truffles) is considered as a delicacy in Mexican cuisine. They are also often used to flavour Mexican dishes like tamales.

8.4 MEDICINAL USES OF FUNGI

Fungi are therapeutically important group of micro-organisms yielding many anti fungal antibiotics. *Penicillium brevicompactum* and *Penicillium notatam* are the most well known sources of Penicillin. They are also a source of immunosuppressive and hypocholesterolemic agents (See Chapter 15).

Many fungi reside in the living tissues of a host plant (endophytic fungus). These produce variety of chemical substances which are potentially useful in medicine such as anticancer compounds, novel antibiotics, antioxidants and anti-diabetics (See Chapter 18).

Agaricus

Agaricus bisporus: Polysaccharopeptides are antineoplastic and immuno stimulating.

Agaricus campestris: is used as tonic, aphrodisiac and laxative; useful in diseases of the nose, eye, pain in the liver, hydrocele, paralysis and weakness.

Agaricus igniarius: is used internally as a bitter tonic and laxative and externally as styptic.

Agaricus ostreatus: paste is applied to gums during excessive salivation and stomatitis. Taken internally to cure dysentery, diarrhoea and haemorrhage.

Agrocybe

Agrocybe parasitica: is used for treating fever and cough.

Aspergillus

Aspergillus awamori: A gene for lactoferrin from human milk has been engineered into the fungus *Aspergillus awamori*. The recombinant lactoferrin is being tested for use in the treatment of oral, intestinal, ear and eye infections. It is also added to infant food formula to prevent opportunistic infections and as a preservative in pharmaceuticals such as eye drops, contact lens solutions and cosmetics.

Aspergillus fumigatus: a sesquiterpenoid fumigillin isolated from the fungus is found to be a inhibitor of angiogenesis (formation of new blood vessels). It is considered as promising therapy for treating breast, cervical, lung and renal cancers (Stradler *et al.,* 1999).

Auricularia

Auricularia auricula judae: is antineoplastic and immuno stimulatory.

Auricularia sambucina: is used as emetic, purgative, anti-inflammatory and for treating sore throat.

Boletus

Boletus crocatus: paste is prepared and applied to gums in excessive salivation. It is also useful in curing diarrhoea and dysentery.

Cantharellus

Cantharellus species are used to prevent eye inflammation and night blindness.

Clavatia

Clavatia species are used in curing infertility in women, chronic hiccups and to relieve pain after scalding.

Cordyceos

Cordyceos robertsi: is used for asthma.

Coriolus

Coriolus versicolor: is used in treating lung and liver infections, and is immuno stimulant and anticarcinogenic.

Daldinia

Daldinia concentrica: is used for curing skin diseases, stomach upset, stomach ulcer, whooping cough and easy child birth.

Fomes

Fomes fomentarius: is used for treating epilepsy, chest pain and mental disorders.

Ganoderma

Ganoderma species are used in Japan as food for inducing sleep and for treating lung and heart diseases, to increase resistance to infections and to promote strength. It is used as an adjunct for treating cancer and chronic fatigue syndrome. The Chinese called it Ling Zhi (Herb of spiritual potency) and the Japanese called it Mannentake (10,000 year old mushroom).

The polysaccharide and triterpenoids in species of *Ganoderma* exhibit immunomodulatory properties and inhibit cholesterol synthesis, allergenic responses and are anti inflammatory, anti-fungal, antiviral, anti-aging and stamina enhancing. The drug by name PSK obtained from this mushroom is an approved anticancer drug in Asia for stomach cancer.

Ganoderma callosum: is used in treatment of cancer, hepatitis, bronchitis, asthma and hemorrhoids; also known to be anti microbial (Ofodile *et al.,* 2005).

Ganoderma lucidum: is used for treating fatigue, asthma, cough, liver and lung ailments and as a hepatoprotective and immunostimulant.In an effort to identify and develop a new anti HIV drug without adverse side effects and viral resistance, *Ganoderma lucidum* extract has proved to be a safe and valuable crude drug. HIV human subjects showed a decrease in the viral load (Mohan *et al.,* 2005). The treatment of various ailments through ganotherapy has been initiated recently in India and is now under study. Capsules of *Ganoderma* are given as alternative medicine for curing diseases and maintaining health. Ganotherapy is mainly intended for treating low blood pleasure, liver and stomach disorders, obesity, prostate cancer, diabetes, paralysis, uterine tumor, gangrene, gastritis, gout, sinusitis, otitis, piles, rheumatism, kidney disorders and bloody dysentery (Rahman *et al.,* 2005).

Grifola

Grifola frondosa: an extremely popular edible mushroom in Japan is used as food and in making tea. It is reported to enhance and stimulate the immune system in cancer patients.

Linnotus

Linnotus obliquus: is used as a tonic, blood purifier, anodyne and restorative. It is useful in stomach ache, gastritis, ulcer, tuberculosis of the bone and glandular diseases.

Lentinula

Lentinula edodes: is used to regulate the immune system and to lower cholesterol level and as an anti-viral agent.

Morchella

Morchella esculenta: is used as aphrodisiac and narcotic.

Phellinus

Phellinus linteus (Heart rot fungus): exhibits highest anti tumor and anti - allergic activity (Nakamura *et al.,* 2005).

Pleurotus

Pleurotus tuberrigium: is used medicinally for treating whooping cough, dysentery, malarial fever and as an energy providing food supplement to pregnant women. Along with *Ocimum gratissimum* it is used to treat ear infection in children (Osagalekhor and Okhuyia, 2005).

Polyporus

Polyporus anthelminticus: is used as anthelmintic.

Polyporus officinalis: is used as diuretic, laxative, expectorant and as a nervine tonic.

Poria

Poria cocos: is used in treating insomnia, vitiligo, and as an invigorate to the spleen and is anti inflammatory, antitumor, anti depressent, antiemetic and antidiarrhoeal. It is a known medicine of immortality by Chinese. It is used in Japanese medicine for enhancing immunocompetency and water overloading and various disorders of women (Plotnikoff, 2005).

Schizophyllum

Schizophyllum is used for treating epilepsy and chest pain (Sumba, 2005).

Xylaria

Xylaria is known to be anti bacterial against *Staphylococcus aureus, Klebsiella pneumoniae, Pseudomonas aeruginosa, Salmonella enteritidis, Aspergillus niger* and *Candida albicans* (Quang, *et al.*, 2005).

8.4.1 Nutriceutically Important Edible Mushrooms

Nutriceuticals are foods that have a medicinal effect on human health. It describes compounds which are potential dietary supplements used as health food and is of therapeutic value. Mushrooms were used extensively in Chinese medicine either alone or in combination with other herbs. Mushrooms such as *Ganoderma lucidum, Lentinus edodes,* and *Grifola frondosa* are rich in Vitamin D_2 (Ergosterol) which regulates gene expression including the immune system. Vitamin D_2 content is reported to be antitumor in activity as it can stabilize chromosomal structure, prevent DNA breaks and induce apoptosis.

Sun exposed dried mushrooms can be an excellent dietary supplement for those with Vitamin D deficiency and those who face the risk of osteoporosis. (Stamets and Plotnikoff, 2005). Preliminary studies involving use of mushrooms in the treatment of Kaposi's sarcoma - an opportunistic skin cancer which affects AIDS patients showed promising therapeutic results. A product by name GACOCA containing mushrooms in Tanzania is used for treating diabetes, high blood pressure and bedwetting.

Following is the list of nutriceutically important edible mushrooms. *Agaricus basianulosus, Agaricus bisporus, Agaricus bitorquis, Agaricus blazei, Agaricus campestris, Agaricus placomyces, Cantharellus cibarius, Cantharellus minor, Cantharellus tropicalis, Clitocybe dealbata, Ganoderma lucidum, Lactarius deliciosus, Lentinula edodes, Lentinus cladopus, Lentinus edodes, Lepiota longicauda, Lepiota mimica, Lepiota procera, Lycoperdon oblongisporum, Pleurotus cretaceous, Pleurotus florida, Pleurotus ostreatus, Pleurotus sajor-caju, Scleroderma aurantium, Scleroderma texense and Volvariella volvacea.*

Health products from mushrooms

1. Trimyo-GenTM (TRI): *Cordyceps sinensis, Ganoderma lucidum, Lentinula edodes.*
2. Reishiplus (Rex): *Ganoderma lucidum, Lentinula edodes.*
3. Mycoplex-7TM: *Agaricus blazei, Cordyceps sinensis, Coriolus versicolor, Ganoderma lucidum, Hericium erinaceus, Poria cocos, Schizophyllum commune, Tremella fuciformis.*

8.4.2 Medicinal Uses of Ergot

Ergot is a fungal disease caused by *Claviceps purpurea* affecting wheat, oat, barley and rye. Ergot alkaloids are produced by various *Claviceps* strains. Two main classes of ergot alkaloids that have been recognised are clavinet alkaloids (clavines) and lysergid acid (ergo) alkaloids.

Ergotism is a type of poisoning caused in humans and animals by consumption of ergot affected plants. The disease affects the grains in the ears and replaces the grains with fungal sclerotia, commercially known as ergot. It is a natural source of hallucinogen LSD. It is employed to control migraine and haemorrhages. The major producers of ergot alkaloids are *Claviceps fusiformis, Claviceps gigantea, Claviceps paspali* and *Claviceps purpurea.* Ergot alkoloids are used to treat various diseases (Table 8.17).

Table 8.17 Therapeutic uses of ergot preparations

Ergot alkaloids	Therapeutic uses
Ergotamine	Uterotonic, stimulates labor; antimigraine, vasoconstrictor
Ergometrine	Uterotonic, oxytoxic
Dihydroergotamine	Antimigraine, sympatholitic, vasoconstrictor
Dihydrotoxine	Cerebral and peripheral vasodilator
Dihydroergocristine	Sympatholitic, peripheral vasodilator
Dihydro-α-ergokryptine	Antiparkinsonian, prolactin inhibitor, cerebral vasodilator
Bromokryptine	Dopamine agonist, antiparkinsonian, prolactin inhibitor
Nicergoline	Cerebral vasodilator
Metergoline	Serotonine antagonist, antimigraine, prolactin inhibitor
Methylergometrine	Uterotonic, oxytoxic
Methylsergide	Serotonine anatagonist, antimigraine
Lisuride	Serotonine antagonist, prolactin inhibitor, antiparkinsonian
Terguride	Dopamine agonist, prolactin inhibitor, antiparkinsonian
Pergolide	Dopamine agonist, prolactin inhibitor, antiparkinsonian
Cabergoline	Dopamine agonist, prolactin inhibitor, antiparkinsonian

Source: Mehta *et al.*, 2004

8.4.3 Fungi in The Treatment of Cancer and Tumour

A number of fungi are used in traditional medicine for treatment of cancers and tumours. (Table 8.18).

Table 8.18 Fungi in treatment of cancer/tumour

Species	Type of cancer/tumour
Agaricus bisporus	Indurations of liver, spleen and cancerous tumour.
Amanita muscaria	Glandular and cystic tumours; purulent cancerous ulcers; uterine cancer
Aspergillus flavus	Neoplastic disease; cancer of skin, breast, uterus, cervix and stomach
Boletus edulis	Cervical cancer
Boletus laricis	Tumors of the throat and gullet.
Claviceps purpurea	Polypi of the uterus; tumours; carcinomatous scirrhus and non malignant indurations; cancer of the uterus; fibroids in the uterus and carcinoma of the cheek

Fomes annosus	Cancer
Fomes applanatus	Cancer and tumours
Fomes fomentarius	Cancer of breast and uterus
Fomes ignarius	Breast cancer and tumours
Fomes nigricans	Malignant tumours and neoplasma; sarcoma; cancer of the ear and uterus.
Fomes officinalis	Tumours of sinews, eyelids tendons, anus and genitals; indurations of breast; carcinoma; malignant ulcers; induration of the spleen.
Helvella sp.	Tumours
Lactarius acris	Warts
Lactarius piperatus	Warts
Lysurus mokusin	Cancerous ulcers
Penicillium notatum	Neoplastic diseases cancer of skin, buccal cavity, breast, uterus, cervix and stomach
Peziza auricula	Uterine cancer
Phallus indusiatus	Indurated tumours
Pleurotus dryinus	Warts
Podaxon sp.	Carcinomatous ulcers and cancers
Poria obliqua	Cancer of the lung, lip, mammary gland, skin, rectum, stomach and genitalia.
Rhizopus sp.	Cancer

Source: Hartwell, 1982

8.4.4 Medicinal Uses in Homoeopathy

Agaricus

In the homoeopathic system of medicine, *Agaricus* is used for treating vertigo from sunlight, dull headache after prolonged headache, neuralgia with icy cold head, double vision, twitching of the eye balls, itching of the nose, spasmodic sneezing, watery discharge, twitching of the facial muscles, herpes on lips, tongue pain, constant thirst, dryness of pharynx, difficulty in swallowing, hiccups, unnatural hunger, distention of stomach and abdomen, scorching pain in the liver, foetid stools, flatulence with diarrhoea, frequent urination, spasm, dysmenorrhoea, leucorrhoea with itching, violent cough, laboured breathing, irregular heart palpitation, hip pain, frozen feeling in the toes, pain in the shin bone, paralysis of lower limbs, paralytic pain in the left arm, pustular and edematous lesions on the skin.

Lycoperdon

Lycoperdon (puff ball) is given in the name of Bovista for treating, headache in the morning, itchy and sensitive scalp, crust in the corners of mouth, chapped lips, bleeding nose and gums, acne due to use of cosmetics, diarrhoea before and during menses, early and profuse menses, metrorrhagia, colic with red urine, chronic diarrhoea in aged, weakness of joints, moist eczema on hand, edema in fractured parts, scurvy, herpetic eruptions, pruritus and urticaria.

Ustilago

Ustilago spores are given for flabby uterus, congestion in various parts, headache due to irregular menses, spermatorrhoea, lumbar pain, irritability, vicarious menstruation, pain in the ovary, profuse menses after miscarriage, hypertrophied uterus, cervical bleeding, lochia, profuse sweat, muscular debility, alopecia, dry eczema, pruritus, copper coloured spots, excess anger, nausea, vomiting; psoriasis and eczema (Boericke, 1991).

8.4.5 Uses in Chinese Medicine

Lycoperdon

Lycoperdon bovista : is used as an expectorant, emmenogogue, astringent, antiphlogistic, antitussive; used in laryngitis, tonsilitis, hematemesis, epistasis; applied externally to wounds; the dry powder arrests hemorrhage; dust mixed with honey is used in curing throat infections.

Poria

Poria cocos: is diuretic and sedative; used in the treatment of insomnia, tachycardia and gastrointestinal atony.

8.5 TOXICOLOGY

Agaricus

Agaricus bisporus (cultivated mushroom) hydrazines are reported to be carcinogenic, causing liver, lung, gall bladder and bile duct tumours (Hirono, 1987). *Agaricus campestris* causes *tridosha*, vomiting, dysentry, fever, bronchitis.

Cladosporium

Cladosporium herbarum - a common air borne fungus causes allergic asthma.

Coriolus

Coriolus versicolor when used in excess causes diarrhoea and nail pigmentation.

Ergot

Ergot preparations when used in excess results in hallucinations, convulsions and excruciating pain or burning known as St. Antony's fire. It is called St. antony's fire or holy fire because affected community exhibited dancing movements. Ergotamine in ergot is a strong vasoconstrictor, which is so powerful as to completely block the flow of blood to all parts of the body creating a conditions known as gangrenous ergotism.

Ganoderma

Excessive consumption of *Ganoderma* as medicine leads to dizziness, dry mouth, stomach upset, nose bleed, skin irritation, constipation and bleeding in pregnant women. *Ganoderma applanatum* causes skin allergy.

8.6 OTHER USES OF FUNGI

Aureobasidium

Aureobasidium pullulans: is used to prevent sugar crystallization and sandiness in ice creams. It is used in formulating powdered cheese. Pullulan is made into polymeric films that are used to pack foods to prevent food exidation.

Mushrooms

In Nigeria mushroom powder is sprinkled around the house to ward off termites. Mushroom hydrazines are employed in herbicides, antioxidants and chemotherapeutic agents.

Rhizomucor

Rhizomucor meihei: Betapol is human milk fat substitute which is incorporated into infant milk formula containing cow's milk. Fat in cow's milk cause reduced fat absorption and hard stools and constipation in bottle fed babies.

Betapol helps in avoiding the adverse effects on bottle fed babies, as it is rich in palmitic and oleic acid. Immobilized *Rhizomucor miehei* mixed with tripalmitin rich vegetable oil and oleic acid results in the desired product.

Schizophyllum

Schizophyllum commune: is used in cosmetic industry, ceramic glazing, printing inks and pesticides.Hydrophobins from *Schizophyllum* species are being used to coat Teflon so strongly that it can withstand treatment with boiling sodium dodecyl sulphate (SDS). They are also used for protein immobilization and as emulsifiers.

Trichoderma

Species of *Trichoderma* produce various enzymes which are used in food processing, baking and detergent industries. They are used as antifouling agents, in composting, waste treatment, biobleaching of paper, recycling of paper and biostoning and biofinishing of denim fabric (stone washing of jeans). They are used as stain removers.

Trichosporum

Trichosporum penicillatum: is used for extraction of pectins from fruits and as a gelling agent.

8.6.1 Uses of Fungal Products

Chitin

Chitosan- a derivative of chitin from many fungi is used in clearing fruit juices and for extracting of beta carotene from carrot juice. Porous bead technology involving chitosan is used in chelating heavy metals, in affinity chromatography, for trapping microbes in gene therapy, as adhesive for drug delivery, as a bandage for corneal wound healing and in hair sprays, nail varnishes, shampoos and hair conditioners (Ng, 2004, b,c).

Carotenoids

Current preference for natural products favours extraction of carotenoids from biological sources. Easy cultivation, simple requirements and rapid growth have favoured the use of fungi in large scale industrial production of carotene. *Aschersonia aleyroides, Aspergillus giganteus, Blackslea trispora, Cercosporidium nicotianae, Mucor circinelloides, Phycomyces blackesleanus, Xanthophyllomyces dendrorhous* are employed in carotenoid production.

Astraxanthin, cantaxanthin, lycopene, torulene, â-carotene are a few examples of carotenoids formed by fungi.

Box 8.2 Antifungal Food Additives

Fungi are the most challenging of organisms responsible for food spoilage. The problem of food spoilage has worried humans throughout history. Earlier humans preserved foods by sprinkling sugar, salt and spices or by exposing the food to wood smoke.

Antifungal food additives are an efficient, cost effective and successful way to control microbial growth in foods and may be added during processing. The various antifungal food additives are sorbic acid, benzoic acid, propionic acid, fatty acids and their esters (Ray and Liewen, 2004).

Cellulase

In 1943, scientists in the United States Army Laboratory isolated a mold (*Trichoderma reesei* from cellulase based equipment deteriorating in storage. This discovery led to the commercial production of cellulase in food and drink industries. Cellulases are widely employed in textile manufacturing. Main producer are *Trichoderma reesei and Trichoderma longibracteatum.*

Laccase

Laccase, a major enzyme produced by white rot fungi such as *Phanerochaete chrysosporium* is used in stabilising wine and beer (Minnussi *et al.,* 1998). Laccase from *Coriolus hirsutus* and *Coriolus zonatus* is used in polyaniline synthesis, an important polymer in light weight batteries and display systems. Laccase from *Coprinus cinereus, Myceliophthora thermophila* and *Rhizoctonia solani* is used in the production of dyes and perfumes (Kulys *et al.,* 2003). A new laccase enzyme Denlite II produced by Novozymes Industries is used to decolourise denim fabrics and to improve the contrast look on fabrics (Kapoor *et al.*, 2005).

Peptides

A novel cyclic peptide which is active against nematodes of veterinary and agricultural importance is synthesized by *Rosellinia.*

Statins

Statins are group of compounds that are competitive inhibitors of 3-HMG-CoA reductase, key enzyme in the biosynthesis of cholesterol. The most important commercially available statins are Lovastatin from *Monascus ruber* and *Aspergillus terreus* and mevastatin from *Penicillium citrinum.*

8.6.2 White Rot Fungi and Their Applications

White rot fungi has immense uses in wood, paper, food, feed, textile and chemical industry. They are also used in bioremediation of wood preservatives.

Saving pulp refining energy

Eriksson and Vallander (1982) and Messner and Srebotnik (1994) reported a reduction in pulp-refining energy by incubating spruce and pine chips with *Phanerochaete chrysosporium*. The fungal pretreatment of wood chips reduces electrical power consumption of mechanical or chemical treatment as they partially disrupt the cell and break down lignin/ carbo hydrate complex of plant cell well (Sawada *et al.*, 1997).

Biopulping

Biopulping is defined as the pretreatment of wood chips with selective delignification of white rot fungi prior to mechanical or chemical pulping. It has the advantage of diffusible lignin and hemicellulose degrading agents excreted by the fungi during incubation of wood chips and also reduce environmental impact due to toxicity of bleach plant effluents from chemical pulping plants.

In paper making, fibres are separated from the plant tissue, bleached and rearranged to form a sheet of paper. The fibres are separated mechanically either by stone grinding whole wood logs or by disc refining of wood chips. Both the processes are energy consuming. In chemi-thermo-mechanical-pulping (CTMP) the refining process is assisted by chemicals and high temperature. Mechanical pulping suffers from low strength and brightness.

Parameters such as amount of delignification after pulping (monitored by Kappa number, an indirect method measuring the amount of permanganate taken by lignin), brightness and whiteness of the paper, tensile strength, tear index etc. decide the quality of paper (Akhtal *et.al.*, 1992).

Biobleaching

Environmental friendly, elemental chlorine free bleaching technology is very much necessary for reducing the toxicity of effluents of wood and paper industry.

Pulp can be bleached with white rot fungi and their ligninolytic enzymes enabling chlorine free bleaching process. *Bjerkandera* sp., strain BOS55, *Polyporus ciliatus, Stereum hirsutum, Phanerochaete cinnabarinus, Phanerochaete chrysosporium, Phlebia radiata, Lentinus tigrinusi, Tramates versicolor,* have been found to be efficient biobleachers (Akhtar *et al.*, 1992).

Chemical industry

Many substances like bisphenol A (BPA), nonylphenols (NP), phthalic esters (PAE) have potential to disrupt the hormonal system and are called endocrine disrupters (EDs). Lignin degrading enzymes have been applied to the detoxification of ED's(Tanaka *et al.*, 2003). Vanillic acid and gallic acid used frequently in the pharmaceutical industries are synthesized by the lignocellulolytic *Phanerochaete chrysosporium* (Howard *et al.*, 2003).

Bioremediation of contaminated sites

Bioremediation using fungi (mycoremediation) is attracting the interest for the past two decades. White rot fungi can withstand toxic levels of most organopollutants which are highly recalcitrant.

White rot fungi have shown potential for bioremediation by their ligninolytic enzymes (Hestbjerg *et al.*, 2003, Vasudev and Kuhad 1995).

The major environmental contaminants are polycyclic aromatic hydrocarbons(PAH) that are generated from burning of fossil fuels, coal mining, oil drilling and wood burning. The efficient degraders of PAH are *Bjerkandera, Cyathus* and *Pleurotus* sp., *Phanerochaete chrysosporium* and *Trametes versicolor* (Bumpus and Tatarko 1994).

Poly-chlorinated biphenyl (PCB) are widely used in dielectric and hydraulic fluids, flame retardants, plasticizers, textiles and printing. *Phanerochaete chrysosporium, Phanerochaete cinnabarinus, Phanerochaete ostreatus, Lentinus edodes, Trametes versicolor,* are major white rot fungi which are capable of mineralizing PCB's.

Trinitrotoleune (TNT) is a major contaminant in many military sites. The white rot fungus *Phanerochaete chrysosporeum* transforms TNT to dinitrotoluene (DNT) and mineralizes it to CO2.

Wood preservatives are used to treat wood and to make the wood free from fungal attack. Most of the preservatives are not natural. White rot fungi produce extra cellular enzymes that oxidize these aromatic organic compounds and burn them down. Creosote and Pentachlorphenol (PCP) are two of the major organic wood preservatives which are also precursors for synthesis of several biocides used worldwide for more than 150 years.

Although wood products treated with creosote and PCP do not represent a threat to the environment during use, improper disposal methods and accidental spillage have resulted in extensive contamination of subsurface and surface soils. Incineration and bioremediation are two alternatives for removal of these pollutants. Creosote consists of mostly polycylic aromatic hydrocarbons (PAH) which have been categorized as priority pollutants by the environmental protection agency (EPA).

The microbial breakdown of creosote like component was described in 1928 by Gray and Thornton. Since then white rot fungi are employed as they utilize coal tar and creosote as nitrogen and carbon source and also degrade phenols and polycyclic aromatic hydrocarbons. Similarly pentalchlorophenol is detoxifiedby the action of enzymes such as laccase, tyrosinase and peroxidase by white rot fungi like *Coniophora puteana*. Though incineration is highly effective, it is expensive and has environmental disadvantage. The process is carried out by heterotrophs or by selected microbes *in situ* or *ex situ.* (McGinnis *et al.,* 1991). Bioremediation techniques such as land farming, bioventing, biopiling and bioslurry treatment.

The preservatives used for the manufacturing of disposal wood make it more difficult as it cannot be placed in a normal municipal landfill. As wood is a natural product, it should be treated as a part of the terrestrial carbon cycle and if possible the contamination should be remedied by natural means and the residue returned to natural environment.

Food and feed industry

White rot fungi cultivated on lignin and cellulose containing substrates such as cotton stalk wheat and rice straw can decompose lignocellulose efficiently that can be utilized for the production of edible mushrooms.

The polysaccharides in lignocellulosic wastes are a potential source of dietary energy for ruminants. Wheat straw which forms a part of the feed as roughage for ruminants is primarily composed of cellulose and hemicellulose with low digestibility and low palpability. *Phanerochaete chrysosporium, Cyathus stercoreus* improve the digestibility of wheat straw as they can degrade celluloses and hemicelluloses. Similarly, digestibility of Bermuda grass is improved by *Phellinus pini* RAB-83-19 and *Phanerochaete chrysosporium* K-3 (Akin *et al.,* 1993). Selective white rot fungi have a potential for upgrading crop residues as animal feed.

Textile industry

White rot fungi such as *Coriolopsis gallica, Polyporus pinisitus and Trametes versicolor* have been used in textile industry for decolorization of azodyes.

8.6.3 Keratinophilic Fungi and Their Applications

Keratinophilic fungi, the natural colonizers of keratinic substrates such as hair, wool, feather, nail, claws, scales, horn, hoof and skin play an important role in the natural decomposition of keratinized residues. Some may even cause human and animal infections. These soil bound fungi are biotechnologically important as they are used in the production of animal feed, fertilizers, anti-microbial agents and biodegradation of tannery wastes.

1. Production of indole compounds by *Microsporum gypseum* is used for the production of plant growth hormones.
2. Keratinophilic fungia are used as tools for making compost from poultry waste.

3. In the production of amylase by *Verticillium tenuipes;* production of catalase, phosphotase, cellulase, urease, pectinase, esterase, lipase, α-galactosidase, α-glucoronidase and α-fucosidase from *Chrysosporium keratinophilum.*
4. In dehairing and debaiting hides (Jain and Jain, 2005).
5. Production of anti-microbial substances (See Table 8.19)

Table 8.19 Production of anti-microbials from keratinophilic fungi

Source species	Anti-microbial substances
Microsporum gypseum	Fusidic acid
	Gypsetin
Trichophyton violaceum	Quinone antibiotics
Trichophyton rubrum	
Trichophyton violaceum	Xanthomagnin
Epidermophyton floccosum	Floccosin
	Flocossic acid
Chrysosporium merdarium	Questin
	Questinol
	Asteric acid
Chrysosporium indicum	L-Arginyl-D-allothreonil-L-Phenylalanine
Chrysosporium pannorum	Cryscandin
Arachniotus aureus	Acetyl aranotin
	Aporaratrotin
	Aranotin-A
Pseudomonas arachinotus	Aranorosinol-A
	Aranorosin

Source: Jain and Jain, 2005

8.7 FUNGI AS PIONEER ORGANISMS FOR GENETIC ANALYSIS

Fungi are among the models for genetic investigation of basic cellular processes. They are ideal systems for construction of efficient vectors for gene transfer and cloning genetic manipulation.

Neurospora crassa, Aspergillus nidulans, Coprinus cinereus are the major model systems for genetic analysis and for uncovering the secret of cell proliferation and exploitation to create new combinations of variations.

8.8 MYCORRHIZAE

The symbiotic association of fungi with the roots of vascular plants transforms the infected roots into unique morphological structures called mycorrhizae, that is 'fungus roots'. Mycorrhizae are formed in cereals, vegetables, flowering plants and gymnosperms (*Pinus*).

There are two types of mycorrhizae that are distinguished by the way the hyphae of the fungi are arranged within the cortical tissue of the roots. They are ectomycorrhizae and endomycorrhizae. In ectomycorrhizae the fungus is a basidiomycete and forms a sheath around the root but does not actually penetrate the host cells. In many gymnosperms such as Pine, Spruce, Larch, ectomycorrhyzae are found. In endomycorrhizae the fungal hyphae penetrate the cells of the root, forming swellings (vesicle) and/or branched haustorial hyphae (arbuscle). The fungus is usually a zygomycete. Most of the orchids, except *Cypripedium, Neottia, Limodorum, Vanilla* and few others have endomycorrhizae.

8.8.1 Arbuscular Mycorrhizal Fungi (AMF)

These are zygomycetous fungi that form specialized structures such as arbuscules and are obligate biotrophs. They are ubiquitous and co-exist with 80% of agricultural and horticulatural plants. A incompatible association between the host and AMF community can lead to serious loss in crop yield. But compatible association increases crop productivity. AMF infection results in the synthesis of phytoalexins in response to the infection.

8.8.2 Vesicular Arbuscular Mycorrhizae (VAM)

VAM fungi are obligate endosymbionts and the hyphae form both vesicles and arbuscules in the cells of the host root system.

8.8.3 Uses of Mycorrhizae

Arbuscular mycorrhizae are a potential source of cheap and risk free control of plant diseases caused by bacteria and viruses. They increase the volume of soil for the plant to explore which is very important in disturbed site. They help in alleviating drought stress and enhance beneficial interaction with other microbes.

Arbuscular mycorrhizae (AM) impart tolerance to salt stress to the plants. AM biofertilizer is known to increase the nitrogen fixing potential of legumes by promoting root nodulation (Xavier and Boyetchko, 2004). Immobilised trace elements such as phosphorus, sulphur, calcium, magnesium, zinc and copper are made available to the plants in the presence of mycorhizae.

VAM fungi improve and maintain soil fertility and they are transporters of minerals such as nitrogen, phosphorus, calcium, potassium, sulphur, iron, zinc, copper and water from the soil to the host.

8.8.4 Producers and formulators of commercial AM inoculum

AgBio Inc, Westminister, Colorado USA
Bio-Organics Supply, Camarillo, California, USA
Bio Scientific, Inc., Avondale, Arizona, USA
Eco Life Corporation, Moorpark, California, USA
Mikro-Tek Inc., Timmins, Ontario, Canada
BioGrow TM, North America
Plant Health Care, Inc., Pennsylvania, USA
Mycor TM VAM Mini Plug TM, North America
T & J Enterprises, Spokane, Washington, USA
Tree Pro, West Lafayette, Indiana, USA
Biological Crop Protection Ltd., Kent, UK
Bio-Organics, Medillin, Columbia
Biorize, Dijon, France
Central Glass Co., Chemicals Section, Tokyo, Japan
Global Horticare, Lelystad, Netherlands
Idemitsu Kosan Co., Sodegaura, Chile
N-Viron Sdn. Bhd, Malaysia
Triton Umweltschuz GmbH, Bitterfeld, Germany
KCP Sugar and Industries Corporation Ltd., Andhra Pradesh, India
Cadila Pharmaceuticals Ltd., Ahemadabad, India.

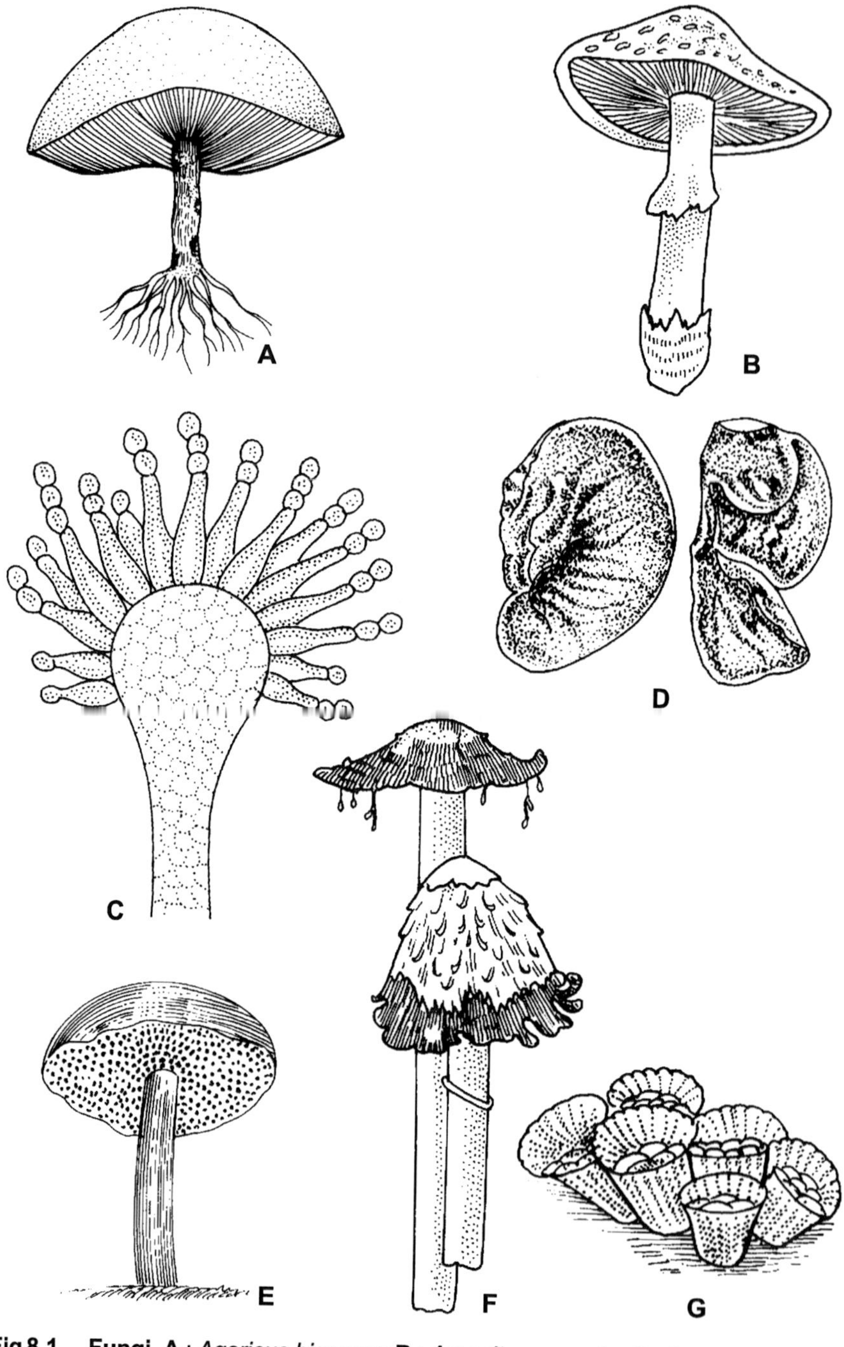

Fig 8.1 **Fungi. A** : *Agaricus bisporus*; **B** : *Amanita muscaria*; **C** : *Aspergillus niger*; **D** : *Auricularia auricula-judae*; **E** : *Boletus* sp.; **F** : *Coprinus atramantarius*; **G** : *Cyathus* sp.

Fig. 8.1 to 8.9 Redrawn from Bessey, E.A. (1950) and Srivastava, J.P. (1962)

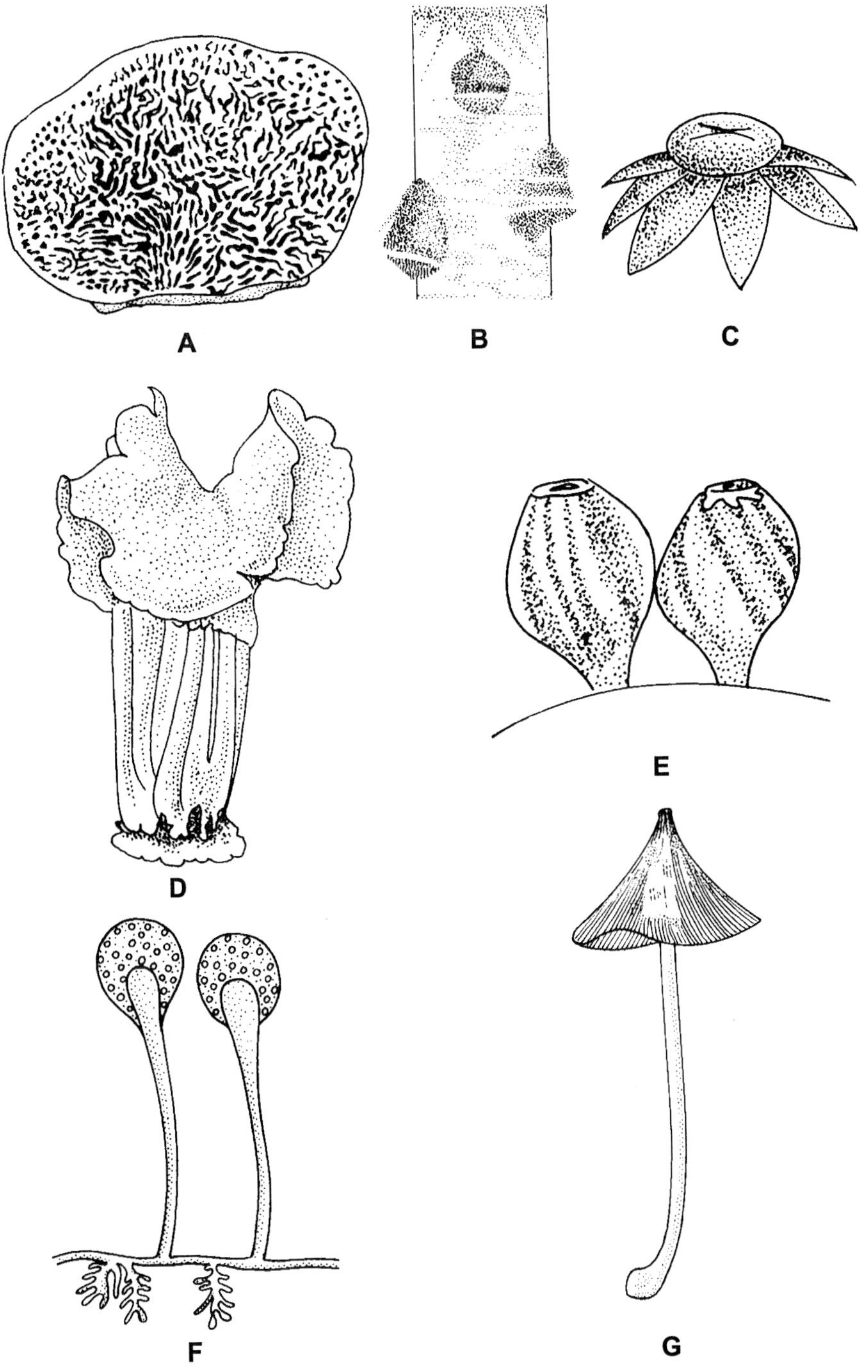

Fig 8.2 **Fungi. A** : *Daedelea* sp.; **B** : *Fomes fomentarius*; **C** : *Geaster* sp.; **D**: *Helvella crispa*; **E** : *Lycoperdon pyriforme*; **F** : *Mucor* sp. **G** : *Mycena* sp.

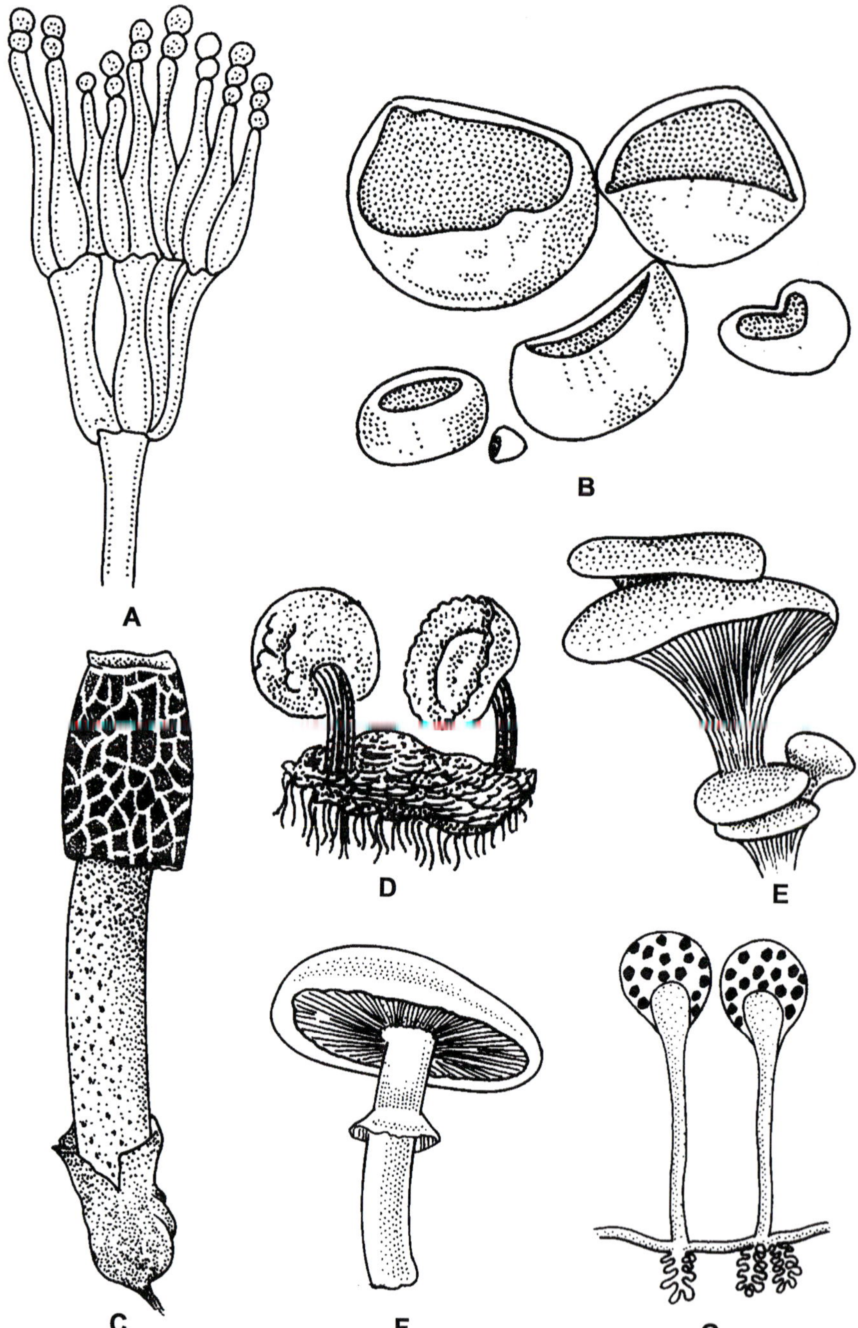

Fig 8.3 **Fungi. A** : *Penicillium notatum*; **B** : *Peziza repanda*; **C** : *Phallus simpudicus*; **D** : *Physarum natans*; **E** : *Pleurotus* sp.; **F** : *Psalliota campestris*; **G** : *Rhizopus nigricans.*

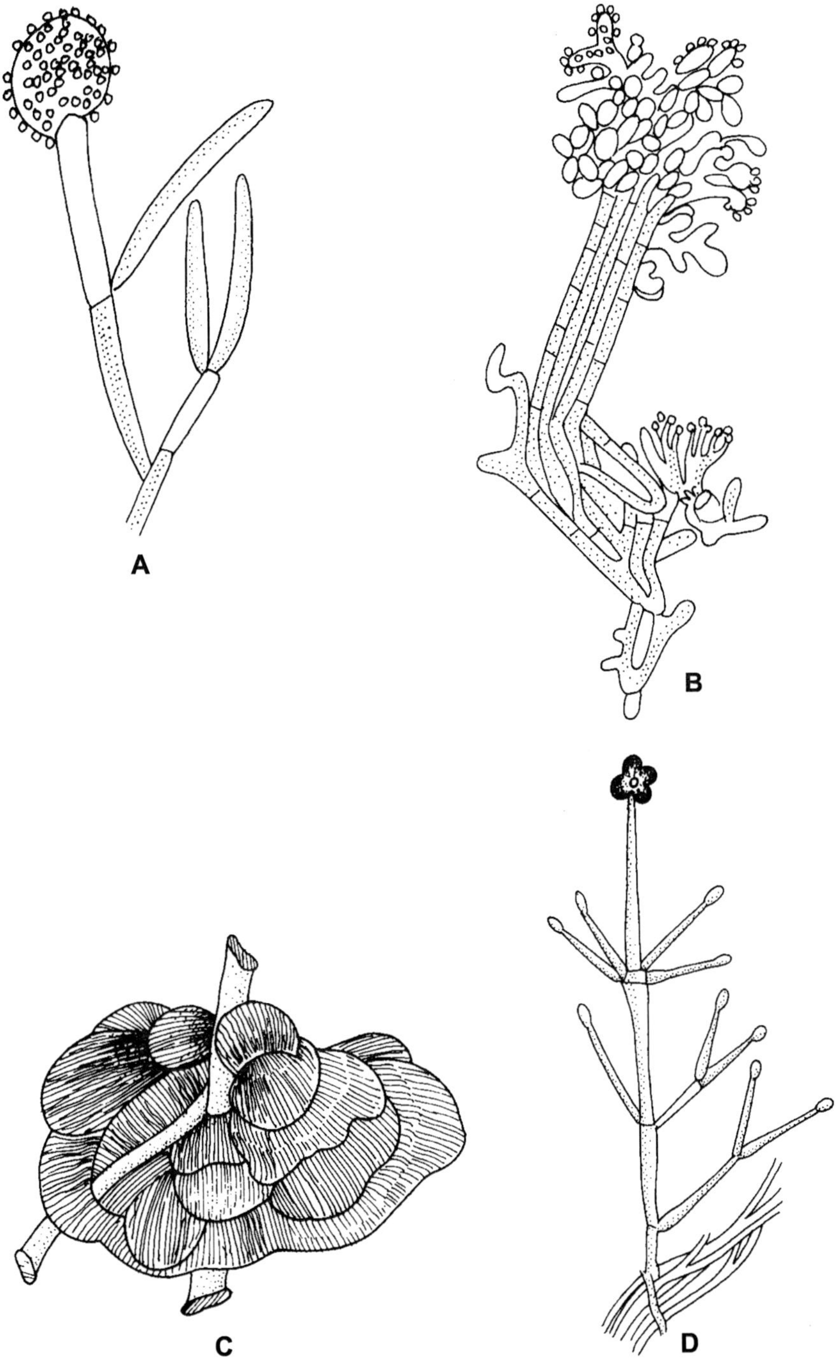

Fig 8.4 **Fungi. A** : *Achlya racemosa*; **B** : *Pellicularia isabellina*; **C** : *Polyporus* sp.; **D** : *Verticillium alboatrum.*

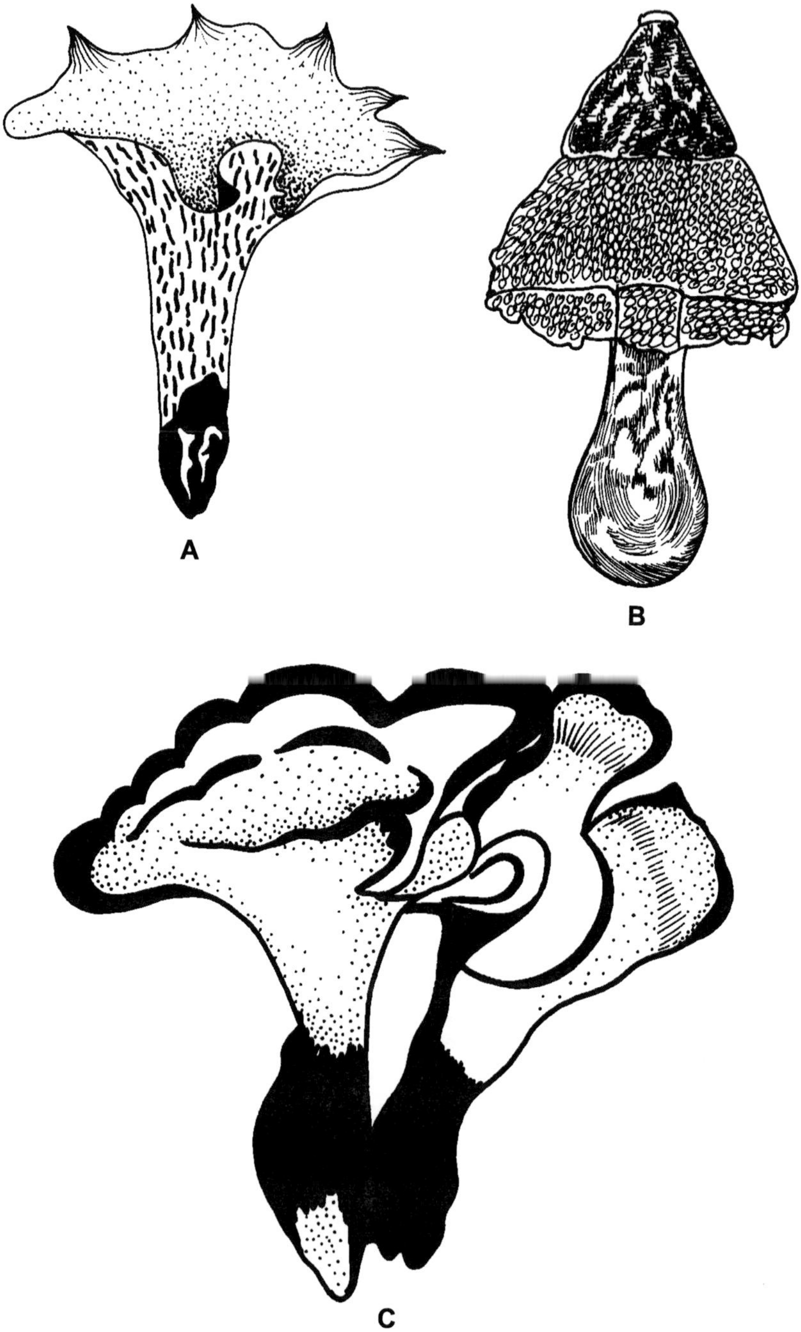

Fig 8.5 **Fungi. A** : *Cantharellus floccosus*; **B** : *Dictyophora duplicata*; **C** : *Grifola berkeleyi.*

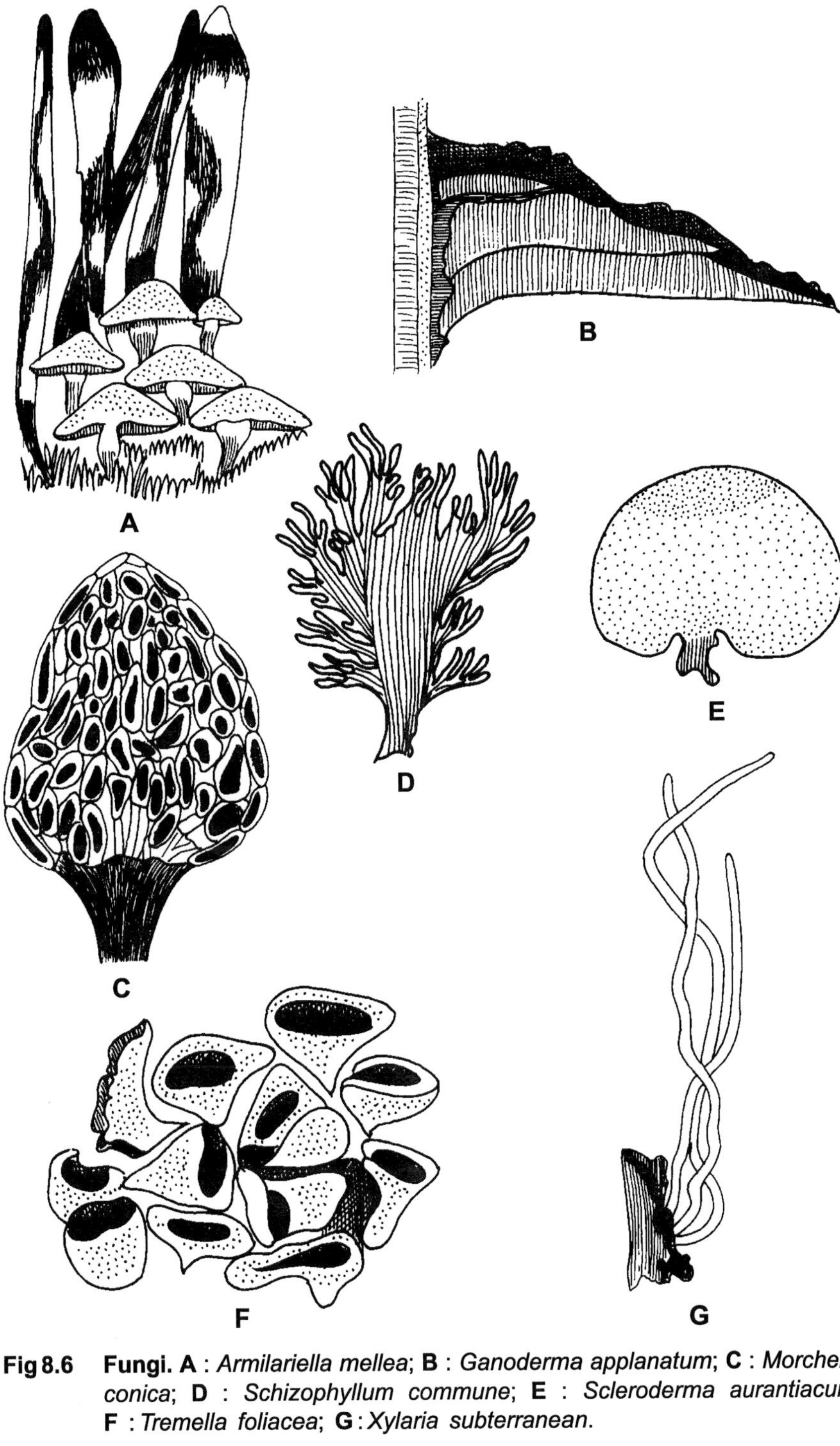

Fig 8.6 **Fungi. A** : *Armilariella mellea*; **B** : *Ganoderma applanatum*; **C** : *Morchella conica*; **D** : *Schizophyllum commune*; **E** : *Scleroderma aurantiacum*; **F** : *Tremella foliacea*; **G** : *Xylaria subterranean*.

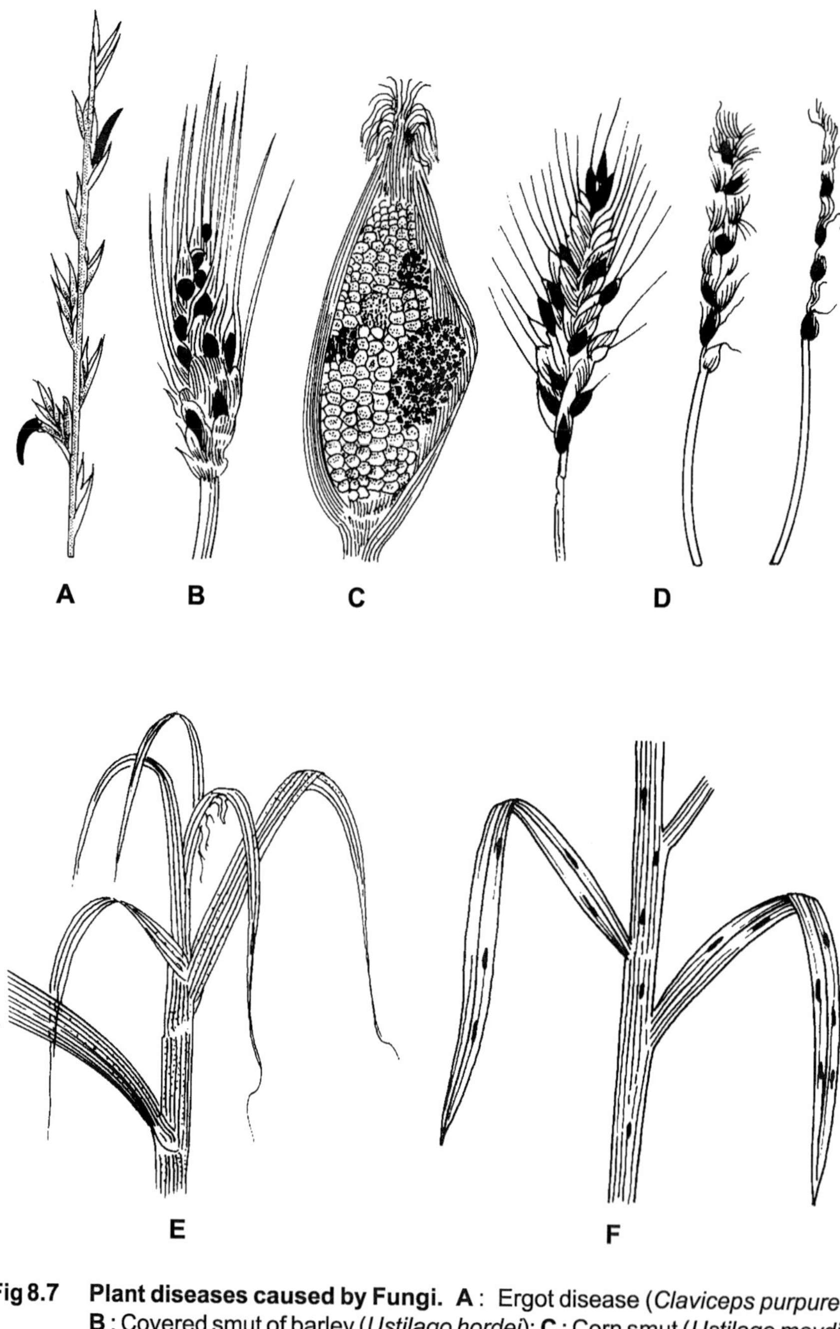

Fig 8.7 Plant diseases caused by Fungi. **A** : Ergot disease (*Claviceps purpurea*); **B** : Covered smut of barley (*Ustilago hordei*); **C** : Corn smut (*Ustilago maydis*); **D** : Loose smut of wheat (*Ustilago tritici*); **E** : Sugarcane smut (Species *of Ustitago*); **F** : Rust of wheat (*Puccinia graminis*).

Fig. 8.7 to 8.9 Redrawn from Agrios, G.N. (1977) and Srivastava, J.P. (1962).

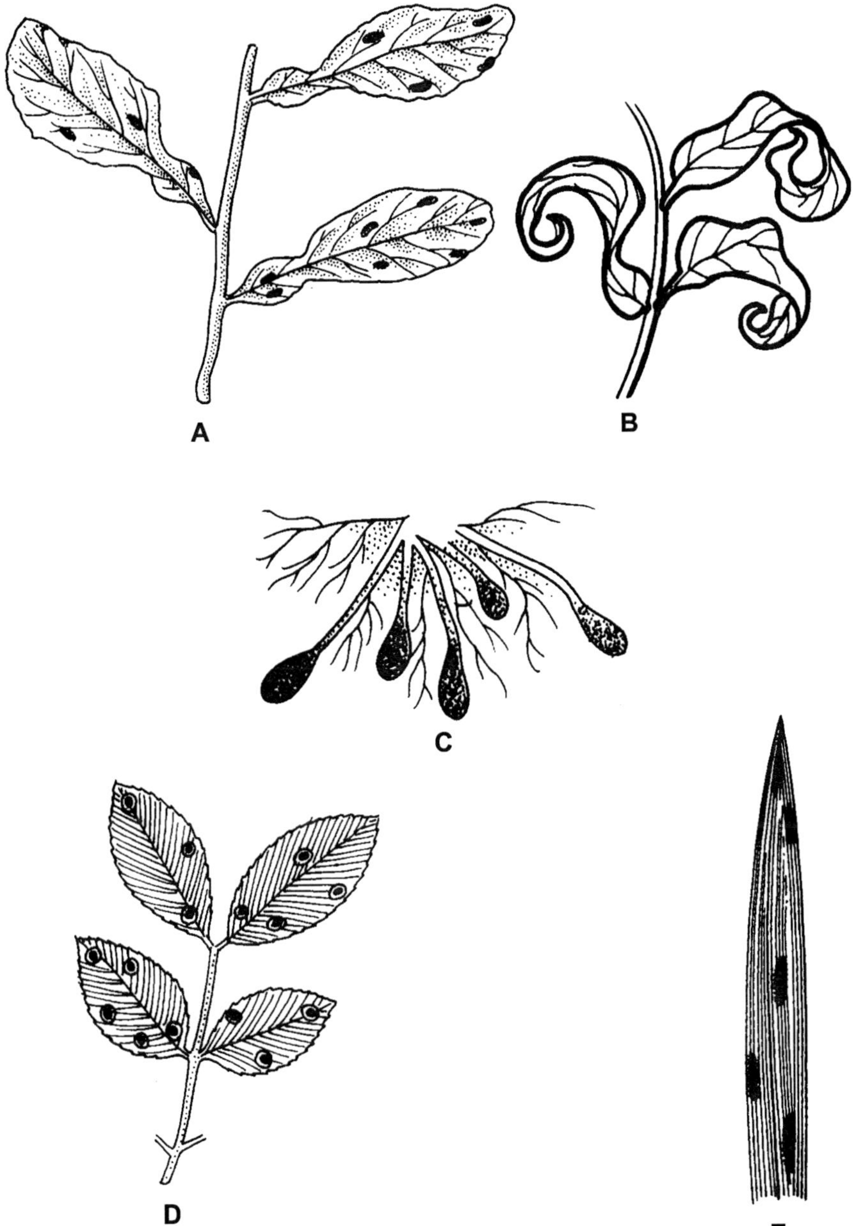

Fig 8.8 Plant diseases caused by Fungi. **A** : White rust of Brassica (*Albugo candida*); **B** : Leaf curl of peaches and plums (*Taphrina deformans*); **C** : Club root of crucifers (*Plasmodiophora brassicae*); **D** : Tikka disease of groundnut (*Cercosporidium arachidicola*); **E** : Downy mildew of jowar (*Perenosclerospora sorghi*).

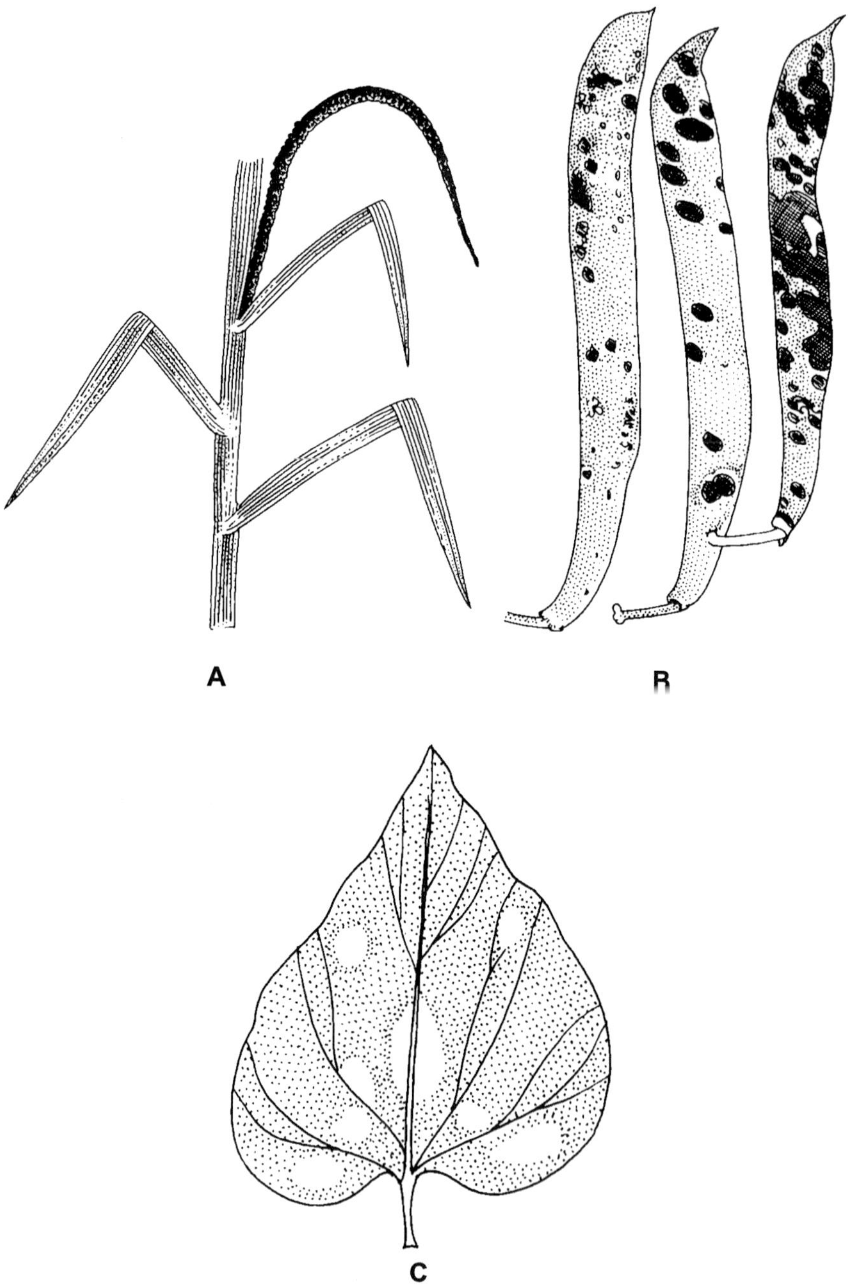

Fig 8.9 Plant diseases caused by Fungi. A : Smut disease (*Ustilago scitaminea*); **B** : Beans affected with *Colletotrichum lindemuthianum* causing anthracnose; **C** : Powdery mildew on cucurbit leaf (*Erysiphe* sp.).

CHAPTER 9

LICHENS

9.1 INTRODUCTION

The word lichen was first used by Theophrastus (371-284 BC) in his book "History of Plants" to describe the extra plant growth on the bark of olive trees. Morision (1699) called them as Musco-fungus and described them as plants with shallow cup like fruits. Dillenius in 1741 used the word Lichenoids for such plants.

Lichens are worldwide in distribution. They grow in wide variety of habitats such as mountain tops to sea shores, in artic regions (*Cladonia rangiferina* = Reindeer moss), on trunks of trees, rocks, walls and soil. They grow equally well on sand and stones. Lichens do not thrive on or near the polluted areas.

Lichen is a thallus of undifferentiated tissue of two symbiont components-algae and fungi. The main skeleton or frame work of the thallus is made by the fungal hyphae that decides the morphological structure and function of protection and reproduction. The algal component performs the function of photosynthesis and assimilates carbon dioxide into carbohydrates as reserve food used by the fungal partner.

Lichens are divided into Ascolichens and Basidiolichens depending upon the group of the fungal component. The fungal component of the majority of lichens is an ascomycetous fungus, but in a few it is a basidiomycetous fungus.

The algal component of Ascolichens is cyanobacteria (blue green algae) (*Scytonema*, *Nostoc*, *Rivularia* and *Gloeocapsa*) while in the Basidiolichens, the algal component is unicellular (*Chroococcus*) or filamentous (*Scytonema*) blue green algae.

On the basis of form and appearance, lichens are divided into foliose (leaf like), fruiticose (shrubby and upright) and crustose (encrusted). Crustose lichens adhere closely to trees, rocks or soils. Foliose lichens have flat, leafy appearance, where as fruticose lichens are upright and have branched appearance.

During hot season, lichens growing on the bark of trees become so dry and inflammable, they act as fuse to carry forest fires to the crown of trees.

Illustration of representative species of lichens are given in Fig. 9.1 to provide an over view of their appearance.

9.2 LICHENS AS SOURCE OF FOOD

Alectoria

Alectoria jubata :is used as a source of carbohydrate in brewing industry (Ambasta, 1986).

Cetraria

Cetraria islandica (Island-moss): has long been eaten by man. It is used in bread and porridge

Evernia

In Egypt, *Evernia* is used in making bread.

Lecanora

In Israel, *Lecanora esculenta* is used used as food (Purohit, 1999).

Parmelia

Parmelia abyssinica and *Parmelia tinctorium*. are used in condiments. In India, species of *Parmelia* are used in preparation of curries and in France they are used in chocolates and pastries.

9.3 LICHENS AS INSECT AND ANIMAL FEED

Lichens are used as food by termites, caterpillars, insects and birds.Species of *Alectoria*, *Cetraria* and *Stereococaulon* are used as animal feed (Rademaker. 2000).

Cladonia alpestris, *Cladonia rangiferina* (Reindeer moss), *Cladonia sylvatica* are used to feed Reindeers. Species of *Alectoria, Cetraria, Cladonia, Roccella montagnei* and *Stereocaulon* are used as fodder for animals.

9.4 MEDICINAL USES OF LICHENS

Lichens have great capacity to survive under extreme harsh environmental conditions by attaining a dormant state and slow metabolic rate, which provides protection against invading viral and bacterial spectrum and animals, insects and nematodes. These metabolites are deposited on the surface of thallus and their antimicrobial properties are employed in traditional medicine for diverse diseases.

Lichens have been used in traditional medicine to treat pulmonary tuberculosis, pain, fever, athelete's foot, dermal lesions and cough.

Cetraria

Cetraria islandica: is used as tonic.

Cladonia

Cladonia alpestris: is used in treating tuberculosis.

Cladonia pyxidata: is useful in whooping cough.

Cladonia rangiferina: hot extract is taken for tuberculosis.

Lobaria

Lobaria pulmonaria: is used to treat lung troubles and given as tonic.

Parmelia

Parmelia kamstchadalis, and *Parmelia perforata*: are used as alexiteric, astringent, laxative, tonic, carminative, aphrodisiac, vulnerant;useful in dyspepsia, amenorrhoea, biliousness, bronchitis, scabies, leprosy, enlarged spleen, burning sensation in the body, bleeding piles, asthma, vomiting, calculi and as antidote for scorpion sting and snakebite.

Parmelia perlata: is astringent, resolvent and diuretic. Is used to relieve pain in renal and lumbar areas.

Parmelia sexatilis: used for treating epilepsy.

Peltigera

Peltigera canina: decoction of this plant is used as deobstruent, purgative and hepato protective.

Roccella

Species of *Roccella* are used to control fever.

Usnea

Usnea barbata: is used as tonic.

Usnea densirostra: In Argentina it is sold as "Barba della piedra" which is used to treat many ailments. Species of *Usnea* is used in the preparation of "Song Lo tea" in China for curing malaria, bronchitis, wounds, snakebites, burns and for detoxification.

Usnic acid from *Usnea* species is a broad spectrum antibiotic used in treating various infections and tuberculosis. It is particularly effective against *Enteroccocci, Mycobacterium* and *Streptococcus aureus,* without perturbing side effects on the beneficial bacterial flora in humans (Cocchietto *et al.*, 2002). It is also used in ointments for burns and wounds.

In clinical trials, a mouthwash containing one per cent usnic acid was administered to volunteers to examine the oral bacterial flora. It was revealed that the growth of *Streptococcus mutans* involved in dental caries was selectively suppressed (Ghione, *et al.,* 1988).

Usnic acid is also known to suppress gram-positive bacteria mainly responsible for body odour and was found to be effective against *Mycobacterium aureum* (Ingolsdottir, *et al.*, 1998) and *Tinea pedis*. It has an inhibitory effect against the pathogenic protozoan *Trichomonas vaginalis* (Wu *et al.*, 1995).

Usnic acid in cancer chemoprevention assay was found to be effective against Epstein Barr virus (Yamamoto, *et al.*, 1995) and Herpes simplex virus type I and Polio virus type I in the infected kidney cells of African green monkey (Perry, *et al.*, 1999). Usnic acid caused moderate inhibition in the murine P388 leukemia assay and exhibited cytotoxic activity against cultured L1210 cells. It also reduced cell counts of leukemic K-562 and endometrial carcinoma cell culture Ishikawa HEC – 50 (Cardarelli, *et al.*, 1997).

Intra-vaginal formulation containing usnic acid as an adjuvant therapy to radio surgical treatment was evaluated in 100 females with genital human papilloma virus. The treatment improved the time of re-epithilization one month after radio surgery (Scrippa, *et al.*, 1999). It also showed anti-inflammatory activity in acute rat paw edema comparable to ibrufen at the same dose level (Vijayakumar, *et al.*, 2000).An oral dose of usnic acid (300 mg/kg) expressed antipyretic activity, determined through lipopolysaccharide induced hyperthermia (Okuyama, *et al.*, 1995).

Xanthoria

Xanthoria parietina: is useful in jaundice.

9.4.1 Lichens in Nanomedicine

The field of medicine has been revolutionized by the application of nanotechnological science useful in cosmetics, dermatological, pharmacological and ophthalmologic fields. Lichens thus are a potential source of nanoparticles to be used in medicine (Shahi and Patra, 2003).

During artificial culture of lichens in synthetic medium for bioactive compounds, *Usnea longissima* was found to be synthesizing bioactive nanoparticle usnic acid in specified medium. This is the first ever report of microbial synthesis of bioactive nanoparticle from lichen in *in-vitro* cultures. *In vitro* analysis of bioactive nanoemulsion was carried out against species of *Epidermophyton, Microsporum* and *Tricophyton* that cause fungal infection of the skin in humans and found to be an effective natural nanomedicine against fungal infections.

9.4.2 Lichens in Treatment of Cancer and Tumour

Lichens were used in treating various types of cancers and tumours (Table 9.1).

Table 9.1 Lichens in the treatment of cancer/tumour

Species	Type of cancer/Tumour
Cetraria islandica	Cancerous ulcers and tumours
Parmelia mollisscula	Indurations of the uterus.
Muscus arboreus	Hard tumours of the liver and inflammatory tumours
Usnea angulata	Warts
Usnea barbata	Warts
Usnea hieronymi	Warts

Source: Hartwell, 1982.

9.4.3 Medicinal Uses in Homoeopathy

Usnea

Usnea is used in homoeopathic system of medicine for congestive headache, and sunstroke (Boericke, 1991).

9.5 TOXICOLOGY

In several short term *in vivo* and *in vitro* animal studies, usnic acid is shown to be an uncoupler of oxidative phosphorylation. It produced hepatotoxicity due to the formation of dysfunctional mitochondria. Since mitochondria play multiple critical roles in tissue haemostasis, toxicity in this organelle may produce a diverse spectrum of clinical syndromes.

Species of *Evernia*, *Pseudoevernia* and *Cladonia stellaris* cause contact dermatitis when used in perfumes.

9.6 OTHER USES OF LICHENS

9.6.1 Tanning and Dyeing

Lichens are a rich source of colouring substances that are used in dyeing and tanning industry.

Cetraria

Cetraria islandica: is used in tanning industry. Lecasnoric acid and erythein, the two coloring principles present in lichens are split into orcin and carbonic acid on treatment with ammonia. Orcin an active principle of orchil-dye is used as a stain in microscopic preparations. These coloring substances yield superior quality dyes ranging from yellow, brown, blue and red, that are valued as insect repellents and are biofriendly.

Cladonia

Cladonia pyxidata: yields a dye used in dyeing industry.

Cladonia stellaris is used in Christmas decorations and flower arrangements and graveyard decoration.

Lecanora

Species of *Lecanora* are the sources of litmus.

Lobaria

Lobaria pulmonaria: yields a brown dye which is used in tanning industry.

Parmelia

Parmelia furfuracea: yields a reddish brown dye.

Parmelia physodes: is used as a substitute for gum arabic in dyeing industry.

Peltigera

Peltigera canina: yields blackish red dye.

Roccella

Roccella: is the source of litmus (Purohit, 1999).

Stricta

Stricta crocata: yields a brown dye.

Teloschistes

Teloschistes flavicans: yields a yellow dye.

9.6.2 Cosmetics and Perfumes

Lichens are rich in aromatic essential oils that are used in cosmetics and perfumery, as for example, *Cetraria, Evernaria* and *Lobaria*. Usnic acid from lichens is used in creams, toothpaste, mouthwash, deodorants and sunscreen products either as a active principle or as a preservative. In the recent years, usnic acid and its salt, sodium usniate are marketed in the US as an ingredient in dietary supplements in weight loss food. It is also a component of brewed tea for raising the metabolic rate, as it acts as a stimulant.

In Italy, usnic acid is used in preparation of vaginal creams, powders and shampoo (Rafanelli, *et al.*, 1995). Ethoxydiglycol of lichens is used as preservative in moisturizing creams (Seifert and Berterm, 1995).

Ramalina

Ramalina calcaris: is used to whiten the hairs-in wigs, as they act as bleaching agents (Purohit, 1999).

Lobaria

Lobaria pulmonaria: is used in perfumery and for cleaning hair (Ambasta, 1986).

Parmelia

Parmelia furfuracea: yields an oleoresin used in perfumery.

9.6.3 Miscellaneous Uses

Lichens are used as floral decoration and as incense.

Cladonia

Cladonia rangiferina: is used in the production of glucose and alcohol.

Cladonia stellaris: is used in christmas decoration and flower arrangements (Brasch and Jacobsen 1991).

Parmelia

Parmelia physodes: is used in making parchment paper and cardboard.

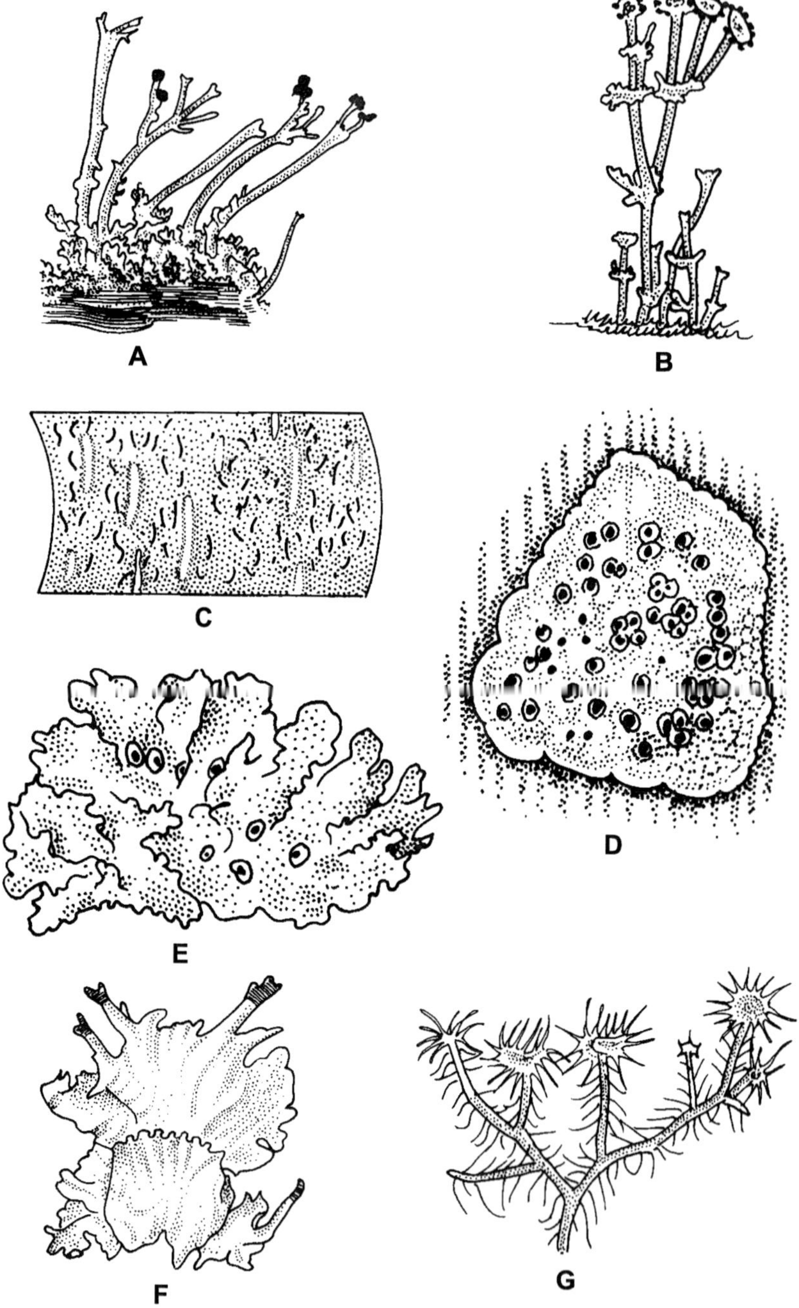

Fig. 9.1 **Lichens.** **A** : Fruticose : *Cladonia flabelliformis* **B** : *Cladonia verticillata*; **C** : Crustose : *Graphis scirpta* **D** : *Haematomma puniceum*; **E** : Foliose : *Parmelia flavicans* **F** : *Peltigera* sp.; **G** : Fruticose : *Usnea barbata*.
Fig. 9.1. Redrawn from Purohit, S.S. (1999)

CHAPTER 10

ALGAE

10.1 INTRODUCTION

Algae, the sea weeds and related forms, are a heterogenous group of chlorophyllous autotrophic plants comprising thirteen major groups called Phyla (Round, 1973). The more important of them discussed in this chapter are the Cyanophyceae (cyanobacteria, blue green algae), Chlorophyceae (green algae), Phaeophyceae (brown algae) Rhodophyceae (red algae), Charophyceae (stone worts) and Bacillariophyceae (golden brown algae).

The algae have a long fossil history and are considered to be the ancestors of more complex groups of cryptogams (non-seed plants), from which the seed plants (Spermatophytes) also arose.

The first reference to algae is found in the early Chinese literature, referring to them as *Tsao,* alluding to their beauty and excellence. The Romans called them *Fucus* and the Greek term for algae is *Phycos*, the sea weeds. Algae initially gained industrial importance when the technology to make agar from algae was discovered in China and Japan, centuries ago.

Algae occur in very diverse habitats. They are primarily aquatic growing in both fresh and salt water (marine). The fresh water forms grow in water bodies such as ponds, pools, lakes, rivers, streams, etc., where as marine forms are found in different zonations of the temperate and tropical seas. Many algae are terrestrial growing on land, and on rocks (lithophytes). Some live on or inside animals (ecto- or endozoophytes), on or inside plants (epiphytes or endophytes), in symbiotic association with lichens, bryophytes, pteridophytes, gymnosperms and flowering plants.

Other unique habitats in which algae grow are ice and snow (cryophytes), hot water springs (thermophilic algae), on and inside soil and sand (edaphophytes). They also inhabit sewage, estuaries, brackish water, salt marshes and swamps. They are the primary colonisers of barren lands.

Algae grow as plankton (free-floating) or as benthons (attached). The planktons belonging to blue-greens and dinoflagellates are the major components of water blooms. *Trichodesmium erythrium* (blue-green alga) causes "red tides" in sea, as the water gets coloured by its abundant growth. Dinoflagellates are also partly responsible for the 'red seas'. Some dinoflagellates exhibit bioluminiscence (emission of light).

Algae exhibit enormous diversity in form, structure, reproduction and life history. The simplest plant body in algae is unicellular (blue green, green and golden brown algae), aggregated unicells (colony) of various shapes and sizes (green algae), branched or unbranched filaments (all groups), siphonaceous (yellow green algae) to multicellular, parenhymatous and macroscopic seaweeds (green, brown and red algae).

Except the Cyanobacteria that are prokaryotic (without a nucleus and other membrane bound organelles), algae are eukaryotic. The cell wall constituents are primarily cellulose and pectin in most algae except in Cyanobacteria that have diaminopimelic acid and muramic acid (similar to bacterial cell wall). In brown and red algae, the cell wall has additional compounds such as carrageenan, algin, furcellaran and funoran. In diatoms the cell wall (frustule) is transparent and silicified, where as the charophytes have lime encrusted cell walls. Cell wall constituents of most algae yield industrially important products.

The algal pigments, mainly chlorophylls and carotenes, occur in specific combinations in different groups of algae and are a major consideration in algal classification. Even the common names of algal groups are based on their most visible colour. The Cyanobacteria, Rhodophyceae and Cryptophyceae possess special protein pigment complexes called phycobilins. The phycobilins present in Cyanobacteria are c-phycocyanin and c-phycoerythrin, where as in the Rhodophyceae and Cryptophyceae they are r-phycocyanin and r-phycoerythrin. The blue-green colour of Cyanobacteria is due to the abundance of phycocyanin compared to phycoerythrin, while it is the reverse in the red algae.

Algae form an important part of human diet in China, Japan and other eastern countries. They are rich in proteins, carbohydrates, fats, fibres, essential minerals such as sodium, potassium, magnesium and vitamins like B_2 (Riboflavin), B_1 (thymine hydrochloride), B_3 (niacin, nicotinic acid), B_5 (pantothenic acid), B_{12} (cyanacobalamine), C (ascorbic acid), D (calcipherols) and M (folic acid), amino acids like alanine, glycine and glutamic acid and carotenoids which are the precursors of vitamin A.

The consumption of seaweeds, which are rich in iodine and bromine, has reduced incidence of goiter (a thyroid disturbance) in the eastern countries.

The third world food survey by FAO draws attention to the shortage of proteins and the need to find alternate sources such as algae, in the developing countries. Single Cell Proteins (SCP) from algae offer a promise to bridge the gap of protein supply, both quantitatively and qualitatively.

Besides SCP, algae are now known to provide a range of therapeutically active compounds, nutriceuticals (such as vitamins, provitamins, amino acids, caretonoids), cosmeceuticals and variety of industrial raw materials, making algae a very rich biological resource.

Illustrations of representative species of algae are given in Fig. 10.1 to 10.31 to provide an overview of their appearance.

10.2 ALGAE AS SOURCE OF FOOD

Agardhiella

Agardhiella: is boiled with sugar and spices and eaten as a sweet.

Alaria

Many species of *Alaria* are edible and sold as 'Saruman'. The reproductive parts of *Alaria,* sold as 'Kausam', are eaten by Alaskan Indians.

Alaria bifida: is called 'Green Kombu' and sold as 'Neko Ashi Kombu'.

Alaria crassifolia: is edible and is sold as 'Chigaiso'.

Bostrychia

Bostrychia is consumed in Burma.

Caulerpa

Caulerpa peltata, Caulerpa racemosa, Caulerpa scalpelliformis, Caulerpa taxifolia: are eaten raw or boiled.

Chlorella

Chlorella pyrenoidosa: is regarded as a rich source of carbohydrates, proteins, fats, vitamins A, B and C, potassium, magnesium, sulphur, iron, calcium, manganese, copper and zinc. It has highest chlorophyll content among all the green algae. Dried *Chlorella* powder is used in making biscuits. *Chlorella* is considered as a good source of Single Cell Proteins.

Chondrus

Chondrus crispus: is one of the most important algal foods. It is used in soups, salads, chocolate milks, puddings and ice creams. A bread called 'Pain des Algues' is prepared from this alga.

Codium

Codium adherens and *Codium tomentosa:* are eaten raw or boiled.

Codium divaricatum, *Codium fragile* and *Codium muelleri:* are used in soup, or eaten as salad.

Durvillea

Durvillea is dried and sold as 'Cachiyugo', for use in salads and soups in Japan and South America.

Enteromorpha

Species of *Enteromorpha* are eaten raw in salads

Fucus

Fucus stipes (stem like part) are eaten with salt, tomato, pepper and onion.

Furcellaria

Furcellaria fastigiata: is used in jams and jellies and as fish preservative.

Gelidiella

Gelidiella species are used in salads.

Gracilaria

Gracilaria coronopifolia and *Gracilaria verrucosa:* are utilized for garnishing soups; with sugar and coconut milk they are eaten as sweet dishes.

Grateloupia

Grateloupia filicina: is used as salad and as vegetable with fish and soybean sauce.

Griffithsia

Griffithsia is boiled with sugar and coconut milk and eaten as a dessert.

Gymnogongrus

Gymnogongrus vermicularis: is used in different food preparations. It is eaten with fish and soy sauce. But before using, it is soaked in water for one day to remove the bitter iodine taste; also used as a thickening agent in chicken broth.

Heterochordaria

Heterochordaria abietina: is salted and eaten. It is also used to preserve and pack mushrooms.

Laminaria

Laminaria is sold as 'Kombu'; cooked with pork is an expensive and exotic dish.

Laminaria japonica: fronds coated with sugar are called 'Kwashi or 'Sweet Cake Kombu'. Powdered kombu is used in soups, tea and curry and also eaten as a vegetable with boiled rice.

Laminaria longipedalis: fronds are soaked in wine-vinegar mixture, dried and the surface layer is removed, leaving a white core called 'Kuru Tororo' which is made into delicate sheets called 'Oboro' or 'Filmy Kombu', used in various preparations.

Laminaria saccharina: stipes are eaten raw in Scotland as it is sweet and has the flavour of peanuts and used in preparation of jellies.

Laurencia

Laurencia obtusa: is used as food.

Laurencia papillosa: used as a condiment.

Macrocystis

Macrocystis: is eaten as salad along with other vegetables. It is made into porridge with cow's milk.

Mesogloea

Mesogloea decipiens: is eaten with soy sauce.

Monostroma

Monostroma species are used as food in Japan. They are made into soup with pepper, olive oil and vinegar.

Padina

Padina australis: is made into a gelatin like sweet meat.

Palmaria

Palmaria palmata: is used as vegetable.

Porphyra

Porphyra tenera, Porphyra umbilicalis, Porphyra kunieda, Porphyra pseudo-linearis, Porphyra okamurai, Porphyra dentata and *Porphyra ungustata* are an important part of Japanese cuisine.

Porphya is sold in the form a paste by the name 'Nori' in Japan, which is used as a sandwitch spread, and in puddings and ice creams.

Dried and salted *Porphyra* is sold as 'Sishi' in Japan, as 'Sloke' in US, 'Tsatsai' in China, and 'Laver' in Korea. It is fried in butter and mixed with lemon juice, olive oil, pepper, salt and eaten with bread or meat.

Porphyra is eaten raw in the far East and Pacific islands. The processed plant is known locally as laver bread in Welsh. In Newzealand *Porphyra columbina* is steamed and eaten as a delicacy on special occasions.

Sargassum

Many species of *Sargassum* are boiled and eaten along with other vegetables.

Sargassum enerve: is eaten with soy sauce in Japan.

Scenedesmus

Scenedesmus species are used as source of food.

Spirulina

Spirulina is regarded as a high value, low calorie diet and is promoted as slimming food in US. *Spirulina* protein contains most of the amino acids important in human nutrition, equalling the nutritive value of milk. It is a very good source of β-carotene, which is a precusor of vitamin A, and has 15 times more pro-vitamin carotenoids than carrot. β-carotene prevents night blindness and increases body resistance. *Spirulina* is also a good source of vitamin C, E and B_{12} and carotenoids which reduce the risk of cancer. In fact, *Spirulina* is the only food from plants containing B_{12}.

Spirulina is an important source of SCP. The Central Food Technological Research Institute, Mysore and the Murugappa Chettiar Research Centre, Chennai, are pioneers in the field of food technology, developing nutritive recipes using this alga. *Spirulina* is marketed in the form of tablets and capsules. Cosmetics are also made from *Spirulina*

Turbinaria

Turbinaria ornata: is pickled and eaten.

Ulva

Ulva lactuca: is used in the preparation of curries and also eaten raw in salads.

Undaria

Undaria pinnatifida: dried and cut into strips and eaten as chips ('Wakame chips'). Rhizoids are dried and eaten with sauce in a preparation called 'Mehibi'.

10.2.1 Single Cell Proteins From Algae

Single cell proteins - a product of unicellular algae such as *Chlorella pyrenoidosa, Scenedesmus acutus, Spirulina fusiformis and Spirulina maxima* is considered as a rich source of vitamins, amino acids and minerals as an alternate source of nutritious human food supplement.

10.3 MEDICINAL USES OF ALGAE

Ascophyllum

Ascophyllum nodosum: is useful in treating sprains and rheumatic pain as an external application.

Chlorella

Chlorella pyrenoidosa: Fibromyalgia syndrome is a chronic musculoskeletal disorder of unknown etiology. The available therapy is disappointing where patients are helped by a combination of medicines. *Chlorella* given as a food supplement produced improvement in the clinical and functional status of the disease. The alga was administered in the form of commercially available products 'Wakasa gold' and 'Sun Chlorella' (Merchant *et al.*, 2000).

Potential of *Chlorella pyrenoidosa* as a dietary supplement in the treatment of fibromyalgia, ulcerative colitis, hypertension, lowered serum cholesterol level is well established. It accelerated wound healing and enhanced immune functions (Merchant and Andre 2001).

Chlorellin, an antibacterial substance is extracted from *Chlorella.*

Chondria

Chondria armata: is anthelmintic

Chondria sanguina: has antiulcerogenic properties.

Chondrus

Chondrus crispus: decoction is used as cough syrup and in the preparation of anticoagulant.

Cutleria

Cutleria multifida: extract is given for stomach ulcers.

Digenea

Digenea simplex: is anthelmintic.

Delesseria

Delesseria sanguinea: is used as anticoagulant

Dictyopteris

Dictyopteris polypodioides: is used in curing lung diseases and scrofula.

Dunaliella

α-carotene from the alga which is a precursor of vitamin A exerts a protective effect against asthma through its antioxidative property (Newman *et al.*, 1999).

Durvillea

Durvillea is used as vermifuge, for scabies and as an antifungal agent. Kelp ash from *Durvillea* is made into pills and sold as *Aethiops vegetabilis* for use against goitre.

Fucus

Species of *Fucus* are given in the form of infusion for goitre.

Fucus distichus, Fucus nodosus, and *Fucus vesiculosus :* used in rheumatism, scrofula and goitre.

Fucus helminthocostus: is used in curing tumours and glandular indurations.

Gelidium

Gelidium cartilagineum : useful in lung diseases and scrofula.

Gigartina

Gigartina stellata: is anticoagulant and is employed in treating diarrhoea, dysentery and inflammation.

Gracilaria

Gracilaria edulis: is antibacterial, antitumour, anticholesterolemic and diuretic.

Gracilaria lichenoides: is used as emmolient, demulcent and alterative.

Hypnea

Hypnea nidifica: is used in stomach troubles in Hawai; is antibacterial, antitumour, anticholesterolemic and diuretic.

Laminaria

Laminaria powder is given in France to strengthen the body. Dried stems of *Laminaria* (Sea girdle) are used as cervical dilator. A strong thread is attached to one end with a collar to prevent the migration of the stem into the cervical canal. As the stem is hygroscopic, it absorbs water and swells 3-5 times its original size within 12 hours. The hollow stipe of *Laminaria* was used as tent for uterine discharges and also to administer antibiotics into uterus.

Laminaria was given as food supplement to patients as a part of treatment for pulmonary tuberculosis. The treatment resulted in improvement of lymphocyte activity (Lozovskaia, 2005).

Laminarin, a polysaccharide of the cell wall forms a viscous, colloidal solution or gel in water. This property has led to its use in laxatives and as an anticoagulant. Sodium laminarin sulphate from *Laminaria* is used as a blood coagulant similar to heparin.

Laminaria bracteata: is used in China for menstrual troubles.

Laminaria hyperborea: short pieces of stipes are employed in surgery for widening wound entrances and fistula. It is based on its property to swell 3-4 times its original size when moistened.

Laminaria saccharina (Sugar wrack): used against syphilis in India.

Nereocystis

In earlier days, *Nereocystis* was used in traditional medicine to cure headache due to congestion. Stipes of *Nereocystis* were placed in the ear, the other end with a bulbous part (pneumocyst) was placed on a hot stove. Steam generated from the bulb passes through the stipe into the ear and clears congestion .

Polysiphonia

Polysiphonia is antibacterial.

Porphyra

Species of *Porphyra* are eaten to control blood cholesterol level. It is a preferred source of red pigment which is used as a fluorescent tag in medical diagnosis.

Porphyra umbilicalis: is used as demulcent, alterative and useful in scrofula.

Rhodymenia

Rhodymenia is used in inducing sweating during fever and when the alga is boiled in water and taken with butter, acts as purgative.

Sargassum

Decoction of species of *Sargassum* is used to treat cancer in China (Yamamoto *et al,* 1995).

Sargassum bacciferum: is employed in South America as a cure for kidney related diseases.

Sargassum linifolium: is used for bladder disorder.

Sargassum vulgare: is an efficient vermifuge.

Spirulina

Spirulina is used to heal wounds, cirrhosis, hepatitis, cancer, heart ailments, atherosclerosis, and anaemia. It is beneficial in restoring body strength, and to combat free radicals in the human body and to act as a stress reliever. Because of high iron and folic acid content, it improves haemoglobin in the blood. Zinc, vitamins A, B_{12}, C and E in *Spirulina* help in delaying aging and increase the body's immune system. It has a role to play in regulating immunoglobulin E mediated allergy.

Spirulina has gamma linolenic acid which prevents cholesterol accumulation in human adolescent studies with hyperlipidaemic nephrotic syndrome (Sammuel *et al,* 2002).

Spirulina diet fed to animal models has exerted an inhibitory action on oral carcinogenesis. It also reversed leukoplakia lesions (Mathew *et al.*, 1995).

In Vietnam, *Spirulina* is recommended for increasing lactation in nursing mothers. It promotes healing process.

10.3.1 Vermifugal Algae

The extracts of the following red algae are used as vermifuge: *Alsidium helminthochorton, Corallina officinalis, Chondria, Digenea simplex, Durvillea, Hypnea musciformis, Jania rubens,* and *Rhodymenia* species.

10.3.2 Algae as Substitutes for Heparin

Heparin is a life saving drug in cardiovascular surgery and in valve replacement. It is conventionally obtained from lungs and intestine of cow and buffalo. The following species exert heparin like activity:

Acanthophora spicifera, Chaetomorpha linum, Chaetomorpha media, Champia parvula, Grateloupia filicina, Gelidiella aurosa, Gracilaria auriculata, Gracilaria corticata, Gracilaria edulis, Hypnea musciformis, Sargassum myriocystum and *Ulva lactuca.*

10.3.3 Algae in Treatment of Cancer and Tumour

Algae are used in traditional medicines for treating cancers and tumours. (Table 10.1).

Table 10.1 Algae in treatment of cancer and tumour

Species	Type of cancer/Tumour
Dunaliella sp	Lung, stomach and colon cancer.
Enteromorpha	Tumours
Fucus vesiculosus	Cirrhous tumours.
Laminaria digitata	Tumours and cancer.
Laminaria japonica	Tumours; swellings; nasopharyngeal, gastric, thyroid and lymph node cancers.
Laminaria saccharina	Tumours and swellings
Laminaria utilis	Tumours
Nostoc commune	Cancerous ulcers
Sargassum siliquastrum	Nasopharyngeal, gastric, thyroid and lymph node cancers.

Source: Hartwell, 1982

10.3.4 Medicinal Uses in Homoeopathy

Fucus

In homoeopathic system of medicine, *Fucus vesiculosus* is used in treating goitre, obesity, obstinate constipation and thyroid enlargement (Boericke, 1991).

10.4 OTHER USES OF ALGAE

Chara

The lime-encrusted thallus is added to acidify soil and to give it a good crumb structure. Charophytes are well suited to experimental biosystematic investigations as they are easy to grow with easily countable chromosomes.

Chlamydomonas

Chlamydomonas species are sewage disposers as they decompose the sewage bacteria and by their photosynthetic activity incerease the oxygen content of sewage water.

Chlorella

Chlorella pyrenoidosa: can be converted to a hydrocarbon stock and is considered for coal hydrogenation. It can be used as a petroleum substitute. *Chlorella* is a good research material.

Diatoms

Diatoms have a potential for the production of materials that is of great benefit to microfabrication. they are being used in the design and manufacture of materials of micrometer proportion. They can be used in the production of low cost flexible and programmable systems in nanotechnology (Parkinson and Gordon, 1999). Because of the fine markings on the frustule they are used to test the resolution power of microscope lenses. Diatoms are considered as the primary source of crude oil that yields petroleum products.

After the death and decay of diatoms, their silicified shells sink to the bottom of the water body or the sea bed, and form an indestructible material called 'diatomaceous earth' or 'diatomite' or 'kieselguhr', which is highly valued in industries. It is used in making bleaching powder, as filtering aid in sugar and brewing industry, as insulators for boilers, furnaces and refrigerators due to its low heat conductivity, in polishing and cleaning metals, as an absorbent in metal and silver polish, as a cleansing agent in tooth paste, in storing and transporting strong acids and nitroglycerine, sprinkled on the walls and floor of coal mines to reduce the chance of secondary explosion, to manufacture sodium silicate (water glass) which is an industrial catalyst. Diatomite bricks are used in construction. The 100 m tall Saint Sophia church in Constantinople was built from diatomite in 532 BCE. The bricks are also used in the construction of sound proof rooms.

In research, Kieselguhr is used as a support for the stationary phase in partition chromatography.

Dunaliella

Dunaliella is a source of glycerol and an ideal alga for biosolar energy conversion in arid zones.

Durvillea

Durvillea (Bull kelp) is split and used as a bag to store water by the natives of South America. In Newzealand it is burnt and mixed with water as a substitute for Chinese ink.

Laminaria

Laminaria is dried and burnt to obtain soda and potash used in making soap, glass and porcelain. (Chapman and Chapman, 1980).

Laminaria radicosa : used in floral decoration.

Lessonia

Dried stipes of *Lessonia* are hard as bone and are used as knife handles.

Nereocystis

Nereocystis, after drying and stretching, is used as fishing lines.

Sargassum

Sargassum enerve: used in floral decoration.

Sargassum tenerrimum: mixed with cow dung produces methane gas (Sarma,*et al.,* 1984).

Spirulina

Spirulina with its high degree of biomass production is an efficient converter of solar energy and carbon dioxide. It is cultivated by using biogas slurry and can be dried using solar driers for feeding livestock.

10.4.1 Algae in Soil Fertility

Phycocolloids present in brown and red algae when added to soil provides a good crumb structure to the soil. Dried and burnt, the ash of brown algae is a good fertilizer. *Turbinaria* is used as fertilizer in India for palms and coconut trees. Species of *Alaria, Corallina, Fucus, Laminaria, Lithophyllum, Macrocystis, Nereocystis, Palmaria,* and *Sargassum* are commonly used as fertilizers.

10.4.2 Algae as Fodder

Many of the sea weeds such as *Ascophyllum, Fucus, Laminaria,Rhodymenia, Sargassum* and *Turbinaria* are used as animal feed for their high micronutrient content.

10.4.3 Biofuels from Algae

The dead algal scum of *Botryococcus* contains 70 per cent hydrocarbons closely resembling crude oil. The hydrocarbons accumulate as globules in the outer walls and cytoplasm. Another alga, *Chlorella pyrenoidosa* can be converted to a hydrocarbon stock and is considered for coal hydrogenation, for use as petroleum substitute.

Immobilized strains of *Chlorella, Microcystis, Oscillatoria, Scenedesmus* and *Synechococcus* are used in the production of hydrogen. *Dunaliella, Sargassum tenerrimum* and *Spirulina maxima* are employed in the production of methane.

10.4.4 Algal Bioremediation

Phosphorous and nitrogen residues are removed by biofilteration from sewage and oil of industrial wastes, using *Oscillatoria annae* mixed with sand. *Oscillatoria pseudogeminata* is successfully used in treating paper mill effluents. (Kumar, 1994).

Sewage is a mixture of domestic and industrial waste with dissolved and suspended organic and inorganic substance, and pathogenic organisms.

Sewage treatment involving chemicals, is highly expensive. Using algal system sewage is successfully treated before its disposal. Species of *Ankistrodesmus, Chlamydomonas, Chlorella, Euglena* and *Scenedesmus* are being employed in complete oxidation of sewage and decomposition of bacteria. Saprophytic diatoms clear the water by feeding on organic waste. Growth of *Chara* clears turbid water.

10.4.5 Ecological Effects of Extracellular Products of Algae

Extra cellular products are defined as soulble chemical compounds liberated from healthy algal cells. Many of them are the by products of metabolism.

The major extracellular products secreted by algae are amino acids, amines, peptides, glycolic acid, oxalic acid, pyruvic acid, carbohydrates, glucose, xylose, galactose, rhamnose, arabinose, vitamins, antibiotics, enzymes, hydrocarbons, nitrogenous substances, sterols, fatty acids, phenolic compounds, carbonyl compounds and hormones.

These compounds secreted by various classes of algae have a profound ecological effects on water chemistry, pH, salinity and phytoplankton periodicity. Many of the dinoflagellates secrete neurotoxins, hepatotoxins, and saxitoxins into water which affects the fauna and flora of water and cause allergic disorders in humans.

The non toxic cellular products are utilized as nutrients by other aquatic organisms. The antibiotics secreted by algae inhibits the growth of saprophytic bacteria, and fungi.

10.5 APPLICATIONS OF CYANOBACTERIA

10.5.1 Secondary Metabolites from Cyanobacteria in Medicine

Cyanobacteria are rich in secondary metabolites that have several applications in medicine.

Extracts of Cyanobacteria inhibit various enzymes such as trypsin, chymotrypsin, thrombin, plasmin, elastase and collagenase (Patterson, 1996). Thrombin inhibitors are useful agents in the treatment of stroke and coronary artery-occlusion. It prevents conversion of fibrinogen to fibrin by which blood clotting ceases.

The phycobiliproteins from Cyanobacteria are used as anticancer agents and as immunostimulants. They are used in treating ulcers and haemorrhoidal bleeding (Richmond, 1990).

Tolytoxins, tantazoles (Carmeli *et al.*, 1993) mirabazoles (Pattenden and Thom, 1993) are modified heterocyclic peptides of cytotoxins in *Scytonema mirabile*. These are tumour selective cytotoxins. Haplosin—a novel cyclic depsipeptide from *Haplosiphon welwitschii* reverses p-glycoprotein mediated multidrug resistance in tumour cells (Stratmann *et al.*, 1994).

Cryptophycin—I, a new microtubule depolymerizing agent (Smith *et al.*, 1994), showed excellent activity against multi drug resistant tumours (Trimurtulu *et al.*, 1994). The extract of *Lyngbya majuscula* yields a lipid curacin A that is antiproliferative in action, which inhibits microtubule assembly (Gerwick *et al.*, 1994).

Antitumour galacto lipids were isolated from nitrogen fixing *Anabaena flos-aquae* (Shirahashi *et al.*, 1996).

10.5.2 Antimicrobial Compounds in Cyanobacteria

A number of Cyanobacteria exert bioactivity against viruses, bacteria, yeasts and fungi (Table 10.2).

Table 10.2 Bioactive compounds in cyanobacteria

Cyanobacteria	Active against	Active compound
Anabaena laxa	*Aspergillus oryzae, Candida albicans, Penicillium notatum, Saccharomyces cerevisiae, Trichophyton mentagrophytes*	Laxaphycin

Calothrix fusca	*Aspergillus oryzae, Candida albicans, Penicillium notatum, Saccharomyces cerevisiae, Trichophyton mentagrophytes*	Calophycin
Dichothrix baueriana	Herpes Simplex Virus (HSV)	B-carbolines
Fischerella ambigua	*Aspergillus oryzae, Candida albicans, Penicillium notatum, Saccharomyces cerevisiae, Trichophyton mentagrophytes*	Ambiguines
Hormothamnion enteromorphoides	*Bacillus subtilis* *Pseudomonas aeruginosa* *Salmonella hyphimurium* *Staphylococcus aureus* *Candida albicans*	Hormothamin Hormothamnion A
Lyngbya majuscula	*Mycobacterium smegmatis, Streptococcus pyogenes*	Malyngolide
Lyngbya lagerheimii	HIV-1 (T-cell lines)	Sulpholipid
Haplosiphon hibernicus	*Aspergillus oryzae , Candida albicans, Penicillium notatum, Saccharomyces cerevisiae, Trichophyton mentagrophytes*	Ambiguines
Nostoc ellipsosporum	HIV-1	Cyanovirin-N
Nostoc sphaerium	Herpes Simplex virus	Indolocarbazoles
Phormidium tenue	HIV-1	Cyanovirin-N
Schizothrix sp. (TAU Strain IL-89-2)	*Bacillus subtilis*	Schizotrin A
Scytonema ocellatum	*Candida albicans, Penicillium notatum, Saccharomyces cerevisiae, Trichophyton mentagrophytes*	Tolytoxin
Tolypothrix tjipanensis	*Candida albicans*	Chloroindolocrazole
Westiellopsis prolifica	*Aspergillus oryzae, Candida albicans*	Ambiguines
Westiellopsis sp.	*Penicillium notatum, Saccharomyces cerevisiae, Trichophyton mentagrophytes*	Tolytoxin

Source: Kulik, 1995.

10.5.3 Other Uses of Cyanobacteria

Cyanobacteria are used in biological assays as they can be grown in abundance. They are used in limnographic and oceanographic studies and as a system for producing isotopically labelled metabolites. *Spirulina* and *Anacystis* are capable of growing at high cell density. When grown on substrates such as carbon dioxide enriched with ^{13}C or ^{14}C, the resulting compounds are utilized as tracers in biology, analytical chemistry and diagnostic medicine (Chorney *et al.*, 1960; Behrens *et al.*, 1994).

Production of renewable fuel employing biosystems is a priority research through out the world. Cyanobacteria such as *Anacystis* and *Oscillatoria* are the potential source of biofuels. Many of the Cyanobacteria possess nitrogenase and hydrogenase that can be utilized in producing hydrogen biosynthetically (Smith *et al.*, 1992). A sustained and substantial production of hydrogen without the co-production of oxygen is reported in *Synechococcus* species, Miami BG043511 strain (Kumazawa and Mitsui, 1994).

10.5.3.1 Cyanobacteria in bioremediation

Cyanobacteria are capable of abating various pollutants in soil and water. They can degrade aromatic hydrocarbons such as naphthalene, n-alkanes and crude oil (Gerniglia *et al.*, 1980). *Phormidium valderianum* is capable of adsorbing cadmium, copper, and cobalt (Karna *et al.*, 1999).

Synchocystis Pcc6803 can metabolise herbicide like glyphosate (Powell *et al.*, 1990) and *Lyngbya major, Gloeocapsa atrata, Calothrix parietina* and *Scytonema pascheri* can metablise organochlorine and carbamate pesticides (Sahu *et al.*, 1992). Organophosphorous pesticides were degraded by *Aulosira fertilissima*.

10.5.3.2 Cyanobacteria in agriculture and soil fertility

Nitrogen fixing blue green algae increase soil fertility by providing nitrogenous compounds and organic matter on their death and decay.

Blue green algae are important for the fertility of rice fields. They release fixed nitrogen, increase soil phosphates, prevent leaching of ammonia and nitrates, and provide humus. *Aulossira fertilissima,* fixes highest amount nitrogen.

Cyanobacteria improve water holding capacity of the soil and also prevent soil erosion due to the presence of mucilage around the cells. They also reclaim the salinity and alkalinity of the soil and add good crumb structure and prevent clumping of soil particles.

Scytonema crassa helps in withering of rocks by their enzymes and result in formation of very thin layer of soil.

10.6 ALGAL PRODUCTS IN INDUSTRY

The pigments and cell wall constituents of algae are extracted and processed about 10 and the liquor is filtered while hot. A solution of calcium chloride is added to the filtrate and the precipitated calcium alginate rises to the top and is into products of diverse application in industry. These wall constituents and pigments are mainly from blue green, brown and red algae.

10.6.1 Algal Pigments

10.6.1.1 Carotenoids

The carotenoids present in cyanobacteria are used as non-toxic colouring agents in food industry. They are used as dietary additives in poultry feed to get rich orange colour to the meat and eggs. (Hirschberg and Chamoritz, 1994).

10.6.1.2 Phycobilins

Phycocyanin and phycoerythrin, the phycobilin proteins which are the characteristic cyanobacterial pigments are commercially available as 'Lina blue' from Dainippan Ink and Chemicals Inc., Japan, which are used for coloring candy, ice cream, yoghurt, dairy products and soft drinks (Cohen, 1986).

Phycocyanin present in *Spirulina* is being used in Japan in making bio-lipsticks, bio-eyeliners, bio-eye shadows, creams and soaps. It is also used in coal tar dyes. *Haematococcus* is being used in making red coloured lipsticks (Biolipsticks). Phycobilins are also used as tools for basic research in fluorescent microscopy, immunoassays and as phycofloures (Glazer and Stryer, 1984; Kronick, 1986)

The phycocyanins and phycoerythrins in *Spirulina* have unique excitation absorbance which permits application of fluorescence based immunoassays, and as non-radioactive markers for DNA probes.

10.6.2 Cell Wall Constituents of Algae

10.6.2.1 Algin

Algin was first discovered by Stanford in 1881. Krefting (1896) purified it and called it as 'tangsaure' meaning sea weed acid. Algin occurs as alginic acid which is the cell wall constituent of brown algae. Alginic acid is a polymer of â-1,4-linked D-mannuronic acid and L-guluronic acid units. It is insoluble in water but can absorb water.

Preparation of algin

The dried sea weeds are washed with faintly acidulated water to remove most salt content and are chopped and then bruised in a hammer mill. The sea weed thus prepared is digested in solution of sodium carbonate (soda ash) at pH of removed. By treatment with hydrochloric acid, alginic acid is precipitated. Neutralization of the purified alginic acid by soda gives sodium alginate.

Uses of algin

Algin in food industry

Algin is used in confectionary as filler for candy bars, whipping creams, cheese, salad dressing and as icing for decorating fancy cakes, for stabilizing beer and sherbets, for creaming and as a suspending agent in jellies and milk shakes and to prevent crystal formation and to give smooth surface to ice creams. Algin is mixed with milk powder and coco powder to make them readily soluble.

Algin in pharmaceuticals

Algin is used as a stablilizing agents in the preparation of surgical jellies, ointments, suppositories, bulk forming laxatives, haemostatic powders, antacid tablets, sulpha suspensions, terramycin and aureomycin tablets, dental and orthopaedic impressions.

Algin in cosmetics

Algin is used as a base for creams, shampoo, hair sprays, hair dyes, toothpastes, shaving creams and calamine face creams. Alginic acid derivatives, triethanolamine alginate and propylene glycol alginate are used in face creams as dust protectors.

Other uses of algin

Algin is used in glazing paper and ceramics and as a stabilizing agent in the production of rubber and tyres; is used in shoe polishes and as a thickening agent for fabric colours and in printing; as a hardner and adhesive for joining threads during weaving by mixing with starch or tragacanth gum; as a sizing material for paper and cardboard; as an adhesive in the manufacture of briquettes of brown coal or lignite; is used in asphalt paints. Paints mixed with algin provide smooth surface, better flow of the paint and stabilization of the color; when mixed with copper it is used as an antifouling agent to paint the undersurface of ships and as water proof material.

Algin is used in the production of artificial fibre by purifying the alkaline extract and forcing it through fine holes to make viscous threads which can be spun in a mixture of furfurol, caustic soda and formalin.

Algin mixed with resin yields a rubber like end product, which is treated with sodium carbonate, formalin and glue or resin for paneling boards. Algin mixed with formalin forms a gel that is used in sealing holes or cracks in oil drilling machines.

The insoluble calcium salt of algin is mixed with tannic acid and emulsified to form a thin transparent film similar to cellophane. When mixed with silica and concentrated sulphuric acid it is an efficient color absorbent for decolorizing liquids. Soaps made of algin give more lather in sea water or hard water.

Alginates are used in the production of insecticides such as DDT and fungicides such as Bordeaux mixture. They are used as natural hydrogel for cell immobilization. Entrapment of cells in beads of calcium alginate is the most widely used method for immobilizing living cells for e.g. immobilizing bone marrow cells for regeneration of bone like tissue (Ho *et al.*, 2005); used as binder for printers ink and as centrifuge primers and non-splinter glass. Ammoniated aluminium alginate and ammoniated copper alginate are employed in making water proof cloth, wagon covers, plastics, imitation leather and linoleum. Calcium alginate with an alkali is marketed as 'Tragaya' which is used as a substitute for gum tragacanth. Yarn treated with calcium alginates and chromium yields an alkali resistant yarn. Cotton yarn treated with calcium alginate removes the normal twist of the yarn to yield untwisted yarn, used to stitch open wounds. It is used to treat canvas to prevent mildew formation. In textile industry, it is an excellent polishing material. Sodium peralginate is used as a bleaching agent. Bricks coated with 2 per cent alginates are weather proof. Copper and nickel salts of alginates are used in spinning industry for giving a natural green color to the yarn. Cobalt alginate gives a red color, while chromium alginate gives a blue color.

Box 10.1 Japanese Silk Yarn.

Alginic acid extracted from *Sargassum* is mixed with ammonium salt (3%) and coagulated in a bath of sulphuric acid (10%) and immersed in ammonium sulphate (10%) and finally rinsed in a weak solution of lead acetate yields a yarn called as Japanese silk yarn.

10.6.2.2 Agar-Agar

Agar or Agar-Agar, is a colloid obtained from red algae such as species of *Acanthopeltis, Ahnfeltia, Gelidium, Gigartina, Gracilaria, Phyllophora* and *Pterocladia* (agarophytes).

Preparation of Agar

The traditional method of preparation utilized extraction by boiling for several hours. Before this, the raw material had to be cleaned and bleached naturally (in the sun), with several washings in fresh water to facilitate bleaching. The boiled extract is acidified and concentrated and then cleaned of waste material by repeated freezing and thawing after which it is dried and marketed in the form of cakes or flakes.

Another method of extracting agar is by heat and pressure in autoclaves where agar is decolorized and deodorized by activated charcoal, filtered under pressure and evaporated under reduced pressure. Further purification is done by freezing.

The quality of the product varies considerably, depending upon the species from which it is obtained, the collection, drying and bleaching procedures, and the manufacture techniques.

Uses of Agar

Agar in food industry

Agar is used as a thickening agent in soups and sauces, and as a clarifying agent in brewing of beer and wine, and a stabilizer in coffee and chocolate drinks. It is extensively used in fruit jellies, ice creams, candies, cream, cheese, custard, jams, marmalade and mayonnaise sauce. Agar is cooked in Japan with rice to make 'agar rice'. Doughnuts are glazed using agar.

Agar in pharmaceutical industry

Agar is used in preparation of pills, suppositories, dental impression, and as a thickener and stabilizer in lotions, creams and ointments. 'Agarol', 'Normacol' and 'Regulin' are agar based laxatives.

Other uses of agar

Agar is used as the medium to culture bacteria and fungi and as a molecular sieve in chromatography. It is used in sizing of silk fabrics; in dyeing and printing paper and textiles; in making waterproof paper; in photographic industry in making films; in the leather industry to impart stiffness and gloss; as a constituent of high grade adhesive and pesticides; in the manufacture of plywood as glue, as a lubricant for electrical fittings; in the manufacture of storage batteries for submarines; as an ingredient in sound and heat proof material, and water based paints.

10.6.2.3 Carrageenan

Carrageenan (carrageenin) is a gelatinous cell wall polysaccharide, extracted from red algae. It is a complex mixture of D-galactose-3—6-anhydro-L-galactose esterified with sulphate. There are two major types of carrageenan, a branched chain polymer, called the Kappa fraction (*k*-carrageenan) and a linear chain polymer, called the Lambda fraction (λ-carrageenan). Another form, the *i*-carrageenan used in chocolate milk, puddings and ice creams, occurs in small quantities in *Chondrus crispus.*

Chondrus crispus, Chondrus ocellatus, Gigartina acicularis, Gigartina pistillata, Gigartina radula, Gigartina stellata and some species of *Gymnogangrus canaliculata, Hypnea musciformis, Endocladia denticulata, Rhodoglossum,* and *Iridophycus* are the important sources of carrageenan.

Preparation of carrageenan

The red sea weeds are dried and washed with fresh water to remove salt and then boiled in water for a few hours. The boiled material is filtered to remove excess water and dried in a rotary vacuum evaporator to obtain carrageenan.

Medicinal uses of carrageenan

Intravenous injections of carrageenan reduces serum lipids and prevents development of atherosclerosis. It is sudorific, stimulant, demulcent, emollient, anticoagulant, pectoral in cough and bronchitis; used in inflammations of the skin and intestinal ailments (Ambasta, 1986).

Carcinogenecity of carrageenan

Carcinogenecity of carrageenan has been studied in both native and degraded forms. Oral administration or subcutaneous injections induced carcinomas, ulcerative lesions of the large intestine, colorectal metaplasia, squamous cell carcinoma and adenomatous colon polyps in laboratory animals (Hirono, 1987).

Other uses of carrageenan

Carrageenan is a hydrocolloid employed in industries for its properties of viscosity, stabilization, suspension and foaming capacity. It is used for taking dental impressions, as a clarifying agent in beer, honey, wine, coffee and in soups and sauces. Carrageenan is employed in hand lotions, creams, tooth paste, cough syrup, pills, tablets, in shoe polish, paints, cosmetics and shaving creams.

Carrageenan is used in leather, paper and printing industry to impart gloss and as a stiffening agent in the textile industry; as an emulsifying agent for cod-liver and other fatty oils employed in spermicidal jellies; is used for cell immobilization as a natural hydrogel.

10.6.2.4 Eucheuman

The product from the *Eucheuma cottonii and Eucheuma spinosum* is called eucheuman, which is intermediate between agar and carrageenan. It has the same uses as agar-agar and carrageenan.

10.6.2.5 Funori

Funori is chemically similar to agar where galactose residues may be sulphated or methylated. Funori, obtained from *Gloiopeltis furcata, Gloiopeltis tenax, Gloiopeltis complanata* and *Gloiopeltis coliformis* is used in textile and paper industry as sizing agent and in decorating porcelain in Japan; is also used as glue.

10.6.2.6 Furcellaran

Furcellaran from *Furcellaria fastigiata* is used in toothpaste, ointments and lotions. In pharmaceuticals it is used as a jelling and thickening agent and as stabilizer and preservative.

10.6.2.7 Mannitol

Mannitol (a sugar alcohol) is a reserve food product of brown algae. Species of *Laminaria* contain more mannitol than other brown algae. It is used as an alternative sweetener, and is used in pharmaceuticals, paints, leather, lacquer and plastic. When nitrated, mannitol forms nitromannite, a very powerful explosive similar to nitroglycerine.

10.7 NEGATIVE IMPACT OF ALGAE

During rainy season and under humid conditions algae in general (mostly unicellular and colonial forms) cause deterioration of fabrics, paints, walls of the building, paper and leather.

Species of *Amphora, Cladophora, Enteromorpha, Macrocystis, Nereocystis, Polysiphonia, Porphyra, Palmaria, Sargarsum, Turbinaria* and *Ulva* cause fouling of substratum by their colonization. Ship hulls and boats are easy targets for fouling by algae. Fouling results in loss of speed, higher fuel consumption and increase in frictional resistance. They also cause imbalance in water chemistry and disturb oil and gas extraction operation. Their excessive growth of algae in water results in algal blooms (water blooms).

10.7.1 Water Blooms

Water blooms are due to a luxuriant proliferation of the algae. Initially this may enhance the oxygen levels in the water, but in the later stages, water blooms change the colour, odour and taste of water, making it unfit for drinking and swimming. Water blooms also increase the biological oxygen demand, severely threatening the flora and fauna in the water, causing an ecological imbalance and various diseases in humans.

Blooms can occur both in fresh water and sea, either as a temporary phenomenon (temporary blooms) during the rainy season and the immediate following period in fresh water habitats or as a lasting phenomenon in sea water (permanent blooms).

Water blooms are caused by several groups of algae, though Cyanobacteria and dinoflagellates predominate (Table 10.3).

Illustration of representative species of algae causing water blooms are given in Fig. 10.27 to 10.31 to provide an overview of their appearance.

Table 10.3 Algal groups that cause water blooms

Algal groups	Species
Cyanobacteria (blue green algae)	*Anabaena majuscula, Aphanothece, Gloeotrichia, Lyngbya majuscula, Microcystis aeruginosa, Oscillatoria, Trichodesmium erythreum*
and	species of *Nostoc* and *Schizothrix.*
Chlorophyceae (green algae)	Species of *Cladophora, Closterium, Cosmarium, Eudorina, Gonium, Pandorina, Pleudorina* and Volvox.
Bacillariophyceae (diatoms)	Species of *Fragilaria, Nitzchia* and *Chaetoceras.*
Dinoflagellates	Species of *Dinoclonium, Euglena, Dinophysis, Gonyaulax, Gymnidium* and *Synura.*

10.7.1.1 Effects of water blooms on human health and welfare

Water blooms spoil drinking water by imparting bad odour and taste. They reduce oxygen content, change the pH and mineral content in water, cause slime formation in water reservoirs, ruin recreational facilities like fishing, swimming and boating, and clog water pipes.

The cyanobacterial and dinoflagellate algal blooms release biotoxins (Table 10.4) like hepatotoxins, neurotoxins, cytotoxins and haemolytic toxins into water resulting in various skin diseases, neurological and gastro intestinal disorders and respiratory disorders, in swimmers and fishermen exposed to such water. Biointoxications can also be caused by dinoflagellates. They are mainly paralytic shell fish poisoning (PSP), diarrhoetic shellfish poisoning (DSP), venerupin poisoning (VP) and amnesic shellfish poisoning (ASP) (Viriani, 1992). Different types of cyanobacterial toxins are given below:

A. Neurotoxins

1. Anatoxins : *Anabaena flos-aquae*
2. Homoanatoxin: *Oscillatoria formosa*
3. Aphanotoxin: *Aphanizomenon flos – aquae*
4. Saxitoxin: *Lyngbya wolleri*

B. Hepatotoxins

1. Microcystin: *Microcystis aeruginosa, Nostoc rivulare, Oscillatoria agardhi*
2. Cylindrospermin: *Cylindrospermopsis raciborskii, Umezakia natans*

C. Cytotoxins

1. Scytophycin: *Scytonema pseudohofmanni*
2. Cyanobacterin: *Scytonema hofmanni*
3. Haplolindole: *Haplosiphon fontinalis*
4. Acutiphycin: *Oscillatoria acutissima*
5. Lyngbyatoxin: *Lyngbya majuscula*

Table 10.4 Algae in water blooms causing specific human ailments

Species	Ailment
Anabaena circinalis	Saxitoxins and aphanotoxin cause shell fish poisoning.
Species of *Anabaena*	Asthma and gastrointestinal disorders; paralysis of skeletal and respiratory muscles, impaired movement and respiratory failure
Species of *Cladophora*	Dermatitis
Fragilaria striatula	Dermatitis
Nitzchia pungens	Amnesic (memory loss) shell fish poisoning and silicosis-a lung and respiratory disorder
Dinoflagellates	Neurotoxins cause paralytic shell fish poisoning resulting in respiratory paralysis
Lyngbya majuscula	Swimmers' itch (skin rashes, blisters and pustular folliculitis)
Microcystis aeruginosa	Microcystin, a hepatotoxin causes water accumulation in lungs, swollen blood vessels, intra-hepatic haemorrhage, haemorrhagic shock and gastrointestinal disorders
Oscillatoria nigroviridis	Contact dermatitis
Species of *Oscillatoria*	Asthma and gastrointestinal disturbance
Prototheca zopfii and *Prototheca wickerhamii*	Protothecosis
Species of *Schizothrix*	Gastrointestinal disturbance

Source: Kumar, 1990.

10.7.2 Plant Diseases Caused by Algae

A few of green algae cause plant diseases. *Cephaleuros coffeae* causes coffee rust and *Cephaleuros viresence* causes red rust of tea. They cause severe economical loss also in palm, magnolia, citrus and pepper plantations.

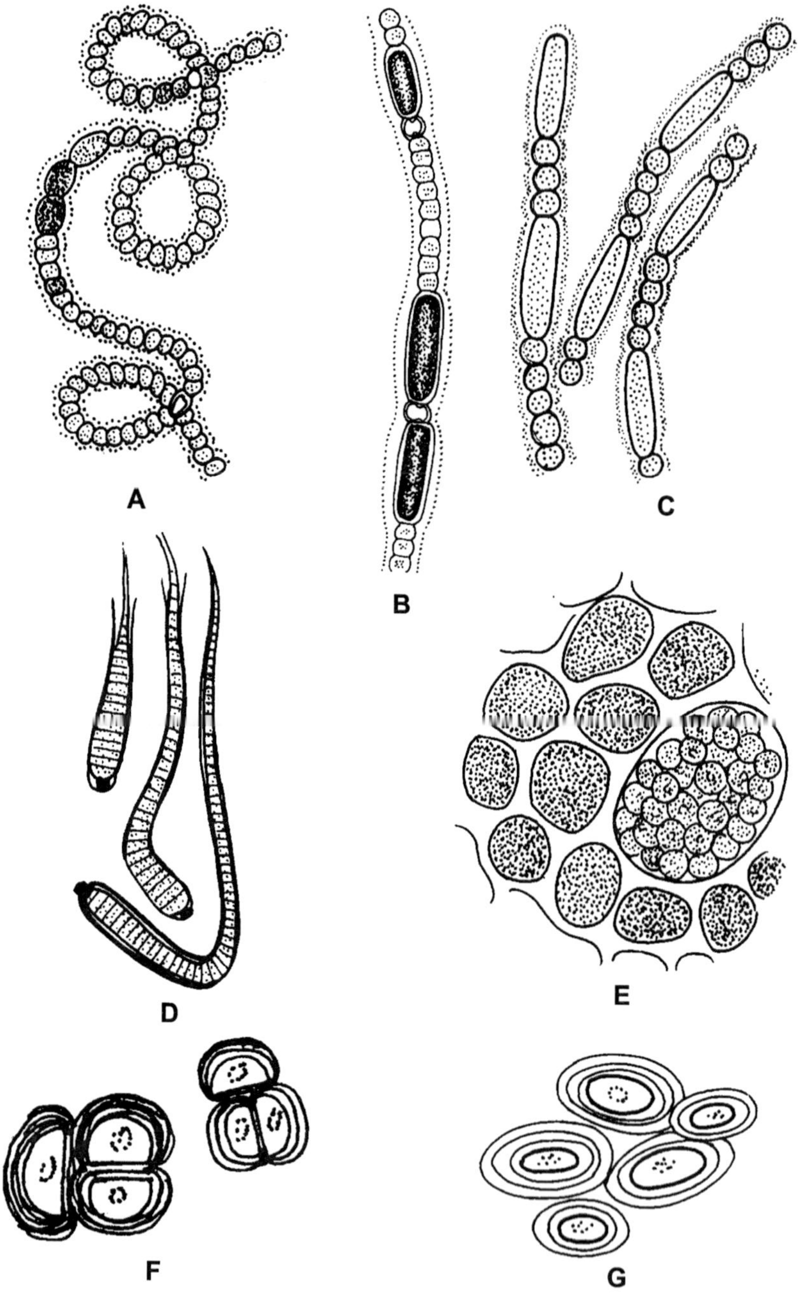

Fig. 10.1 Cyanobacteria. A : *Anabaena circinalis*; **B** : *Anabaena oscillarioides*; **C** : *Aulosira laxa*; **D** : *Calothrix parietina*; **E** : *Chlorococcum humicolum*; **F** : *Chroococcus montana*; **G** : Gloeocapsa montana.

Fig. 10.1 to 10.31 Redrawn from Bhatia, K.N., (1984) Dixon, P.S., (1973) Fritsch, F.E. (1935), Kumar, H.D.(1990), Penta cost, A. (1984) and Smith, G.M. (1951, 1955)

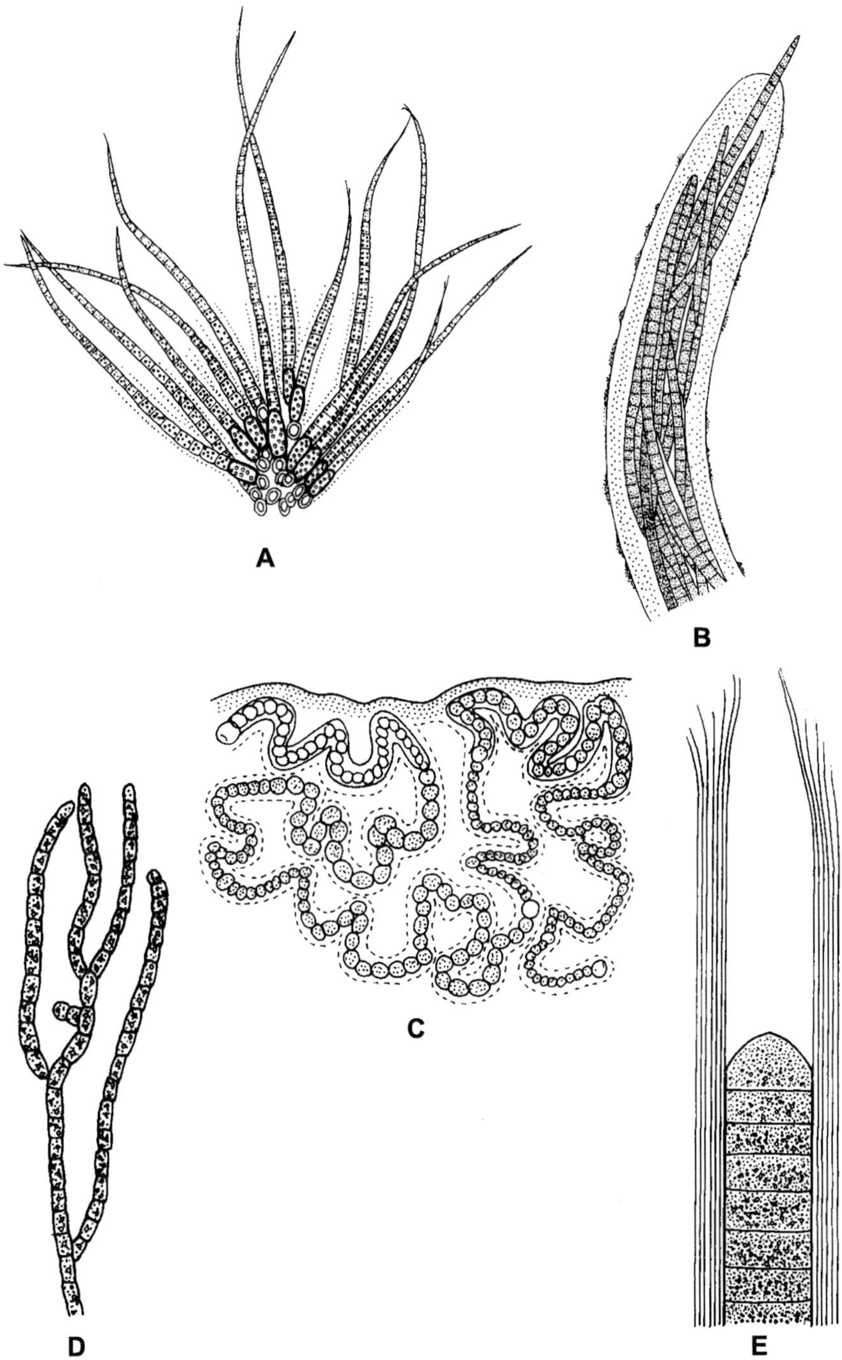

Fig. 10.2 **Cyanobacteria. A** : *Gloeotrichia echinulata*; **B** : *Microcoleus vaginatus*; **C** : *Nostoc linckia*; **D** : *Nostocopsis lobatus*; **E** : *Porphyrosiphon notarisii*.

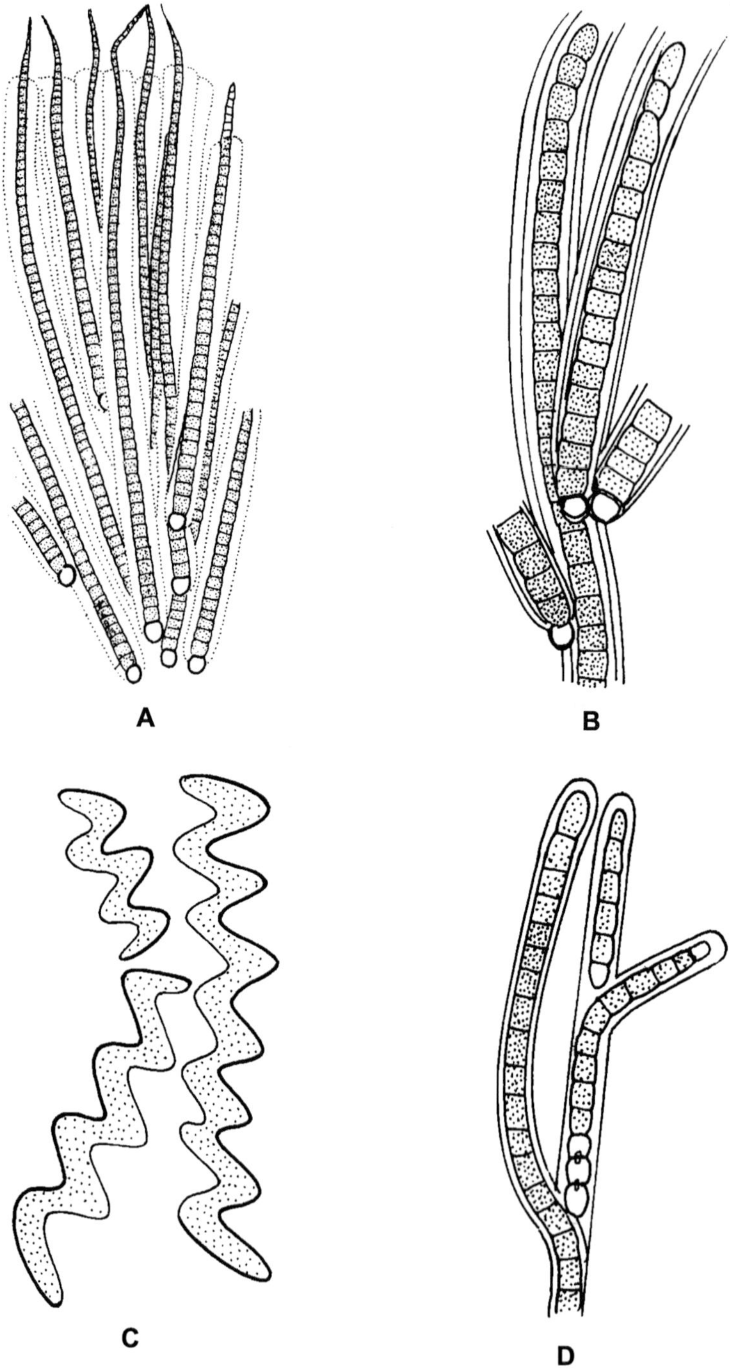

Fig. 10.3 **Cyanobacteria.** **A** : *Rivularia dura*; **B** : *Rivularia minutula*; **C** : *Spirulina* sp.; **D** : *Tolypothrix tenuis*.

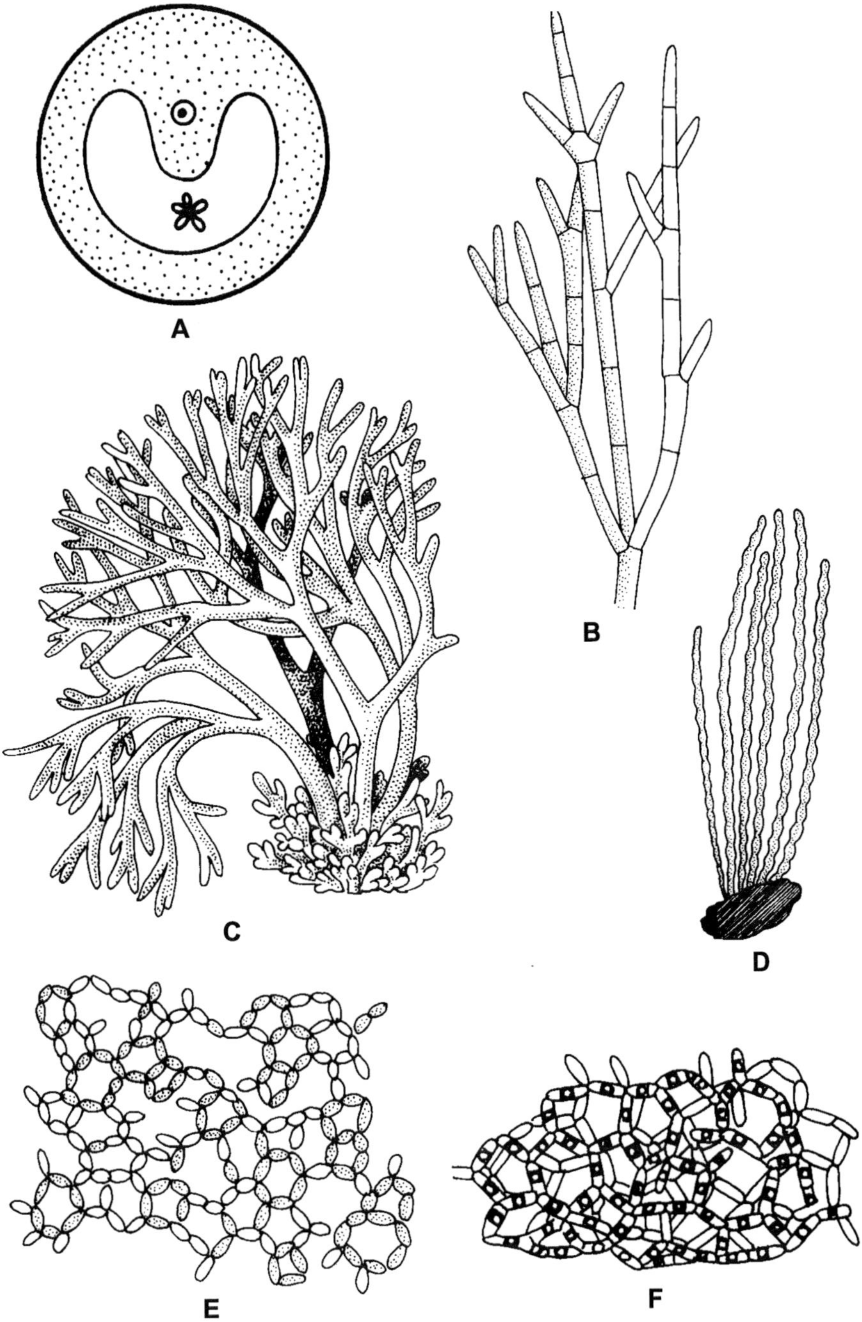

Fig. 10.4 **Green algae.** **A** : *Chlorella vulgaris*; **B** : *Cladophora glomerata*; **C** : *Codium fragile*; **D** : *Enteromorpha intestinalis*; **E** : *Hydrodictyon africanum*; **F** : *Hydrodictyon reticulatum*.

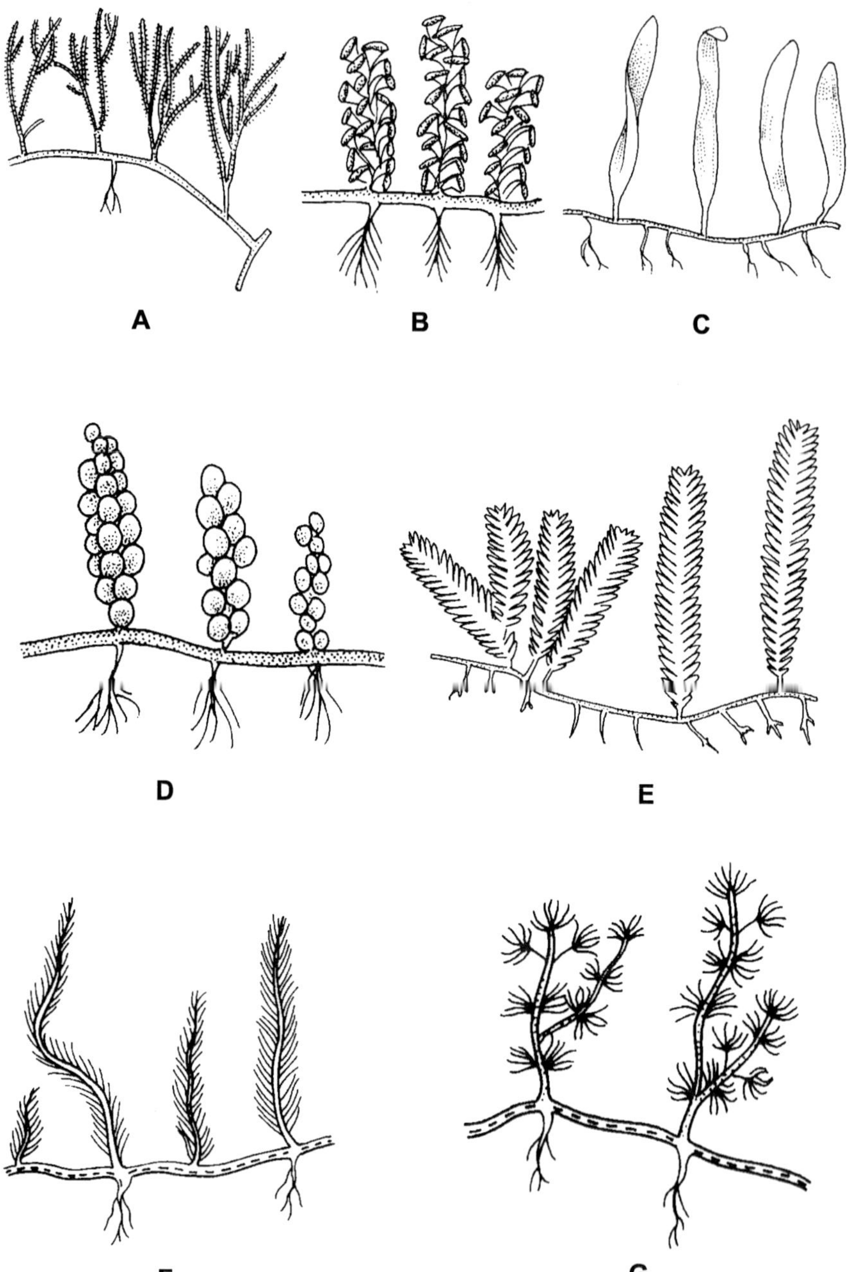

Fig. 10.5 **Green algae. A** : *Caulerpa cupressoides*; **B** : *Caulerpa peltata*; **C** : *Caulerpa prolifera*; **D** : *Caulerpa racemosa*; **E** : *Caulerpa scalpelliformis*; **F** : *Caulerpa taxifolia*; **G** : *Caulerpa verticillata*.

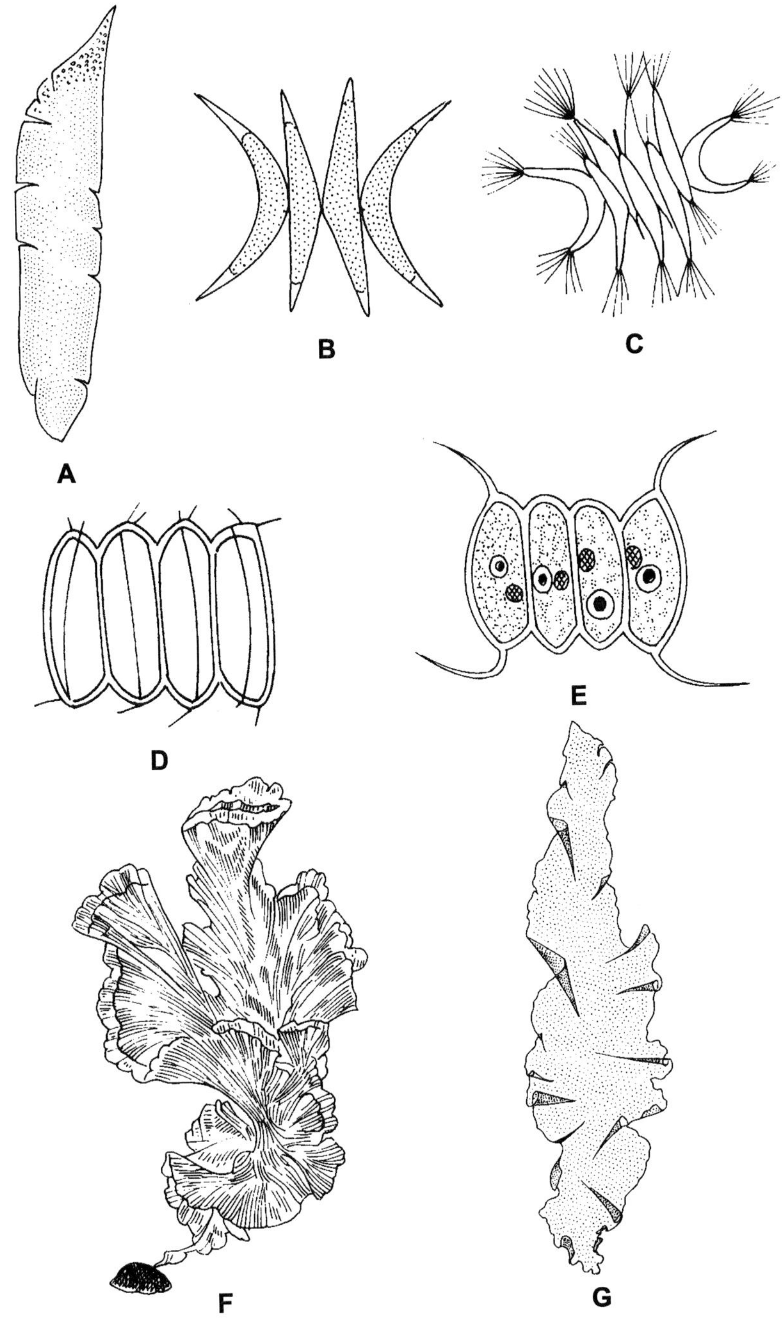

Fig. 10.6 **Green algae. A** : *Monostroma membranaceum*; **B** : *Scenedesmus acuminatus*; **C** : *Scenedesmus acuminatus* showing bristles; **D** : *Scenedesmus brasiliensis*; **E** : *Scenedesmus quadricauda*; **F** : *Ulva lactuca*; **G** : *Ulva stenophylla*.

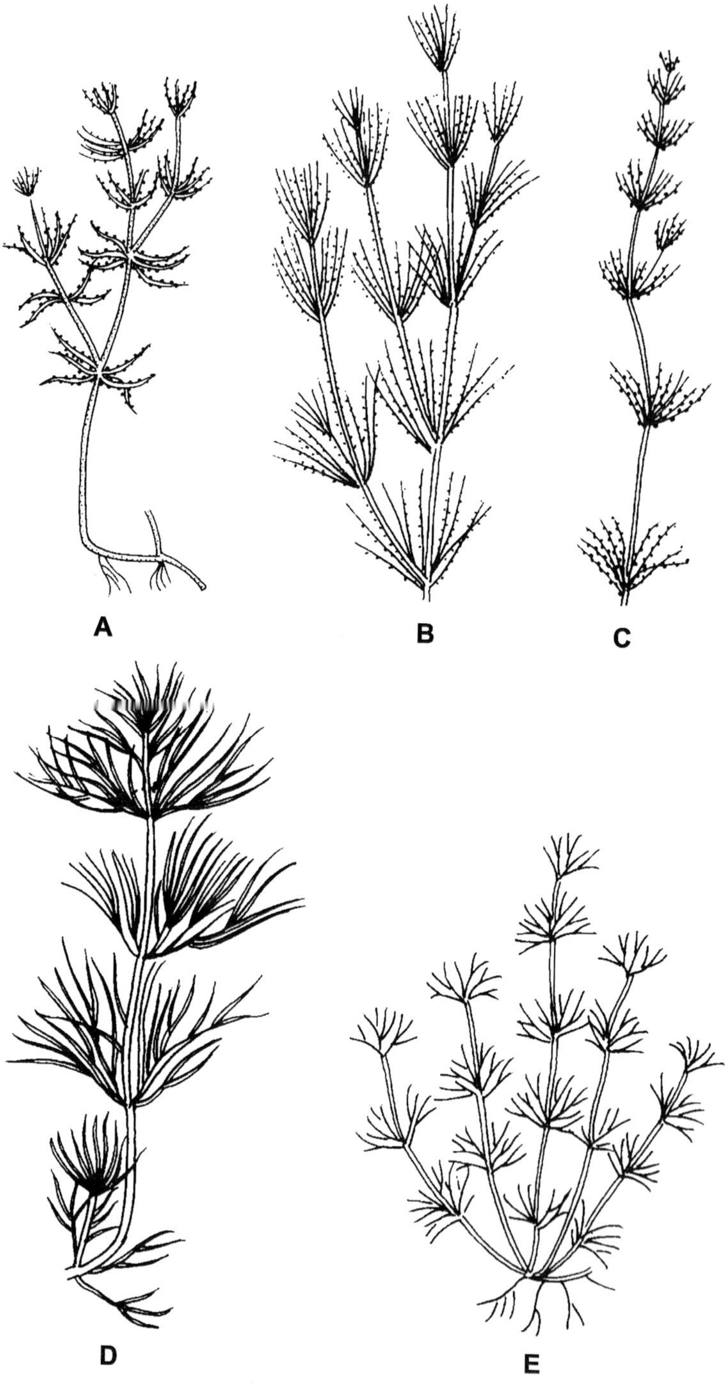

Fig. 10.7 **Stoneworts. A** : *Chara foetida*; **B** : *Chara hispida*; **C** : *Chara vulgaris*; **D** : *Nitella batrachosperma*; **E** : *Nitella tenuiissima*.

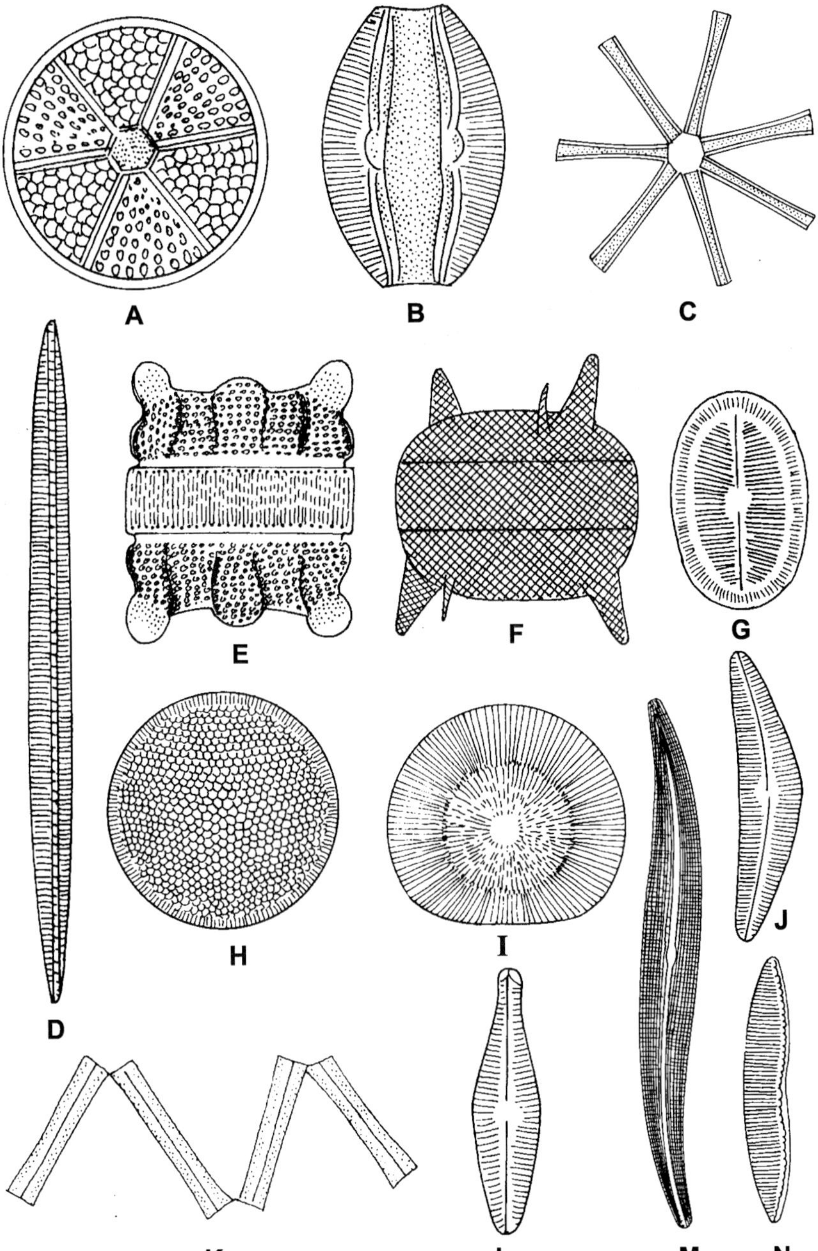

Fig. 10.8 **Diatoms.** **A** : *Acinoptychus undulata*; **B** : *Amphora ovalis*; **C** : *Asterionella* gracillima; **D** : *Bacillaria paradoxa*; **E** : *Biddulphia pulchella*; **F** : *Biddulphia smithii*; **G** : *Cocconeis placentula*; **H** : *Coscinodiscus excentricus*; **I** :*Cyclotella comta*; **J** : *Cymbella cistula*; **K** : *Diatoma* sp.; **L** : *Gomphonema ventricosum*; **M** : *Gyrosigma attenuatum*; **N** : *Hantzschia amphioxys*.

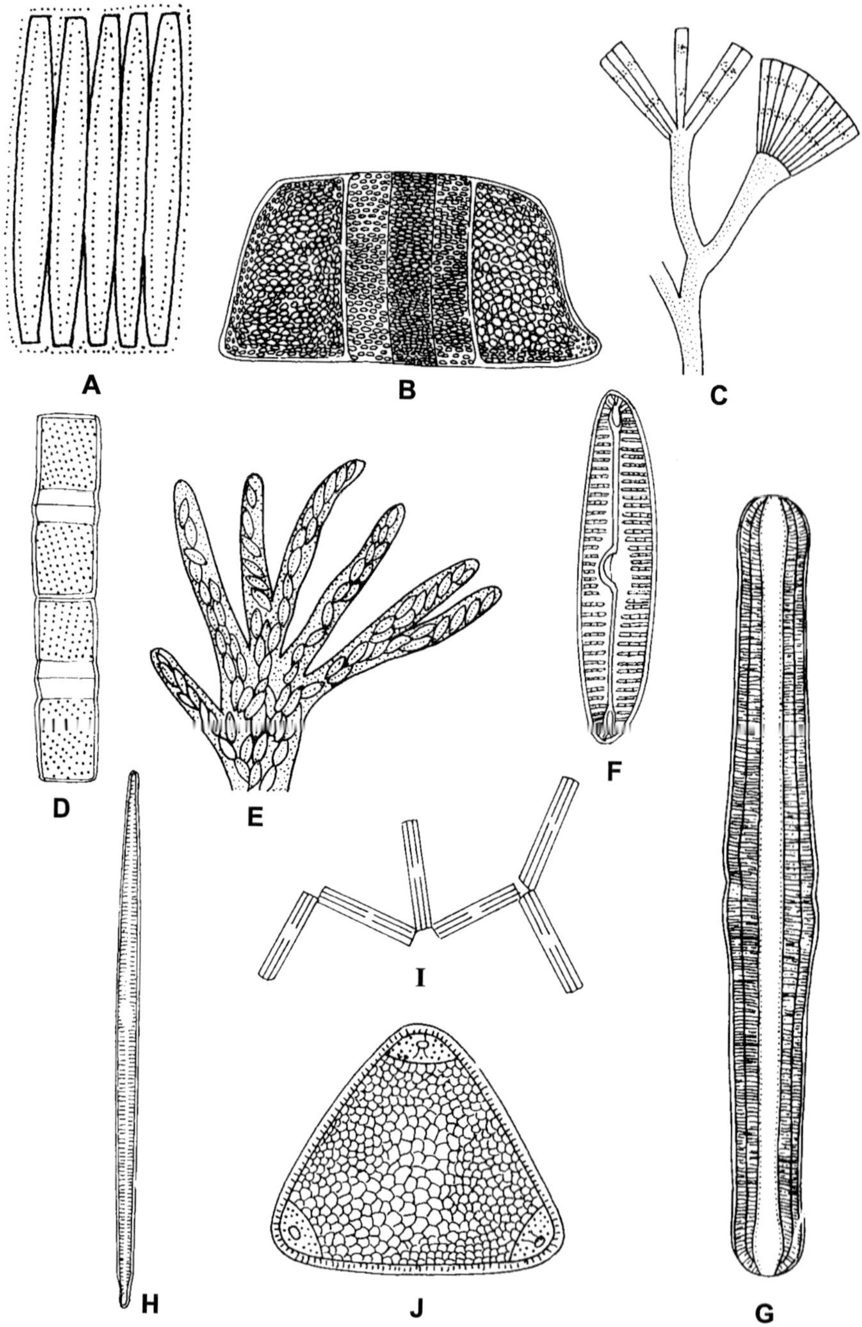

Fig. 10.9 **Diatoms. A** : *Fragilaria crotonensis*; **B** : *Isthmia enervis*; **C** : *Licmophora flabellata*; **D** : *Melosira varians*; **E** : *Navicula* sp.; **F** : *Pinnularia streptoraphe*; **G** : *Rhopalodia gibba*; **H** : *Synedra ulna*; **I** : *Tabellaria* sp.; **J** : *Triceratium distinctum*.

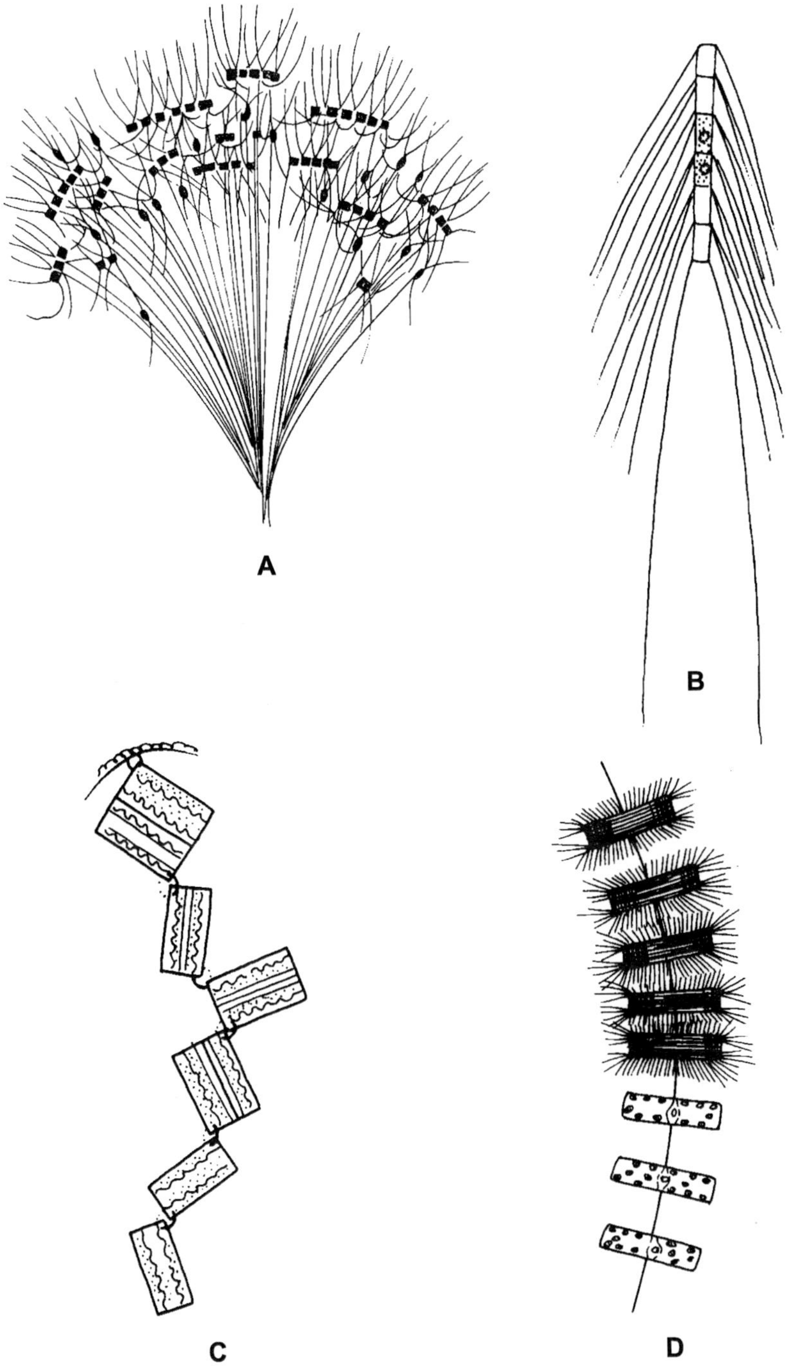

Fig. 10.10 **Diatoms. A** : *Chaetoceras socialis*; **B** : *Chaetoceras subtilis*; **C** : *Grammatophora serpentina*; **D** : *Thalassiosira gravida*.

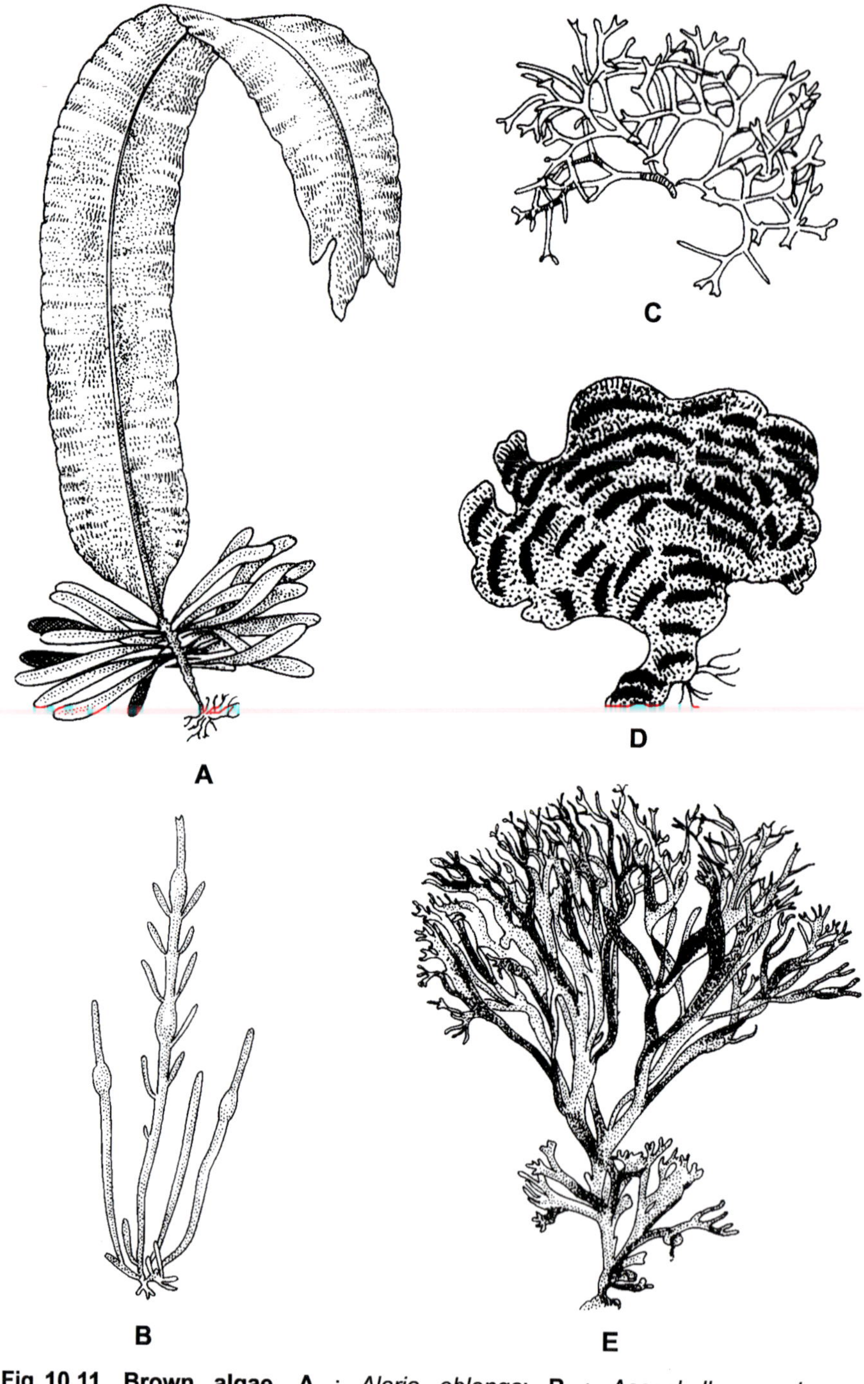

Fig. 10.11 Brown algae. A : *Alaria oblonga*; **B** : *Ascophyllum nodosum*; **C** : *Ascophyllum nodosum* var. *mackaii*; **D** : *Cutleria adspersa*; **E** : *Cutleria multifida*.

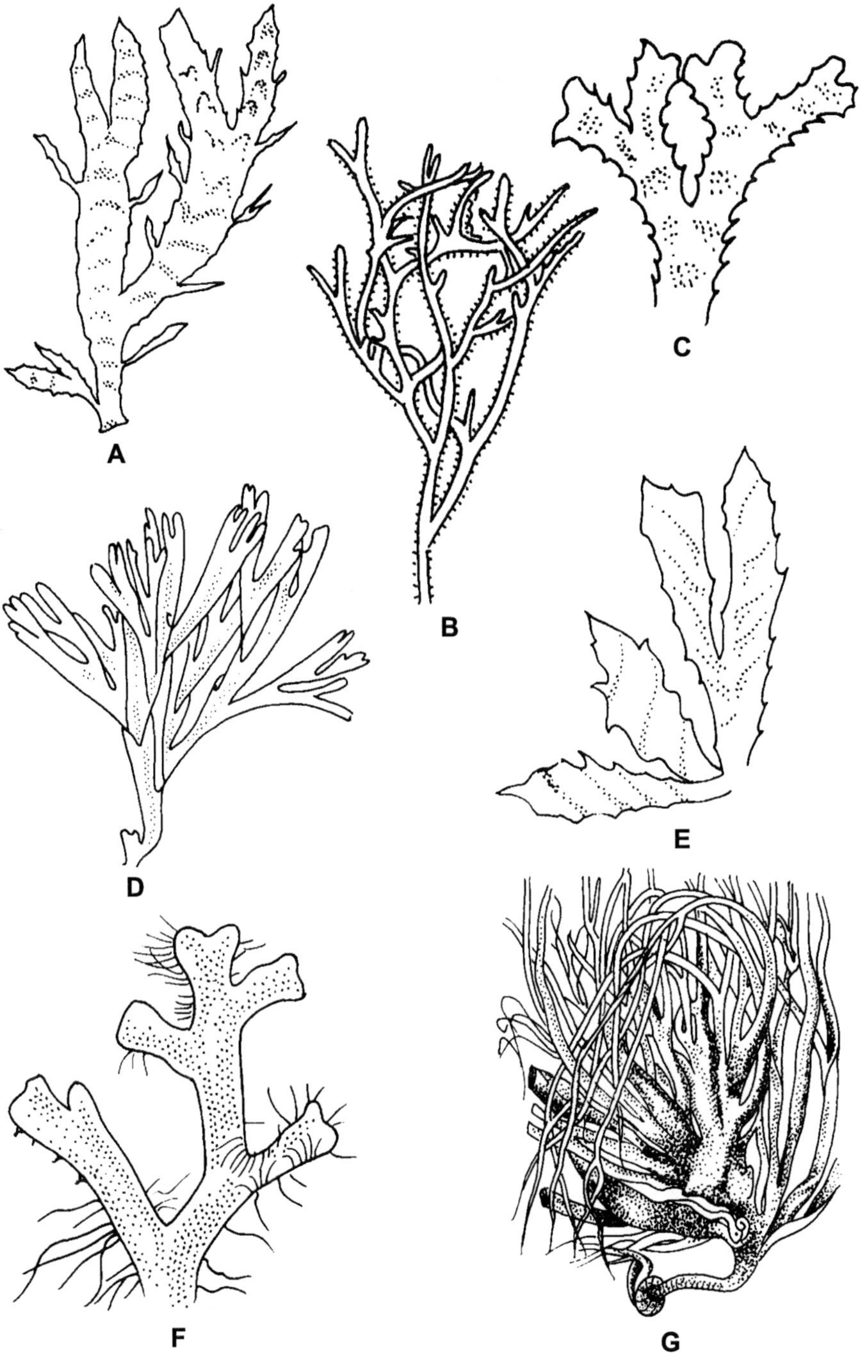

Fig. 10.12 Brown algae. A : *Dictyota atomarica*; **B** : *Dictyota cervicornis*; **C** : *Dictyota ciliata*; **D** : *Dictyota dichotoma*; **E** : *Dictyota maxima*; **F** : *Dictyotopsis propagulifera*; **G** : *Durvillea antarctica.*

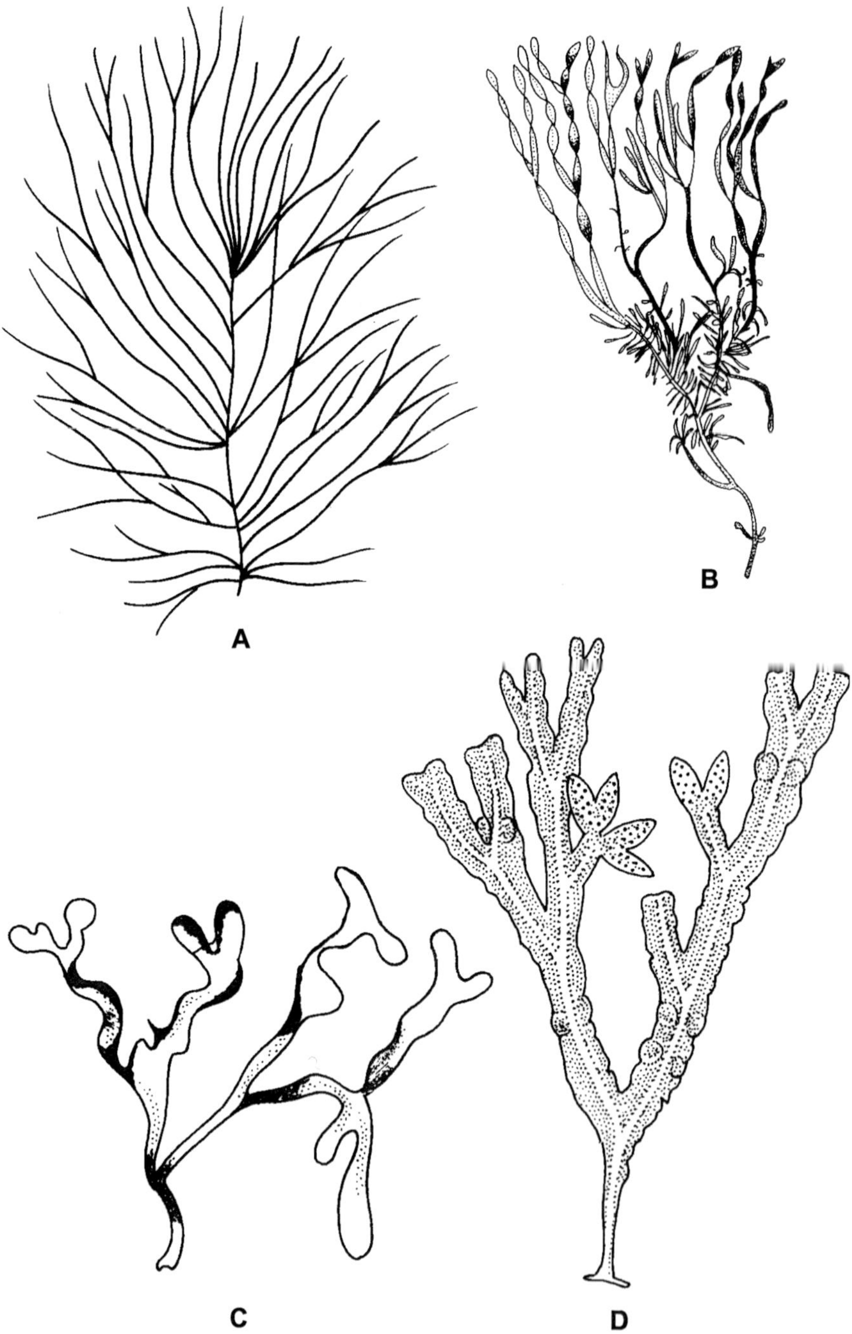

Fig. 10.13 Brown algae. A : *Chordaria flagelliformis*; **B** : *Fucus lutaricus*; **C** : *Fucus spiralis*; **D** : *Fucus vesiculosus*;

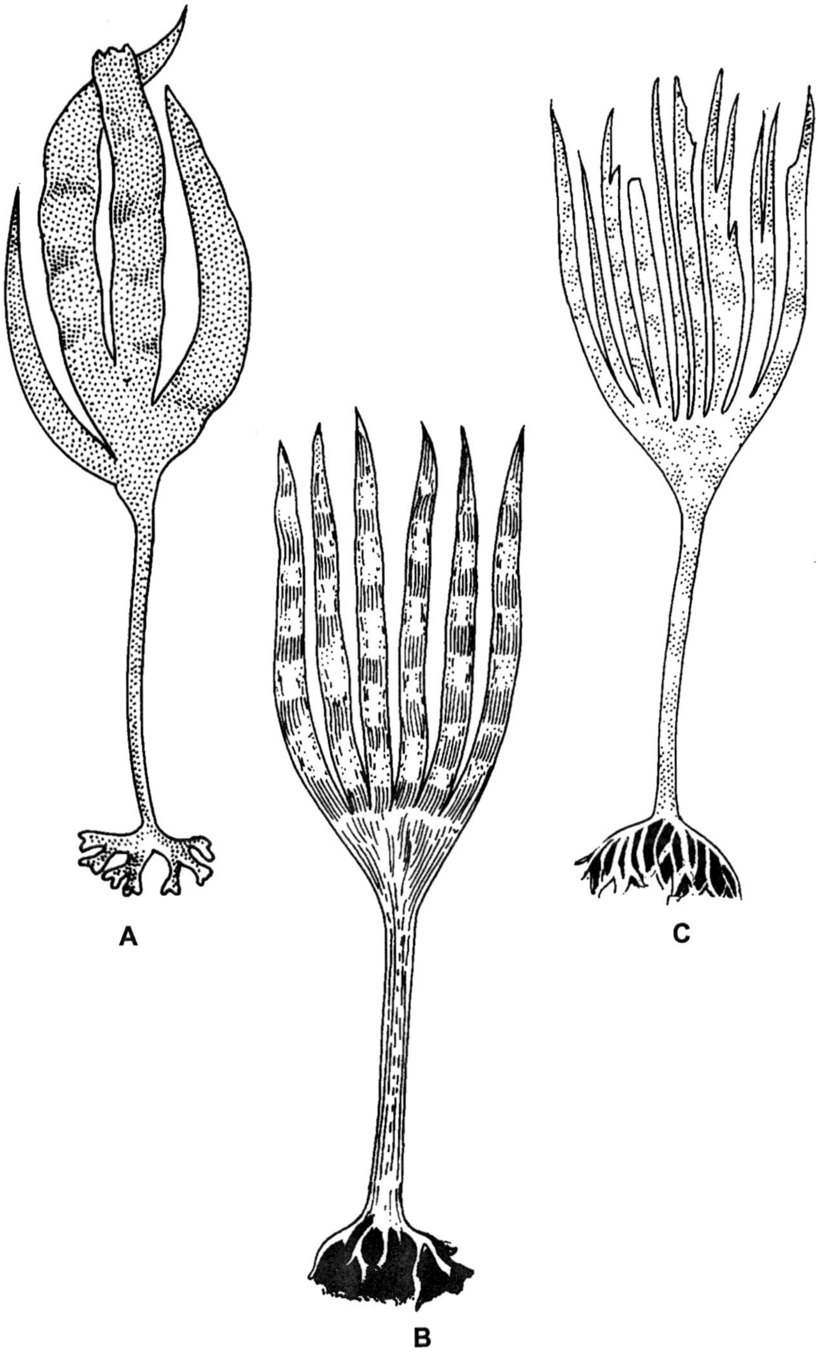

Fig. 10.14 Brown algae. A : *Laminaria andersonii;* **B** : *Laminaria digitata*; **C** : *Laminaria hyperborea*.

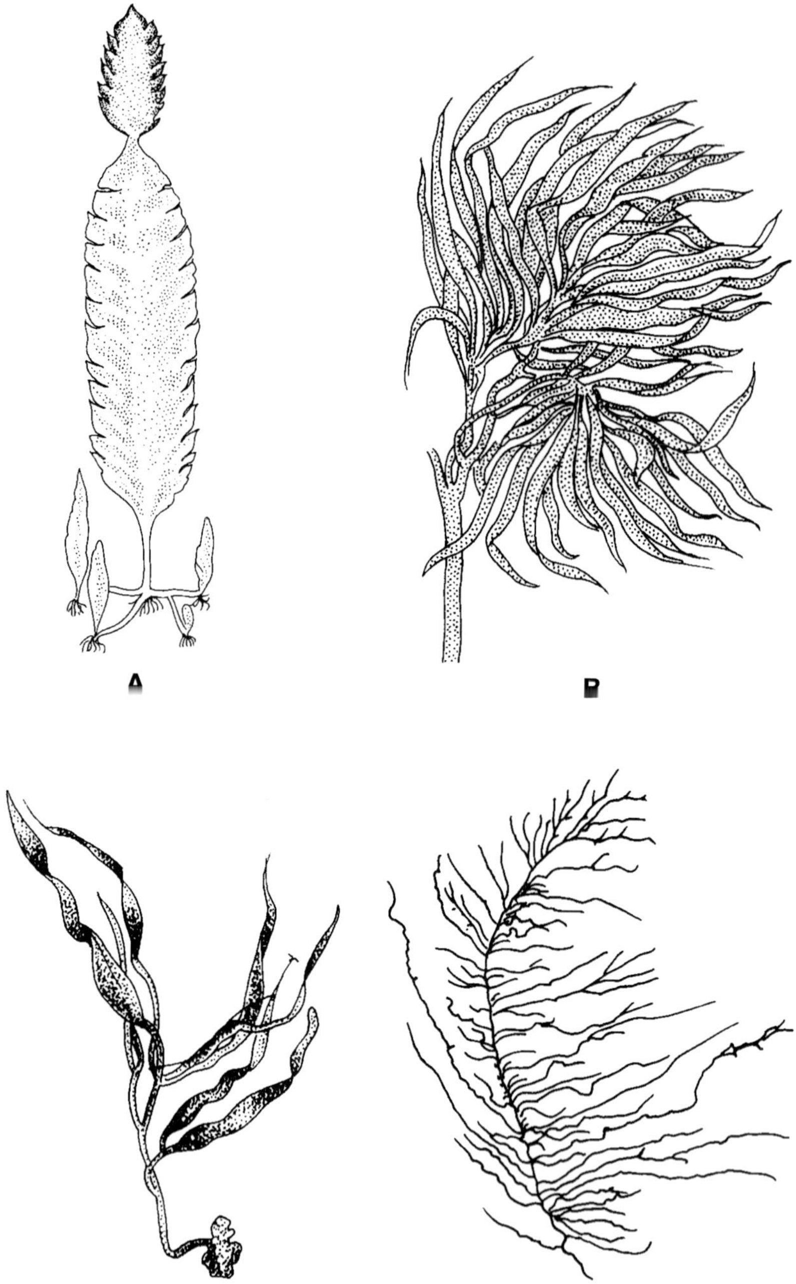

Fig. 10.15 Brown algae. A : *Laminaria saccharina*; **B** : Lessonia *flavicans*; **C** : *Lessonia nigrescens;* **D** : *Mesogloea vermiculata*.

Fig. 10.16 Brown algae. A : *Macrocystis pyrifera*; **B** : *Nereocystis leutkeana*; **C** : *Padina commersoni*; **D** : *Phymatolithon calcareum*.

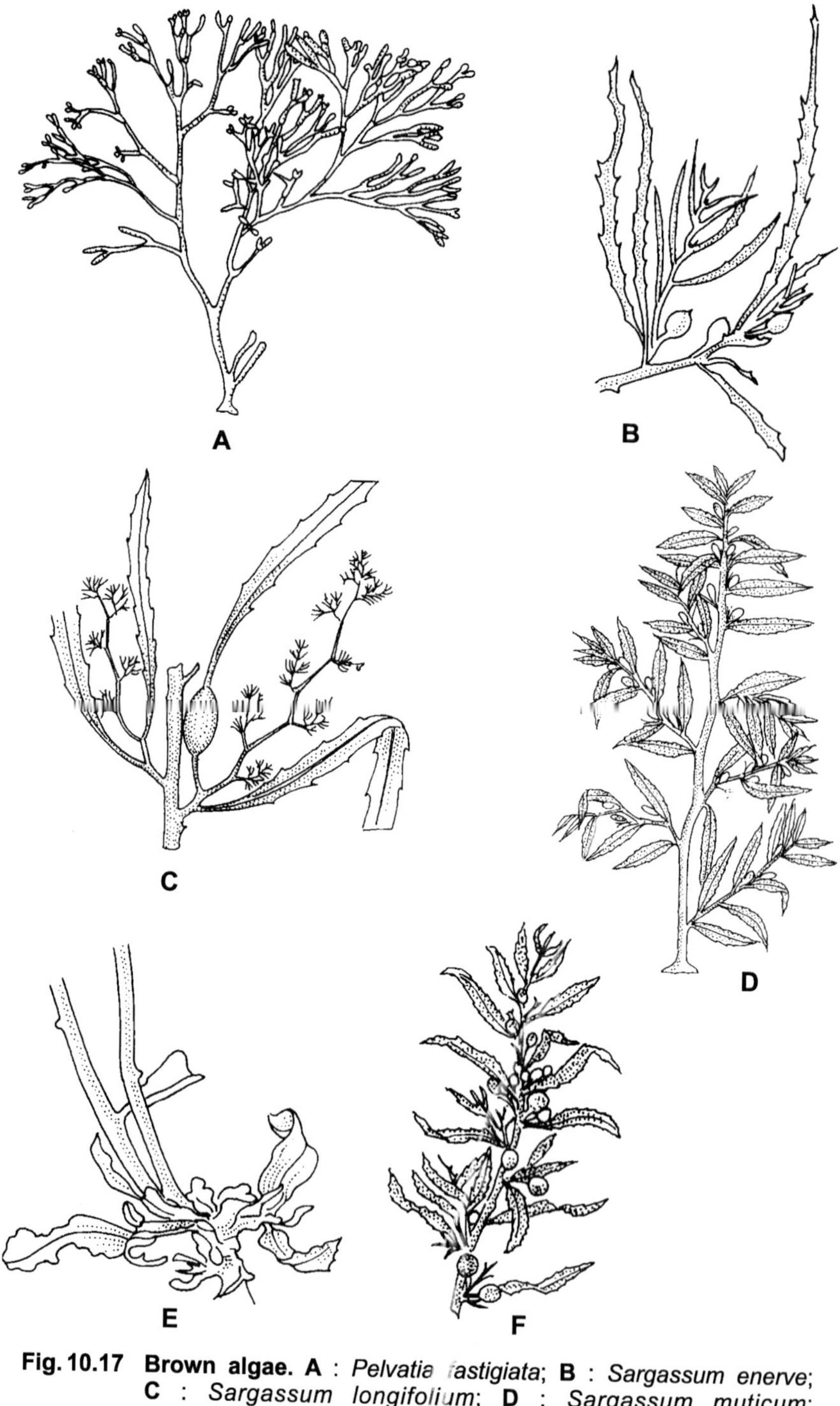

Fig. 10.17 **Brown algae. A** : *Pelvatia fastigiata*; **B** : *Sargassum enerve*; **C** : *Sargassum longifolium*; **D** : *Sargassum muticum*; **E** : *Sargassum vachellianum*; **F** : *Sargassum vulgare*.

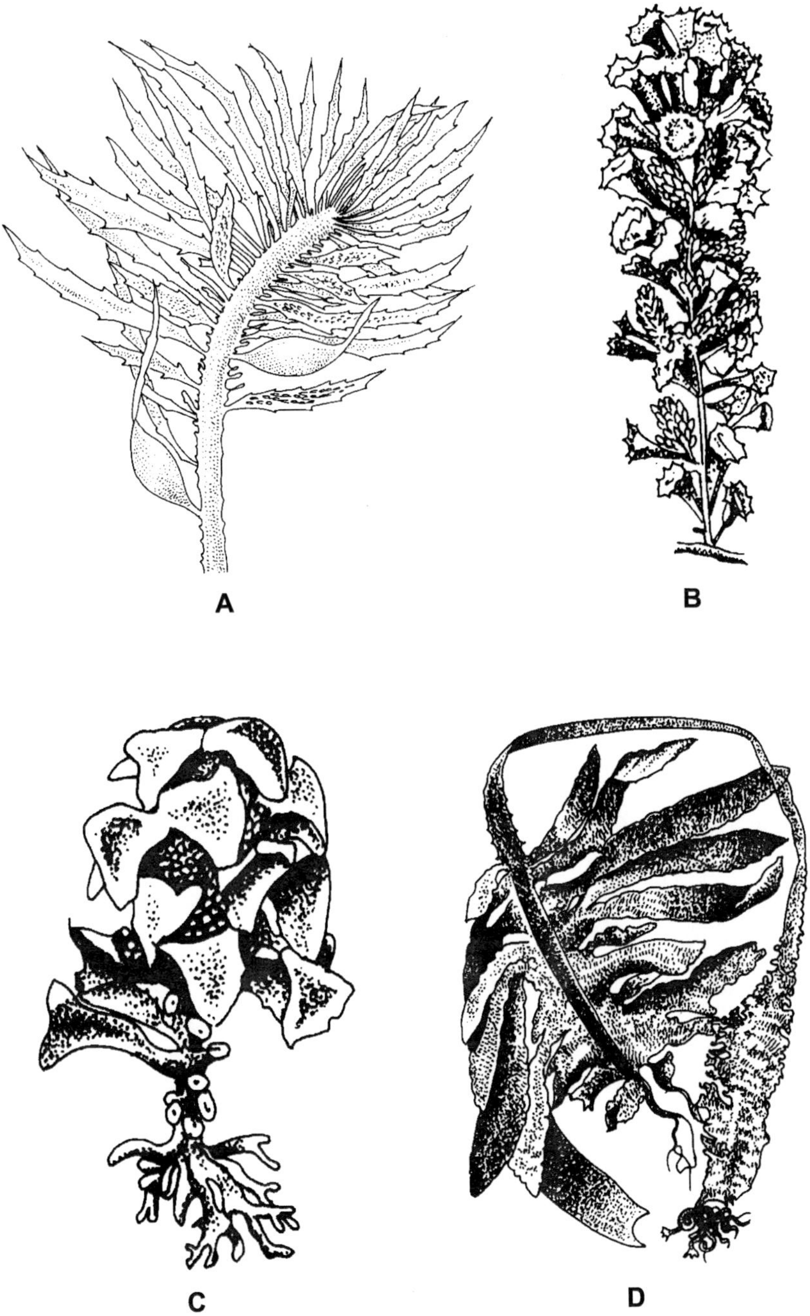

Fig. 10.18 **Brown algae. A** : *Phyllospora comosa*; **B** : *Turbinaria dentata*; **C** : *Turbinaria murrayana*; **D** : *Undaria pinnatifida*

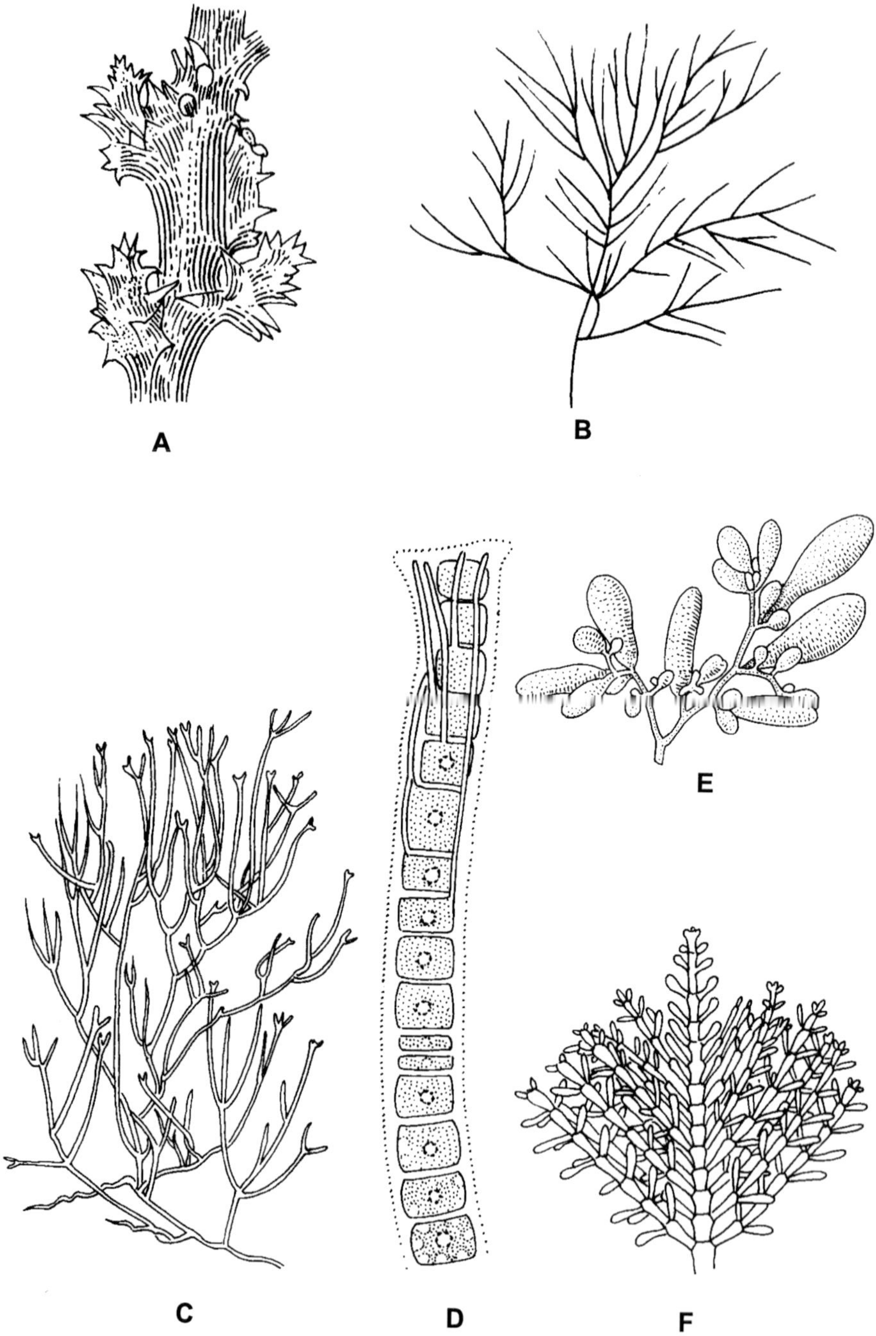

Fig. 10.19 **Red algae. A** : *Acanthophora spicifera*; **B** : *Agardhiella subulata*; **C** : *Ahnfeltia plicata*; **D** : *Bangia fuscopurpurea*; **E** : *Botryocladia pseudodichotoma*; **F** : *Choreonema officinalis*.

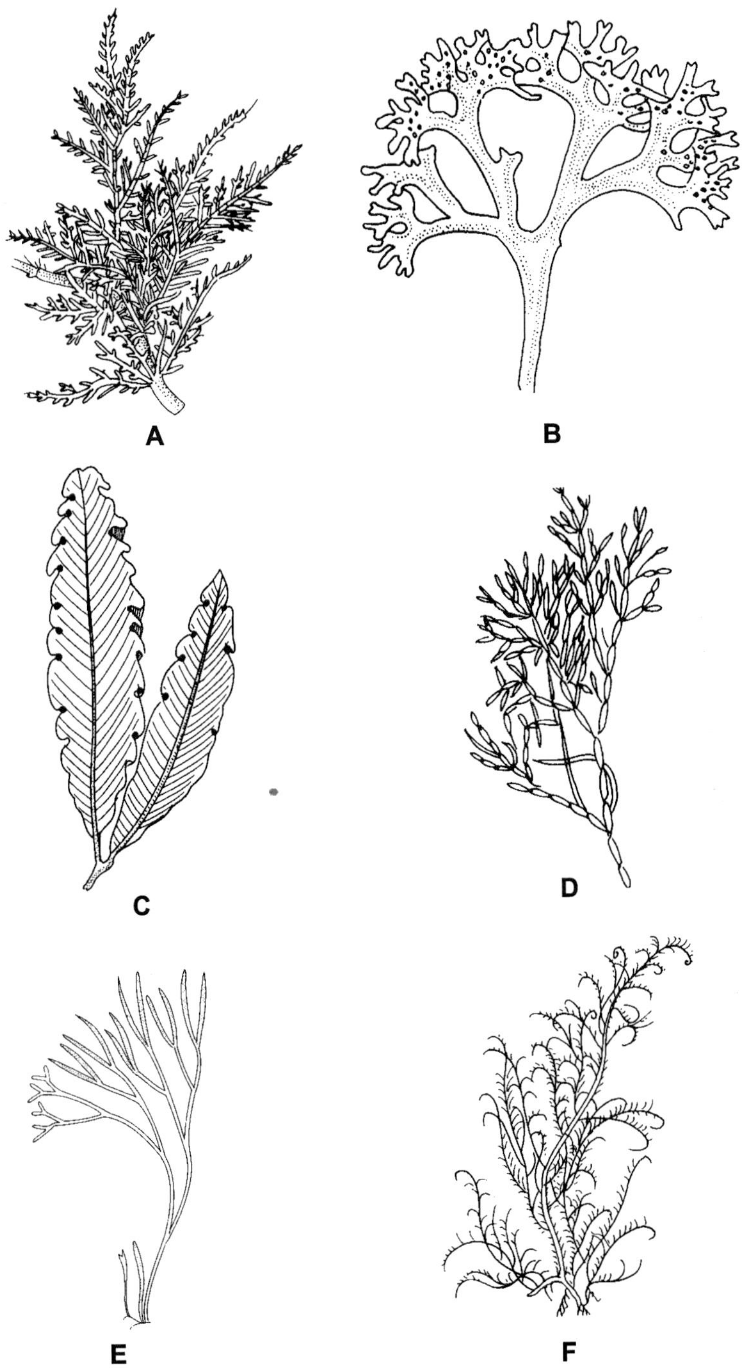

Fig. 10.20 Red algae. A : *Chondria dasyphylla*; **B** : *Chondrus crispus*; **C** : *Delesseria sanguinea*; **D** : *Erythroclonium mulleri*; **E** : *Furcellaria fastigiata*; **F** : *Hypnea musciformis*.

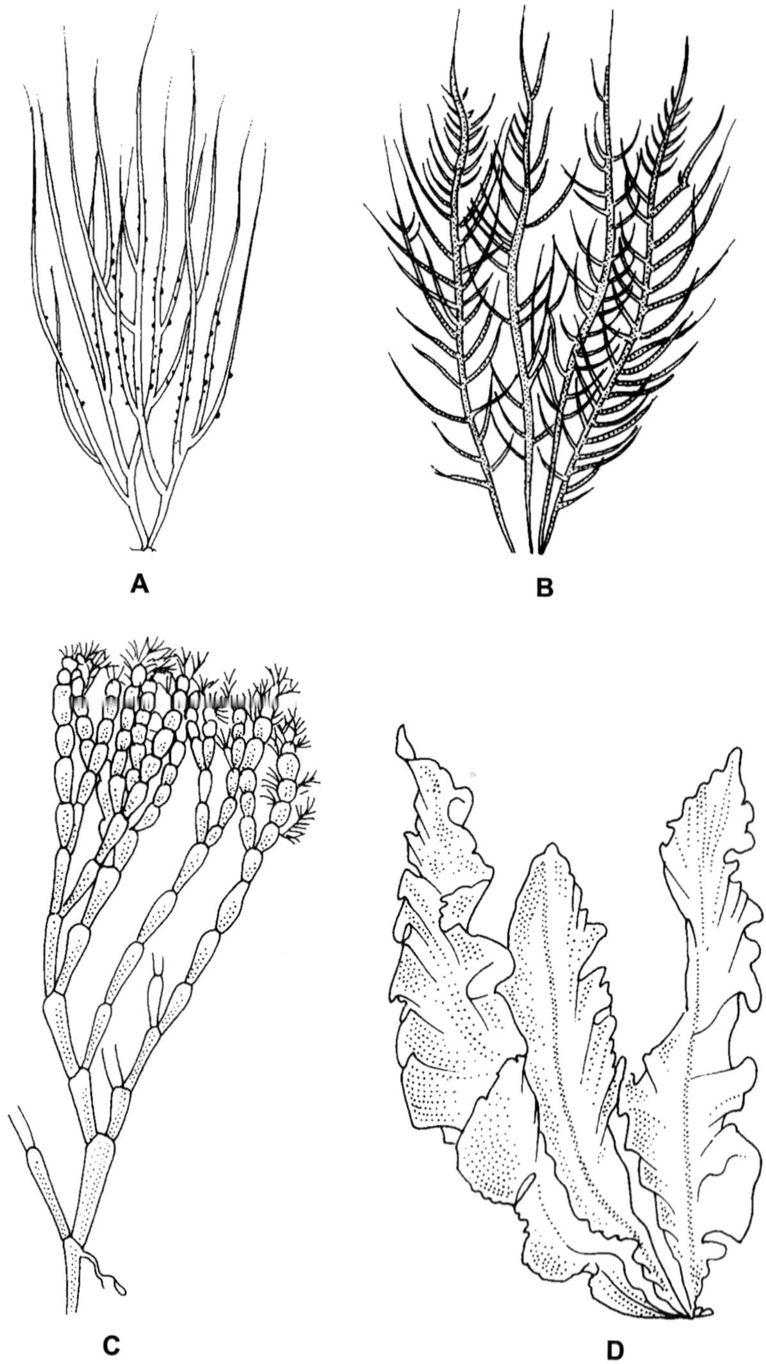

Fig. 10.21 Red algae. A : *Gracilaria confervoides*; **B** : *Grateloupia filicina*; **C** : *Griffithsia corollina*; **D** : *Grinnellia americana*.

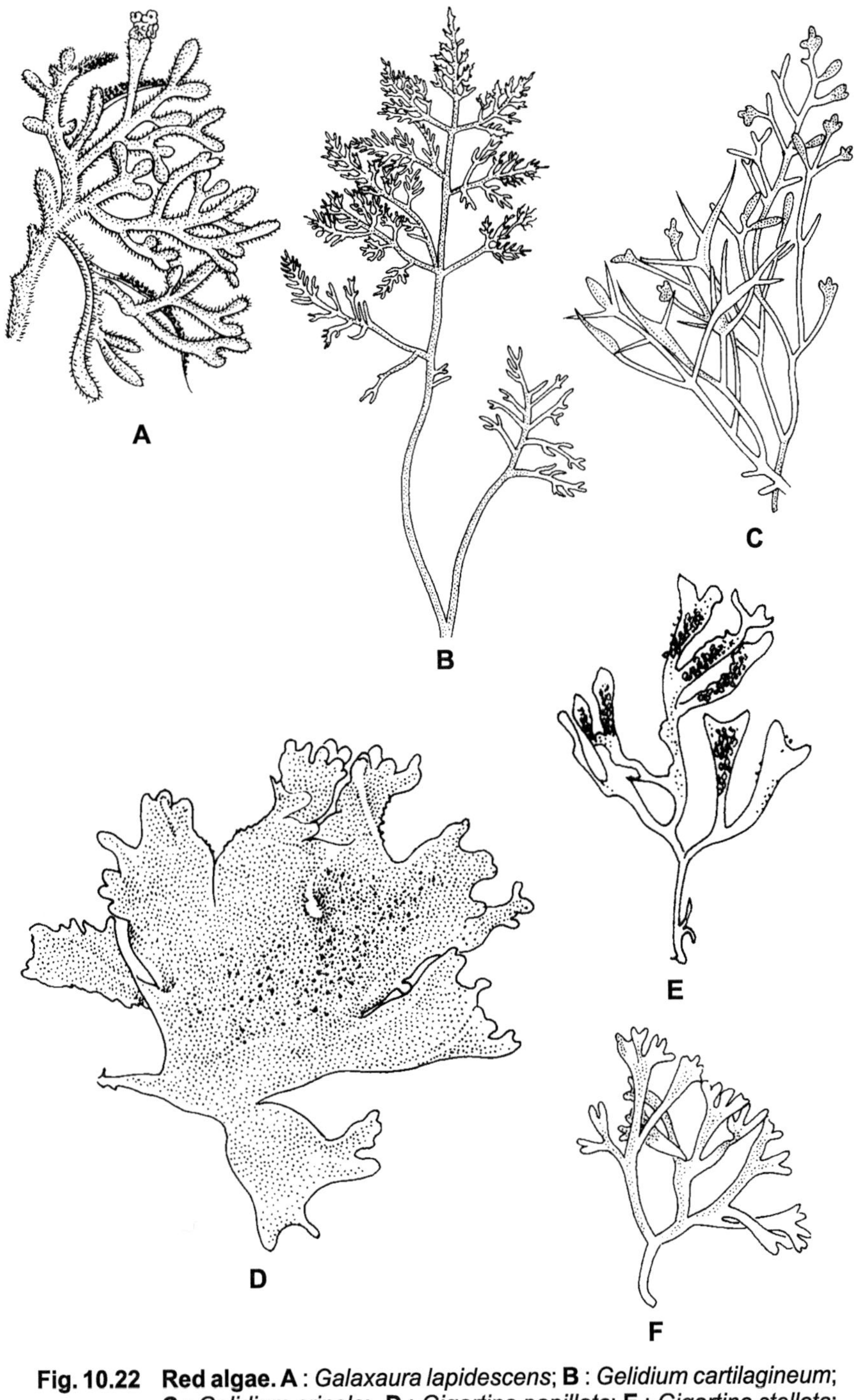

Fig. 10.22 **Red algae.** **A** : *Galaxaura lapidescens*; **B** : *Gelidium cartilagineum*; **C** : *Gelidium crinale*; **D** : *Gigartina papillata*; **E** : *Gigartina stellata*; **F** : *Gymnogongrus norvegicus*.

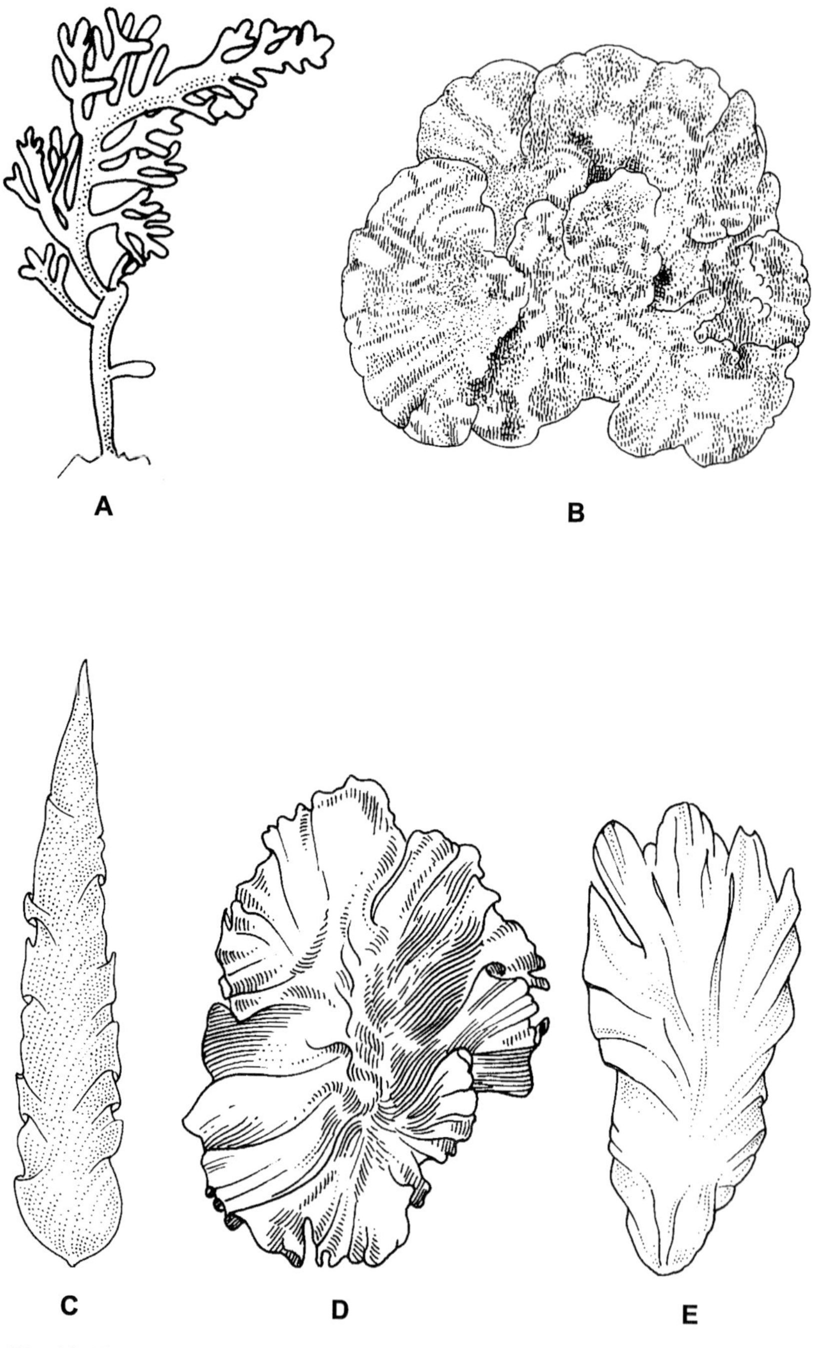

Fig. 10.23 **Red algae.** **A** : *Laurencia pinnatifida*; **B** : *Lithophyllum expansum*; **C** : *Porphyra occidentalis*; **D** : *Porphyra umbilicaulis*; **E** : *Porphyra yezoensis*.

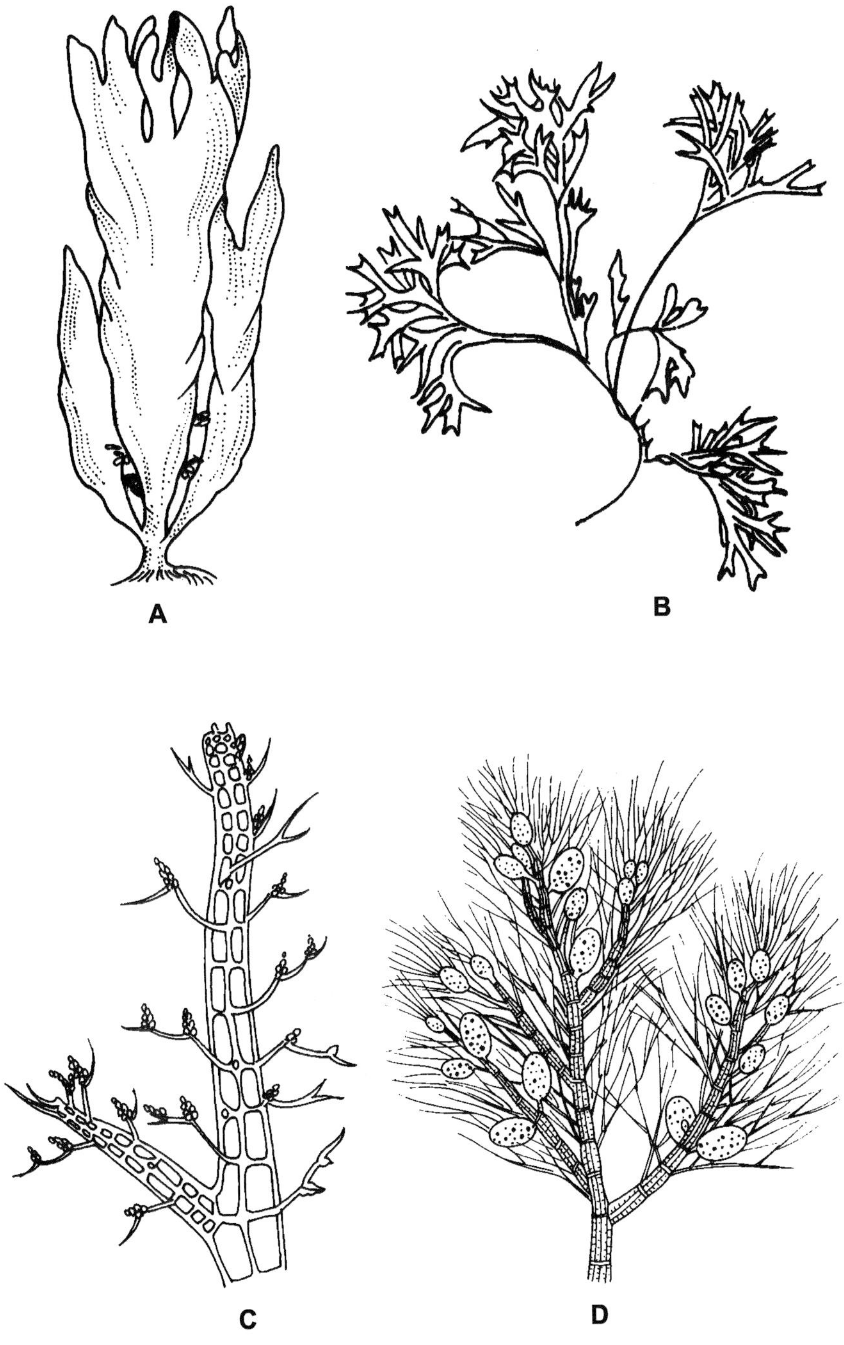

Fig. 10.24 **Red algae. A** : *Palmaria palmata*; **B** : *Phyllophora membranifolia*; **C** : *Polysiphonia ferculacea*; **D** : *Polysiphonia* sp..

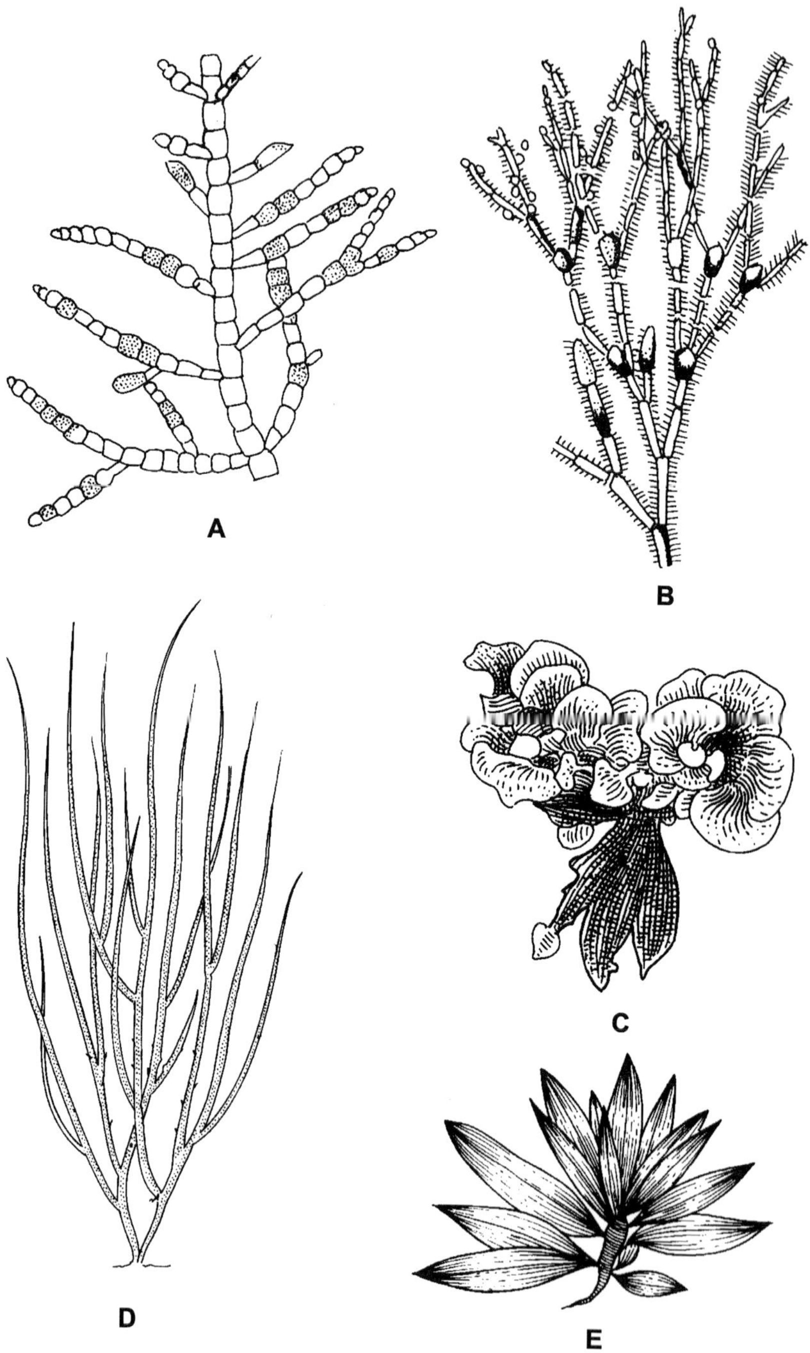

Fig. 10.25 Red algae. A : *Champia parvula*; **B** : *Corallina rubens*; **C** : *Rhizophyllis squamariae*; **D** : *Rhodophyllis bifida*; **E** : *Rhodymenia palmata*.

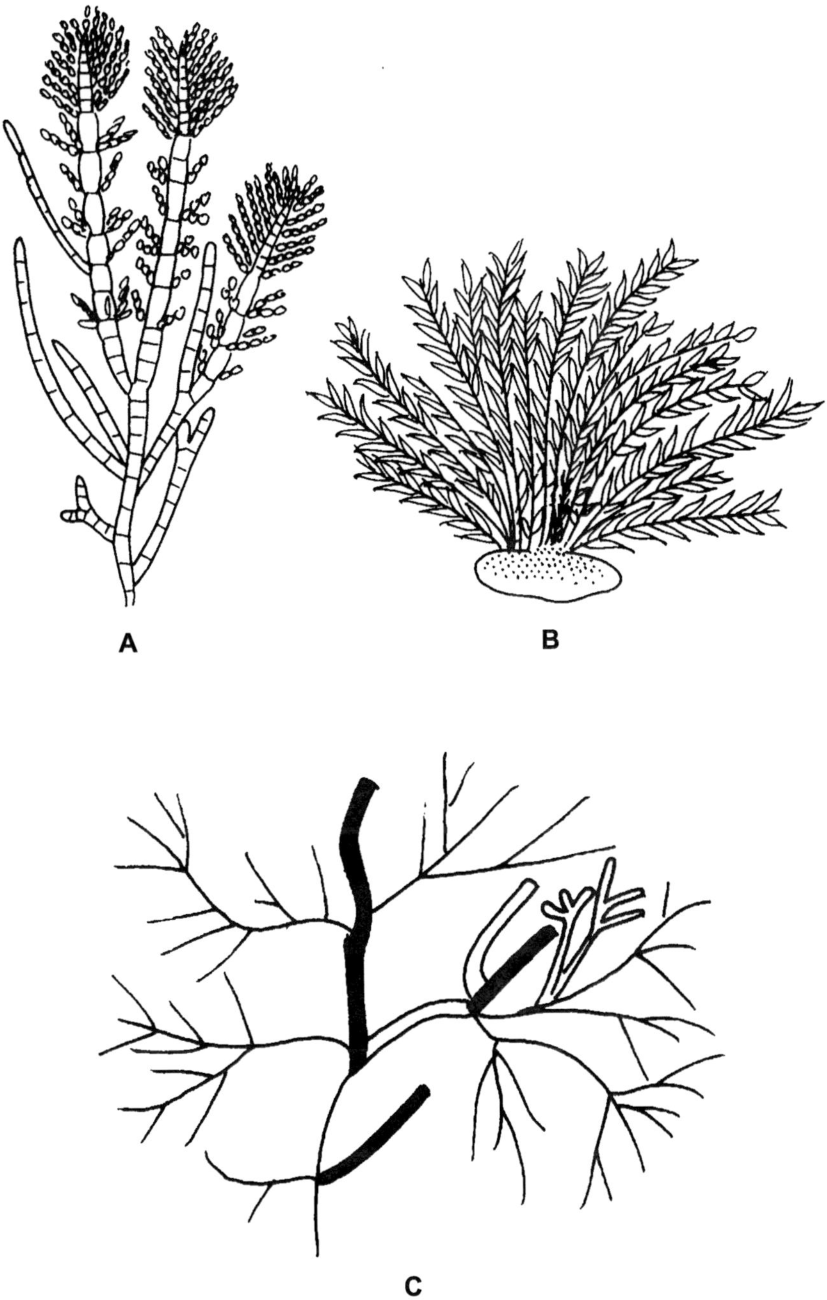

Fig. 10.26 Red algae. A : *Batrachospermum moniliforme*; **B** : *Eucheuma isiforme*; **C** : *Farlowia compressa*.

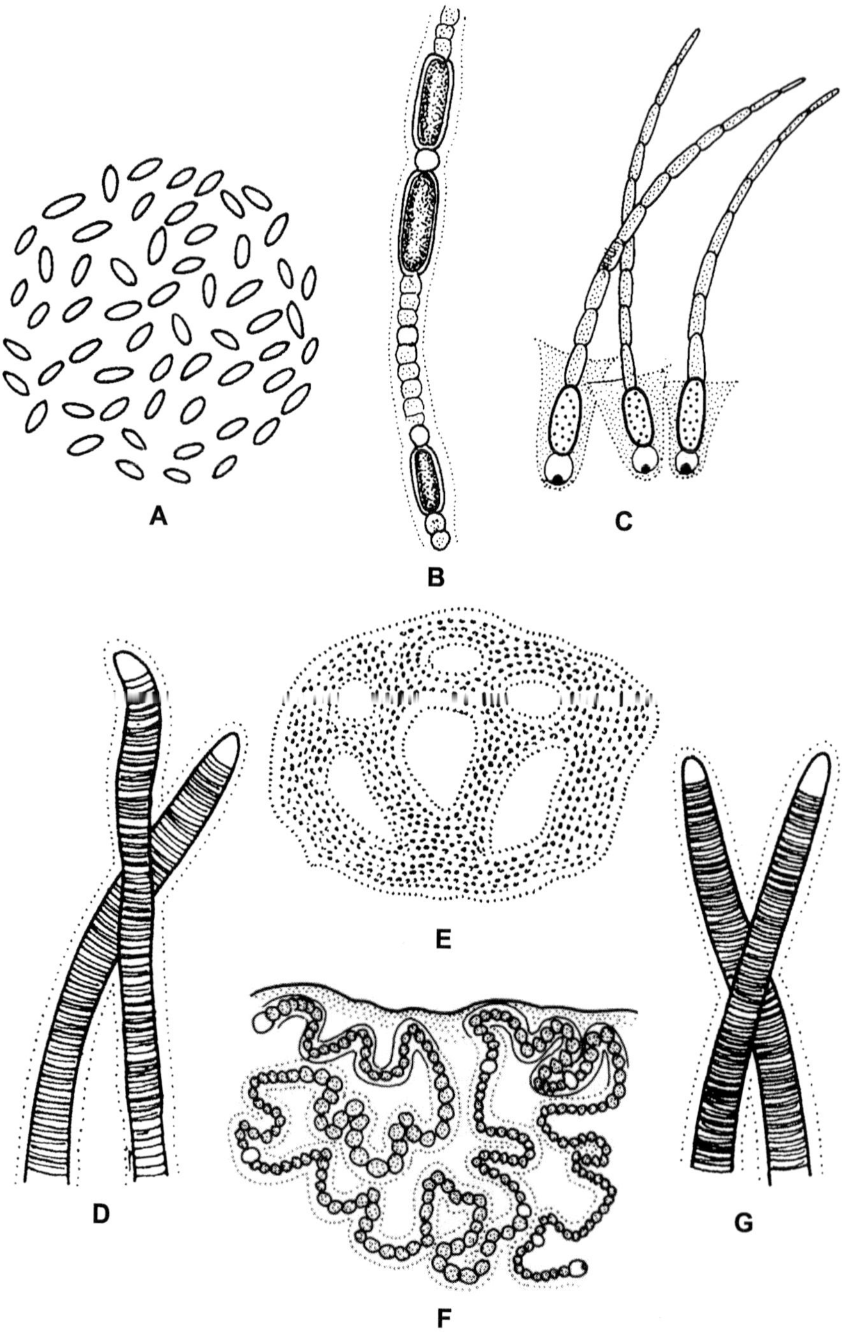

Fig. 10.27 Cyanobacteria in water blooms. A : *Aphanothece* sp.; **B** : *Anabaena oscillarioides*; **C** : *Gloeotrichia pisum*; **D** : *Lyngbya majuscula*; **E** : *Microcystis aeruginosa*; **F** : *Nostoc linckia* **G** : *Oscillatoria* sp.

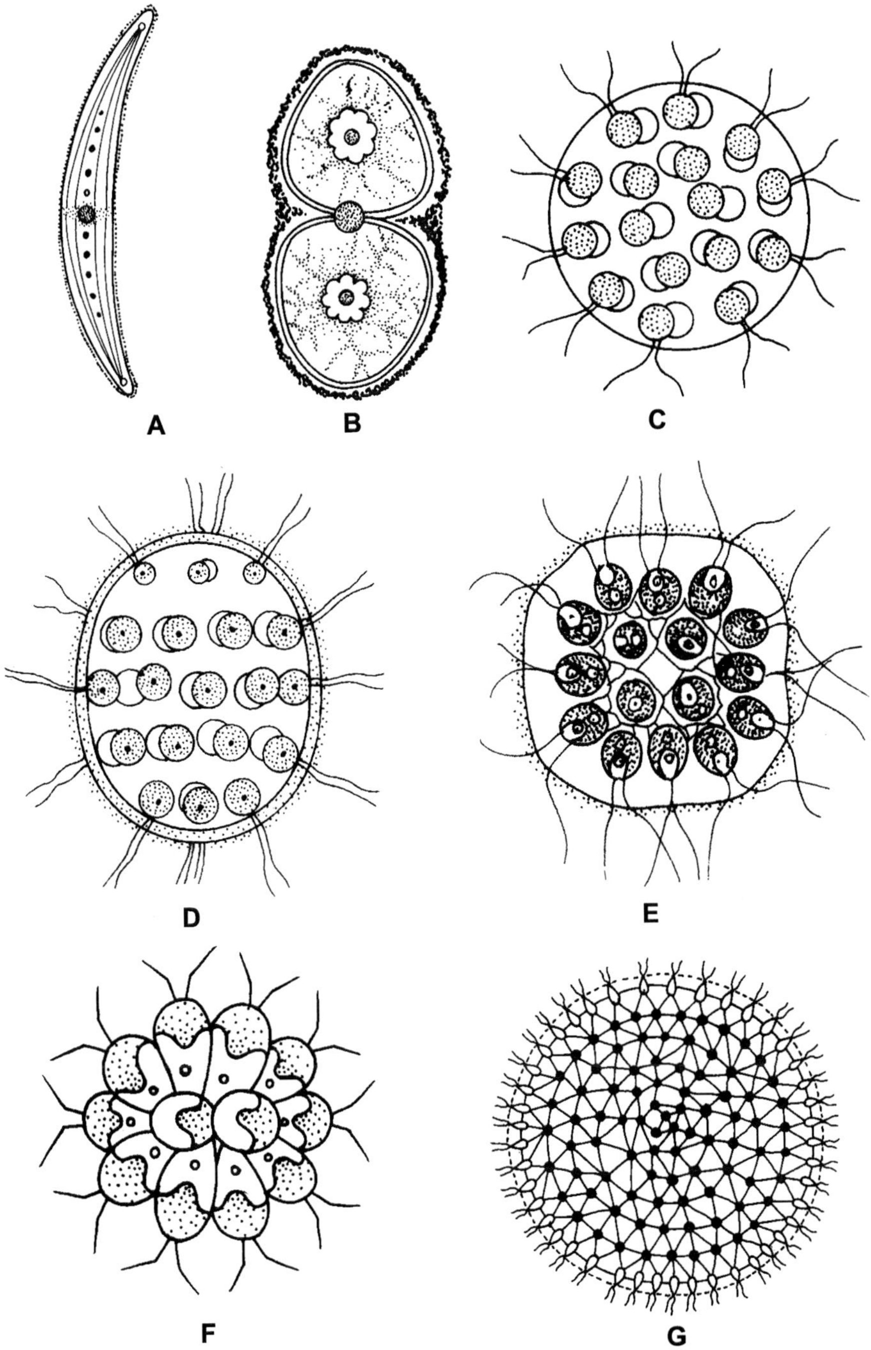

Fig. 10.28 Green algae in water blooms. A : *Closterium* sp.; **B** : *Cosmarium pseudopyramidatum*; **C** : *Eudorina elegans*; **D** : *Eudorina illinoiensis*; **E** : *Gonium pectorale*; **F** : *Pandorina morum;* **G** : *Volvox* sp.

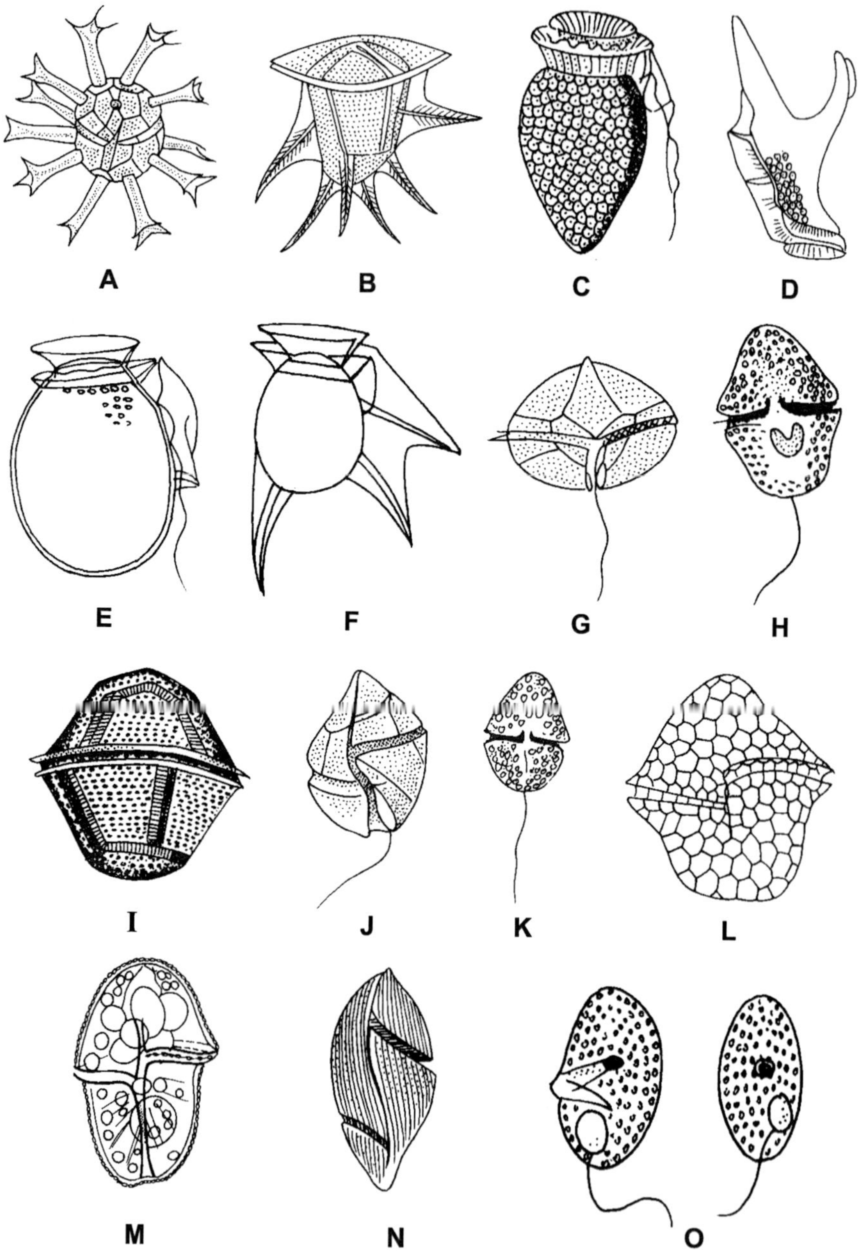

Fig. 10.29 Dinoflagellates in water blooms. A : *Acanthodinium caryophyllum*; **B** : *Ceratocorys horrida*; **C** : *Dinophysis acuta*; **D** : *Dinophysis miles*; **E** : *Dinophysis recurva*; **F** : *Dinophysis schuttii*; **G** : *Diplopsalis lenticula*; **H** : *Glenodinium cinctum*; **I** : *Goniodoma acuminatum*; **J** : *Gonyaulax fragilis*; **K** : *Gymnodinium aeruginosum*; **L** : *Gymnodinium hiemale*; **M** : *Gymnodinium pachydermaticum*; **N** : *Gyrodinium spirale*; **O** : *Hemidinium nasutum*.

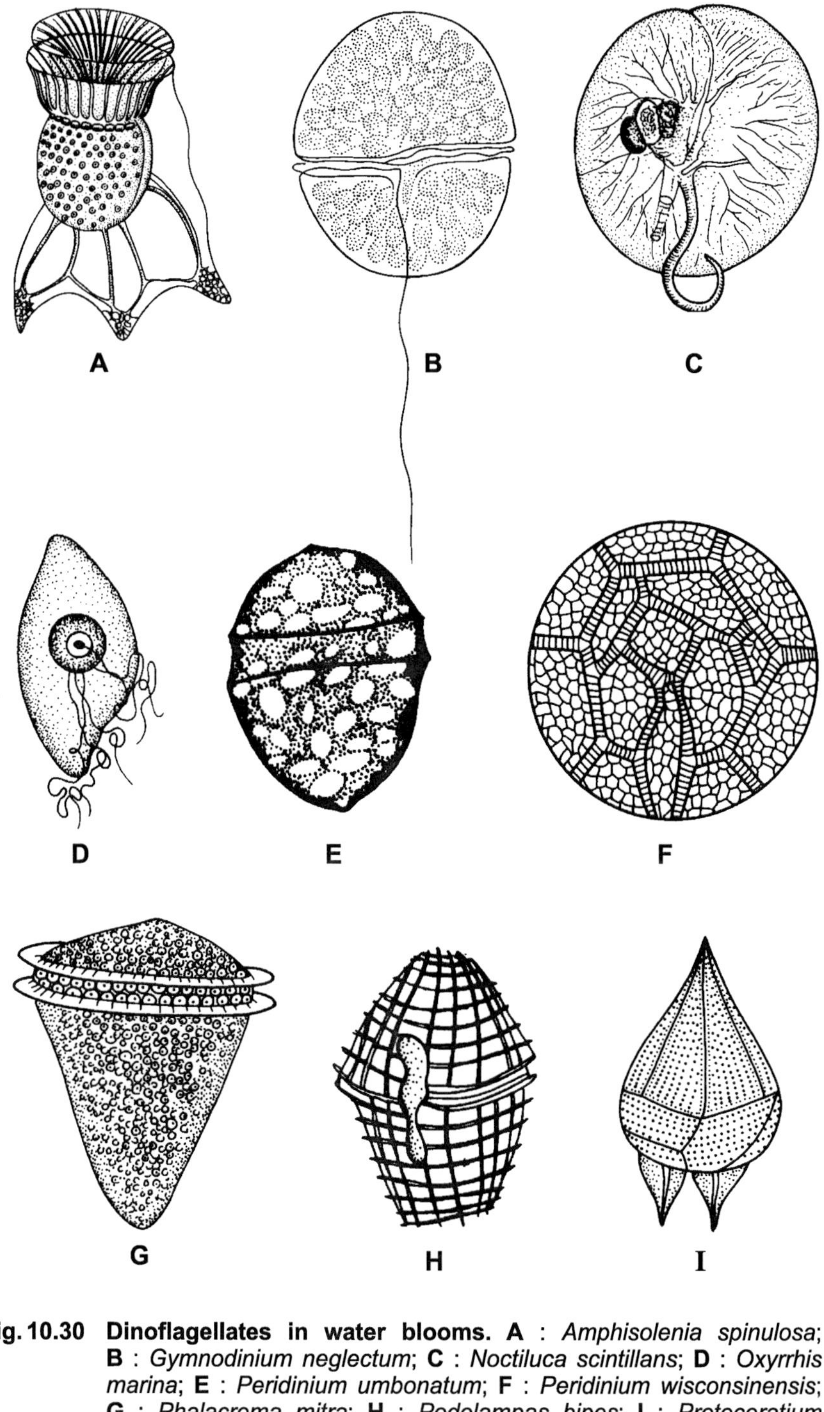

Fig. 10.30 Dinoflagellates in water blooms. A : *Amphisolenia spinulosa*; **B** : *Gymnodinium neglectum*; **C** : *Noctiluca scintillans*; **D** : *Oxyrrhis marina*; **E** : *Peridinium umbonatum*; **F** : *Peridinium wisconsinensis*; **G** : *Phalacroma mitra*; **H** : *Podolampas bipes*; **I** : *Protoceratium reticulatum*.

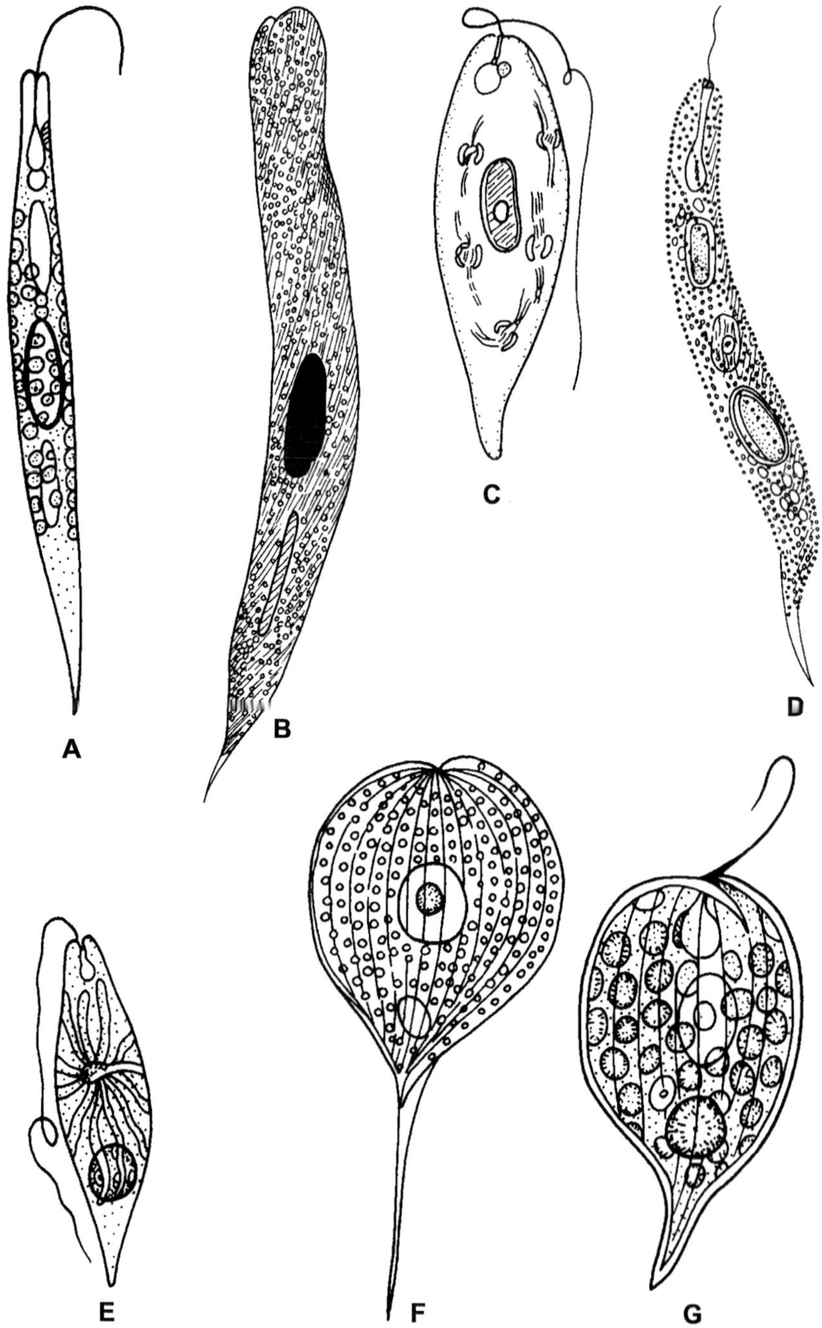

Fig. 10.31 Euglenas. A : *Euglena acus*; **B** : *Euglena oxyuris*; **C** : *Euglena polymorpha*; **D** : *Euglena spirogyra*; **E** : *Euglena viridis*; **F** : *Phacus anomala*; **G** : *Phacus pleuronectes*.

CHAPTER 11

BRYOPHYTES

11.1 INTRODUCTION

Bryophytes the simplest of land plants are restricted to moist and shady habitats. They grow on banks of streams or on shaded cliffs, rocks and hill slopes while a few are aquatic (*Ricciocarpus*). Some are epiphytic on fern fronds (*Radula pratense*).

Bryophytes generally fall into three groups. The liverworts with flat plant body that grows along the surface of the substratum (*Riccia, Marchantia*), the hornworts with horn like sporophytes (*Anthoceros, Notothylas*) while mosses (*Funaria, Polytrichum*) are erect and leafy and are attached to the substratum by rhizoids. Few of the mosses (*Sphagnum*) are semi-aquatic and branched, their weak branches floating along the water surface.

The usefulness of bryophytes to man is relatively unknown. They have important applications in pharmaceuticals, horticulture and play a significant role in ecology. Their potential in biomapping of atmospheric precipitation and pollution is enormous.The full potential of applications of bryophytes is yet to be realized.

Illustrations of representative species of bryophytes are given in Fig. 11.1 to 11.5 to provide an overview of their appearance.

11.2 MEDICINAL USES OF BRYOPHYTES

Marchantia

Marchantia convoluta: is used in China for treating fumefaction of skin, hepatitis and fever. The flavonoids present in the plant are a potential source of an anti-inflammatory drug which can replace the synthetic drugs in the market, without various side effects, such as bone marrow depletion, water and salt retention, and gastric intolerance.

Marchantia polymorpha: is used to cure pulmonary tuberculosis and liver affections. It also has anti-tumour activity.

Polytrichum

Polytrichum commune: Tea prepared from the plant dissolves kidney and gall bladder stones.

Polytrichum juniperium: is diuretic and given to remove urinary obstruction and for fever and inflammation (Hutchens, 1991).

Sphagnum (Peat moss, bog moss, quack moss, creeping moss)

Sphagnum decoction is given for treating acute haemorrhages and used as eye wash for eye infections. As a lotion it cures skin diseases and act as antibacterial.

Sphagnum magllanicum, Sphagnum palustre and *Sphagnum papillosum:* were used in place of cotton in first world war, as an absorbent bandage in surgical dressing as it is antiseptic, cooling, soft and non-irritant.

11.2.1 Bryophytes in The Treatment of Cancer and Tumour

Bryophytes are used in traditional medicines for treating various types of tumours and cancer (Table 11.1)

Table 11.1 Bryophytes in the treatment of cancer and tumour

Species	Type of Cancer/Tumour
Marchantia polymorpha	Condylomata; Inflammatory tumours
Marchantia stellata	Condylomatous tumours
Polytrichum commune	Indurations and tumours of the liver

Source: Hartwell, 1982.

11.2.2 Antimicrobials from Bryophytes

Bryophytes have antibacterial and antifungal activity (Tables 11.2 and 11.3)

Table 11.2 Antibacterial activity of Bryophytes

Species	Target organisms
Plagiochasma appendiculatum, Sphagnum portoricense, Sphagnum strictum	*Staphylococcus aureus*
Hyophila involuta, Sphagnum strictum.	*Pseudomonas aeruginosa*
Anomodon rostratus	*Micrococcus flavus*
Minium cuspidatum, Orthotrichum rupestre.	*Micrococcus rubens*
Anomodon rostratus, Orthotrichum rupestre Asterella sanguinea, Campylopus laetus, Hyophila involuta, Plagiochasma appendiculatum, Reboulia hemisphaerica	*Streptococcus pyogenes Salmonella typhii, Vibrio cholerae*

Asterella sanguinea, Asterella angusta, Brachythecium procumbens, Marchantia palmata, Oxystegus cylindricus, Plagiochasma appendiculatum, Reboulia hemisphaerica	*Mycobacterium phlei*
Plagiochasma appendiculatum	*Bacillus subtilis, Sarcina lutae, Klebsiella pneumoniae.*

Table 11.3 Anti fungal activity of Bryophytes

Species	Target organisms
Conocephallum conicum, Dumortiera hirsuta, Anomodon rostratus, Orthotrichum rupestre Staphylococcus aureus	*Candida albicans*
Porella sp.	*Trichophyton rubrum, Trichophyton mentagrophytes.*
Makinoa sp.	*Microsporum gypseum*

11.3 OTHER USES OF BRYOPHYTES

Mosses, such as *Polytrichum commune* are grown to cover rocks and stones in gardens. Moss carpets are used in France to cover lawns. Mosses are used to provide landscape in horticulture particularly in miniature landscaping to give different colours to rocks adding beauty in gardens. These gardens are called as moss gardens.

Mosses are used to make hanging baskets, doors, mats, mattresses, and cushions. They are dried, powdered and sprinkled on cereal grains and millets as insect repellents.

Mosses play a role in bog succession from an open water body to a climax forest. The peat moss especially establish itself on the water bank and grow inwards. Its extension into water results in thick matty growth appearing like solid soil.

The moss mat with humus forms a suitable substratum for the growth of other hydrophytes. Dead and decomposed older plants settle at the bottom of water and raise the level of water, finally depleting the water sheet. The water body is converted into solid soil on which mesophytes grow replacing hydrophytes.

Sphagnum, on account of its remarkable potential to absorb and retain water, is used in seed beds for root cuttings and in air layering (methods of grafting), as packing material to protect the grafted parts from drying and to pack young seedlings and cuttings.

Sphagnum is responsible for the formation of peat. The dead and decomposed *Sphagnum* with other plant materials, accumulates from year to year. It is slowly compressed by increasing super incumbent weight of fresh layers of dead mass above it, which eventually hardens into a compact dark coloured carbon rich substance known as peat.

Peat is formed from *Sphagnum acutifolium, Sphagnum cuspidatum, Sphagnum fimbriatum, Sphagnum plumulosum,* and *Sphagnum squarrosum.* It is used as raw material for the production of acetic acid, alcohol, paraffin and naphtha. It is also used in textile, paper and plastic industry and as fertilizer. It is used in building sound proof wooden houses and for caulking ships.

The thick deposits of peat are cut into blocks, dried and used as fuel. Peat fuel is used commercially in Netherlands. Treatment of peat yields a brown dye used in tanning industry. It is added to clayey soil to improve the texture and porosity of soil. When added to sandy and humus poor soil, it improves the water holding capacity and adds organic matter to the soil. Dried peat is used as garden mulch and helps in maintaining soil acidity required by decorative indoor plants.

Peat is used in lining hiking boats to absorb moisture and bad odour and used as packing and storing material for eggs, fish and fruits during cold storage.

Because of water retaining capacity of *Sphagnum,* its enormous biomass production and continuous accumulation of dead tissue, land may be drained of water. A raise in the land surface is noticed by the accumulation of dead plant debris.

Due to its excessive growth *Sphagnum* converts water bodies into swamps, killing larger trees around. Because of water logging it can even destroy forests in course of time. For this reason, *Sphagnum* is called quacking bog, blanket or creeping bog.Many of the birds use dried *Sphagnum* for constructing their nests.

Many of the bryophytes are rich in bicarbonates. Their death and decomposition builds up calcium deposition in soil, which hardens into calcareous rock like deposits around other plants.

11.4 BRYOPHYTES AS INDICATORS OF POLLUTION

Bryophytes play a significant role as indicators of environmental pollution (LeBlanc & Rao, 1975). They can be employed in developing an index of atmospheric purity (IAP) which is based on the number, frequency coverage and resistance factor of species which can provide a long range effects of pollution in a given area (Rao, 1982).

Bryophytes have a tendency for accumulation of heavy metals from the atmosphere . They show visible symptoms of pollution such as change in the color of the plant body and poor growth. The retention of pollutants is reflected in anatomical changes such as swelling of the thylakoids, degeneration of chloroplast membrane and excess accumulation of starch. Bryophytes have the capacity to absorb and retain pollutants and prevent the recycling into the ecosystem. Some heavy metals that accumulate certain bryophytes are given in Table 11.4.

Table 11.4 Bryophytes as indicators of pollution

Metals	Species
Arsenic	*Sphagnum nemoreum*
Cadmium	*Funaria hygrometrica, Marchantia polymorpha, Sphagnum fimbriatum, Sphagnum nemoreum*
Chromium	*Aerobryopsis longissima*
Copper	*Cephaloziella massalongi, Cephaloziella phyllacantha, Dryptodon stratus, Gymnocolea acutiloba, Gymnocolea inflata, Marchantia alpestris, Merceya ligulata, Mielichhoferia elongata, Mielichhoferia ligulata, Mielichhoferia macrocarpa, Mielicchoferia meilicchofer*
Lead	*Dicranella varia, Grimmia doniana, Hylocomium splendens, Hypnum cupressiforme, Pleurzium schreberi, Rhytidiadelphus squarrosus*
Mercury	*Dicranella, heteromalla, Hylocomium splendens, Hypnum cupressiforme, Tomenthyphum nitens,*
Nickel	*Marchantia polymorpha*
Vanadium	*Lunularia cruciata, Marchantia polymorpha*

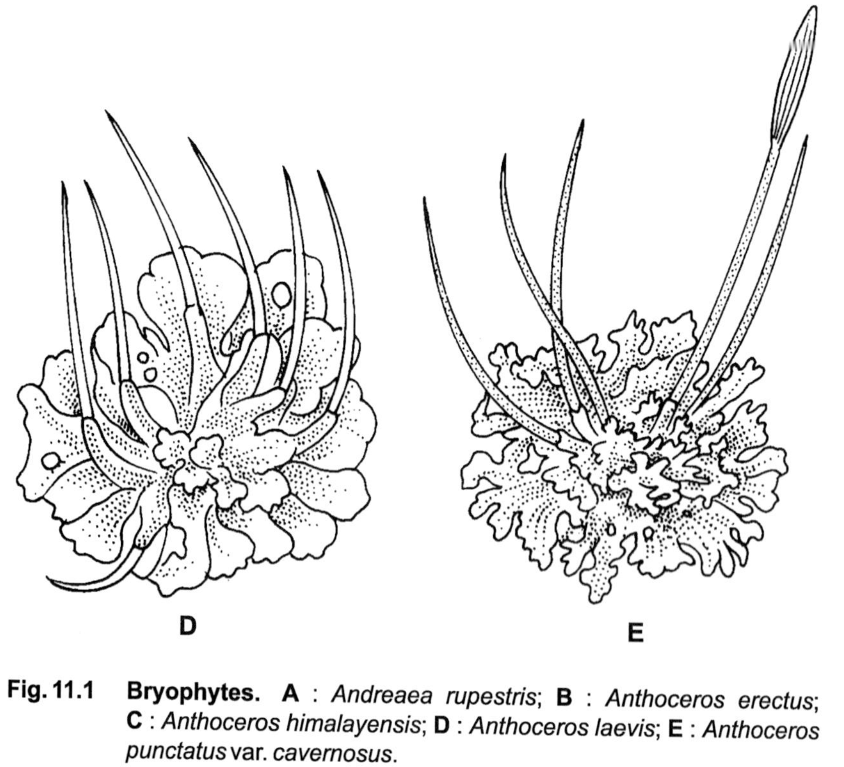

Fig. 11.1 **Bryophytes.** **A** : *Andreaea rupestris*; **B** : *Anthoceros erectus*; **C** : *Anthoceros himalayensis*; **D** : *Anthoceros laevis*; **E** : *Anthoceros punctatus* var. *cavernosus*.

Fig 11.1 to 11.5 Redrawn from Parihar, N.S. (1965)

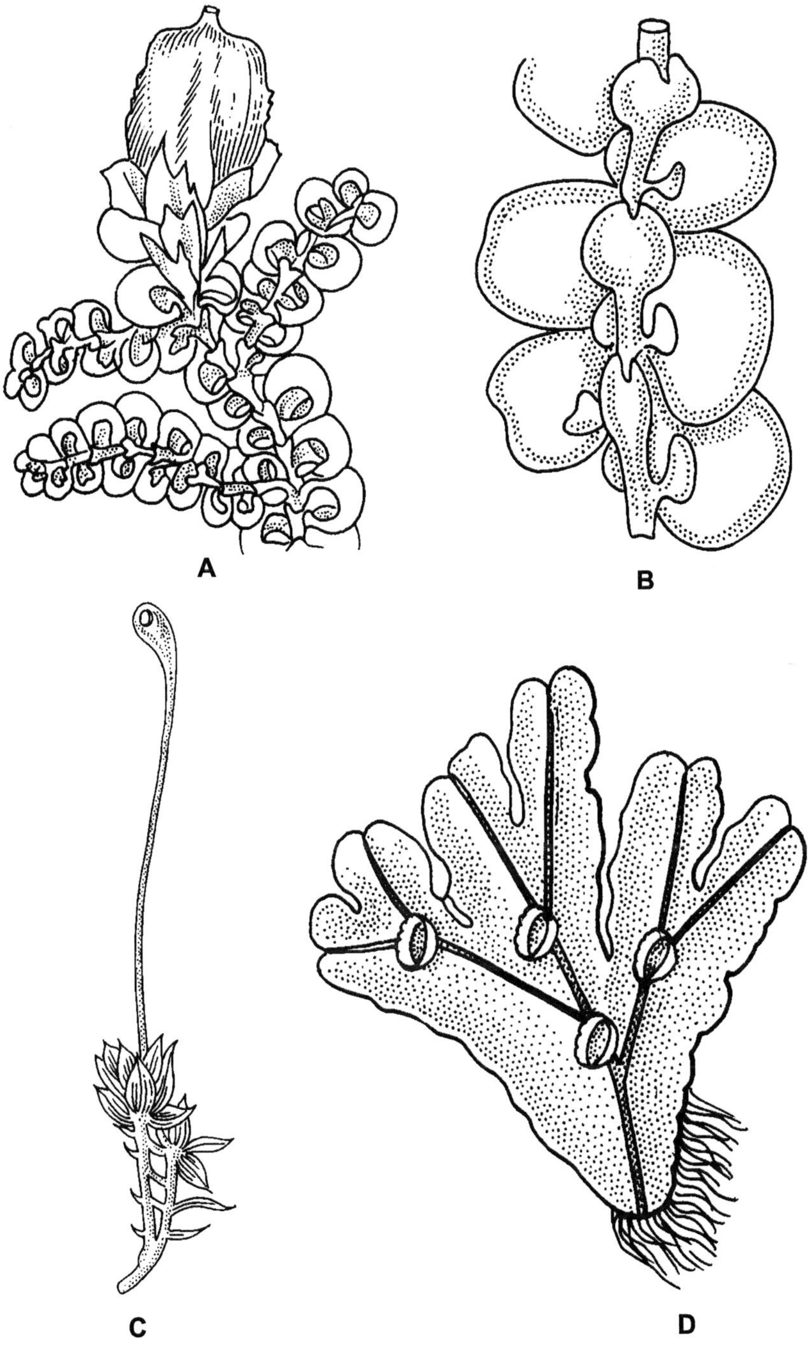

Fig. 11.2 **Bryophytes. A** : *Frullania dilatata*; **B** : *Frullania muscicola*; **C** : *Funaria hygrometrica*; **D** : *Marchantia polymorpha*.

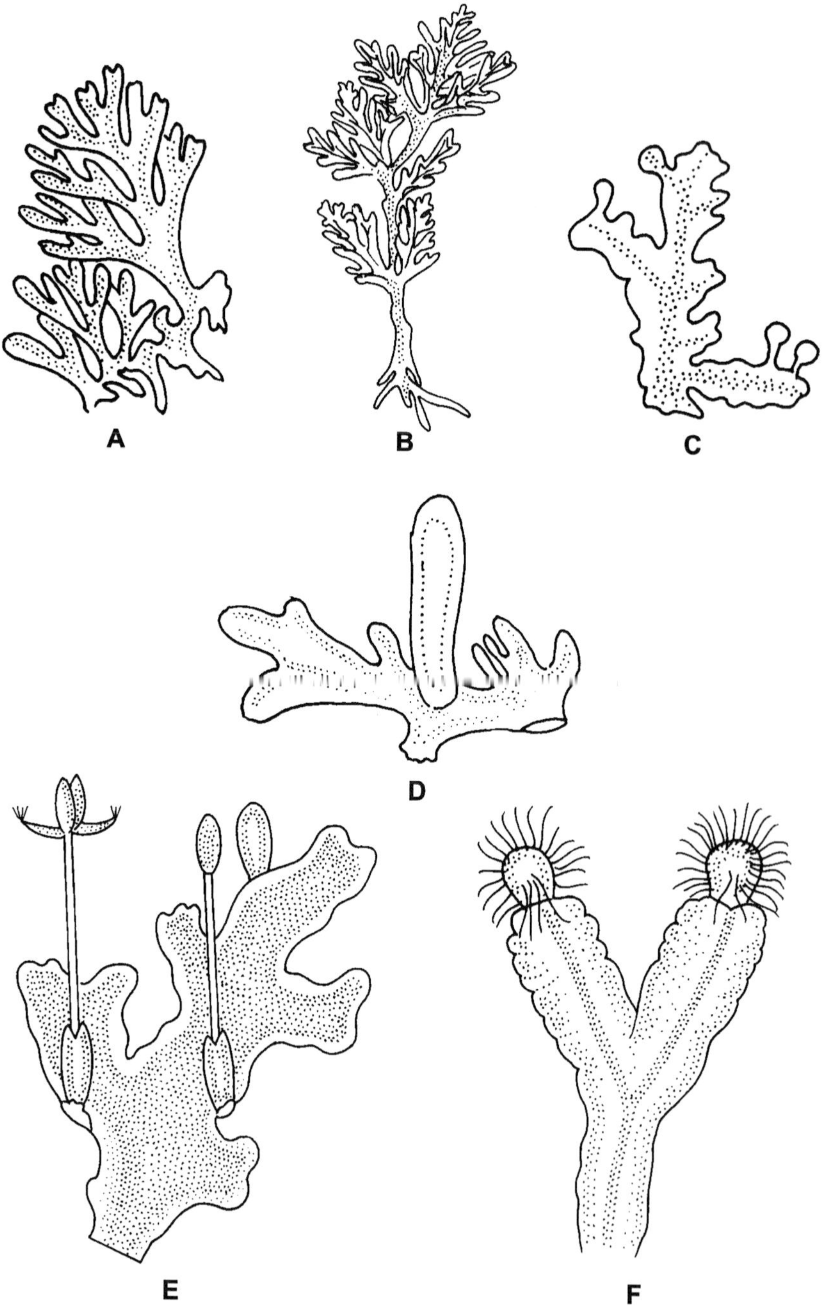

Fig. 11.3 **Bryophytes. A** : *Riccardia bogotensis*; **B** : *Riccardia eriocaulis*; **C** : *Riccardia indica*; **D** : *Riccardia multifida*; **E** : *Riccardia pinguis*; **F** : *Riccia billardieri*.

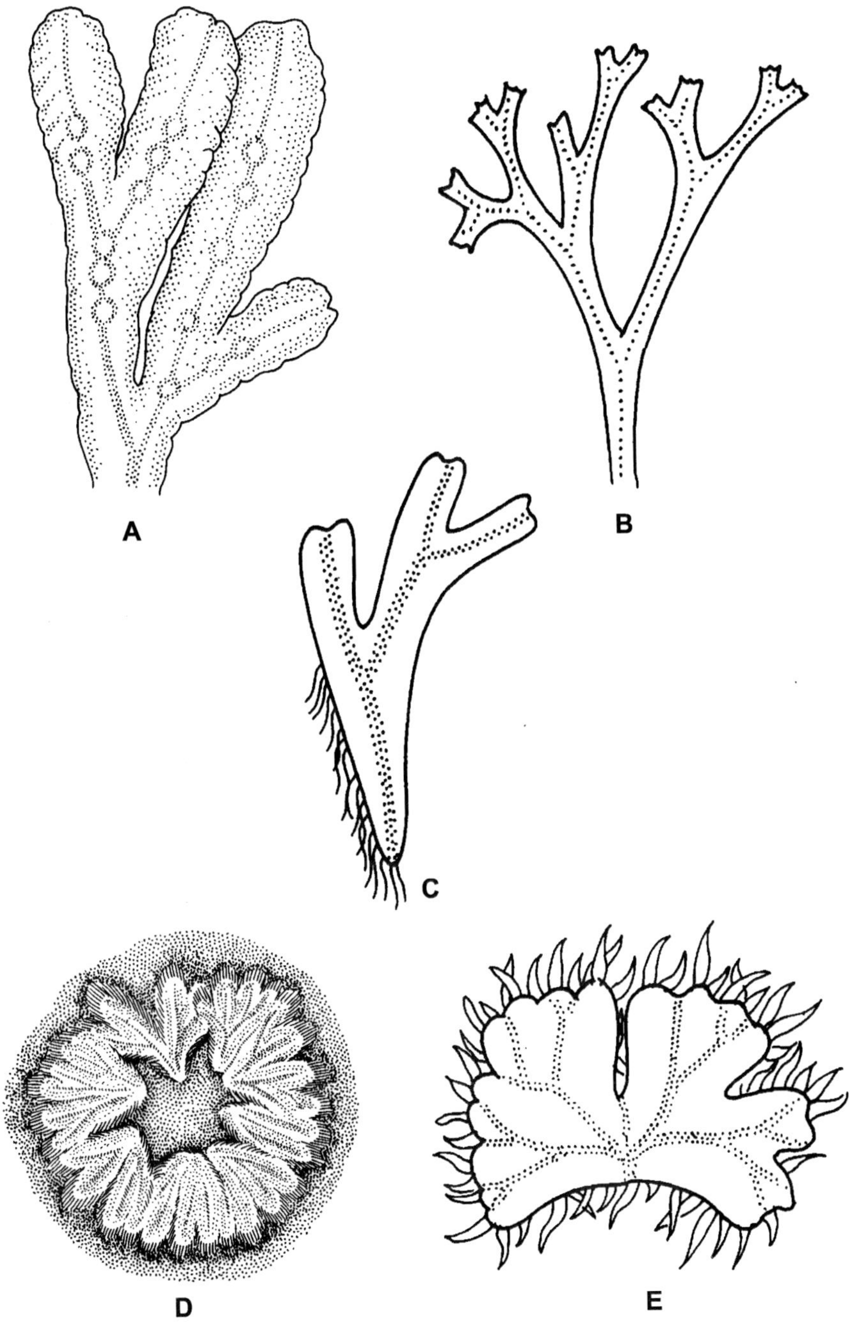

Fig. 11.4 **Bryophytes. A** : *Riccia discolor*, **B** : *Riccia fluitans*-land form; **C** : *Riccia fluitans*-water form; **D** : *Riccia trichocarpa*; **E** : *Ricciocarpus natans*.

Fig. 11.5 **Bryophytes. A** : *Polytrichum commune*; **B** : *Porella platyphylla*; **C** : *Sphagmum acutifolium,*; **D** : *Sphagnum palustre*.

CHAPTER 12

PTERIDOPHYTES

12.1 INTRODUCTION

The name Pteridophyta has a Greek root, as *'Pteron'* means feather, which refers to the feather like leaves (fronds). Pteridophytes grow on tree trunks (*Lycopodium*), on moist damp soil in shady places (many ferns) in the crevices of rocks and walls (*Adiantum*) and on the banks of rivers, streams, pools and ponds and water ferns grow in aquatic habitat.

The plant body of pteridophytes exhibits a great variation in form, size and structure. Some pteridophytes have simple leaves, spores contained in sporangia are formed on the underside of leaves. In a few, the plant body is branched and the branches terminate in spore bearing strobili (*Lycopodium, Selaginella*), cones (*Equisetum*), tassels (*Osmunda*), or spikes (*Ophioglossum*). Some of these pteridophytes, more particularly the horse tails (*Equisetum*) are sometimes found in gardens.

A very large number of ferns are grown as ornamentals for their beautiful general form and leaves. Though most ferns are herbaceous, few tree ferns (*Cyathea*) also occur.

Illustrations of representative species of pteridophytes are given in Fig. 12.1 to 12.35 to provide an over view of their appearance.

12.2 PTERIDOPHYTES AS SOURCE OF FOOD

The fronds of *Acrostichum aureum, Botrychium ternatum, Botrychium verginianum, Ceratopteris siliqulosa, Cyathea gigantea, Dryopteris crenata, Dryopteris macrocarpa, Dryopteris sparsa, Helminthostachys zeylanica, Lygodium microphyllum, Marsilea minuta, Polystichum aculeatum, Pteris ensiformis* and *Selaginella willdenovii* are edible (Ambasta, 1986).

Matteuccia

Matteuccia struthiopteris: young shoot tips are steamed or boiled and eaten.

Nephrolepis

Nephrolepis biserrata (Nephrolepis acuta): shoots and rhizomes are edible.

Ophioglossum

Ophioglossum reticulatum: fronds are eaten as vegetable. It is mixed with other vegetables and consumed raw as salad.

Phymatodes

Phymatodes longissima (Pleopeltis longissima): young fronds which taste like bitter almonds are edible.

Polypodium

Polypodium vulgare: a soup is prepared from the acrial parts.

Polystichum

Polystichum aculeatum (Polypodium aculeatum): fronds are edible, used in curries.

Psilotum

Psilotum nudum: is boiled and eaten.

Selaginella

Selaginella willdenovii: fronds are edible.

Stenochlaena

Stenochlaena palustris: young shoots are edible and used in salads.

12.3 MEDICINAL USES OF PTERIDOPHYTES

Acrostichum

Acrostichum aureum: rhizome paste is applied on boils.

Actinopteris

Actinopteris australis (Adiantum dichotoma), known as Peacock's tail is used as antifertility drug. It is antiseptic, anthelmintic and styptic and used in treating ulcers.

Adiantum

Adiantum aethiopicum (Adiantum emarginatum): is emetic, astringent and emollient. A decoction of the fern is used by the South Africans as an aboritifacient. Useful in cough and chest complaints.

Adiantum capillus-veneris (Maiden hair fern): is used as a emmenogogue, diuretic, febrifuge, pectoral, refrigerant, demulcent, expectorant, tonic; useful in throat affections, bronchial disorders, amenorrhoea, head and chest cold. The fern is boiled with wine and given for curing hard tumours of the spleen and liver. Used in treating cough, cold, nasal congestion, menstrual complaints, catarrh, asthma, influenza and pleurisy.

Adiantum caudatum: Leaves are used to cure cough and fever; leaf paste is applied externally for skin diseases. Fronds are useful in treating skin diseases, diabetes, cough, fever, asthma, bronchitis, hemicrania and is anti-inflammatory.

Adiantum flabellulatum: rhizomes are used for cough and as an anthelmintic.

Adiantum lunulatum : is cooling, alterative, alexiteric; useful in fever, erisipelas, dysentry, ulcers, inflammation and epilepsy.

Adiantum pedatum: rhizomes are used as astringent, stimulant, expectorant, demulcent, tonic and emmenogogue. Fronds burnt with oil is applied to itch and as a pectoral in chronic catarrh and fever.

Adiantum phillippense (Walking maiden hair fern): is used to treat asthma, bone fracture, bronchitis, epilepsy, atrophy, cough, fever, diarrhoea, erysipelas, leprosy, baldness, muscle pain, pimples, sprain and as an antidote for rabid dog's bite.

Adiantum tenerum: is useful as demulcent, expectorant and sudorific.

Adiantum venustum: is used in bronchitis, tumours, biliousness, fever, inflammatory diseases of the chest and ophthalmia. It is tonic, resolvent, expectorant, deobstruent, febrifuge, hydrophobia, colds, headache, diuretic, emmenogogue, astringent, emetic and an antidote for scorpion sting.

Angiopteris

Angiopteris evecta: is useful in leprosy and pain in ribs.

Aspidium

Aspidium polymorphum: is used as anthelmintic.

Asplenium

Asplenium adiantum-nigrum: is bitter and is used as diuretic, anti-inflammatory, expectorant, emmenogogue and laxative. It is useful in treating ophthalmia, diseases of spleen and jaundice. In women, it induces sterility.

Asplenium ceterach: is used as diuretic and astringent. Useful in curing disorders of the spleen.

Asplenium falcatum and *Asplenium parasiticum:* is useful in enlargement of spleen, incontinence of urine, calculus, jaundice and malaria. Rhizome is used as anthelmintic.

Asplenium ruta-muraria: is used as expectorant and deobstruent. Leaves are useful in curing rickets.

Asplenium trichomanes: is laxative and expectorant; used to cure chest cold.

Blechnum

Blechnum orientale: is used as poultice for boils. Decoction of rhizome is given in delirium and is anthelemintic.

Botrychium

Botrychium lunaria (Moon wort): is vulnerary. Used for cuts, wounds and dysentery. Juice of roots and fronds is used for breast cancer.

Botrychium ternatum: is vulnerary, used in dysentery.

Botrychium virginianum (Rattle-snake fern): used in dysentery. Fleshy roots applied to cuts and bruises.

Ceratophyllum

Ceratophyllum demersum: is used as cooling, antiperiodic and is useful in treating biliousness. It is also an antidote for scorpion sting.

Ceratopteris

Ceratopteris siliquilosa (Ceratopteris thalictroides): poultice of the leaves is used for skin complaints.

Cibotium

Cibotium barometz: rhizome is vermifuge; root is given as tonic and used in lumbago; long silky hairs at the base of fronds have the property of rapidly coagulating blood; stem is used as tonic and styptic.

Dicranopteris

Dicranopteris linearis: is used as anthelmintic, antibacterial, expectorant and tonic. It is useful in curing body ache, bone fracture, bowel inflammation, cough, headache, rheumatism, skin diseases, swellings and typhoid fever.

Dryopteris

Dryopteris barbigera and *Dryopteris blandfordii:* rhizomes are used as anthelmentic.

Dryopteris cochleata: is used to cure cholera.

Dryopteris filix-mas, Dryopteris marginata, Dryopteris odontoloma and Dryopteris schimperiana: rhizomes are anthelmintic.

Drynaria

Drynaria quercifolia (Polypodium quercifolium): is used in pthisis, fever, dyspepsia, typhoid fever and cough. Rhizome is bitter and used as astrigent. Fronds are used as poultice for swellings.

Equisetum

Equisetum arvense (Horse tail): is haemostatic, antitubercular, refrigerant, and astringent. Stem is used as a stimulating diuretic in kidney and dropsical disorders. Presence of silica in this plant has been used in eye and skin treatment and to cleanse the system of lead poisoning. Ash of the plant is useful in acidity and dyspepsia. Dried powder of the stem is dusted on bleeding wounds and ulcers (Hutchens, 1991).

Equisetum diffusum: is used to treat acidity, dropsy, dyspepsia, dysentery and as an appetizer.

Equisetum debile: is cooling and useful in gonorrhoea.

Equisetum ramosisimum: is used as a poultice on swellings.

Gleichinia

Gleichinia dichotoma : rhizome is used as anthelmintic. Frond is used to cure asthma and is antibacterial.

Helminthostachys

Heleminthostachys zeylanica: is useful in treating blisters on the tongue, impotency and jaundice. Fronds are used in sciatica and are aperient. Rhizomes are useful in treating whooping cough, dysentery, catarrh and early stages of pthisis. It has intoxicating and anodyne properties.

Hemionitis

Hemionitis arifolia (Mule fern): Juice of the fronds is applied to burns.

Hypolepis

Hypolepis punctata (Phegopteris punctata): fronds are used as poultice for boils.

Lemmaphyllum

Lemmaphyllum carnosum: fronds are diuretic, pectoral and astringent; used in urinary calculi and rheumatism; decoction prescribed to arrest haemorrhages.

Lycopodium

Lycopodium cernuum: decoction of the plant is used as a lotion in Beri-Beri and taken internally for cough and uneasiness in the chest. Embrocation of ashes in vinegar is recommended for skin eruption. It is also used to cure rickets in children.

Lycopodium clavatum (Common club moss): spores are dusted as absorbent in excoriation of skin and used as medicated snuff. It is employed as a covering for pills to prevent adhesion. Is used as diuretic and antiseptic. The decoction of the plant is useful in rheumatism, lung and kidney disorder, skin eruptions, urinary disorders and constipation (Ambasta, 1986).

Lygodium

Lygodium circinatum: is useful in gynaecological disorders and rheumatism. Stem is ground and used as external application for wounds.

Lygodium flexuosum: is used as expectorant. Root boiled with mustard oil is used as poultice to carbuncles, rheumatism, sprains, scabies, ulcers, eczema, cuts and wounds; is used to cure abdominal pain, cholera, constipation, cough, debility, fever, indigestion, jaundice and stomach ache.

Lygodium japonicum: is used as expectorant. Decoction of vegetative parts and spores is used as diuretic and cathartic in veterinary medicine to heal wounds.

Lygodium microphyllum (Lygodium scandens): Leaf decoction is used for dysentery and abdominal pain. Leaf poultice is used for skin diseases and swellings.

Marsilea minuta

An anti epileptic drug Ayush 56 has been formulated and patented.

Microsorium

Microsorium punctatum: is used as an antidote for snake bite.

Nephrolepis

Nephrolepis cordifolia: decoction of fresh fronds is useful in cough, bleeding, congestion, cuts, wounds, nasal blockage, breathing problem, thirst and loss of appetite.

Notholaena

Notholaena eckloniana: dired leaves smoked for relief from cold in the head and chest. (Ambasta, 1986)

Oleandra

Oleandra neriiformis: decoction of frond is emmenogogue.

Oleandra wallichii: rhizome decoction is rejuvenating.

Onychium

Onychium siliculosum: decoction of fronds is used in dysentery.

Ophioglossum

Ophioglossum pendulum: fronds are ground in coconut oil and applied on scalp for promoting hair growth.

Ophioglossum vulgatum: is used as vulnerary, styptic, and antiseptic. Decoction is useful in angina and warm decoction of rhizome is applied to boils, haemorrhoids and wounds.

Osmunda

Osmunda regalis (Royal fern): root is tonic, antibacteral, stimulant, styptic and abortifacient. An aqueous extract of the fern is used for intestinal gripe, rheumatism, dysentery, rickets and muscular debility. Fronds used as diuretic drink and for swelling. Tender shoots are used in balms and as healing plasters.

Pellaea

Pellaea calomelanos : rhizome is used as anthelmintic and infusion of leaves is useful in cold in the head, chest and fronds are useful in asthma.

Pityrogramma

Pityrogramma calomelanus (Gymnogramma calomelanus): decoction is used for kidney troubles.

Phymatosorus

Phymatosorus scolopendria: is useful in dysentery and gonorrhoea.

Pleopeltis

Pleopeltis lanceolata (Polypodium lanceolatum): decoction is useful in curing or treating cold, sore throat and itching.

Polypodium

Polypodium vulgare (Rock polypod, Bracken root, Common polypody, Fern root.) root is pectoral, demulcent, cholagogue, purgative, anthelmintic and vermifuge. Decoction is used as a cure for rickets, pulmonary and hepatic diseases in children. Root and the whole plant decoction is used as a dietary drink which is useful in spleen and other disorders (Hutchens, 1991).

Psilotum

Psilotum nudum (Psilotum triquetrum): tea prepared from the plant is given to children for thrush, and as purgative. Spores are used to treat diarrhoea.

Pteridium

Pteridium aquilinum: rhizome is used as astringent, anthelmintic, and haemostat. Rhizome and frond decoction is given for disorders of spleen.

Pteris

Pteris aquilina: Rhizome is astringent and anthelmintic; decoction is given in chronic spleen disorders.

Pteris biaurita: is useful in healing wounds.

Pteris ensiformis: decoction of fronds is used in dysentery and rhizome juice is applied for glandular swellings on neck.

Pteris multifida (Pteris serrulata): decoction of rhizome and fronds is given to treat dysentery and as vermifuge. Paste is used as a poultice for cutaneous affections.

Pteris vitata : is used as demulcent and tonic.

Pteris wallichiana: is used to heal cuts and wounds.

Pyrrosia

Pyrrosia adnascens: is useful in dysentery and burns.

Schizaea

Schizaea dichotoma: rhizomes is useful in cough, throat troubles and parturition.

Selaginella

Selaginella bryopteris: is used to treat Beri-Beri, dysentery, gonorrhoea, leucorrhoea, spermatorrhoea and other venereal disorders. It is used as diuretic and tonic.

Selaginella wallichii : decoction is used for parturition.

Selaginella willdenovii: is used to cure back ache, fever, lumbago, rheumatism and skin disorders.

Sphenomeris

Sphenomeris chusana (Stenoloma chinensis; Trichomanes chinesnse: is useful in chronic enteritis.

Stenochlaena

Stenochlaena palustris: decoction of fronds is used as febrifuge.

Trichomanes

Trichomanes javanicum: dried fern is mixed with garlic and onions and smoked for headache.

12.3.1 Pteridophytes in The Treatment of Cancer and Tumour

Pteridophytes are used in traditional medicine for treating various types of cancer and tumours (Table 12.1)

Table 12.1 Pteridophytes in the treatment of cancer and tumour

Species	Type of cancer/tumour
Adiantum capillus-veneris	Tumour
Adiantum venustum	Tumour
Botrychium lunaria	Breast cancer
Equisetum hyemale	Leukemia
Lycopodium clavatum	Liver, mouth, eyelid, cancer, epithelioma, polyps
Lygodium flexuosum	Carbuncles
Polypodium lingua	Cancer, carcinomatous ulcers
Polypodium vulgare	Tumour
Pyrrosia lingua	Cancerous swellings
Selaginella doederleinii	Tumour

Source: Hartwell, 1982.

12.3.2 Medicinal Uses in Homoeopathy

Aspidium

Aspidium is used in homoeopathic system of medicine for treating worms, constipation, pulmonary tuberculosis, scrofula, ambylopia, abdominal pain, diarrhoea, vomiting and blue rings around the eyes.

Equisetum

Equisetum (Horse tail) is used in treating anuresis, dysuria, burning pain in the urethra, nightamare, mucous in the urine, involuntary urination, kidney pain, lumbar pain, micturition, catarrhal conditions with offensive, pus-like in discharge of ear, nose and throat, gravel, glandular dicharge, kidney affections, skin disorders, and perspiration in the feet. It strengthens the heart and lungs and is used in treating cystitis, dropsy, anuresis, gonorrhea, haematuria and renal affections,. It is internally taken for bleeding. Externally it is used as a compress or sponge bath for old injuries, putrid wounds, gangrenous ulcers and external bleeding.

Lycopodium

Lycopodium spores are used for atony, malnutrition, catarrh in aged, melancholy, weak memory, sadness, vertigo, pain in the occiput, grey hair, premature baldness and grey sty on the eyelid, ulceration of the eye lid, yellow offensive discharge from the ear, otorrhoea, deafness, ulcerated nostrils, scaly herpes on the face, yellow colour of the skin, tooth ache, swollen cheeks, dry mouth, dry, black and swollen tongue, blisters on the tongue, dry throat, swollen tonsils, inflamed throat, diphtheria, laryngitis, dyspepsia, bulimia, flatulence, bloated abdomen, pain, dropsy, diarrhoea, painful urination, polyuria, erectile dysfunction, delayed menstruation, varicose veins, leucorrhoea, dyspnoea, night cough, neglected pneumonia, catarrh of the chest in infants, bloody, salty phlegm, back pain, chronic gout, profuse sweating of foot, cramps in the calves and toes, sciatica, painful callosities on soles, fever, itching of the skin, chronic eczema, erectile tumours, offensive secretions and psoriasis (Boericke, 1991).

12.3.3 Uses in Chinese Medicine

Cibotium

Cibotium barometz: is tonic and analgesic; used in rheumatism, myospasm and lumbago.

Cyclophorus

Cyclophorus lingua: is used in treating carbuncles and carcinomatous ulcers.

Equisetum

Equisetum hyemale: is diuretic and hemostatic; used in enterorrhagia, hemorrhoidal haemorrhages and as ophthalmic lotion in treating eye affections.

Drynaria

Drynaria fortunei: Used as analgesic in arthralgia, lumbago, myalgia, bruises and contusions.

Lygodium

Lygodium japonicum: is antiphlogistic and diuretic; used in curing cystitis, urethritis, hematuria, dysuria, gonorrhoea and urinary calculus.

Polypodium

Polypodium lingua: is astringent and diuretic; used in treating gonorrhoea, menorrhagia, urethrorrhagia and cystitis.

12.3.4 Uses in Unani Medicine

Adiantum venustum: is used as anodyne

12.3.5 Uses in Ethnomedicine in India

Actiniopteris

Actinopteris australis: used for treating tuberculosis and ulcers.

Actinopteris radiata: used in bronchitis, gynaecological complaints, headache and dysentry; leaf powder used for typhoid.

Adiantum

Adiantum capillus-veneris: used in bronchitis, cold, cough, fever, menstrual complaints and throat disorders.

Adiantum caudatum: used in asthma, bronchitis, burning feeling inside body, cough, fever and skin disorders

Adiantum incisum: used for treating diabetes, skin disorders, cough and fever.

Adiantum philippense (=Adiantum lunulatum): used in asthma, bone fractures, bronchitis, cough, fever, diarrhoea, bite of rabid dog or jackal, erysipelas, leprosy, muscular pain, pimples and sprain.

Adiantum venustum: used to cure fever.

Angiopteris

Angiopteris evecta: used to treat leprosy and pain in ribs.

Asplenium

Asplenium falcatum: used in calculi, jaundice, malaria and urinary complaints.

Asplenium nidus: used for sores and ulcers

Asplenium viride: Fronds are used as poultice on burns and wounds.

Blechnum

Blechnum orientale: used in boils and in delirium

Botrychium

Botrychium ternatum: is anthelmintic, useful in asthma and in epilepsy.

Cyathea

Cyathea albo-setacea: used as poultice for sores and wounds.

Cyathea spinulosa: used as poultice for sores and wounds.

Dicranopteris

Dicranopteris linearis: is anthelmintic; useful in asthma and epilepsy.

Drynaria

Drynaria quercifolia: is anthelmintic, tonic, antibacterial and expectorant; used in body ache, bone fracture, bowel inflammation, cough, headache, rheumatism, skin disorders, swelling and typhoid fever.

Dryopteris

Dryopteris cochleata: used in cholera.

Dryopteris sparsa: is anthelmintic.

Equisetum

Equisetum diffusum: used in acidity, dropsy, dyspepsia and dysentry; used as appetizer.

Equisetum ramosissimum: used in swellings.

Helminthostachys

Helminthostachys zeylanica: used for blisters on tongue, impotency and jaundice; is aphrodisiac.

Lycopodium

Lycopodium clavatum: used in constipation, skin eruptions and urinary complaints.

Lycopodium phlegmaria: used in dropsy, gastric complaints and pulmonary complaints; is an emmenogogue.

Lygodium

Lygodium circinatum: used in gynaecological complaints and for rheumatism.

Lygodium flexuosum: is abortifacient and appetizer; used in cholera, constipation, cough, cuts, delirium, eczema, fever, indigestion, jaundice, rheumatism, scabies, stomach ache, ulcer and wounds; Root decoction with *Zizyphus mauritiana* is used to cure chest pain.

Lygodium microphyllum: useful in dysentry.

Lygodium pinnatifidum: used to treat blood in sputum, fever, stomach upsets, loose motions and to revive taste.

Lygodium scandens: used to treat abdominal pain, skin disorders and swellings.

Marsilea

Marsilea quadrifolia: used for bone fractures, diarrhoea, eye complaints, fever, muscular pain, pimples, snake bites and spleen complaints.

Microsorium

Microsorium punctatum: used as an antidote for snake bite

Nephrolepsis

Nephrolepis cordifolia: checks bleeding; used for treating chest congestion, cuts, wounds, nasal blockage, breathing difficulties, thirst and loss of appetite.

Ophioglossum

Ophioglossum pendulum: used as hair tonic.

Ophioglossum vulgatum: used for haemorrhage and wounds.

Osmunda

Osmunda regalis: is abortifacient.

Phymatosorus

Phymatosorus scolopendria: used in dysentry and gonorrhea.

Pleopeltis

Pleopeltis lanceolata: useful in treating itching.

Pteridium

Pteridium aquilinum (= Pteris aquilinum): as anthelmintic and haemostatic.

Pteris

Pteris biaurita: used for treating wounds.

Pteris ensiformis: useful in treating dystentry and glandular swellings on neck.

Pteris vitata: is demulcent and tonic.

Pteris wallichiana: useful in treating cuts and wounds.

Pyrrosia

Pyrrosia adnascens: in burns and dysentry.

Schizaea

Schizaea dichotoma: used to treat cough and throat disorders.

Selaginella

Selaginella bryopteris: used in beri-beri, dysentery, gonorrhoea, venereal disorders, leucorrhoea and spermatorrhoea; is useful as tonic and diuretic.

12.4 TOXICOLOGY

Dryopteris filix-mas

Large doses of extract are potentially toxic resulting in muscular weakness, coma and temporary or permanent blindness, headache, dyspnoea, nausea, diarrhoea, vertigo, tremors, convulsions, cardiac and respiratory failure (Spoerke, 1980).

Equisetum (Horse tail)

Horsetail has been listed as an herb of undefined safety by the FDA (Der Marderosian and Liberti, 1988). Ingestion of large amounts of the fern may be toxic. Children have been reported to be poisoned by using the stems as blow guns or whistles. In animals, the ingestion of horse tail produces muscle weakness, ataxia, weight loss, abnormal pulse rate, cold extremities and fever. (Spoerke, 1980).Hay composed of 20% or more *Equisetum arvense* produced these symptoms in 2-5 weeks. Cattle appear to recognize the odour of horse tail and have been documented to have refused hay contaminated with horse tail. (Kamphues, 1990). Seborrheic dermatitis has been reported to be induced by horse tail (Sudan, 1985).

Pteridium aquilinum (Bracken fern)

Bracken fern is used as food in Japan and also as cattle feed. Bracken fern poisoning in cattle develops slowly after the ingestion and the affected animal exhibit depressed bone marrow activity. Live stock like sheep, goat and cow when fed with bracken fern developed, multiple tumours, adenomas, chronic haematuria, characterized by urinary bladder haemorrhages, thromobocytopenia and leucopenia (Bracken hair poisoning).

In laboratory trials, the incidence of urinary bladder carcinoma was significantly high in rats fed on a diet of bracken fern.

Indirect human exposure to bracken fern through milk and dairy products is likely to occur (Hirono, 1987).

12.5 OTHER USES OF PTERIDOPHYTES

Acrostichum

Acrostichum aureum: dried fronds are used in thatching and for making brooms.

Azolla

Azolla pinnata: used in rice fields as green manure. It is also used as green forage and poultry feed.

Equisetum

Equisetum arvense: is used in polishing hard wood, ivory and brass. Also used in cosmetics.

Lycopodium

Lycopodium cernuum: dried plant is used for stuffing pillows.

Lygodium

Lygodium microphyllum: old stems are dried and made into baskets.

Onychium

Onychium japonicum: is used for indoor decoration.

Osmunda

Osmunda claytoniana: rhizomes are ground and used as a rooting medium for growing epiphytic orchids.

Pteridium

Pteridium aquilinum (Pteris aquilina): employed as feed for pigs and green fronds are used as fodder. It is also used as manure with cow dung (Ambasta, 1986).

Stenochlaena

Stenochlaena palustris: rhizomes are used as cordage for binding fishing traps, anchor ropes, caulking boats and making baskets (Ambasta, 1986).

Pteridophytes as ornamental plants

Species of *Adiantum, Equisetum, Lycopodium, Nephrolepis, Platycerium, Pteris* and several others are grown in gardens.

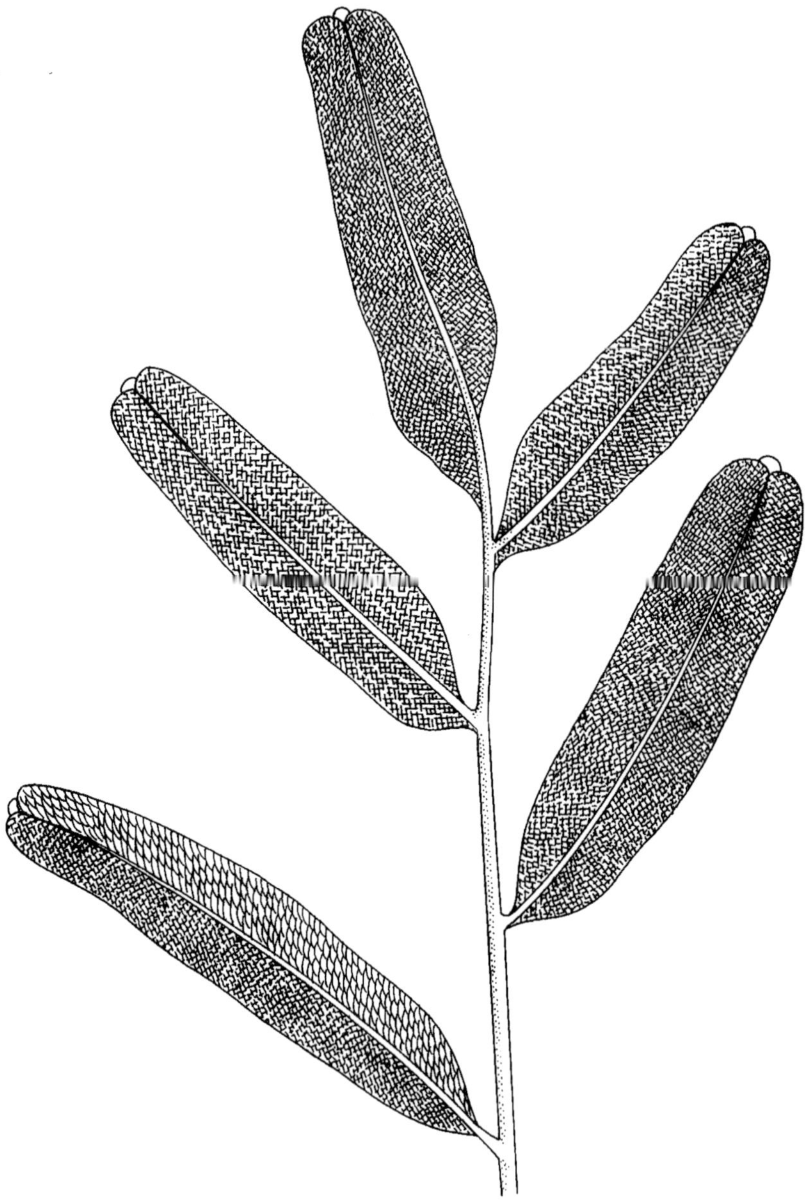

Fig 12.1 Pteridophytes : *Acrostichum aureum.*
Fig. 12.1 to 12.35 Redrawn from Beddome, R.H., (1970) Parihar, N.S. Vasishta, P.C. (1971) and Sporne, K.R. (1974).

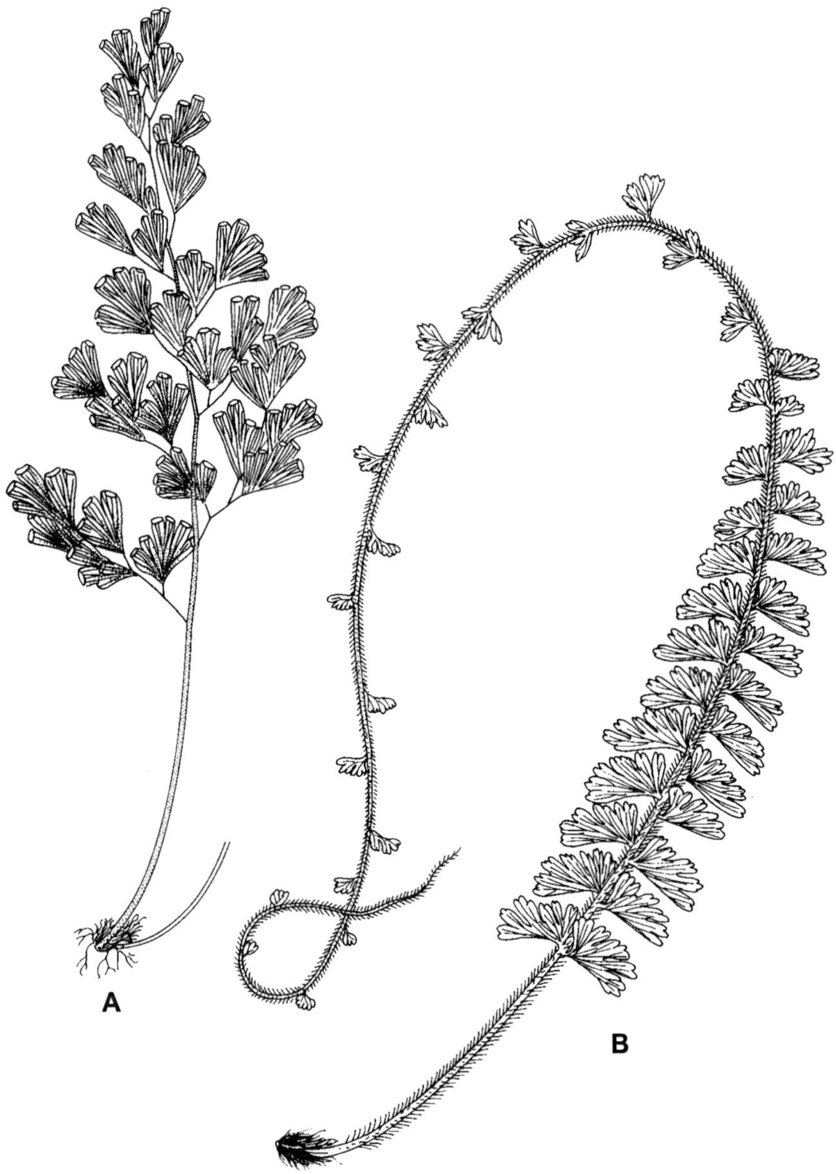

Fig 12.2 Pteridophytes. A : *Adiantum capillus veneris*; **B** : *Adiantum caudatum*

Fig 12.3 Pteridophytes. *Adiantum lunulatum*

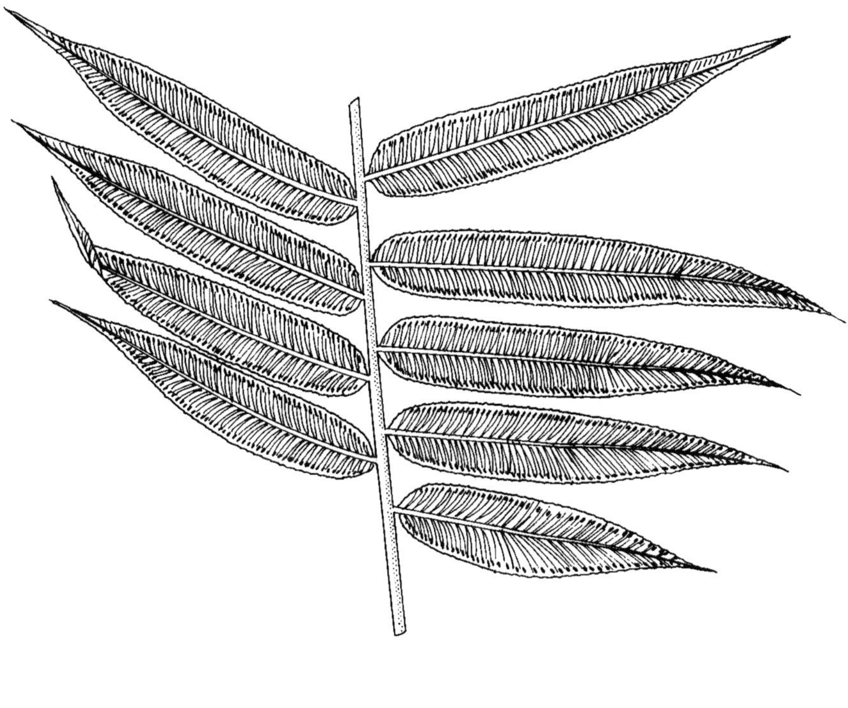

Fig 12.4 **Pteridophytes.** *Angiopteris evecta*

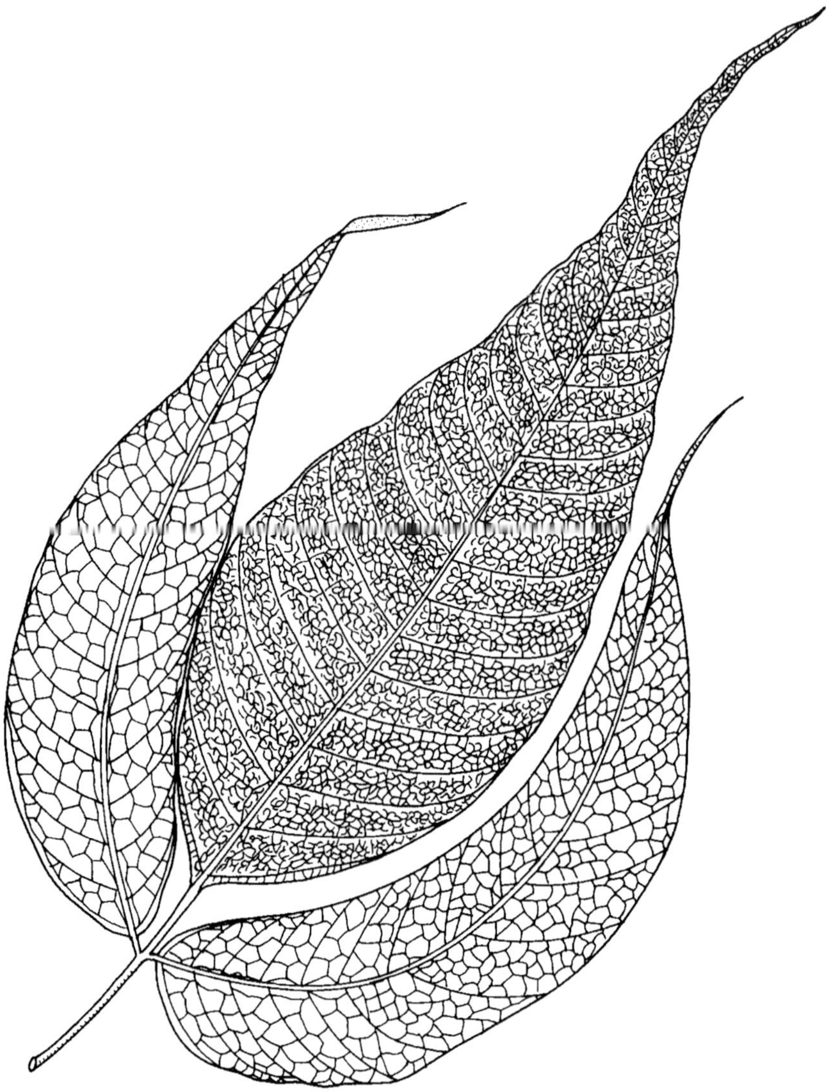

Fig 12.5 Pteridophytes. *Aspidium polymorphum*

Fig 12.6 Pteridophytes. *Asplenium falcatum*

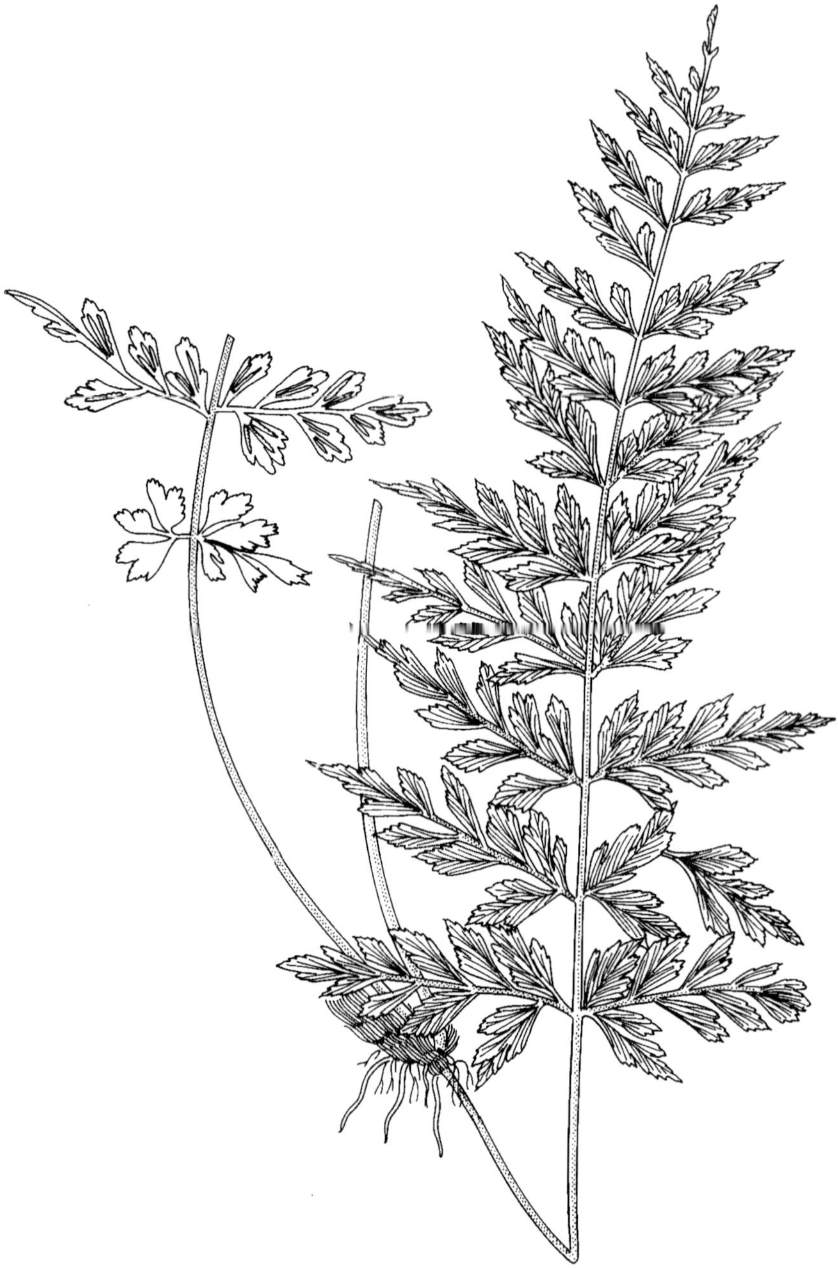

Fig 12.7 Pteridophytes. *Asplenium furcatum*

Fig 12.8 **Pteridophytes.** *Asplenium trichomanes*

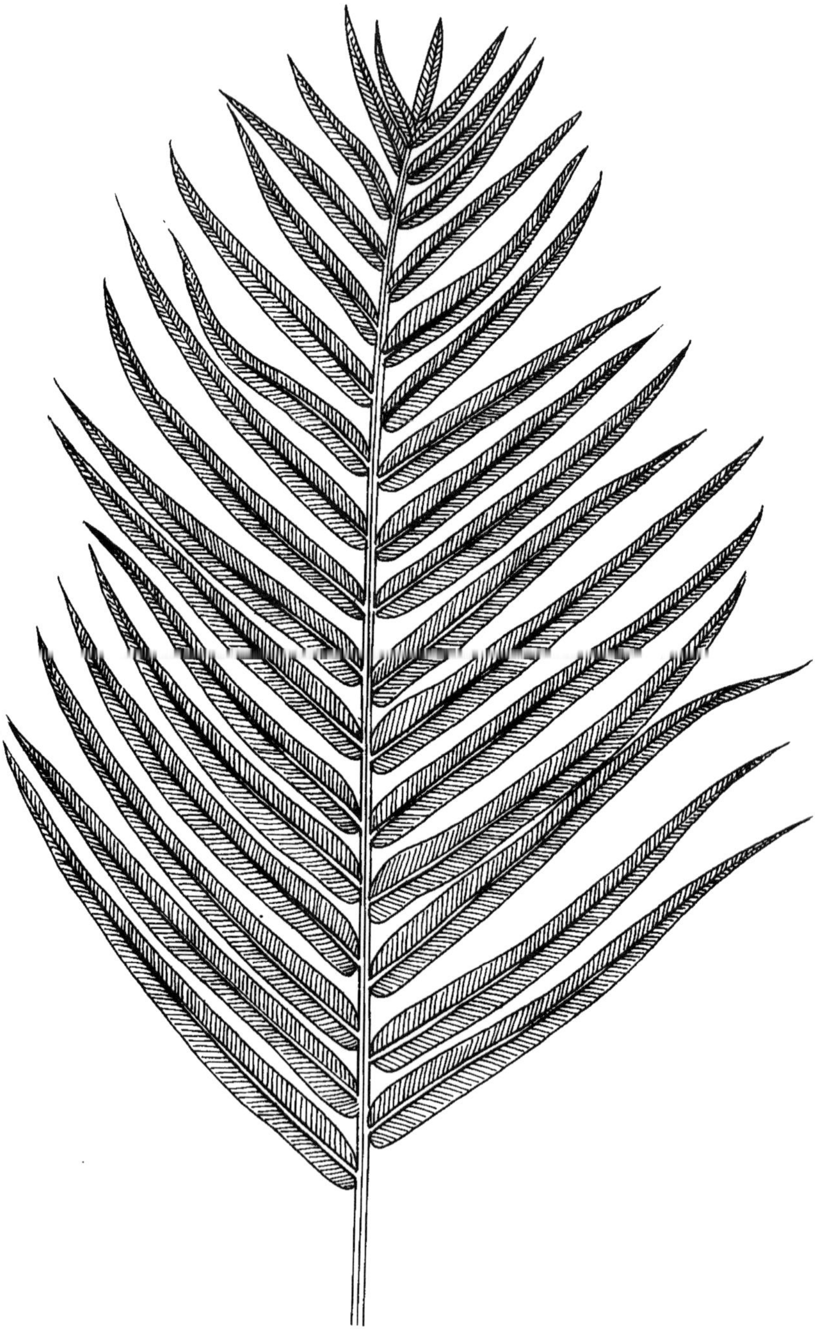

Fig 12.9 Pteridophytes. *Blechnum orientale*

Fig 12.10 Pteridophytes. *Botrychium verginianum* var. *lanuginosum*

Fig 12.11 Pteridophytes. *Cyathea spinulosa*

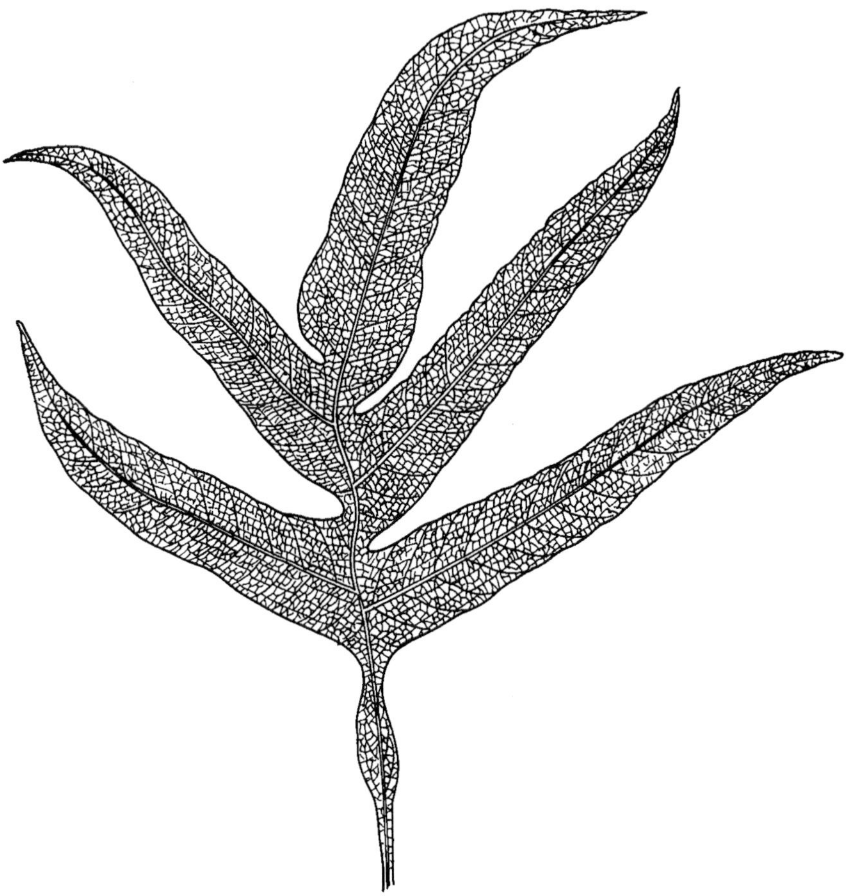

Fig 12.12 Pteridophytes. *Drynaria quercifolia*

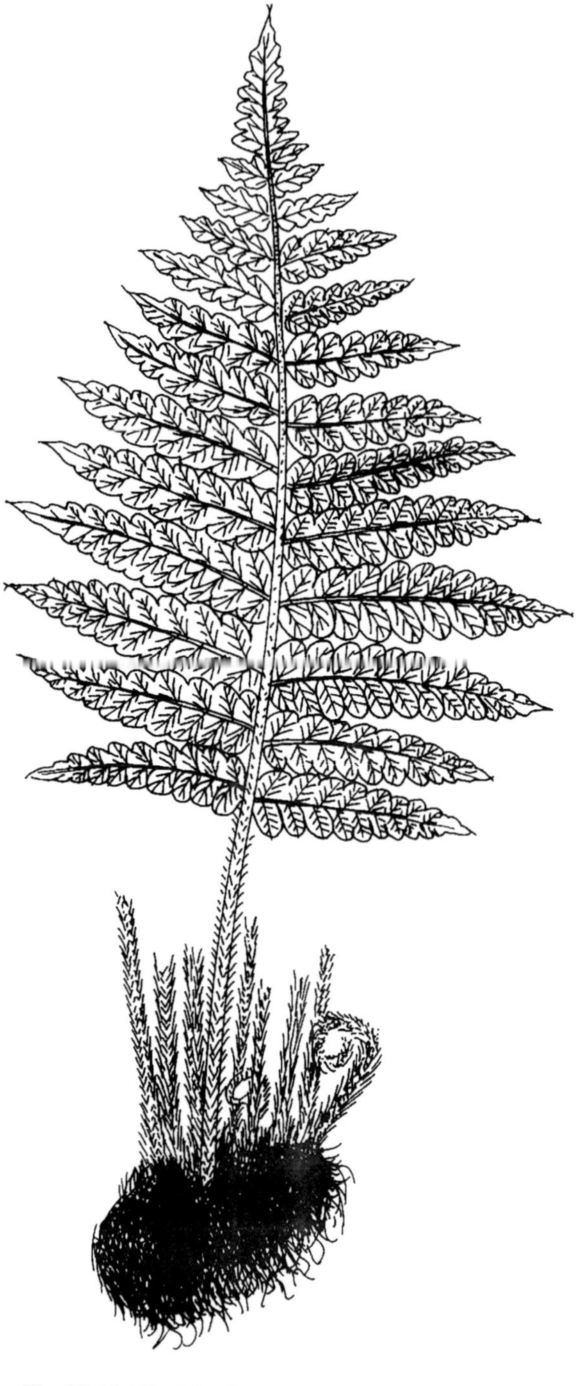

Fig 12.13 **Pteridophytes.** *Dryopteris chrysocoma*

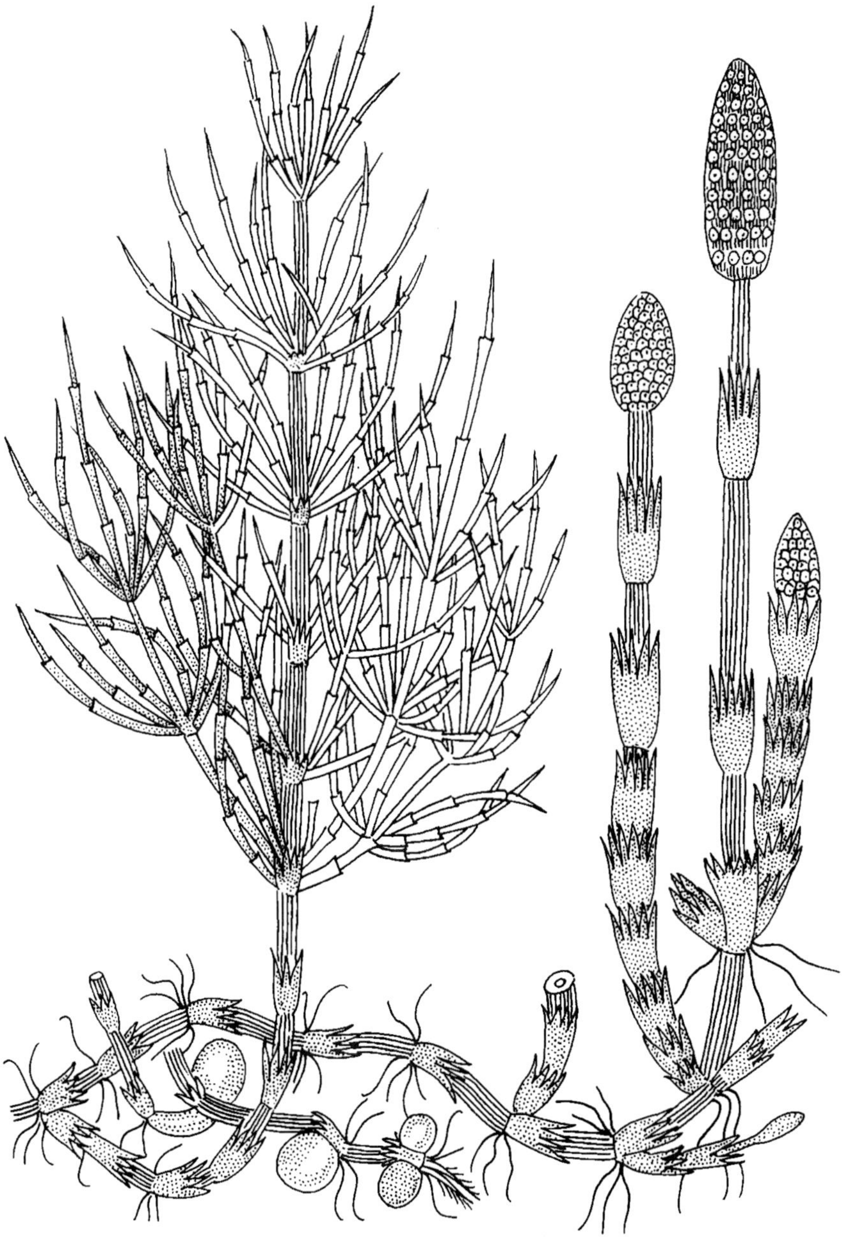

Fig 12.14 Pteridophytes. *Equisetum arvense*

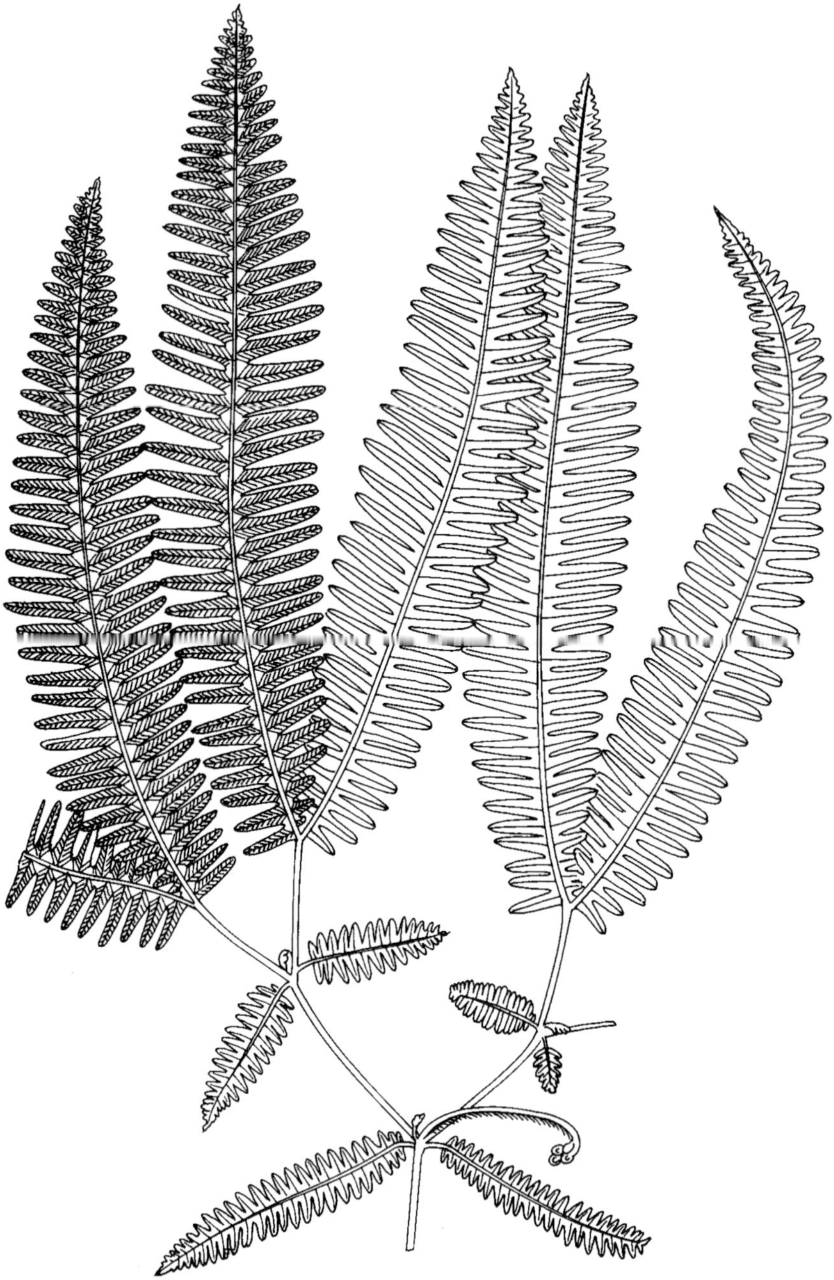

Fig 12.15 Pteridophytes. *Gleichenia dichotoma*

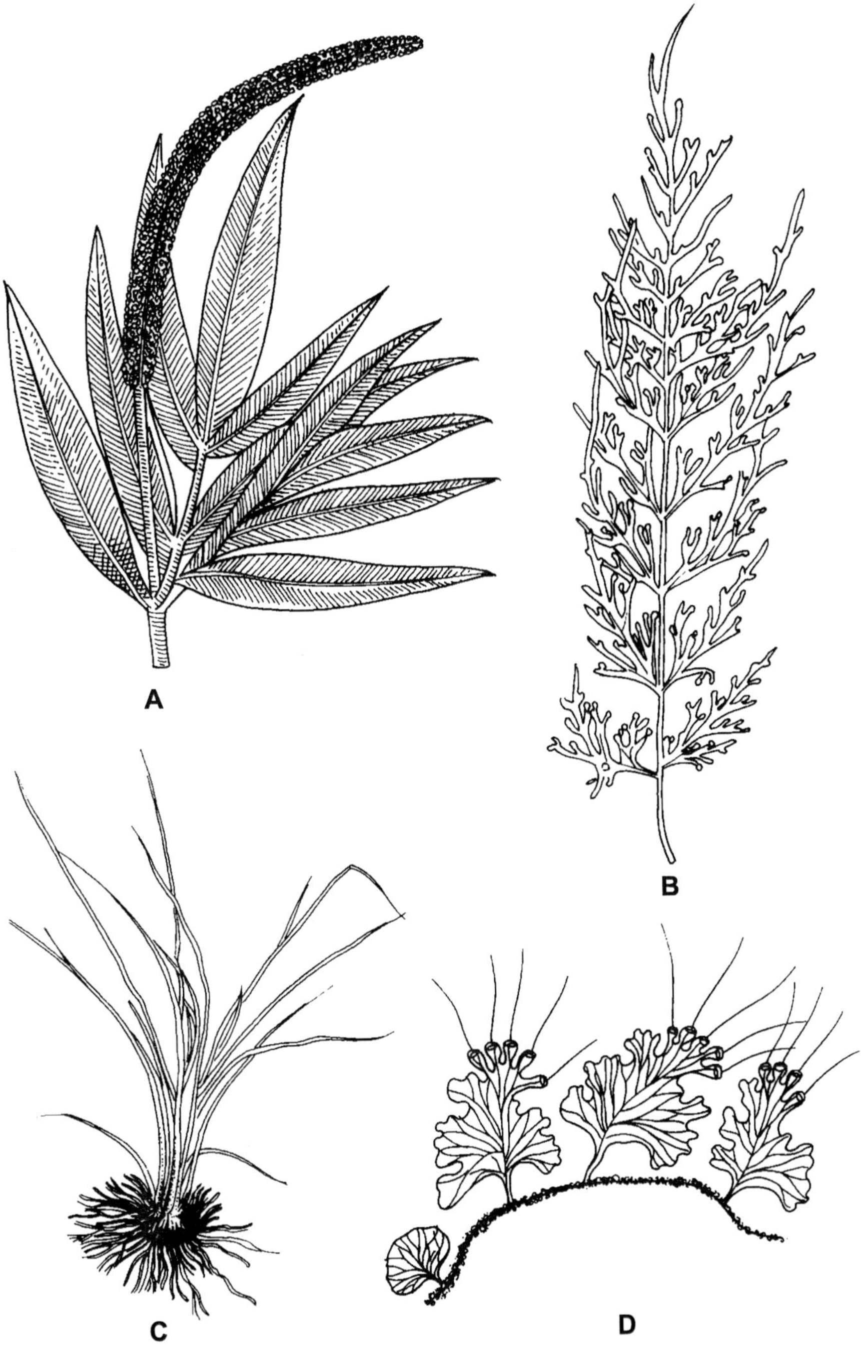

Fig 12.16 Pteridophytes. **A** : *Helminthostachys zeylanica;* **B** : *Hymenophyllum flabellatum*; **C** : *Isoetes indica*; **D** : *Trichomanes cuspidatum.*

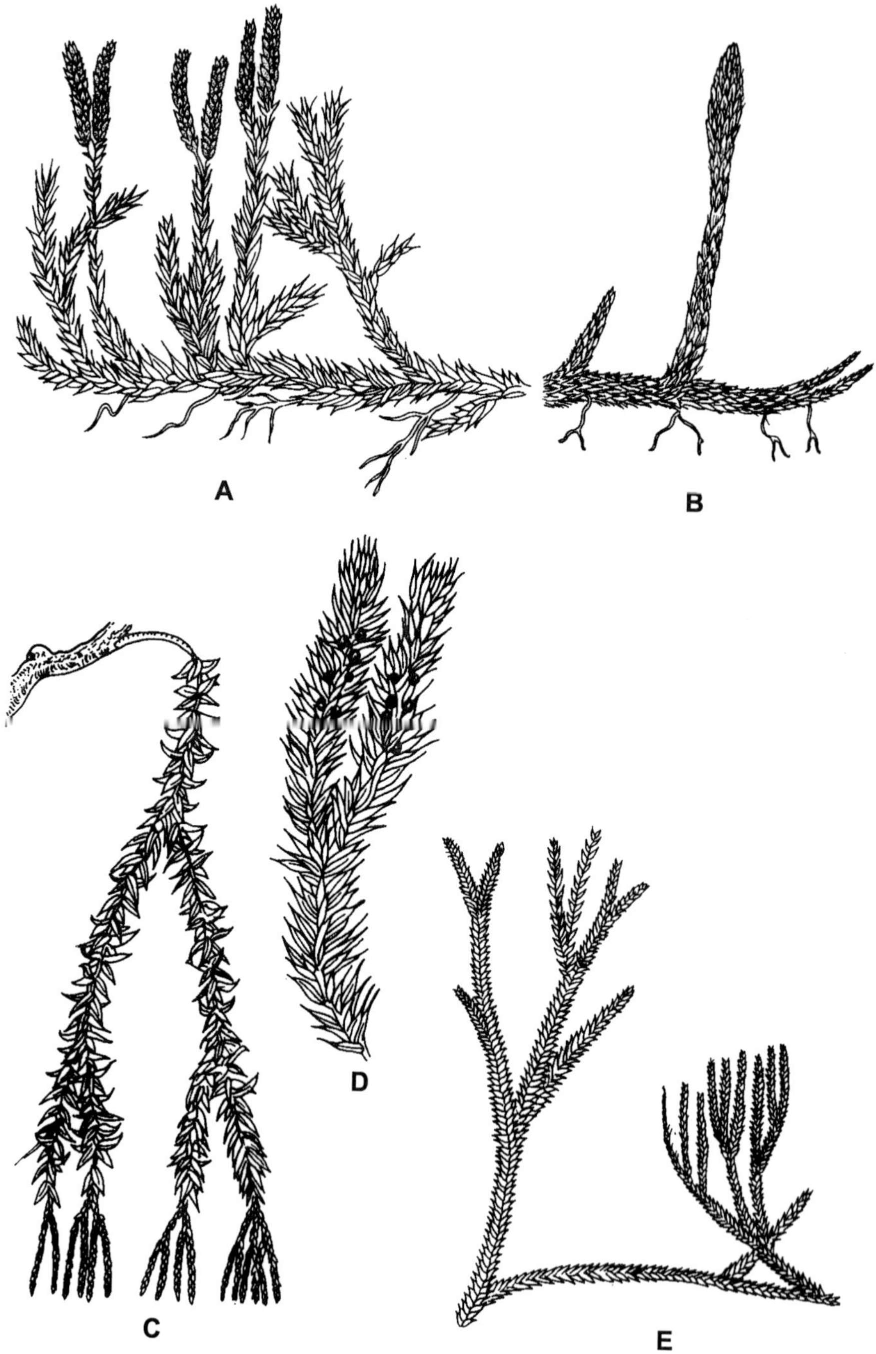

Fig 12.17 **Pteridophytes.** **A** : *Lycopodium clavatum*; **B** : *Lycopodium inundatum*; **C** : *Lycopodium phlegmaria*; **D** : *Lycopodium selago*; **E** : *Lycopodium volubile*

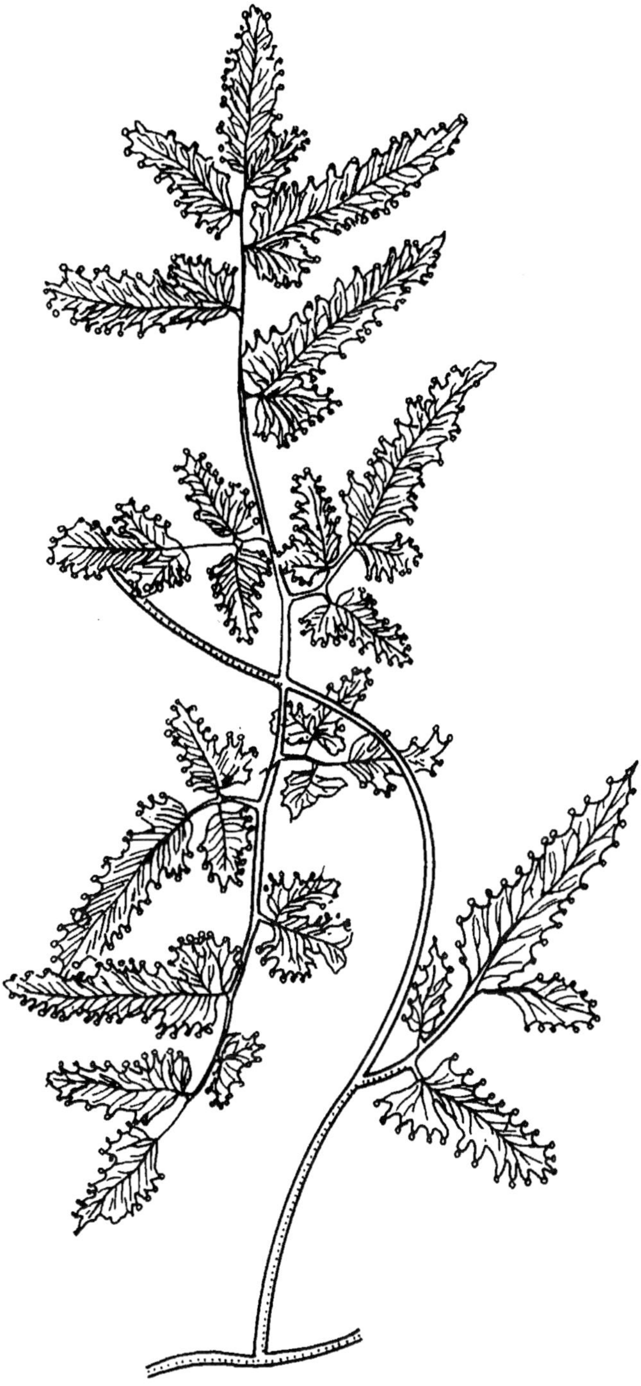

Fig 12.18 Pteridophytes. *Lygodium flexuosum*

Fig 12.19 Pteridophytes. *Lygodium japonicum*

Fig 12.20 Pteridophytes. *Nephrolepis tuberosa*

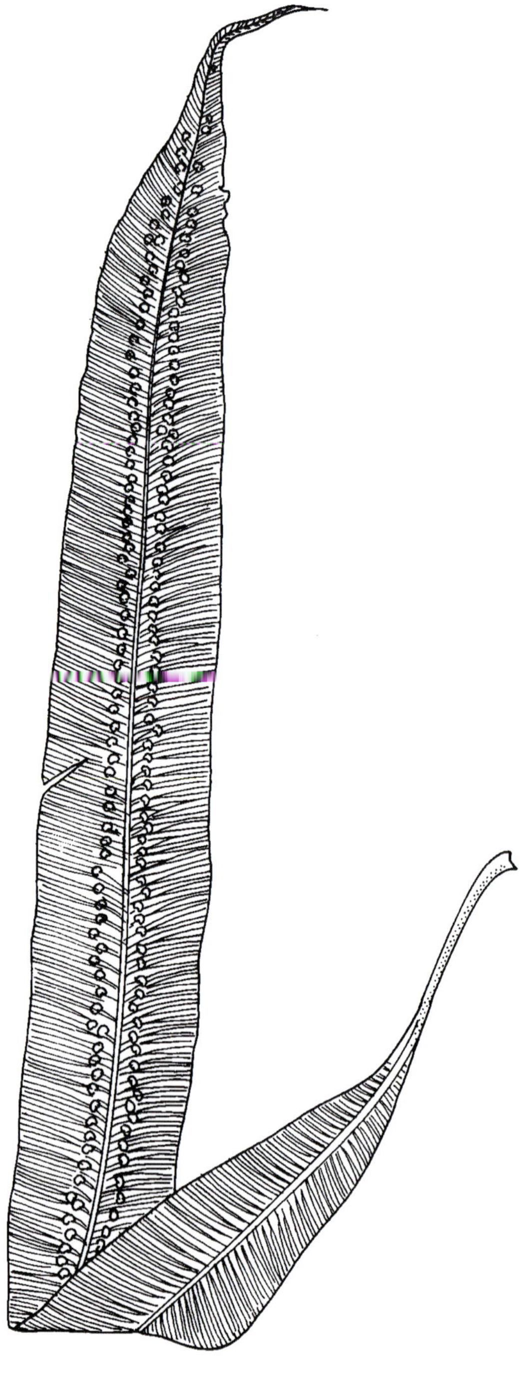

Fig 12.21 Pteridophytes. *Oleandra neriiformis*

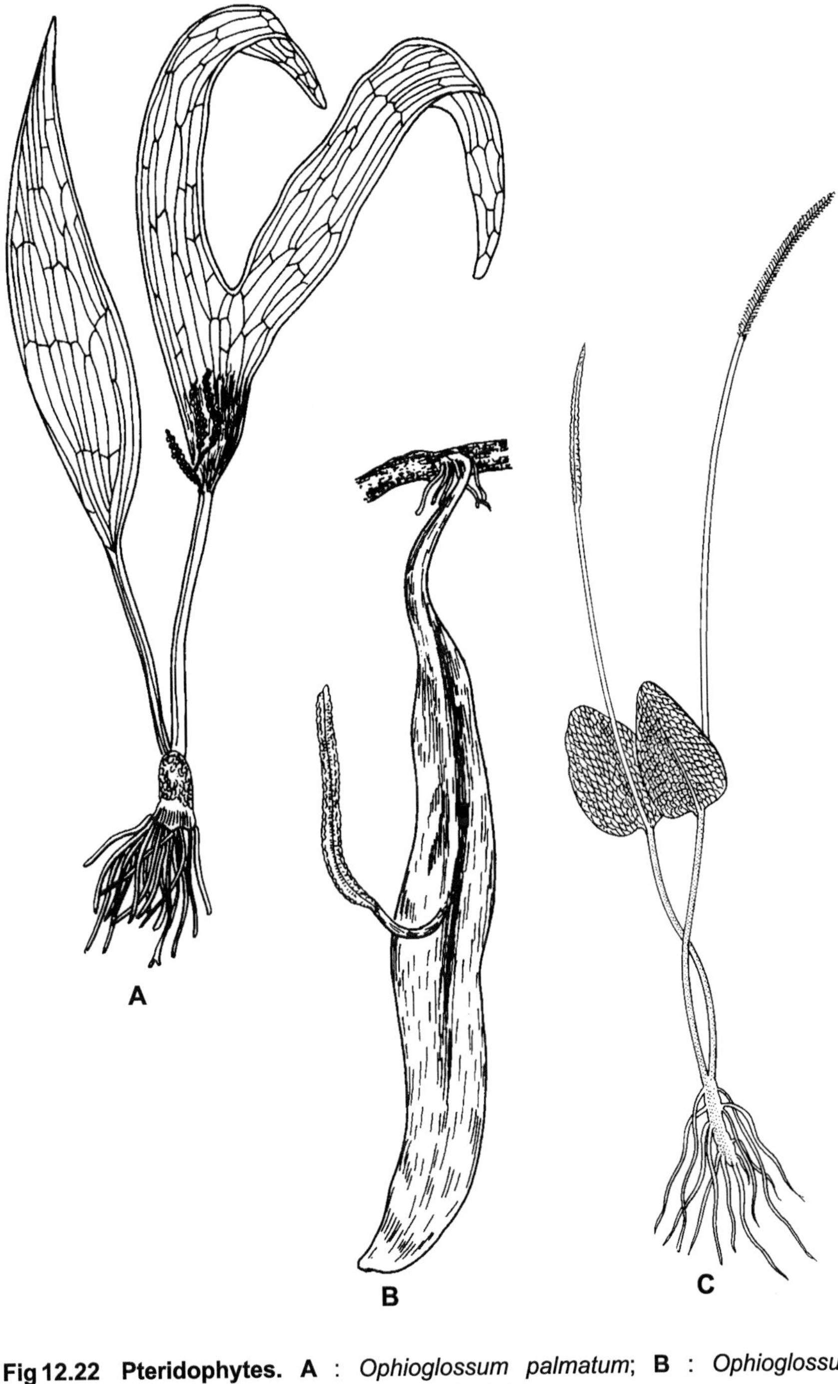

A

B

C

Fig 12.22 Pteridophytes. A : *Ophioglossum palmatum*; **B** : *Ophioglossum pendulum*; **C** : *Ophioglossum reticulatum.*

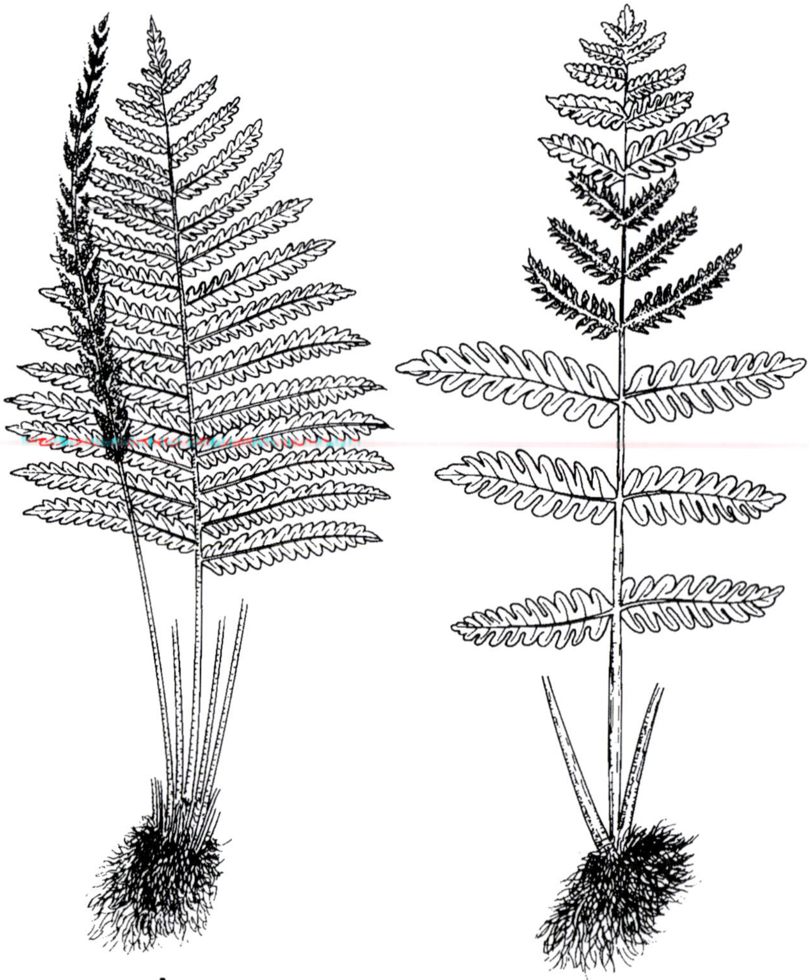

Fig 12.23 Pteridophytes. A : *Osmunda cinnamomea;* **B** : *Osmunda claytoniana.*

Fig 12.24 Pteridophytes. *Osmunda regalis*

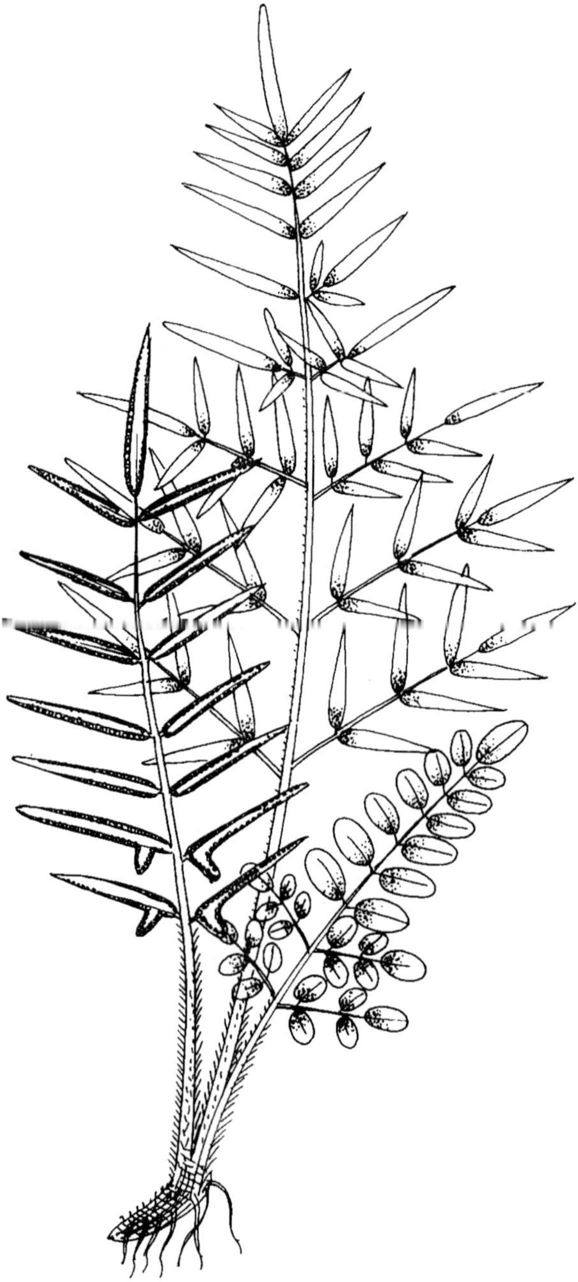

Fig 12.25 Pteridophytes. *Pellaea atropurpurea*

Fig 12.26 Pteridophytes. *Platycerium holtummi*

Fig 12.27 Pteridophytes. A : *Polypodium decorum;* **B** : *Polypodium glandulosum;* **C** : *Polypodium parvulum*

Fig 12.28 Pteridophytes. *Polypodium obliquatum*

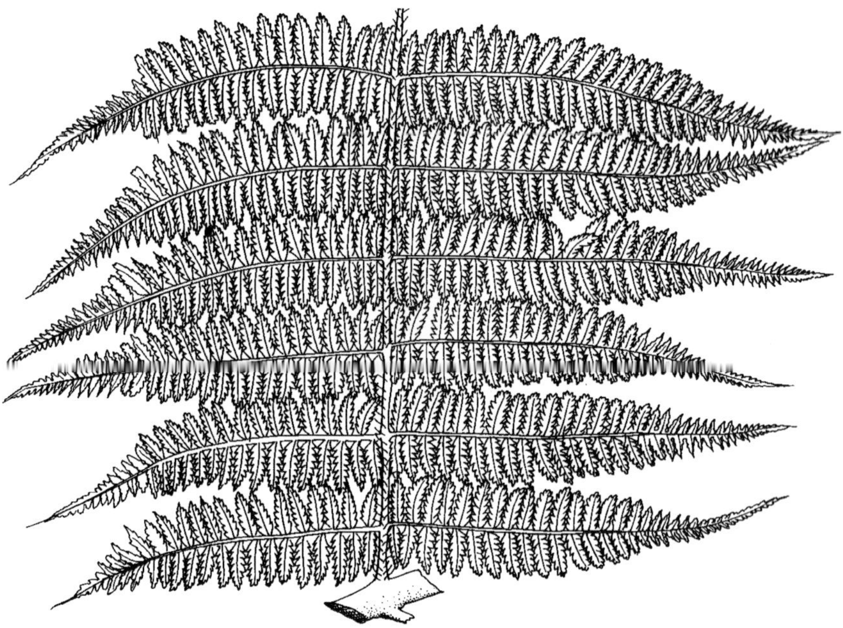

Fig 12.29 Pteridophytes. *Polypodium ornatum*

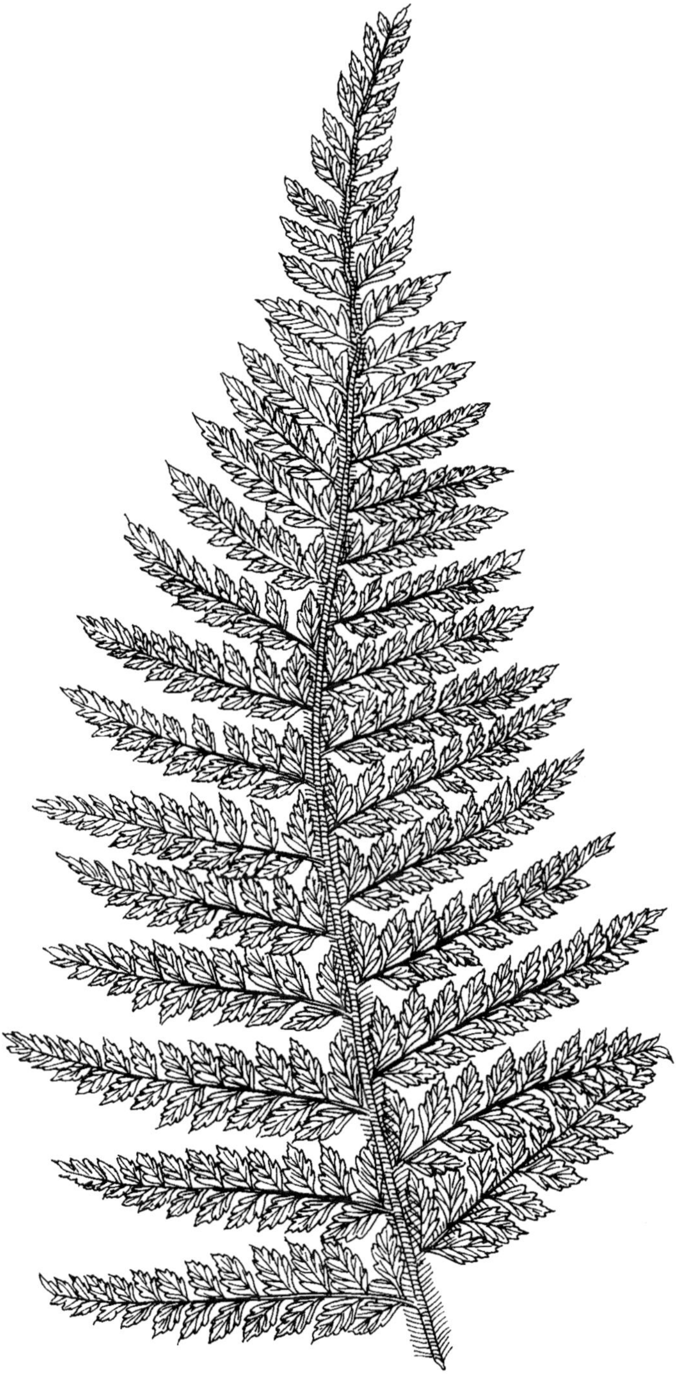

Fig 12.30 **Pteridophytes.** *Polystichum aculeatum*

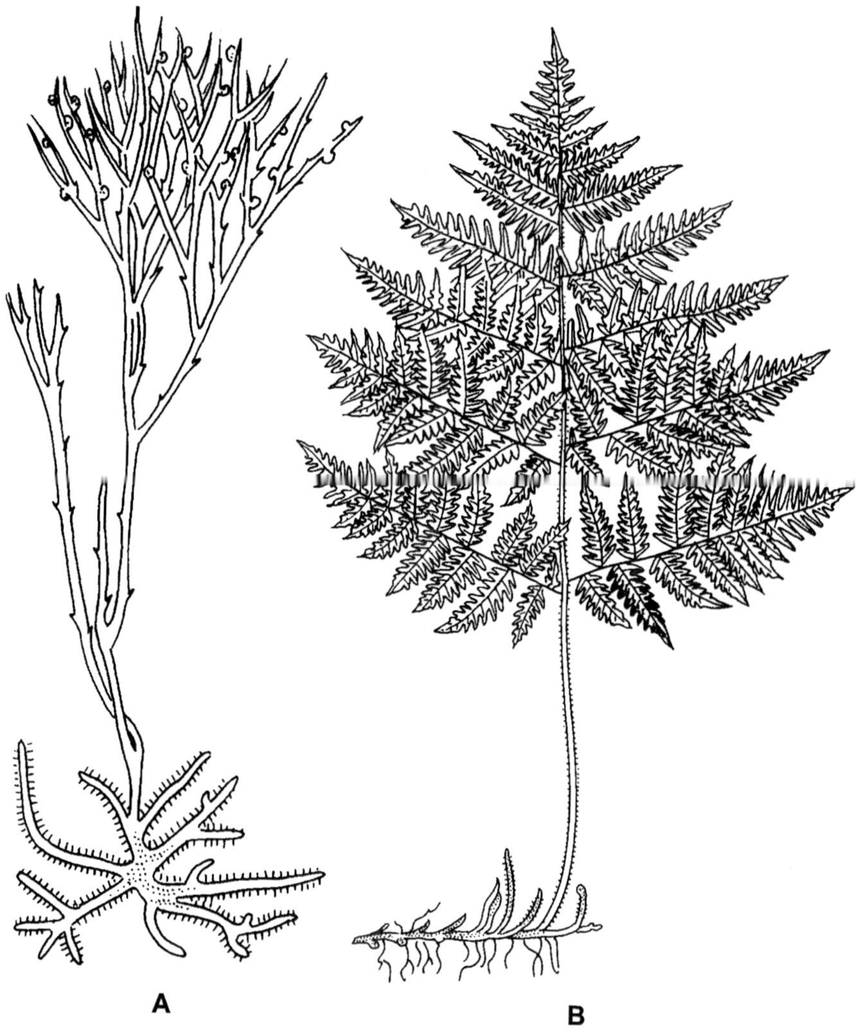

Fig 12.31 Pteridophytes. **A** : *Psilotum nudum*; **B** : *Pteridium aquilinum*

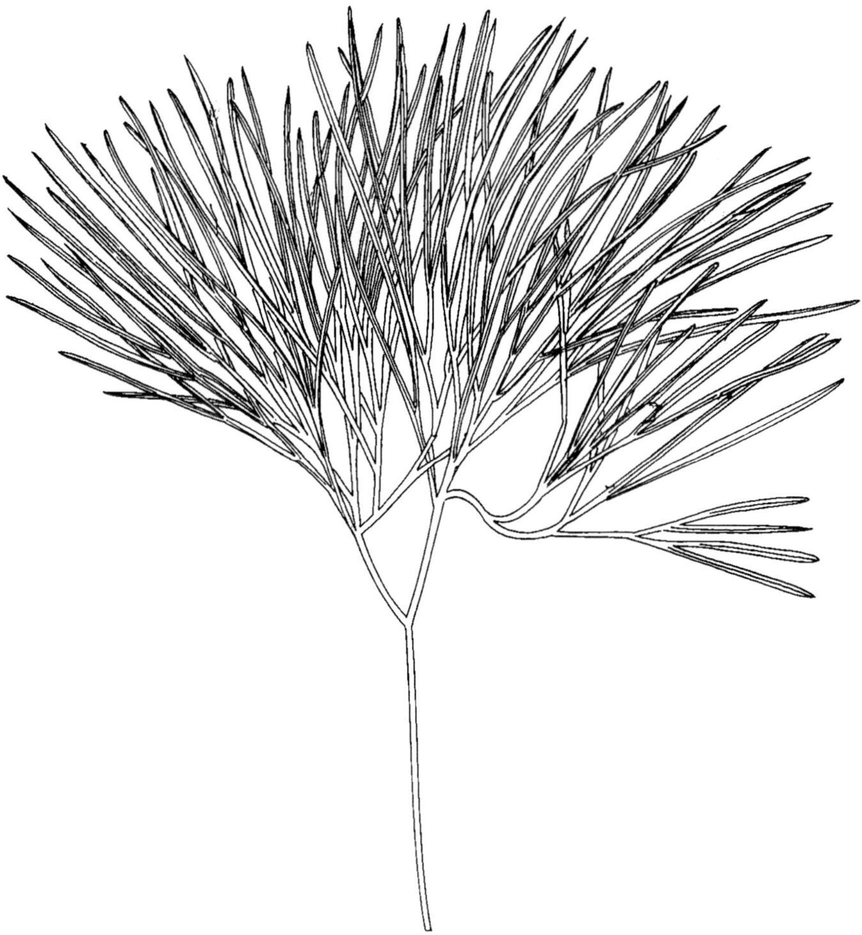

Fig 12.32 Pteridophytes. *Schizea dichotoma*

Fig 12.33 Pteridophytes. *Seleginella chrysocaulos*

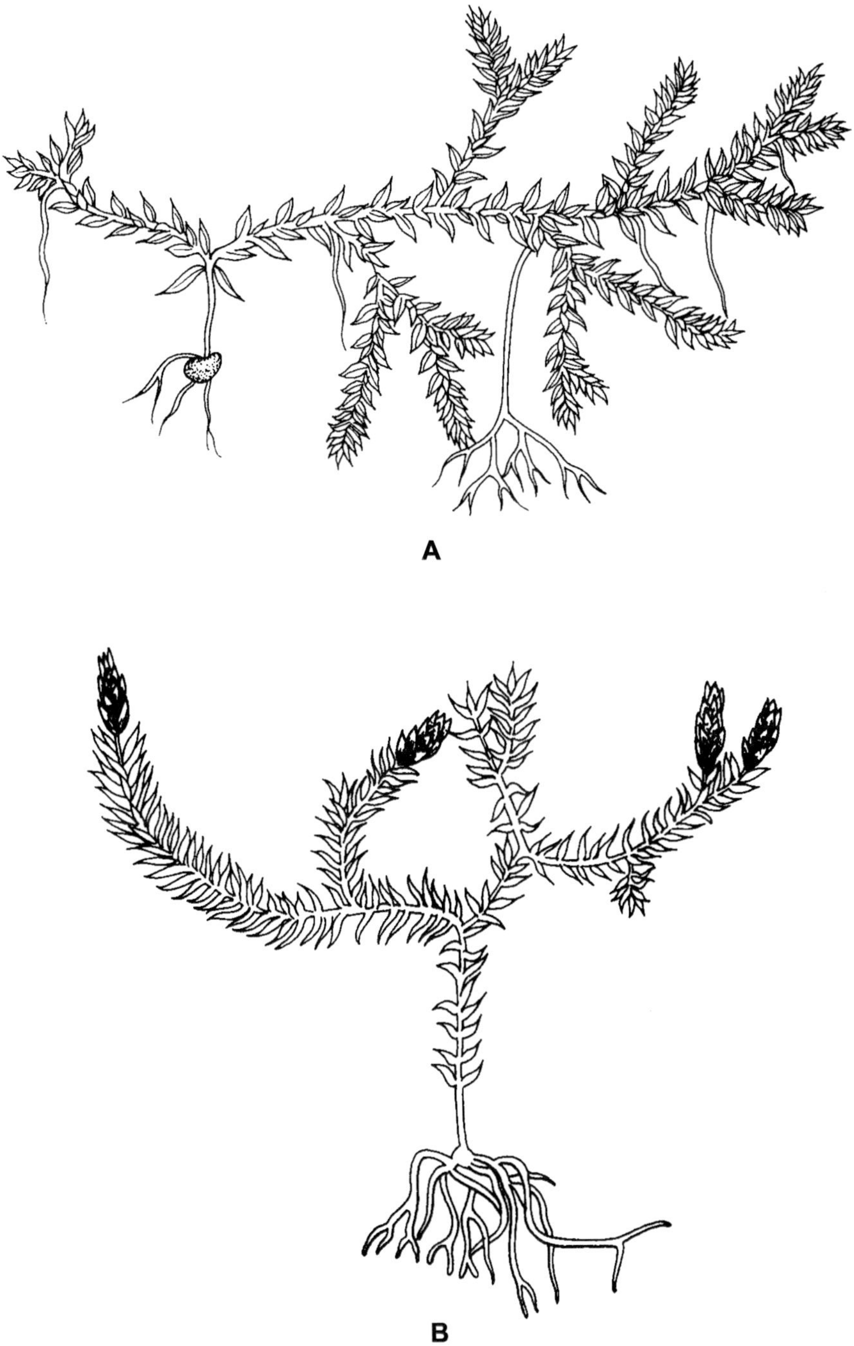

Fig 12.34 Pteridophytes. A : *Selaginella kraussiana*; **B** : *Selaginella selaginoides*

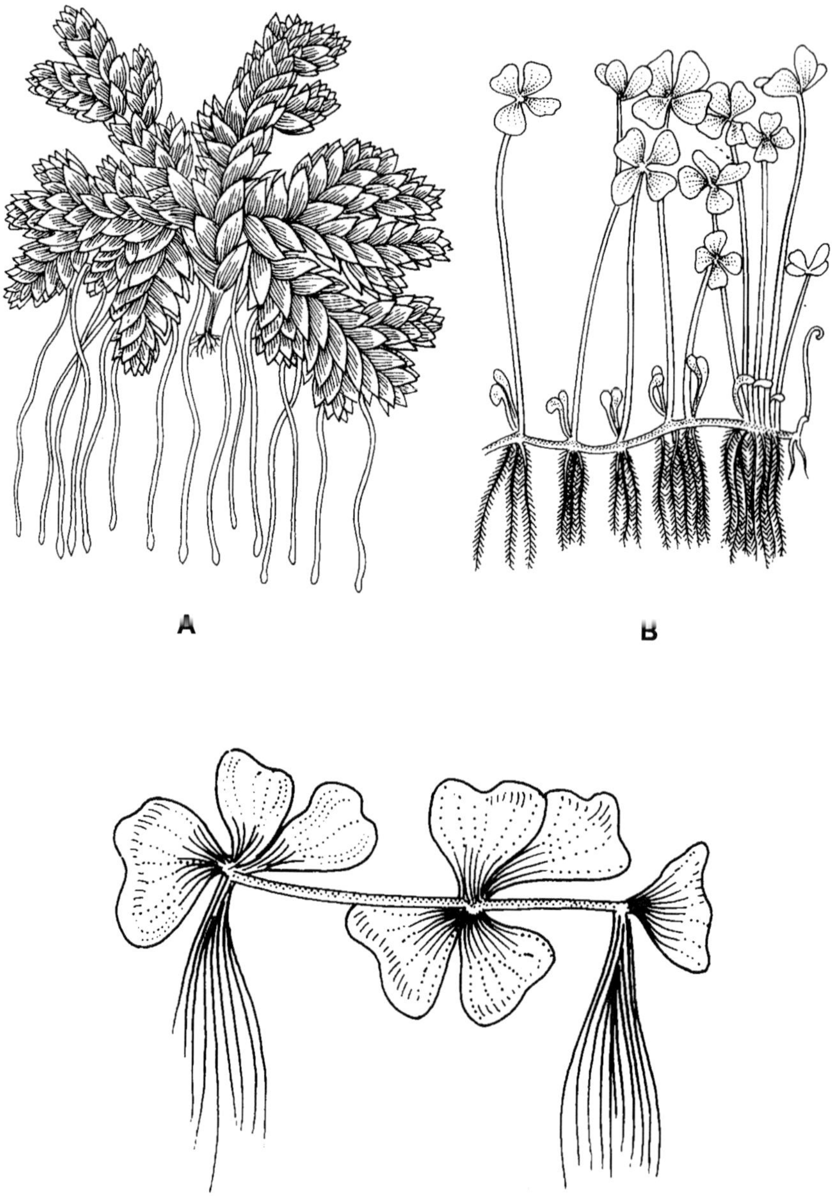

Fig 12.35 Water ferns. A : *Azolla filliculoides;* **B** : *Marsilea minuta;* **C** : *Salvinia auriculata.*

CHAPTER 13

GYMNOSPERMS

13.1 INTRODUCTION

Theophrastus first used the term 'Gymnosperms' in his book "Enquiry into plants" in 300 BCE, referring to those plants whose seeds are naked, as they are not enclosed in a fruit.

The seed plants (Spermatophyta) are divided into two major groups, the gymnosperms and the angiosperms, on the basis of protection offered to the ovule, before or after fertilization. In the angiosperms the ovules that become the seeds later on are enclosed in an ovary that would become the fruit. In the gymnosperms there is no ovary and the ovules and seeds are exposed, and so the gymnosperms form seeds but not fruits.

There are three major groups of living gymnosperms. The cyacads (*Cycas, Zamia, Encephalartos)* resemble the palm trees in general form and leaf architecture. The cone bearing species (*Pinus, Juniperus, Cupressus, Thuja)* called the conifers constitute the largest group. The third group Gnetales (*Gnetum, Welwitschia, Ephedra)* is the smallest group. The species of this group do not resemble each other. They also do not resemble the cycads or the conifers but resemble the angiosperms more, though this likeness is rather superficial.

Gymnosperms are evergreen trees and shrubs with graceful habit and attractive leaves and cones. The conifers among the gymnosperms are the most valued by the garden lovers and landscapists. They are also economically valued as a source of timber, resins, oil, paper and charcoal. Gymnosperms are also important as food and in medicine such as ephedrine and taxol. The conifers are the dominant components of forests in the temperate and semi-temperate regions.

Gymnosperms have some of the tallest trees known in the plant kingdom. *Sequoia sempervirens* is the tallest tree in the world reaching a height of 112 meters with a girth of 15 meters. But the oldest among the present day gymnosperms is *Sequiadendron giganteum,* (General Sherman tree) which is the largest living plant in the Sequoia National Park, California.

The living gymnosperms are trees or shrubs, but a few like *Gnetum ula* are woody climbers (lianas). While most of the gymnosperms are evergreen,

Taxodium and *Larix* shed leaves in a season (deciduous). The plants have a well-developed tap root system. In *Cycas,* the roots harbour the Cyanobacteria species of *Anabaena, Calothrix* and *Nostoc* and become the much branched coralloid roots while in some like the species of *Pinus* the roots are associated with mycorrhiza.

The stems are erect and aerial in most but underground in *Welwitschia,* which with only one species (*Welwitschia mirabilis*) is a bizarre gymnosperm endemic to the Namib desert in South West Africa. The stems may be erect and unbranched (*Cycas*), or branched (*Pinus, Thuja, Araucaria*). The branches bear two types of shoots, one with unlimited or indeterminate growth (long shoots) and the other with limited or determinate growth (short shoots).

The leaves are mostly simple or compound (cycads) and vary in size from minute scale leaves (*Ephedra*) to compound green, thick leaves as in *Agathis* or the cycads. The arrangement of leaves may be opposite, decussate (*Gnetum*), whorled (*Cedrus*), or spiral (*Podocarpus, Taxus*). The leaves show great variation in shape and size; needle like in *Pinus,* fern like in *Cycas,* fan shaped in *Ginkgo,* and strap shaped in *Welwitschia.* The venation may be reticulate (*Gnetum*) or parallel (*Welwitschia*) or open dichotomous (*Ginkgo*).

The male and the female reproductive structures are separately aggregated into cones (conifers) or strobili (*Ephedra*).

Illustrations of representative species of gymnosperms are given in Fig 13.1 to 13.12 to provide an over view of their appearance.

13.2 GYMNOSPERMS AS SOURCE OF FOOD

Araucaria

Araucaria angustifolia (Brazil pine) and *Araucaria bidwillii:* seeds are roasted and eaten.

Bowenia

Bowenia spectabilis: tuberous stem is cooked and eaten.

Cycas

The species of *Cycas* are widely used as food. The young leaves and seeds are consumed by man. The cycad seeds are still used as a source of starch and as a constituent of the bean paste 'miso' in Japan. Seeds are boiled with soy sauce and eaten.

Cycas circinalis, Cycas revoluta and *Cycas rumphii:* leaves are cooked and eaten. The stem starch obtained from species of cycas is called as sago, which is dried and made into flour, and used in various preparations.

Cycas media: seeds are made into paste and eaten as cake in Nicobar islands.

Cycas pectinata and *Cycas siamensis:* young unfolded leaves are cooked and eaten.

Cycas revoluta: seeds are used in making wine in Japan.

Dioon

Dioon edule and *Dioon spinulosum:* seeds yield starch which is used in making porridge.

Encephalartos

A beverage is prepared from the soft core of the stem. It is also used in Africa to make "Kaffir bread".

Ephedra

Mormon tea, a stimulant, is prepared from the North American species of *Ephedra* which is free from the alkaloid ephedrine.

Ephedra intermedia: ash from the burnt leaves is mixed with chewing tobacco in Pakistan.

Ginkgo

Ginkgo biloba: the endosperm of the seed is roasted and eaten.

Gnetum

Gnetum gnemon, Gnetum montana and *Gnetum ula:* seeds are roasted and eaten. Young leaves are cooked and eaten as vegetable.

Juniperus

Fruits are an important ingredient of gin and are also used in flavouring food, in pickling meat and for seasoning.

Macrozamia

Macrozamia spiralis: Stem is boiled and given as animal feed.

Pinus

Pinus cembra, Pinus cembroides, Pinus edulis, Pinus koraiensis, Pinus monophylla, Pinus parryana and *Pinus pumila:* seeds are edible. They are used in the preparation of candies, sweets and desserts. *Pinus* pollen is consumed as health food in various countries.

Pinus gerardiana (Chilgoza pine): the edible dry seeds are sold as 'Chilgoza'. The young cones are used as vegetable. Seed kernels are used in desserts, chocolate, soups, ice creams, pine butter and marzipan.

Pinus pinea: seeds are used in soups, desserts and sweets.

Stangeria

Stangeria paradoxa: the fleshy testa of the seeds is edible.

Taxus

Taxus baccata and *Taxus brevifolia*: fleshy aril is edible.

Torreya

Torreya nucifera: the seeds yield an edible fatty oil.

13.3 MEDICINAL USES OF GYMNOSPERMS

Abies

Abies balsamea (Silver fir tree): canada balsam obtained from the plant is applied externally for dermatitis, wounds and rheumatic pains. The decoction is applied on sore muscles. Taken internally, the decoction cures rheumatism, kidney disorders, inflammation of the bladder, typhoid fever and capillary bronchitis.

Abies webbiana (*Abies spectabilis*): leaves are used as carminative, expectorant, stomachic, tonic, astringent; useful in treating cough, cold, phthisis, asthma, bronchitis and catarrh of the bladder. Leaf juice is used as antiperiodic.

Cedrus

Cedrus atlantica: oil is used in treating bronchitis, tuberculosis, ulcers, skin diseases, gonorrhoea and as diaphoretic.

Cedrus deodora: wood decoction is used as diaphoretic, diuretic, and carminative. It is useful in treating fever, flatulence, pulmonary and urinary disorders, rheumatism, piles, gravel in kidney and as antidote for snake bite. Bark is astringent and useful in treating fever, diarrhoea, dysentery, piles, rheumatism and tuberculosis. The oil is applied for skin diseases, inflammations, dyspepsia, urinary discharges, bronchitis, itching, elephantiasis, leucoderma and piles. Resin is used for ulcers and skin diseases.

Cephalotaxus

Cephalotaxus harringtonia var. *drupacea:* homoharringtonine from the plant is in clinical trials against leukemia. (Luo *et al.*, 2004)

Cupressus

Cupressus cones are used as astringent in treating diarrhoea, dysentery, bleeding haemorrhoids, haemorrhages, menstrual disorders and spitting of blood. The decoction is used as mouth wash for spongy gums and in pyorrhea.

Cupressus sempervirens: wood is used as astringent and anthelmintic.

Cycas

Cycas circinalis: bark and seed paste mixed with coconut oil is used as poultice for sores and swellings. Seed flour is given for haemorrhoids and in bowel complaints. Juice of tender leaves is useful for flatulence, vomiting and skin diseases. Leaflets are crushed in rice water and mixed with clay for dressing ulcerated wounds and boils. Seed decoction is emetic and purgative. Paste of megasporophylls (young seed bearing structures) is applied on loins for nephritic pain.

Cycas revoluta: is used as expectorant and tonic. Seeds are pounded into paste and mixed with coconut oil for dressing sores, wounds and skin diseases. Seed tincture is used for headache, giddiness and sore throat.

Cycas rumphii: resin is applied to malignant ulcers. The scales are used as anodyne, galactagogue and stomachic. In Indonesia seed decoction is given for asthma.

Dacrydium

Dacrydium franklinii: yields an oil, which is used in treatment of cuts, wounds, burns and *Tinea* infection.

Ephedra

Ephedra: has a long history of use in oriental medicine. It is used in treating hay fever, chills, headache, edema, syphilis, rheumatism, nasal congestion, liver diseases, joint pain, cold and cough. It improves respiratory function and is anti-inflammatory. Root is used in rheumatism.

The plant contains the alkaloids ephidrine and pseudoephidrin which work as central nervous system stimulants. They also stimulate heart muscles, increase blood pressure and heart beat rate and are bronchodilators. They are diuretic, decrease blood sugar level and correct calcium phosphorus imbalance, stimulate contraction of uterine muscles.

Ephedra gerardiana: tincture is used as cardiac and circulatory stimulant; useful in headache and hepatic disorders. Decoction of the stem and roots is used in rheumatism and syphilis. The juice of the berries cures respiratory affections.

Ephedra intermedia: some believe it to be the source of the drink 'Soma', which was consumed by sages in ancient times to promote alertness, energy and as appetite depressant and astringent.

Ginkgo

Ginkgo biloba: is protective against cardiac ischemia, thrombosis and peripheral vascular disorders. It increases blood vascular elasticity in patients with pathologic blood flow disorders and also useful in treating obliterative arterial diseases of the lower limbs by improving walking performance. It is used as a nutritional supplement and in the treatment of dementia, circulatory disorders, asthma, cerebral inactivity, anxiety, stress, mood swings, hearing disorders, lack of memory, cerebral ischemia, and peripheral nerve damage. (Issing *et al.,* 2005; Bedzan *et al.,* 2005; Zeng *et al;* 2005.)

The fresh leaves which are antibacterial and antifungal are used in treating asthma and chill blains. Dried leaves are useful in treating stress and anxiety, to arrest fibrosis, promote hair growth, and cure impotency.

Seeds are used as digestive and in the treatment of allergic reactions and inflammations. *Ginkgo* is also known to reduce clastogenic factors, deterioration of lipid profiles, functional and morphogenetic retina impairments and inhibits platelet activation (Boelsma *et al.,* 2004).

Evaluation of clinical studies has proved the efficacy of EGB 761 (special extract of *Ginkgo*) in treating peripheral arterial occlussive disease (PAOD). (Horsch and Walther 2004); used in the treatment of diabetic nephropathy, regulation of blood lipids, impaired renal function, thrombogenic coagulation in chronic peritoneal dialysis and pulmonary interstitil fibrosis (Kim *et al.*, 2005; He *et al.,* 2005) and to decrease the symptoms of acute mountain sickness (AMS) (Chow *et al,* 2005).

Gnetum

Gnetum gnemon: is useful in gynaecological disorders and as antidote for snake bite.

Gnetum scandens: stem and root are used as antiperiodic.

Juniperus

Juniperus is used as carminative and in treating arthritis, bronchitis, kidney disorders, dropsy, urinogenital disorders and cutaneous affections. Bark is expectorant, astringent and used in checking excess mucous secretions, scurvy, tonsillitis, laryngitis and rheumatism.

Resin from the plant is used in treating cold, cough, tuberculosis, influenza and indigestion. Stem decoction is used in curing piles and bleeding gums.

Berry is used as diuretic, anthelmintic, carminative and antiseptic. Used in treating sluggish liver, cough, gout, convulsions, dropsy, epilepsy, sciatica, breathing disorders, muscular cramps and to strengthen weak nerves. It increases urine output and cures vaginal inflammation and stomach ulcers. An infusion of berries is used as an antidote for snake bite and bee stings.

Root and leaf decoction is taken to relax uterine muscles during childbirth and given to strengthen the uterine muscles after child birth.

Juniperus communis: Plant is used as appetizer, carminative, anthelmintic, alexipharmic, laxative; useful in diarrhoea, abdominal pain, strangury, diseases of the spleen, ascites, piles, bronchitis, indigestion, constipation and vaginal discharges;berry and oil used as diuretic, carminative, and stimulant. Useful in curing dropsy, gonorrhoea, gout, leucorrhoea and skin disorders.

Juniperus excelsa: smoke from the branches is used in delirium during fever.

Juniperus macropoda: berry and oil used as diuretic, carminative, and stimulant. Useful in curing dropsy, gonorrhoea, gout, leucorrhoea and skin disorders.

Juniperus oxycedrus: oil is used in curing eczema and other skin diseases.

Juniperus recurva: wood smoke is used as emetic. Decoction is used in curing kidney diseases.

Juniperus sabina: is used as emmenogogue and anthelmintic.

Larix

Larix americana: is useful in melancholy, lung disorders, profuse menstruation, diarrhoea, dysentery, rheumatism, bronchitis, and asthma. It is an effective antidote for poisonous insect bites. Decoction of the bark is used as a wash in cleaning ulcerated sores and as a poultice it is used for haemorrhoids, abscesses, and gangrene. Decoction taken internally cures bronchitis, kidney and bladder trouble (Hutchens, 1991). Arabinogalactans in the plant enhance vascular permeability and decrease inflammation (Benedict, 1999).

Picea

Picea abies: resin is used as stimulant and rubifacient.

Picea excelsa: burgundy pitch, an oleo resin obtained from the plant is used for dressing wounds.

Pinus

Pinus: wood decoction is used as stimulant, thermogenic, emollient, haemostatic, anthelmintic, digestive, liver tonic, rubifacient, diuretic, and diaphoretic. Useful in inflammation of the body, cough and ulcers. It is also useful in ophthlmia, otopathy, pharyngopathy, foul ulcers, haemorrhages, haemoptysis, dyspepsia, flatulence, strangury, bronchitis, inflammation, skin diseases, pruritus and giddiness. As it is antibacterial, decoction is used to wash maggot infested wounds, abscesses and gonorrhoeal sores. Oleoresin from the plant is used as antidote for snake bite and scorpion sting.

Turpentine derived from species of *Pinus* is used to relieve tooth ache, musclar pains, rheumatic pains, cough and cold.

Pinus gerardiana: seeds are used as anodyne and stimulant. Seed oil is used as poultice for wounds and ulcers.

Pinus griffithii, Pinus merkussii: resin is used as stimulant and stomachic. Useful in treating gonorrhoea. As a plaster it is applied to buboes and abscesses for suppuration.

Pinus excelsa and *Pinus longifolia*: Plant is useful in treating diseases of the eye, ear, throat, blood, skin, bronchitis, giddiness, ulcers, inflammation and itching; gum is purgative, carminative, aphrodisiac, diuretic, anthelmintic and analgesic. Useful in diseases of uterus, vagina, eye; given for curing dyspepsia, ulcers, scabies and inflammation; Wood is used as stimulant and diaphoretic and given in cough, ulcers and burning of the body.

Pinus roxburghii: is used in boils, bone fracture, cracks in sole, leprosy, skin disorders, sprain, swellings, ulcers and in urinary disorders. It yields an oleoresin useful in treating chronic bronchitis, flatulence, colic, haemorrhages in nose and gums, otalgia, odontalgia, hepatopathy, spleenopathy, urethrorrhoea, gonorrhoea, scabies, epilepsy, lumbago, haemorrhoids and tuberculous glands. It is externally applied as rubifacient in deep seated inflammations and rheumatic pains and as plaster applied to buboes to hasten suppuration.

Pinus strobus: yields a resin which is used for sore throat. Heated resin is used as an application for sores, boils, swellings, insect bites, sciatic pain, muscular soreness and to relieve chest congestion in pneumonia. Bark is useful in bronchitis, catarrh, rheumatism, scurvy, chest affections, tonsillitis, laryngitis, influenza and tuberculosis (Hutchens, 1991).

Pinus wallichiana: is useful in curing abscesses and ulcers.

Podocarpus

Podocarpus neriifolius: is used in rheumatism.

Taxodium

Taxodium resin is carminative, diuretic and vulnerary. Bark, leaves and roots are used in diarrhoea and bronchitis. Application of resin topically heals wounds and ulcers (Ambasta, 1986).

Taxus

Taxus baccata and Taxus brevifolia: leaves are prescribed in hysteria, asthma, rheumatism, epilepsy and nervousness (Nadkarni and Nadakarni, 1991). Used as antispasmodic, emmenogogue, astringent, aphrodisiac and stomachic. Tincture of the leaves is used to treat rheumatism, liver and urinary tract disorders. It is also used in treating breast, lung and ovarian cancer. A tincture of young shoots is used for headache, failing pulse, biliousness, diarrhoea and giddiness. Seeds are used as sedative (Chopra *et al.,* 1956). Fleshy aril is useful as carminative, expectorant, diuretic and laxative. The Japanese use the needles to treat diabetes and chest complaints. Leaf and berry with clarified butter cures cancer and tumours.Wood decoction is used in hydrophobia.

Taxol (an anti-cancer compound) occurs in the leaves of *Taxus* species. It is demonstrated to be effective against Koposi's sarcoma, bladder, prostate, head, neck, cervical and endometrial cancer (Kowalski *et al.,* 1997).

Thuja

Thuja occidentalis: is useful in treating muscular aches, chronic cough, fever and menstrual pain. It is stimulant, diuretic, expectorant, emmenogogue and anthelmintic. Oil of *Thuja* is used as tonic, antipyretic and diuretic.

13.3.1 Gymnosperms in The Treatment of Cancer and Tumour

Gymnosperms are used in traditional medicine for treating various types of cancer and tumours. (Table 13.1)

Table 13.1 Gymnosperms in treatment of cancer and tumour

Species	Type of cancer/tumour
Abies balsamea	Cancer; tumour; cancerous ulcers
Cedrus deodora	Abdominal tumours
Cedrus libenensis	Cancer
Cupressus sempervirens	Polyp of the nose; tumour of the penis and testes; cancer of the uterus, breast polyp; eye tumour; induration of the joints.
Cycas rumphii	Malignant ulcers
Ginkgo biloba	Cancer
Gnetum gnemon	Tumours
Juniperus bermudiana	Warts and tumours
Juniperus chinensis	Cancer
Juniperus communis	Cancer of kidneys and legs
Juniperus excelsa	Cancerous ulcers and nasal polyps
Juniperus oxycedrus	Malignant ulcers and hard tumours
Juniperus sabina	Indurations of the joints; venereal warts; hardness of genitalia; nasal polyps; condylomata and cancer.
Juniperus squamata	Cancer
Juniperus virginiana	Cancer
Pinus edulis	Finger tumours
Pinus monophylla	External cancers
Pinus palustris	Abdominal tumours and cancer
Pinus pinea	Indurated breast and condylomata; tumours of kidney, bladder, throat and uterus.
Pinus roxburghii	Parotid tumours; tumour of the testicles; malignant ulcer; cancer of the liver, spleen, uterus, breast; cancerous sores, tumours.

Pinus sylvestris	Anal condylomata; nasal polyps and tumours of the feet.
Taxus baccata	Breast, cervix, uterus and liver cancer
Taxus brevifolia	Tongue cancer
Taxus cuspidata	Cancer
Taxodium distichum	Tumours
Taxodium mucronatum	Leg and feet tumours
Tetraclinus articulata	Cancer and hard tumours

Source: Hartwell, 1982

13.3.2 Medicinal Uses in Homoeopathy

Abies

Abies canadensis: is used for affections of mucous membrane, canine hunger, distention of stomach, flatulence, uterine displacement, chills, night sweat, and debility.

Abies nigra: is used for gastritis, dyspepsia, pain in the stomach, cough, choking sensation in the throat, cutting pain in the heart, bradycardia, rheumatic pain, bone pain and chronic intermittent fever.

Cupressus

Cupressus berries and leaves as tincture are used in treating keloid tumours and warts.

Juniperus

Juniperus communis: is used for catarrhal inflammation of kindney, dropsy, suppressed urine, swollen and painful testicle, scrotal itching, suppression or delayed menstruation, hysteria, loss of appetite, constipation, menstrual colic, breathing problems, heart palpitation, back pain radiating towards thighs, joint pain, numbness in the limbs and fever.

Pinus

Pinus needles are used for treating weak ankles, scrofula, emaciation of lower extremities, stiffness, gouty pain in joints, cramps in calves, nettle rash and itching of the nose.

Turpentine obtained from *Pinus* species is used against tympanitis, inflammation of the kidney with haemorrhage, Bright's disease, drowsiness, strangury, chilblains, pain in the head, sore nostrils, neuralgia of the eye, burning tongue, nausea, vomiting, heat in the epigastria, urethritis, purpural peritonitis, metrorrhagia, haemoptysis, bloody phlegm, kidney pain, acne, erythema, vesicular eruption, urticaria, purpura, dropsy, sore muscles, fever, profuse cold, typhoid, stupor and delirium.

Taxus

Taxus baccata: is used for pustular diseases of the skin and night sweats, supra orbital and temporal pain, lacrimation, dilated pupils, puffy face, feeling of pins and needles in the pit of stomach, itching pustules and erysipelas.

Thuja

Thuja twigs are used in treating distended abdomen, angina pectoris, anal fistula, fissures, asthma, balanitis, chorea, condylomata, constipation, convulsion, diarrhoea, dysmenorrhoea, ear polyps, anuresis, epilepsy, eye-tumour, glandular inflammation, flatus, frontal sinus, ganglion, gonorrhoea, haemorrhage, haemorrhoids, hair fall, headache, hernia, herpes zoster, levitation, myopia, neck polyps, paralysis, post-natal catarrh, rheumatism, rickets, sciatica, seminal emissions, syphilis, teeth caries, tongue ulcers, tooth ache, tumours, vaginitis, warts and whooping cough. It is a leading remedy for cancers and tumours (Boericke, 1991).

13.3.3 Uses in Chinese Medicine

Digenia simplex: is used as anthelmintic and laxative.

Ephedra sinica: Stem and root are used in treating bronchial asthma and hay fever.

Thuja orientalis: is used as astringent, antipyretic and emmenogogue; seeds are used as sedative in neurasthenia, palpitation and insomnia.

Torreya grandis: is used as vermifuge.

13.3.4 Uses in Ethnomedicine

Cedrus deodora: is anthelmintic; used for rheumatism and ulcers.

Cycas rumphii: is galactagogue; used for stomach ache.

Ephedra gerardiana: is a blood purifier; used to treat asthma, headache, hepatic disorders and rheumatism.

Ephedra saxatilis var., *sikkimensis*: used to treat asthma.

Gnetum gnemon: used for gynaecological complaints and as an antidote for snake bites.

Juniperus recurva (=Juniperus macropoda): used to treat kidney disorders.

Pinus roxhburghii: used for boils, bone fractures, cracks in sole of feet, leprosy, skin disorders, sprain, swellings, ulcers and urinary complaints; used as an antidode for snake bite.

Pinus wallichiana: used to treat abscesses, dislocated joints and ulcers.

Podocarpus neriifolius: used for rheumatism.

Taxus baccata: used for swellings; worn as a charm to prevent conception.

Thuja tetragona: used to treat piles.

13.4 TOXICOLOGY

Cycas

Cycasin, a toxic glucoside occuring in the seeds, trunks and leaves of *Cycas* causes intoxication in humans, which is attributed to improper washing and cooking before eating. The symptoms of intoxication include nausea, vomiting and unconsciousness.

Feeding the test animals like mice, rat, guinea pigs and hamster with cycad meal showed damage to intrahepatic bile duct resulting later in bile duct carcinoma, liver tumour, hepato-cellular carcinoma, adenoma, renal carcinoma, adenomatous polyps of the colon and cholangioma (Hirono, 1987).

Ephedra

In large doses, ephedrine (alkaloid from *Ephedra*)causes nervousness, headache, insomnia, palpitation, skin flushing, vomiting, increased blood pressure, restlessness coupled with skin irritation. The n-nitrosamines in the plant are reported to cause lung, stomach and liver cancers.

Ginkgo

Though there are no reports of severe side effects, in some it may cause headache, dizziness, heart palpitation and skin reaction. Contact with fleshy fruit results in blisters, itching and dermatitis. Ingestion of the fruit results in rectal burning and erythema. *Ginkgo* pollen may cause allergy.

Juniperus

Juniperus pollen causes skin and respiratory allergy. Large doses of berry infusion may cause renal damage and convulsions. The oil can cause gastric and kidney irritation.

Pinus

Ingestion and inhalation of turpentine obtained from species of *Pinus* results in headache, insomnia, cough, vomiting, haematuria and albuminuria. It causes skin rashes and irritation in repeated application. *Pinus* pollen cause respiratory and dermatalogical allergy.

Taxus

Excluding the red aril, all parts of the plant are poisonous. Following ingestion, dizziness, dry mouth, abdominal cramps develop along with skin rash, bradycardia, hypertension, dyspnoea, coma and death may occur due to cardiac or respiratory failure.

Docetaxel and paclitaxel, the anticancer drugs obtained from *Taxus* when injected intravenously result in skin allergy, rashes, anaphylaxis, drop in blood pressure, bronchospasm causing breathing problems, fluid build up around the heart, low neutrophil and low platelet count, anaemia, bone pain and hard to control bleeding. Myelosuppression is the most common adverse effect. These side effects may not be that severe if the drug is administered slowly over a period of several hours.

Thuja

Thuja plicata saw dust causes asthma and rhinitis in wood workers.

13.5 OTHER USES OF GYMNOSPERMS

Abies

Abies alba: is an important timber tree of Europe, the wood of which is used in general carpentary. It is white or yellowish white, light, soft with distinct growth rings. It is used in making musical instruments, wooden boxes, paper pulp, planks and boards. Treated with preservatives it is used in making telephone poles. The most pleasant aromatic oil is extracted from the plant which is used in cosmetics and perfumery.

Abies balsamea: wood is light and used in making boxes and paper. It yields a resin Canada balsam which has a very high refractive index nearing that of glass and is the most preferred mounting medium for biological specimens for microscopy. It is also used as cement for lenses in optical work.

Abies firma: wood is used in making wooden boxes.

Abies magnifica: is the best soft grained and easy to handle soft wood used for making furniture, boxes and cigar cases..

Abies sibirica: yields an oil (Siberean fir needle oil) used in scented soaps, bath salts, room sprays and deodorants.

Abies spectabilis: wood is pale yellow, lustrous, straight grained and is used for making camp furniture, plywood, paper pulp and slanting roofs of cottages. It yields a purple or violet dye used in dyeing industry.

Abies amabilis, Abies concolor, Abies grandis and *Abies procera:* wood is used for interior woodwork, boxes and general carpentary.

Agathis

Agathis wood is heavy and strong and used in building boats and railway carriages.

Agathis alba (Amboina pine tree): wood is used in making printing and writing paper.

Agathis australis: wood is fine, hard and even textured with lustrous surface. It is used in making cabinets, boxes, plywood and battery separators.

Agathis robusta: wood is used in making writing, printing and wrapping papers. Yields an oleoresin used in paints, polishes, inks and lacquer and as adhesive.

Agathis vitiensis (Fiji pine): yields wood, which is used for making masts, boats and ships. It is used for flooring houses.

Araucaria

Many species of *Araucaria* are grown as ornamental trees in gardens and the twigs are used as decorations during Christmas.

Araucaria araucana: is the Christmas tree or the Monkey puzzle tree.

Araucaria angustifolia: resin is fragrant. When mixed with wax it is used in making aromatic candles. Wood is suitable for furniture, boxes, cabinets, railway coaches, flooring and agricultural implements.

Araucaria bidwillii, Araucaria columnaris, Araucaria cunninghamii, and *Araucaria heterophylla:* wood is suitable for furniture, boxes, cabinets, railway coaches, flooring and agricultural implements.

Callitris

Callitris wood is olive brown in colour and is used in making decorative furniture.

Cedrus

Cedrus atlantica: yields an oil which is used as odour fixative.

Cedrus deodora : has a strongly scented heart wood that is resistant to insect and fungal attacks. It is used in making window panes, furniture, railway sleepers and carriages. Yields cedar oil which is used in perfumery, soap, pharmaceuticals and as oil for oil immersion lens of the microscope.

Chamaecyparis

Chamaecyparis formosensis: wood has a smooth surface. It is used for making musical instruments.

Chamaecyparis lawsoniana: wood is light, fine and fragrant. It is used in making fancy articles, linen drawers, interior work and furniture.

Chamaecyparis obtusa and *Chamaecyparis pisifera:* yields a strong straight grained fragrant, durable wood which is valued for modeling, panelling, cabinets and furniture. The bark is used in making roofs. The sapwood is dried and plaited into mats and hats.

Chamaecyparis nootkatensis: wood is ideal as battery separators. It is used for making, poles, boats and furniture.

Cryptomeria

Cryptomeria japonica: yields a coarse grained, fragrant, strong, durable, easy to work wood. The wood is employed in panelling, furniture, high-grade paper, ships, bridges, boxes and furniture. The bark is stripped and used for roofing houses. It is grown as an ornamental tree.

Cunninghamia

Cunninghamia lanceolata: yields a durable fragrant yellow wood which is used for building wooden houses, indoor work, boxes and coffins as it is excellently preserved. Bark is used for roofing in wooden houses and paper making.

Cupressus

Cupressus glauca (Mexican cypress): leaves yield an aromatic oil used as a fixative.

Cupressus lusitanica: wood is used for paper making.

Cupressus macrocarpa: wood is yellowish brown, and durable. It is used in making furniture, fences and general carpentary. As the wood is insect repellent and does not exude resin, it can be painted and polished. It is burnt as incense in temples.

Cupressus sempervirens (Mediterranean cypress): An oil obtained from the wood is used in perfumery and soap. Wood is suitable for making boxes, drawers and furniture.

Cupressus torulosa (Himalayan cypress): wood is used for construction of door panels, window frames, pen holders, flooring and cheap grade pencils.

Many species of *Cupressus* are grown as ornamental plants.

Cycas

Cycas wood is used to make small decorative boxes, penholders, bowls, and bottle stands. Seeds are used in making fancy snuff boxes. Fine fibres from the base of the leaves are used for stuffing pillows and mattresses. The leaves are woven into hats, mats, thatches and broom. The bark yields a strong fibre used in making twine threads. Dried leaves are used as manure.

Cycas circinalis: seeds are used as fish poison. Ropes and twines are made from the bark. The gum obtained from the trunk is used as adhesive.

Cycas circinalis and *Cycas revoluta:* gum from the injured stems and megasporophylls is used as adhesive. Dried leaves are used as decorative material for floral wreaths. It is used to make artificial palm trees.

Species of *Cycas* are grown as ornamental plants.

Dacrydium

Dacrydium wood is highly resistant to insects and fungi. It is used in building wooden houses, planks, boards, packing cases, doors and window frames.

Dacrydium biforme, Dacrydium cupressinum and *Dacrydium intermedium:* wood is valued for furniture and railway sleepers. It is highly resinous and tannin rich. The bark yields a red dye used in dyeing.

Dacrydium elatum and *Dacrydium franklini:* wood is used in making packing cases, furniture and boxes.

Encephalartos

Encephalartos leaves are dried and woven into hats, mattresses and made into brooms.

Gingko

Gingko biloba: is grown as an ornamental tree in gardens. It is worshipped in temples of China and Japan. Wood of *Ginkgo* is used in making chess boards and trays (Ambasta, 1986).

Gnetum

Gnetum gnemon and *Gnetum ula:* yields a fibre of high tensile strength, which is used in making ropes, fishing nets and papers.

Juniperus

Juniperus bermudiana and *Juniperus procera:* wood is cedar scented, durable and fine textured. It is used in pencil making, ship building and furniture. Cabinets made from the wood are highly priced.

Juniperus chinensis (Chinese juniper): wood is used in cosmetics and in making furniture.

Juniperus communis: bark is useful in tanning. The wood is used for veneering. Fruit yields dye. Young twigs are burnt as incense.

Juniperus excelsa: yields a durable wood preferred for building posts, railway sleepers and furniture.

Juniperus macrocarpa: wood is used in house building, in making walking sticks, pencils, cups and bowls. It yields good charcoal. Twigs are burnt as incense.

Juniperus macropoda: wood is suitable for making pencils and wall plates. It is burnt as incense and yields good charcoal.

Juniperus recurva: wood is used in making cabinets, pencils and furniture.

Junipreus virginiana: wood is used for making superior quality pencils. Because of its strong aroma it is valued in making cigar boxes, linen chests, and coffins. It is a good indicator of hydrogen fluoride pollution.

Juniperus wallichiana (Black juniper): wood is used as fuel and burnt as incense in Buddist temples. It is also used in the preparation of perfumes, cosmetics, soaps, room sprays, deodorants, insecticides, lubricating grease and floor polish. Oil from the berries is used for preserving catgut ligatures.

Many species of *Juniperus* are used in developing Bonsai plants.

Larix

Larix wood is heavy and strong scented. It is used in making furniture, doors and windows.

Larix decidua: yields venetian turpentine and a gum called orenburg gum, which is used in varnishes, lithography, histological work and veterinary medicine.

Larix larcina: wood is heavy, hard and durable and used for construction work. The roots are used for weaving baskets.

Larix leptolepis: wood is hard and used for making transmission poles, ships, boats, houses, railway sleepers, and bowls.

Libocedrus

Libocedrus decurrens: wood is soft, fragrant, decay resistant and is chocolate brown in color. It is used for making pencils, venetian blinds, and fancy articles.

Macrozamia

Macrozamia: seed paste is used in laundrying. The fine fibres at the base of the leaves used in stuffing pillows.

Phyllocladus

Phyllocladus rhomboidalis and *Phyllocladus trichomanoides:* wood is used in flooring, railway sleepers and furniture.

Picea

Picea species are the source of rayon, photographic film and lacquer.

Picea abies: wood is light in weight, soft, long fibred, straight grained and is used for plywood, carpentary, kitchen tables, musical instruments and carving. Ropes and fishing lines are made from the pliable roots.

Picea alba: bark yields tannin used in tanning industry.

Picea excelsa: Burgandy pitch obtained from the plants used in polishing lenses, leather, in audio mobile and photographic industry, rubber, perfume and dyeing intustries; in engraving and stonc cutting; in asphalt emulsion and furniture making.

Picea glauca, Picea mariana and *Picea rubens:* wood is used in making packing cases, sound proof boards, organ pipes, paper pulp and artificial fabric pulp; yields an oil which is an ingredient of room sprays and bath salts.

Picea sitchensis (Sitka spruce): yields non-resinous wood, which is pinkish brown, long fibred with satin finish. The plywood from it is used in making lamination inside aircrafts, gliders, and musical instruments.

Picea smithiana (West Himalayan spruce): is white, straight and even grained which is used in making slanting roofs of house, flooring, packing cases, furnitures and plywood. It also yield excellent quality paper pulp. It is one of the best box woods available in India.

Pinus

Pinus wood scrapings and saw dust of *Pinus* species are used in making plastics, artificial wood, linoleum, insulated bricks. Seed oil is used in pharmaceutical and detergent industry. Seed testa is used for colouring fishing nets.

Pinus cembra: seeds yield an oil used in paint industry.

Pinus halepensis (Aleppo pine): is grown as windbreak and soil binder.

Pinus merkusii, Pinus pinaster (Cluster pine), *Pinus roxburghii* (Himalayan long leaved pine), *Pinus thunbergii* (Japanese black pine) and *Pinus wallichiana* (Blue pine): wood is useful in construction of telegraph poles, railway sleepers, ships, boxes, packing cases, drawing boards, camp furniture, battery separators and as source of charcoal.

Pinus roxburghii: pine oil from the wood is used in paints, varnishes, furniture and metal polishes,shoe polish, floor waxes,printing inks and pharmaceuticals.

It is used as a wetting agent in textiles, degreasing agent in leather industry and as synergist in insecticides. The wood is used in making paper pulp.

Pinus species are grown in gardens for ornamental purpose. Pine tar is used in caulking boats and preserving ropes tied to boats. *Pinus* wood is cut longitudinally into small pieces and dried to obtain wood wool, which is used for stuffing pillows and cushions.

Podocarpus

Podocarpus wood exhibits well-marked growth rings and is used for making decorative and fancy furnitures.

Podocarpus elongatus, Podocarpus gracilior, Podocarpus latifolia, Podocarpus taxifolia: yield timber of high quality.

Podocarpus neriifolius: wood is used in making oars, paddles, boat hooks, masts, camp furniture, ladders, mathematical instruments, pen holders, tea boxes and pencils.

Pseudostuga

Pseudostuga menziesii (Green Douglas fir, Oregon pine): wood is used in house and bridge construction, masts, telegraph poles, railway carriages, flooring, furniture, veneers, and boxes. Saw dust is used as insecticide. Timber is used in making logs, scantling and planks. Bark is used as substitute in making plastics, adhesives and explosives and in tanning. Hard wax obtained from the bark is useful in preparation of polishes, lubricants, ointments and soap, and as sizing agent.

Pseudostuga taxifolia: wood yields veneer and plywood which is used for making tanks in breweries and fancy laminated arches.

Sequoia

Sequoia sempervirens: wood is useful in making furniture, fence posts, veneered table tops and tanks. Wood waste is useful in paper industry. The bark yields a fibre, which is useful as insulating material and used as blending fibre with wool and cotton (Ambasta, 1986).

Thuja

Thuja occidentalis (White cedar): is grown as an ornamental, windbreak and hedge plant. The leaf oil is an ingredient of room sprays, disinfectants, insecticides, soaps, hair lotions and as industrial cleansor.

Taxodium

Taxodium distichum: wood is dark coloured and durable. It is used in making doors, windows and pillars.

Taxus

Taxus baccata and *Taxus brevifolia*: extracts of leaves and berries are added to hair lotions and shaving creams. Wood is used for cabinet wok and other fancy articles such as knife handles, combs, wooden articles, inlaying, paraquet flooring, paneling, poles, ploughs, water wheel cogs and cheap grade pencils. Tribals use the leaves as arrows to kill animals. The leafy twigs are worn by tribal girls as ornament. The bark yields an inferior red dye.

Thuja

Thuja plicata: used in making baskets, hats, mats and cordage.

Tsuga

Tsuga canadensis and *Tsuga heterophylla:* yields oil of hemlock, which is used in room sprays, bath salts, deodorants and lacquer. It is also used in making writing paper.

Widdringtonia

Widdringtonia whytei (African cypress): yields strong heart wood, which is used in construction of fences.

13.5.1 Oleoresins from Gymnosperms

Oleoresins are naturally occuring mixture of essential oils and resin extracted from various plants such as pine, balsam fir used in perfumes and medicine. Resin is a hydrocarbon secretion of many plants. Terpenes and non-volatile solids are resin compounds.

Rosin

The pine oleoresin, also called as pine gum or pine pitch is a resin deprived of oil. Rosin is separated from the oleoresin by distillation.

Pinus caribaea, Pinus khasya, Pinus merkusii, Pinus nigra, Pinus palustris, Pinus pinaster, Pinus roxburghii, Pinus sylvestris, Pinus wallichiana are the most important species which yield oleoresin. In ancient times oleoresin was used to smear mummies in Egypt.

Rosin is used in making soap, paper, sealing wax, printing inks, cosmetics, polishes, adhesives, grease, paints, varnishes, rubber, roofing cement, explosives, insecticides, disinfectants and fire works. Rosin oil extracted after distillation is used as an adulterant for rape seed and olive oil and in making printing inks.

Turpentine

A number of species of *Pinus* are tapped for turpentine oil by distillation of oleoresin. *Pinus insularis, Pinus roxburghii* and *Pinus wallichiana* are the chief source of turpentine which is used in preparation of sealing wax, ointments, shoe polish, printing inks and disinfectant. It is used as solvent for paints and varnishes. Also used in preparation of stains, soaps, enamel, wax, polishes, stain removers, crayons, liniments, grease, medicated soaps, insecticides and in the manufacture of camphor. It is used for reclaiming rubber and in ore flotation. Several esters of pinic acid are prepared from turpentine oil, which are used as plasticizers and as lubricants for jet aircrafts.

Turpentine deprived of oil by distillation and the resin that remains is called colophony. It is named after Ionion city of Colophon where it was formerly prepared. It occurs in irregularly shaped pieces of different sizes and is pale yellow, brownish yellow, transparent and brittle. It is used as a stimulant and as an ingredient of plasters.

It is used as frothing agent in separation of metals, laundries and for dyeing in textile industry.

Several kinds of turpentine are obtained from the conifers (Box. 13.1).

Box 13.1 Types of Turpentine

White turpentine: obtained from the sapwood of *Pinus palustris* and *Pinus taeda.* It is yellowish white, pungent with bitter taste.

European turpentine: whitish turpentine obtained from *Pinus sylvestris* and *Pinus pinaster,* commercially called Bordeaux turpentine.

Canada balsam (Canada turpentine): is pale yellow or greenish yellow, transparent and acrid. It is obtained as an oleoresin from *Abies balsamea, and* sold as Canada balsam. Balsam of fir, which is similar to Canada balsam, is obtained from *Abies fraseri* (double fir).

Oregon balsam: a turpentine similar to Canada balsam, obtained from *Abies menziesii.*

Venice turpentine (Terebinthina laricis or Larch terpentine): an oleoresin obtained from *Larix europea.* It is yellowish green, with bitter characteristic taste and odour.

Strasburg turpentine: obtained from *Abies picea.*

Amber

The fossilized resin of *Pinites succinifera,* an extinct Baltic pine, is called amber. Amber is transparent, glossy and aromatic. It is used in *x*-ray therapy. Amber boxes are used to store blood to prevent coagulation. It is used in making fancy cigarettes and cigar holders and smoking pipes. Amber containing fossilized pieces of plants and animals are highly expensive and used in making jewellery.

Copal

Agathis alba, Agathis australis, Agathis brownii, Agathis macrophylla, Agathis ovata and *Agathis vitiensis:* are the most important copal yielding plants. Copal is a a hard resin containing essential oils. It is obtained from both living and fossil forms. The most important and most valued of all copals is the Kauri resin (Kauri gum or Kauri copal) from *Agathis australis*. Fossilized resin is considered superior to fresh resin. Copal is used in varnishes, printing ink, sizing material, plastics, oil cloth, linoleum and lacquer.

Sandarc

Callitris calcarata, Callitris glauca, Callitris quadrivalis and *Callitris verrucosa:* yield pale yellow or orange coloured resin called Sandarc which is used in varnishes, as temporary filling for tooth, in paper and leather industry. It is used as a metal, paper and leather varnish to give luster and as a varnish to preserve old paintings and as incense. It is used in cementing glass and porcelain.

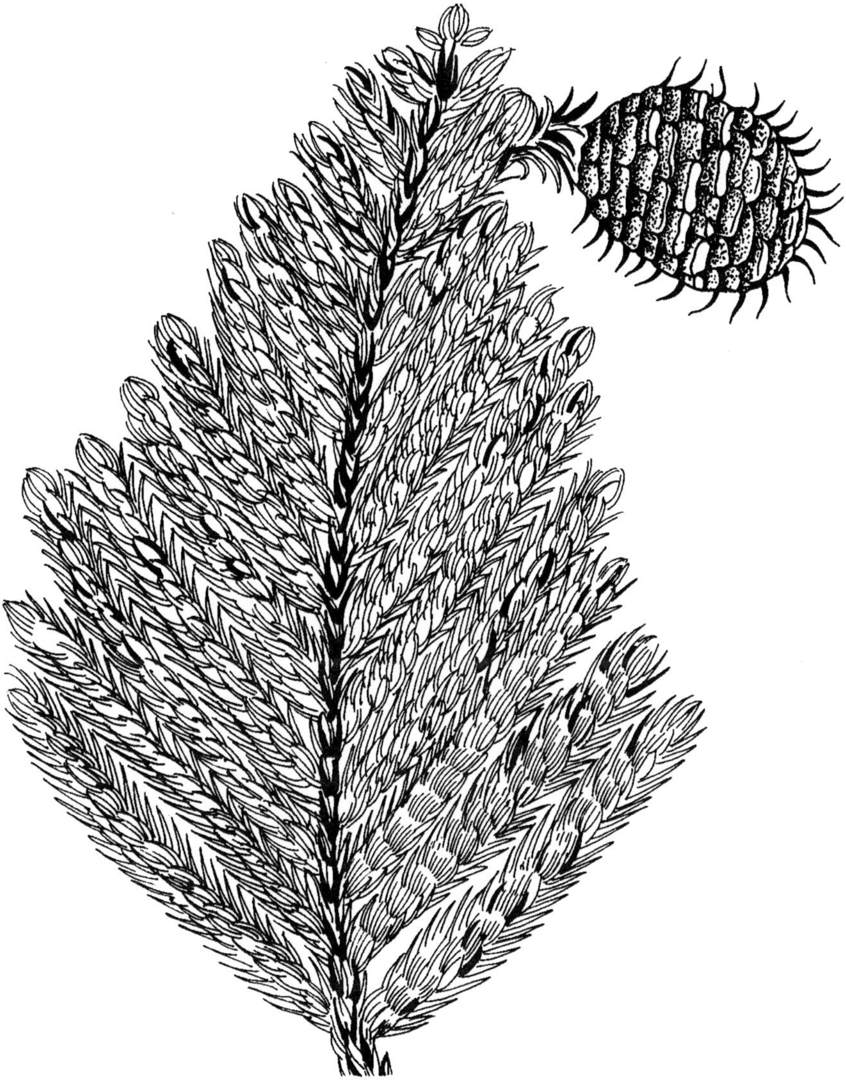

Fig 13.1 **Gymnosperms.** *Araucaria* sp.
Fig 13.1 to 13.12 Redrawn from Bhatnagar, S.P. and Moitra, A., (1996) Chamberlain, C.J., (1986), Sporne, K.R. (1965), and Vasishta P.C. (1996)

Fig. 13.2 Gymnosperms. A : *Agathis alba*; **B** : *Agathis australis*; **C** : *Cedrus deodora*.

Fig 13.3 Gymnosperms. A : *Cycas circinalis*; **B** : *Cycas revoluta*.

Fig. 13.4 Gymnosperms. *Cycas rumphii.*

Fig. 13.5 Gymnosperms. A : *Bowenia spectabilis*; B : *Dioon spinulosum*;

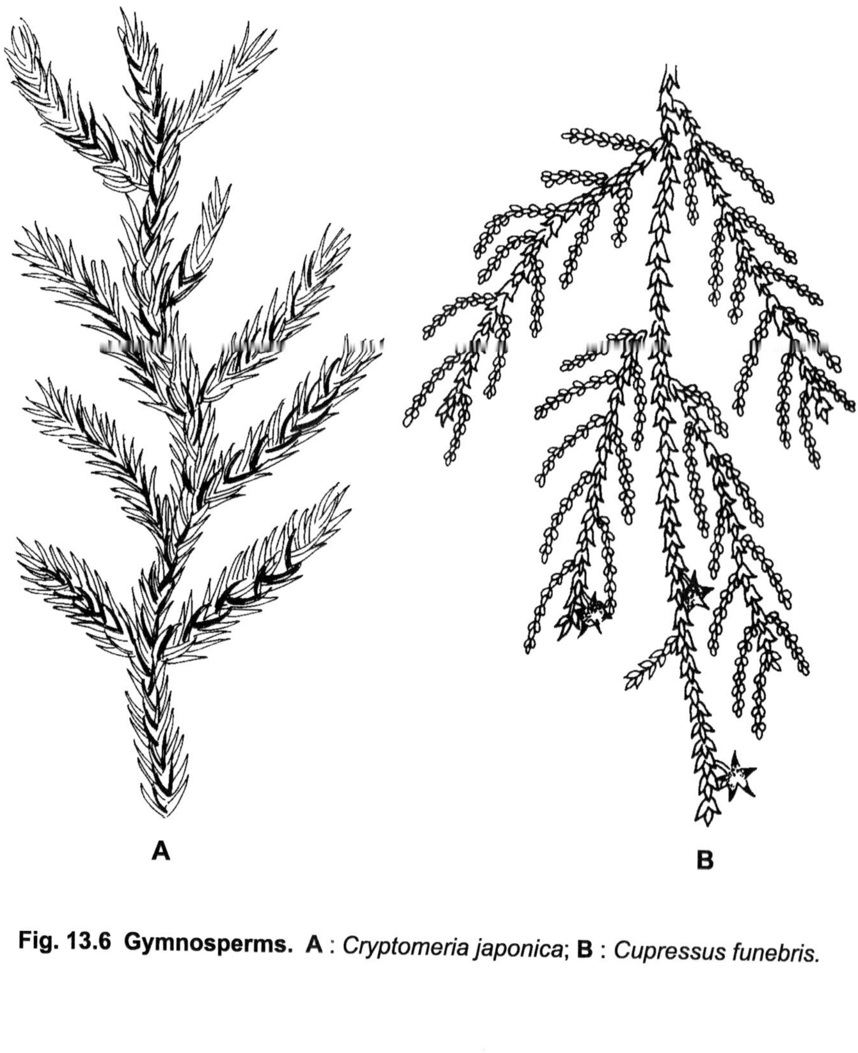

A B

Fig. 13.6 Gymnosperms. A : *Cryptomeria japonica*; **B** : *Cupressus funebris.*

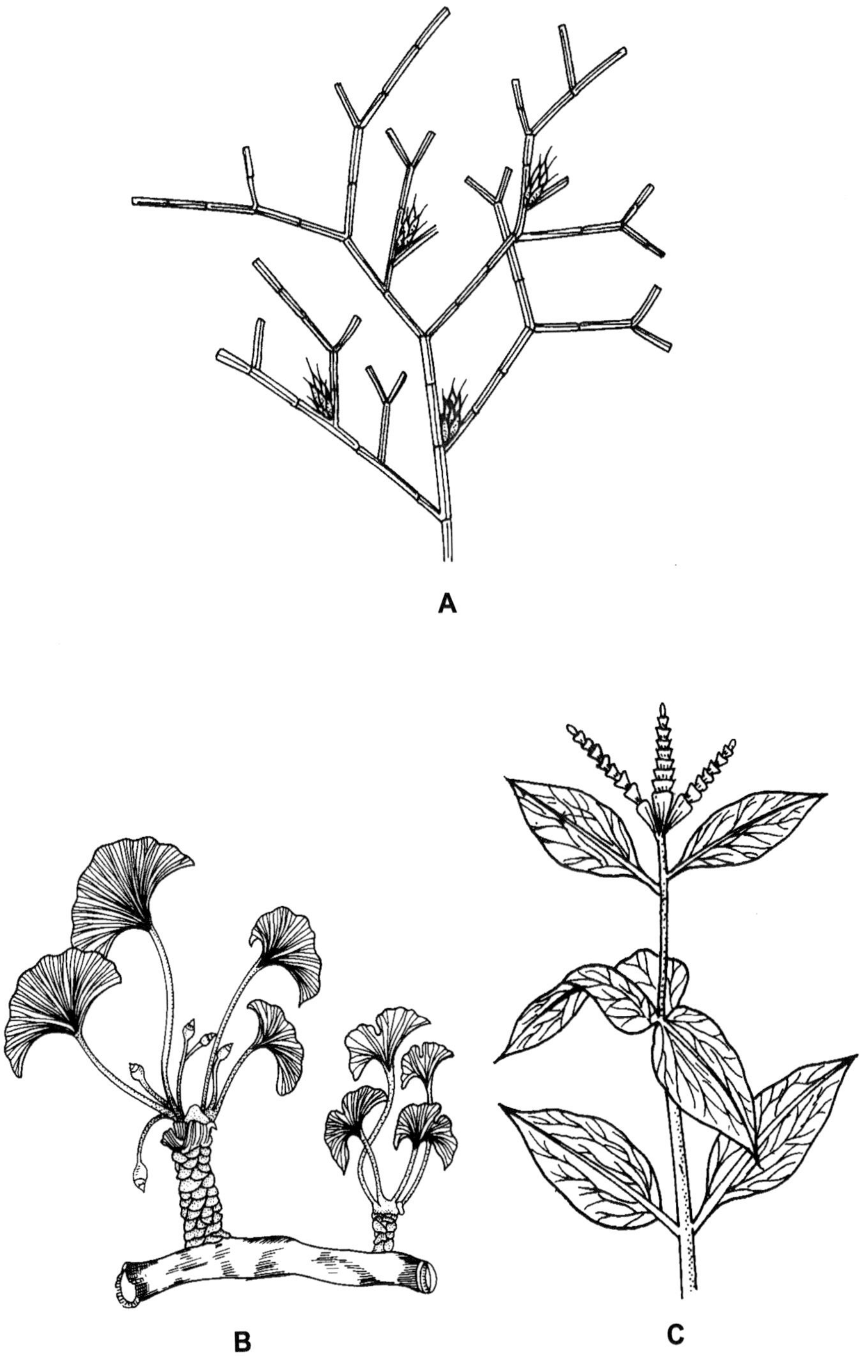

Fig. 13.7 Gymnosperms. A : *Ephedra foliata*; **B** : *Ginkgo biloba*. **C**: *Gnetum montana*

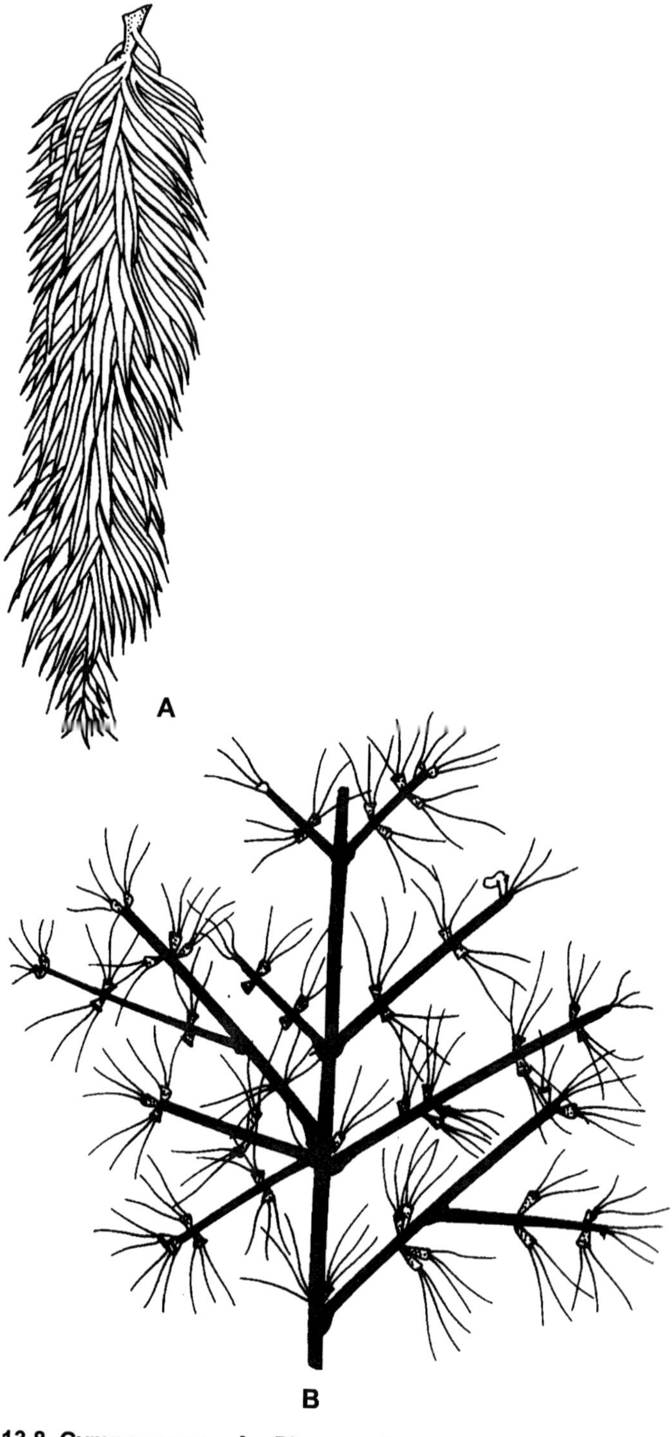

Fig. 13.8 Gymnosperms. A : *Picea* sp.; **B** : *Pinus triphylla.*

Fig. 13.9 Gymnosperms. *Podocarpus gracilior*

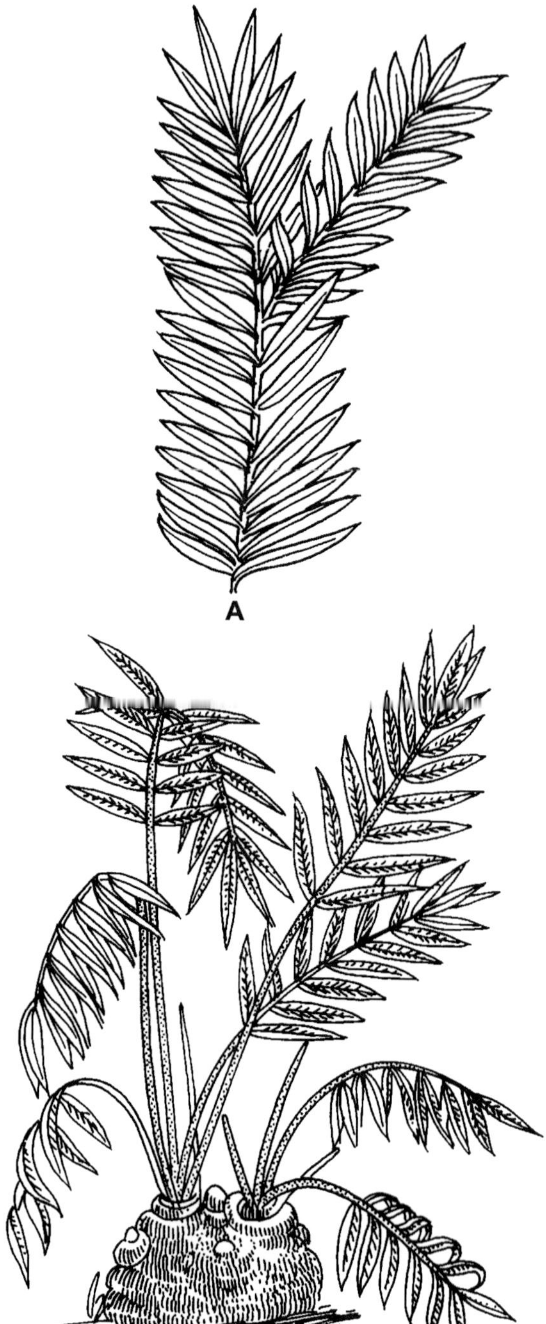

A

B

Fig. 13.10 Gymnosperms. A : *Sequoia sempervirens*; **B** : *Stangeria paradoxa*.

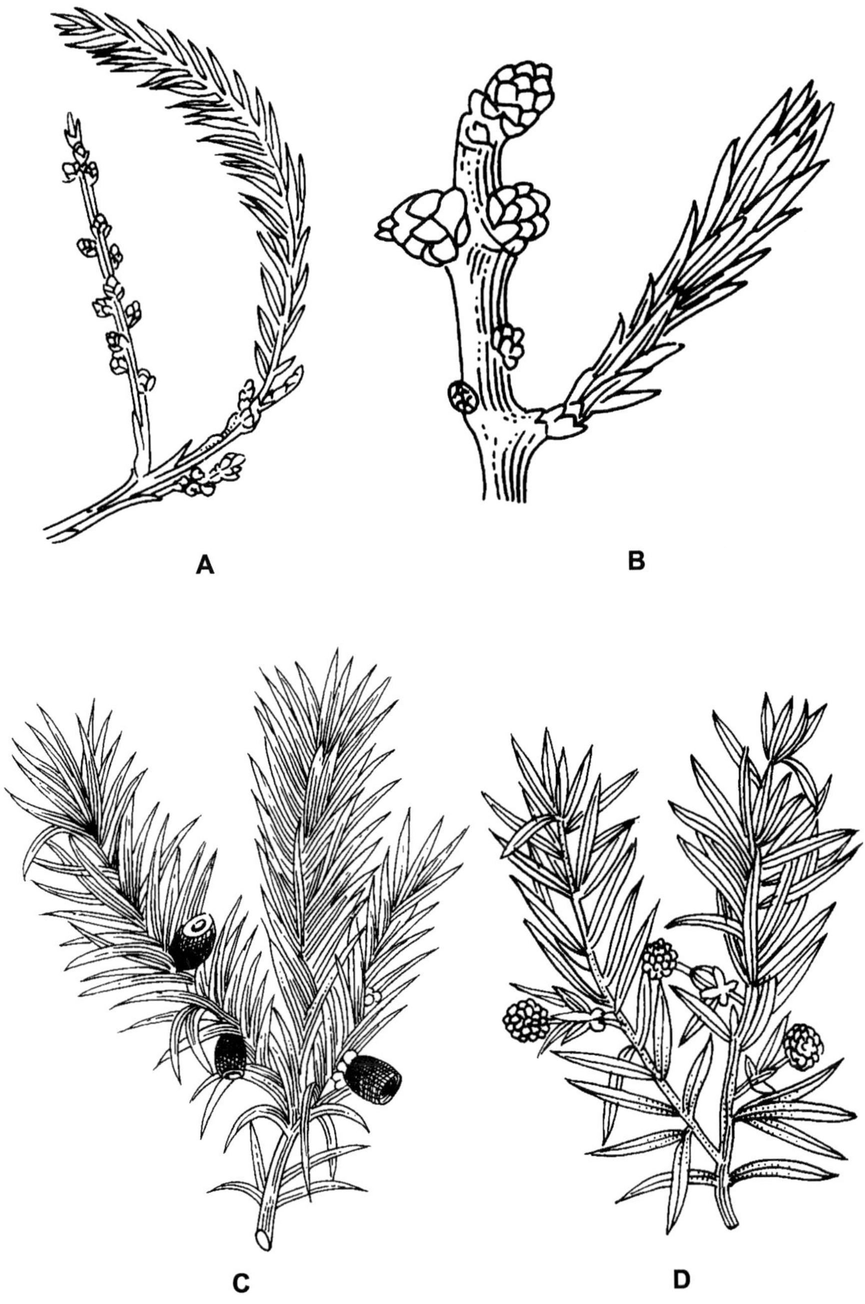

Fig. 13.11 Gymnosperms. A : *Taxodium mucronatum* male cones; **B** : *Taxodium mucronatum* female cones; **C** : *Taxus baccata* female strobilus; **D** : *Taxus baccata* male strobilus.

Fig. 13.12 Gymnosperms. A. *Welwitschia mirabilis* **B.** *Zamia* sp.

CHAPTER 14

NON-FLOWERING PLANTS IN BIOTECHNOLOGY

14.1 INTRODUCTION

Biotechnology is the large industrial scale use of organisms or their products. In this sense, the whole of this book is about biotechnology.

Biotechnology is an old technology of about 7,000 years, adopted in the form of agriculture, animal husbandry, traditional fermentation products such as wine, vinegar, and many others. The industrial production of antibiotics and vaccines (see Chapters 15 and 16) that has an immensely important role in preventing and managing infectious diseases, started about the middle of the last century. Modern biotechnology differs from traditional biotechnology in impressively innovative concepts, tools and techniques, and value added products and services.

14.2 MODERN BIOTECHNOLOGY

Most of modern biotechnology involves genetic engineering (GE) such as insertion of useful genes from one organism into the genome of another organism, irrespective of genetic relationship of the two. Such a thing was not possible a few decades ago. This is transgenic or recombinant DNA technology that results in transgenic organisms and the genes transferred are called transgenes. GE may also involve alteration of expression genes, whether enhancement, suppression or even gene silencing. The objective is generation of more beneficial products such as food and medicine or providing efficient services such as environmental remediation.

While non-flowering plants do not figure promiently among the GE organisms, it is the micro-organisms which have contributed handsomely both to production tools and techniques, as well as to products and services.

14.3 MICROBES AS TOOLS IN BIOTECHNOLOGY

Microbial biotechnology can be harnessed as an effective solution for global health problems. Its potential in novel diagnostics, drug development and delivery can be tapped by pharmaceutical and biotechnology companies. Though the advance of microbial biotechnology has led to new approaches in medicine and healthcare, abuse of such genetic information, ethical issues of using nature and environmental abuse by the release of hazardous by products should be considered.

14.3.1 Biotechnological Importance of Bacteria

Bacteria have been manipulated by genetic engineering for use in medicine. They are the most effective candidates in drug development and delivery by pharmaceutical and biotechnology companies. The recombinant human insulin currently available is prepared by cloning and expressing genes coding for proinsulin in *E.coli*. Proinsulin is purified and enzymatically cleaved to form functional insulin.

Magnetostatic bacteria are a class of bacteria that exhibit peculiar ability to orient themselves along the earth's magnetic field. The magnetosomes in the bacteria is sensitive to the earth's magnetic field and these are used as magnets to align themselves to the earth's geomagnetic field. These bacteria are being studied in magnetic domain analysis, in immobilising enzymes in quantification of IgG and to introduce genes into cells.

DNA polymerase, an essential enzyme for DNA synthesis obtained from normal sources. It cannot withstand the high temperature of polymerase chain reaction technique (PCR) and so is denatured by the end of each cycle and has to be replenished. DNA polymerase from thermophilic bacteria particularly the hot spring bacteria can withstand temperature over 90 degree centigrade, has made PCR technique immensely practicable. *Taq*-polymerase from *Thermus aquaticus, Thermococcus littoralis and Pyrococcus furiosus* produce DNA polymerase which is more thermostable and are preferred. DNA topoisomerase produced by *Methanopyrus kandleri* is used to unwind super coiled circular DNA and play a role in understanding the interaction of human Topoisomerase I and cancer chemotherapeutic agents (Margesin and Schinner 2001);

Halothermophiles are obligate bacteria inhabiting hypersaline and high temperature environment and are halotolerants. Their adaptiveness to high salt concentration and temperature make them biotechnologically potential source of industrially valuable enzymes. *Halothermothrix orenii* is capable of producing alpha-amylases when immobilized on alginate beads (Bagai and Madamwar 1997); biosurfactants produced by *Halobacterium salinarum* has potential application in the remediation of oil contaminated sand and water, degradation of hydrocarbons, enhanced oil recovery in petroleum reservoirs and as emulsifiers in petroleum emulsification; *Dichotomicrobium thermohalophilum* is used in the production of oil derived thermoplastics; Bacteriorhodopsin, a membrane bound retinal pigment in micro-organisms isolated from *Halobacterium salinarum* is commercially used as purple membrane patch used as biological material for information processing;

compatible solutes accumulated by halothermophiles can be used as stress protectants of enzymes such as lactate dehydrogenase and phosphofructokinase and whole cells against thermal denaturation and high salinity (Ventosa *et al.*, 1998); halocins, peptide antibiotics produced by *Haloferax mediterranei* is used against ischemia and reperfusion injury in mammals.

Agrobacterium tumefaciens is a natural genetic engineer capable of modifying genetic material of its host through introduction of genetic material into the host chromosome. It is now being used as a vehicle for transferring genes; *Escherichia coli* is a commonly used shuttle vector; *Photobacterium phosphoreum* is a marine bacterium which is successfully immobilized and used as a bioluminiscent indicator of pollutants in marine water.

14.3.2 Biotechnological Importance of Yeasts

Yeasts are safe systems among the eukaryotic organisms which can be engineered to carry out post translational modifications of expressed characters and safe history of use in the production of medically important products.

They are extensively used as hosts to express cloned eukaryotic genes. They are listed by US FDA as GRAS (Generally Regarded As Safe) organisms. A number of proteins including insulin, blood coagulation factor XIIIa has been produced in these hosts. *Hansenula polymorpha, Pichia pastoris* and *Schizosaccharomyces pombe* are used for recombinant protein production.

Oestrogen is an important human hormone. As with many other steroid hormones oestrogen affects gene expression by binding to a cytoplasmically sequestered receptor proteins and this complex then enters the cell nucleus where it binds to promoters containing a specific receptor recognition sequence referred to as the oestrogen receptor element (ERE). Normal yeast cells do not contain either the receptor or ERE sequence. However, yeast cells have been re-engineered with the ERE and these cells when treated with oestrogen , the hormone binds to ERE that is cloned into the promoter, in front of á – galactosidase gene, inducing expression. The level of enzyme in the cells is used as an index of oestrogen induced receptor activity, allowing antagonists to be tested on the cells to identify the cells that inhibit the activity of hormones.

Many recombinant proteins of therapeutic importance such as prevention and treatment of clots, production of peptide hormones and cytokines for treating blood cell deficiency and protein vaccines including hepatitis B vaccine which is already in the market has been successfully made by using *Pichia* (Curran and Bugeja 2005). *Saccharomyces cerevisiae* is engineered to express heterologous proteins used in recombinant DNA technology. Protein expression of *Saccharomyces cerevisiae* can be manipulated to provide *in vitro* tools with which to probe molecular interactions. These sophisticated heterologous protein expression systems re-engineer promoter elements, transcription factors and signal cascade proteins to transducer heterologous molecular intereaction into easily scoreable phenotypes. Such assays enable biotechnologists to screen for molecules in drug development, particularly to examine the molecular biology of expressed heterologous steroid and peptide receptor proteins.

14.3.3 Biotechnological Importance of Fungi

Fungi are among the models for genetic investigation of basic cellular processes. They are ideal systems for construction of efficient vectors for gene transfer and cloning genetic manipulation.

Neurospora crassa, Aspergillus nidulans, Coprinus cinereus are the major model systems for genetic analysis and for uncovering the secret of cell proliferation and exploitation to create new combinations of variations.

Ö (S) – Naproxen is a non-steroidal anti-inflammatory drug with analgesic and anti-pyretic properties. Biotransformation of Naproxen R isomer to S isomer is carried out by fungi like *Cordyceps militaris, Beauvaria bassiana* and *Exophiala wilhansii* which is not possible in chemical synthesis.

Fungal biotechnology has aided in transformation of renewable resources such as terpenes, α – pinene, α – pinene and limonene to valuable metabolites that are in demand by the flavour and fragrance industries which is closely associated with food and cosmetics. (+) limonene is hydrated to terpineol which is a widely used fragrance and when oxidized to perillyl alcohol it is a precursor of artificial sweetener perillaldehyde oxime.

Lactoferrin from human milk has been engineered into the fungus *Aspergillus awamori.* The recombinant lactoferrin is being tested for use in the treatment of oral, intestinal, ear and eye infections. It is also added to infant formula to prevent opportunistic infections and as a preservative in pharmaceuticals such as eye drops, contact lens solutions and cosmetics.

14.3.4 Microbes as Donors of Genes

Several genes have been sourced from microbes. The most widely used pesticidal genes come from the universally occurring soil bacterium *Bacillus thuringiensis,* which produces several proteins that selectively control a number of crop pests. These genes, called Bt genes, have been incorporated into several crop plant species such as Brinjal, Cabbage, Cauliflower, Chickpea, Corn, Cotton, Okra, Pigeon pea, Potato, Rice, Sorghum, Soybean, Tomato. There are also gene stacked transgenics, as for example, cotton varieties with two Bt genes (Cry 1 Ac and Cry 1 Ab), for better pest control.

Genes from a species of *Agrobacterium* strain CP4EPSPS are used to impart herbicide tolerance in Canola, Corn, Soybean, tomato and other crops. Genes for resistance against a number of viral and fungal pathogens are sourced from microbes, to develop disease resistant varienties of rice, virus resistant okra and several other crops.

14.3.5 Microbes as Vectors of Genes

Genes taken from a donor organism require a vector, the vehicle to carry and introduce them into the recipient organism. Though there are physico-chemical and mechanical means of achieving this, *Agrobacterium tumefasciens,* regarded as a natural genetic engineer, is the most widely used gene vector in transgenic technology.

14.3.6 Viral Vectors in Gene Therapy

Gene therapy is being used to correct genetic diseases by supplying cells with missing or defective genes with functional, normal genes. It is accomplished by transferring the desired genes using adenoviruses, herpes viruses and retroviruses as vectors for gene delivery. It also holds a powerful tool for treating incurable maladies arising from genetic disorders such as cystic fibrosis, haemophilia, Sickle cell anaemia and Lesch Nyhan syndrome.

Although many clinical trials have demonstrated non-toxicity of viral vectors for gene therapy, public risk assessment and safety issues are being still being assessed.

14.3.7 Gene Promoters

Transgenes require a mechanism, the promoters, to express in the new cell environment. Such promoters of gene expression, influencing the expression of gene in nature are isolated and used in transgenic technology. CaMV 35S, from Cauliflower Mosaic Virus infecting turnips, is a standard gene promoter widely used in plant transgenic technology.

14.3.8 Markers of Genetic Transformation

When a transgenic organism is developed, a genetic marker is needed to confirm the transfer of the genes into the recipient organism. Several microbes contain naturally occurring genes for resistance to antibiotics. Genes for resistance to antibiotics like Kanamycin and Hygromycin are incorporated as markers into transgenics along with the genes for desired benefits.

14.3.9 Elicitors

Specially designed techniques and equipment such as fermenters and bioreactors are used for the biosynthesis of economically important microbial products. However, such technologies often result in poor yield of the products. Elicitors are organisms that induce or enhance production of a compound in fermenters and bioreactors, as illustrated by the following examples of yeasts and fungi:

Saccharomyces cerevisiae: enhanced production of glyceollin by *Glycine max,* and berberine by *Thalictrum rugosum.*

Rhizopus arrhizus: trebled diosgenin production by *Dioscorea deltoidea.*

Verticillium dahlieae: production of morphine and codeine by *Papaver somniferum* increased by 18 times.

14.4 BIOTRANSFORMATION OF PRODUCTS

Micro-organisms have the ability to chemically modify a wide variety of organic compounds. Such changes are called transformation or generally referred to as bioconversion/biotransformation and are aided by enzymes. The end products of transformation are usually extracellular.

14.4.1 Transformation of Steroids and Sterols

Naturally occurring steroids such as androgens, estrogens and progesterone are used therapeutically as sedatives, contraceptives, antitumour agents. Cortisone is anti-inflammatory and used in treatment of rheumatism, arthritis and skin diseases. Bacteria and fungi are used to produce commercially important steroids and amino acids, (Table 14.1 and Table 14.2)

Table 14.1 Commercially produced steroids

Steroids	Source species
Progesterone	*Rhizopus nigricans*
Cortisol	*Curvularia lunata*
Hydroxycortisol	*Streptomyces roseochromogenes*
Prednisolone	*Arthobacter simplex*
Dehydrotestolactone	*Cylindrcarpon radicicola*
Androstadiendione	*Mycobacterium fortuitum*
Triendiol	*Streptomyces affinis*

Source: Sebek and Perlmn, 1979

Table 14.2 Microbial production of amino acids from biotransformation

Amino acid	Source species
D, L- Alanine	*Microbacterium ammoniaphilum*
L-Alanine	*Pseudomonas dacunhae*
L-Arginine	*Serratia marcescens* AT428
L-Aspartic acid	*E.coli* KI-1023

L-glutamine	*Corynebacterium glutamicum*
L-Histidine	*Serratia marcescens* L120
L-Isoleucine	*Brevibacterium lactofermentum* *Bacillus sphaericus*
L-Lysine	*Brevibacterium lactofermentum AJ11204* *Brevibacterium flavum* *Achromobacte obae* *Cryptococcus laurentii*
L-Methioinine	*Corynebacterium glutamicum*
L-Ornithine	*Corynebacterium glutamicum* *Rhodococcus sp.,* *Lactobacillus confuses* *Lactobacillus casei*
L-Phenylalanine	*Brevibacterium lactofermentum* *Bacillus sphaericus* *Rhodotorula glutinis*
L-Proline	*Corynebacterium acetoacidophilum*
L-Serine	*Brevibacterium lactofermentum* *Klebsiella aerogenes*
L-Threonine	*E.coli* VL344
L-Tryptophan	*E.coli* JP4114 *Bacillus subtilis* *Candida utilis* *Claviceps purpurea* *Hansenula anomala* *Klebsiella aerogenes*
L-Tyrosine	*Corynebacterium glutamicum Pr -20+* *Klebsiella aerogenes* *Erwinia herbicola*
L-Valine	*Brevibacterium lactofermentum 487*

Source: Crueger and Crueger 2005

14.4.2 Transformation of Morphine Alkaloids

Transformation of morphine alkaloids by microbes are employed in the production of semi-synthetic opiate drugs. The morphine alkaloids are the major alkaloid components of *Papaver somniferum* and provide some of the most important pain killing drugs in clinical use. Semi-synthetic opiates are derived from the naturally occurring morphine, codeine and thebaine. Like many bioactive compounds, these alkaloids possess a variety of functional groups that require protection during chemical transformation resulting in poor yield.

Biotransformation offers advantages over chemical route to yield several important therapeutic compounds.

The morphine molecule is highly complex as it has abundance of functional groups. This has made transformation of morphine difficult using conventional chemical techniques. Industrial scale synthesis is complex, expensive and also results in production of toxic wastes. In addition, product yield is moderate to low. Microorganisms offer the obvious biological alternatives to contend with current procedures.

Since morphine alkaloids are naturally occurring compounds, the environment in which they are found also possess microbes capable of their degradation. Selective enrichment of samples from waste streams of an opiate processing plant yielded a microbial culture of *Alcaligenes xylosoxidans, Methylobacterium* and *Pseudomonas putida* M10 (Bruce *et al.,* 1990). When these species of bacteria were incubated with morphine, a large number of transformation products were identified, including morphinone, hydromorphone, 14 α – hydroxymorphinone and di-hydromorphine. (Long *et al.,*1995). Similarly, when codeine was used as a substrate, codeinone hydrocodone, 14 α-hydrocodeinone and di-hydrocodeine were identified as transformation products (Lister *et al.,* 1999) which have significant commercial value. Hydromorphone and hydrocodone are widely used as analgesics, dihydromorphine and 14 – hydroxymorphine and 14 – hydroxycodeine are used as intermediates in the synthesis of analgesic compounds.

An NADP+ dependant morphine dehydrogenase was shown to initiate degradation of morphine in *Pseudomonas putida* and this enzyme has proven to be important as a biorecognition component in the development of a highly sensitive biosensor for the detection of illicit heroin (Holt *et al.,* 1996).

Biological synthesis of hydromorphone and hydrocodone on an industrial scale is unsuitable due to unwanted side reactions.

14.4.3 Bioresolution

One of the relatively recent development in the area of biotransformation is the resolution of enantiomers of the synthetic racemates by microbial cells or the enzymes derived from them. Some of the commonly available ester hydrolases used for bioresolution are obtained from *Aspergillus niger, Aspergillus oryzae, Candida cylindracea, Pseudomonas fluorescens and Rhizopus arrhizus.*

14.5 BIOPHARMING

Biopharming is the interface of biopharmaceuticals and agricultural biotechnology, aimed at producing therapeutically important agents in genetically engineered crops and farm animals, to bypass technical and other difficulties involved in the use of conventional technologies. Examples of such transgenic organisms involving microbial genes that are in advanced stages of development are given in Table 14.3.

Table 14.3 Transgenic crops producing pharmaceutical compounds

Golden Rice:	two genes for β-carotence (precursor of vitamin A) from the bacterium *Erwinia uredevora*
Transgenic rice:	a) pest tolerant varieties, mostly with Bt genes, b) pathogen tolerant varieties with genes from microbes
Transgenic potato:	antigens of cholera and diarrhoeal pathogens and hepatitis B vaccine
Transgenic maize:	AIDS antigens
Transgenic fruits and vegetables:	a) bananas, melons, brinjals and tomatoes with subunit vaccines against Rabies, b) AIDS antigens in tomato
Transgenic tobacco:	human antibodies against hepatitis B virus
Transgenic chicken lines:	human antibodies against hepatitis B virus

14.6 GENETICALLY ENGINEERED TREES

Genetic engineering of trees is a more difficult task than with crops, due to the long time required between seed germination and the plant reaching the reproductive phase. Genetic engineering of trees is designed to impart pest and disease tolerance, herbicide tolerance, tolerance to abiotic stress like drought, faster growth, higher yield, reduction of lignin in the wood, or to potentiate the trees to clean up toxic elements from the soil. Pest tolerance will enable the trees to withstand pest and pathogen pressure while herbicide tolerance will result in weed-free plantations. The weeds are killed using herbicides without affecting the trees. Genes for such traits come from microbes as with the crop plants.

Tree genetic engineering is still in very formative stages although several patents have been filed and some approved in the US.

14.7 GENETICALLY MODIFIED (GM) MICROBES

Genetically engineered microbial strains are already a component of current fermentation. Enrichment, isolation and modification of microorganisms will continue to be a source of biological material for production of enzymes, industrial solvents and pharmaceuticals. In agriculture, amino acids such as lysine, threonine and tryptophan can be used to enhance the nutritive value of animal feeds. Genetically modified organisms are used for the production of proteinase, cellulase, glucanase, arabinoxylanase and phytase. Genetically modified *Lactobacillus* is used in silage production to control aerobic and anaerobic fermentation.

Consortia of microbes under defined conditions are industrially very important. They are used in recycling waste products of various industries; in upgrading or modification of wood waste, pulp and paper manufacturing, petroleum refining, agricultural and mining wastes.

Use of engineered microbial inoculants with improved ability can enhance growth, nutrient and water absorption and stress tolerance in plants. Engineering the plant to produce specific chemo-attractants and at the same time engineering the bacteria to recognize these signals and grow in response to them would allow researches to establish a defined plant microbe interaction in the soil. This engineered symbiosis can enhance phyto-remediation of toxic chemicals and metals, to bolster the presence of disease suppressive microbial population and to enhance nutrient uptake by encouraging the growth of microorganisms that mobilize and absorb nutrients.

14.8 BIOSYNTHESIS OF THERAPEUTIC PRODUCTS

Micro-organisms have been manipulated by genetic engineering for use in medicine. With the advent of recombinant DNA and protein engineering technique a large number of genetically engineered therapeutic agents are being constructed and marketed.

Genetically engineered bacteria, fungi and yeasts are used as mini-factories for the commercial synthesis of a number of therapeutically important products (Table. 14.3 and 14.4).

One of the most dramatic impacts of recombinant human insulin gene into *Escherichia coli* is its successful production. *Escherichia coli* became the first genetically engineered therapeutic agent to be approved for clinical use. In addition, engineered organisms produce human proteins used as therapeutic agents.

Human growth hormone (hGH) is a protein naturally made by the pituitary gland. The inadequate secretion hGH in children results in dwarfism. hGH was prepared from pituitaries removed from cadavers which was cost prohibitive and some patients injected with pituitary hGH developed Jakob-Creutzfeldt syndrome. Now hGh is produced in genetically engineered *Escherichia coli* in large amounts and at relatively less cost.

Human tissue plasminogen activator (tPA) a proteolytic enzyme with an affinity for fibrin clots is an other therapeutic agent made available in large amounts by using recombinant DNA technology involving *Escherichia coli.* tPA is a life saving drug in myocardial infraction for rapid dissolution of blood clots (damage to heart muscle due to arterial blockage).

These novel drugs include hormones for pituitary dwarfism, insulin for diabetes mellitus, erythropoietin for renal failure and anaemia; growth factors such as insulin like growth factor I for Laron dwarfism, brain derived neurotrophic immunomodulators, immunostimulators, cytokines such as interferon for hairy cell leukemia, chronic myeloma, hepatitis B and C; stem cell factor, granulocyte colony stimulating factor for bone marrow failure, interleukins I and 6 for cancer, tumour necrosis factor, hematological products such as tissue plasminogen activator for thromboembolic lysis, coagulation factor VIII for haemophilia, α- I antitrypsin for emphysema, encephalin as an analgesic, antibacterial agents such as lactoferrin, microbial vaccine and cancer vaccines.

Table 14.4 Therapeutically important products synthesized by genetically engineered bacteria and yeasts

Genetically engineered organism	Product	Use
Escherichia coli, Saccharomyces cerevisiae	Human insulin	For diabetic patients as a better substitute of animal insulin
Escherichia coli	Somatostatin	Human hormone
Escherichia coli	Human growth hormone(hGH)	For treatment of growth deficiency in children suffering from hypopituitary dwarfism
Escherichia coli	Gamma interferon	Therapy for chronic granulomatous disease
Escherichia coli	Tumour necrosis factor (TNF)	Used to disintegrate tumour cells in cancer therapy
Escherichia coli	Interleukin	Used in cancer treatment to stimulate the immune system
Escherichia coli	Relaxin	To ease physical and psychological tensions during child birth
Escherichia coli	Taxol	Used in the treatment of ovarian and lung cancer
Saccharomyces cerevisiae	Hepatitis B surface glycoprotein	Vaccine against hepatitis B virus
Escherichia coli, Saccharomyces cerevisiae	Alpha interferon	Therapy against viral diseases and cancer
Escherichia coli, Saccharomyces cerevisiae	Prourokinase	An anticoagulant used in the treatment of heart attack.

Pichia pastoris, Saccharomyces cerevisiae	Superoxide dismutase	Antioxidant
Streptomyces avermitilis	Avermectins	Anthelmintic
Sporomiella intermedia, Leptodontium elatius	Zaragozic acid	Cholesterol lowering

Source: Banerjee and Banerjee, 2006

Table 14.5 Examples of human proteins cloned in *Escherichia coli* and their therapeutic use.

Human protein	Function	Therapeutic use
α-antitrypsin, calcitonin	Protease inhibitor Influences phosphate metabolism	Threatment of emphysema in osteomalacia
Epidemal growth factor	Epithelial cell growth	Wound healing
Factor VIII and Factor IX	Blood clotting factor	Blood clotting in haemophilics
Interferons *(α, β-γ)*	Low molecular weight proteins	Antitumour
Interleukins	Stimulators of immune functions	Antitumour and treatment of immune disorders
Lymphotoxin	A bone resorbing factor produced by leukocytes	Antitumour
Somatomedin	Sulphate uptake factor	Growth promotion
Serum albumin	Plasma protein	Plasma supplement
Superoxide dismutase	Decomposition factor of free radicals	Cardiac treatment and organ and transplantation
Tumour necrosis factor	Product of mononuclear phagocytes	Antitumour
Urogastrone	Control of gastro-intestinal secretion	Antiulcerative
Urokinase	Plasminogen activator(tPA)	Anticoagulant

Source: Glazer and Nikaido, 1995

14.9 MICROBE CONTAINING PRODUCTS

14.9.1 Biofertilizers

Biofertilizers are preparations containing beneficial micro-organisms in a viable state and are used as soil or seed application to improve soil fertility. It is also intended to increase the number and biological activity of the microflora in the plant root environment (rhizosphere). Biofertilizers are ecofriendly unlike chemical fertilizers and do not pose the problems of degradability, accumulation in the soil, contamination of water and entry into food chain. Nobbe and Hiltner in 1885 were the first to produce laboratory grown *Rhizobia* under the brand name 'Nitragin' for about 17 legumes. Rhizobial incoculants are used mainly to enrich soil of pasture legumes crops.

The most commonly used biofertilizers are non-symbiotic nitrogen fixing bacteria like *Azatobacter* and *Azospirillum.* The cyanobacterial (blue green algae) inoculants such as species of *Aulosira, Anabaena* and *Nostoc* are being promoted for use in rice cultivation.

The seaweeds *Ascophyllum, Rhodymenia* and *Sargassum,* and the water fern *Azolla* are being employed as biofertilizers (Sullia and Shantharam, 2005).

14.9.2 Combinatorial Biosynthesis

Genetic modification or gene transfer to create organisms that produce hybrid molecules, a technique known as combinatorial biosynthesis, has been the basis for many new products. Genes for enzymes that facilitate synthesis of several natural products have been introduced into non-producing strains to obtain hybrid antibiotics.

Combinatorial biosynthesis is proposed to generate "unnatural" natural products that use genetic information and genetic recombination technique to alter the biosynthetic pathway of micro-organisms to produce the designed chemical structure and used to diversify the chemical library of microbial metabolites.

Gene transfer from *Streptomyces* strains producing Granaticin and Mederomycin, into non-producing strains led to the discovery of new antibiotic derivatives such as Mederrhodin A.

Genetically engineered *Pseudomonas fluorescens* produces a toxin that is normally produced by *Bacillus thuringiensis*, which kill European corn borer.Xiaoquiang *et al.* (1997) cloned genes like cryIVA, cryIVD and P20 from *Bacillus israelensis* into *Anabaena* species strain Pcc7120 by a shuttle vector pRL-488p. A Bt toxin gene is cloned into cyanobacteria for biocontrol of mosquito larvae.

14.10 ULTRASENSITIVE DEVICES

14.10.1 Biofilms and Biofilters

Both free and immobilised bacterial, cyanobacterial and fungal biomass are used in biofilms and bioreactor filters. However, several species occur in a state where the cells are closely packed and firmly attached to each other forming a solid surface film, the biofilm. Biofilm is a complex aggregation of micro-organisms whose excretions form a protective and adhesive matrix.

Biofilms are characterized by surface attachement, diversity, genetic diversity, community interactions and the extra-cellular matrix of synthetic polymers. Biofilms have found several uses in the industry and are widely used to perform otherwise difficult operations.

Biofilters are columns packed with an inert support coated with biofilms. Organic compounds from industrial effluents are degraded when passed through the biofilms. *Pseudomonas, Staphylococcus* and *Streptococcus* species are used in biofilms while species of *Nitrobacter* and *Nitrosomonas* are used in biofilters.

14.10.2 Biotraps

Biotechnology has been successfully exploited as a remedy for organic wastes and *in situ* bioremediation of sediments at contaminated sites. Many micro-organisms act on toxic metal ions and alter them in such a way that renders them less harmful. These biological materials are therefore termed biotraps (Table 14.6) Fungi and yeast by biosorption, uptake, transformation by oxidation and reduction can precipitate the metal ions forming insoluble salts (see Chapter 23).

Table 14.6 Precipitation of metal ions by biotraps

Species	Metal ions
Trichoderma viride	Copper, Magnesium.
Rhizopus arrhizus	Copper, Nickel, Uranium.
Aspergillus niger	Zinc.
Aspergillus carbonarius	Copper and Chromium
Fusarium flocciferum	Cadmium, Nickel, Copper.
Kluyveromyces marxianus	Uranium
Saccharomyces cerevisiae	Copper, Chromium, Cadmium, Zinc, Nickel and Cobalt
Saccharomyces uvarum	Lead.

14.10.3 Biosensors

A biosensor is a compact analytical device incorporating a biological or biologically derived sensing element either integrated within or intimately associated with a physico-chemical transducer. All the components are integrated on a single microchip, using semiconductor fabrication technology. Even natural products like bacteriorhodopsin from species of *Halobacterium* are being used to develop computer chips, in an attempt to replace silicon chips, as light-driven bacteriorhodopsin chips seem to process information faster.

The aim of a biosensor is to produce either discrete or continuous digital electronic singnals that are proportional to a single analyte or a related group of analytes. For example, bacteria are harnessed as the biological element in microbial electrodes for the measurement of alcohol. A wide variety of devices exploiting enzymes, nucleic acids, cell receptors, antibodies and intact cells, in combination with electrochemical, optical, piezoelectric and thermometric transducers have been developed, to solve numerous analytical problems in health care, food and drink industries, environmental monitoring, defense and security.

Different species of bacteria have different lipid profiles that work like signatures of the species. They also have characteristic mass spectral signatures. Biosensors have been developed to identify individual bacterial species using these principles.

Transgenic bacteria, with a transgene for the enzyme luciferase was employed in a biosensor to produce characteristic morphological resonance of fluorescent light to identify individual species of bacteria. Such bacteria can function as 'bioreporter' used to monitor the progress in their metabolizing toxic environmental components such as toluene and trichloroethylene.

Microbes have been a component of biosensors to monitor urea, alcohol, hypoxanthin, inosine, glucose and carbon dioxide. Immobilized *Escherichia coli* are a component of a sensor to measure Vitamin B_{12}.

Thermophilic bacteria were used in sensors that measures biological oxygen demand and carbon dioxide output.

Microbiosensors can be implanted in the human body for *in vivo* measurement of several metabolic parameters simultaneously, from exteremely small samples.

A number of biosensors containing laccase have been developed for immunoassays and determination of glucose, aromatic amines and phenolic compounds (Minussi *et al.,* 1998). A carbon paste biosensor modified with crude extract of *Phanerochaete ostreatus* as laccase source for determination of catecholamines in pharmaceutical formulations. Catecholamines play an important role in the central nervous system as neuro transmitters (Leite 2003).

Biosensors have innumerable applications in various fields and innovations are growing at a very rapid rate. Biosensors are currently expensive, but when mass-produced, they would become affordable and more commonly deployed.

14.10.4 Nanodevices

Nanotechnology involves the use of extermely small cells such as bacteria or molecules such as DNA or elemnts like carbon, in producing ultrasensitive physical micro-devices. Bacterial S-layer is being used as functional molecules or templates in the formation of ordered arrays of nanoparticles, which are requied for nanoelectronics and nonlinear optics. S-layer can also be used as stabilizing and supporting structures for lipid membranes as an alternative to soft polymer cushions.

14.11 BIOTECHNOLOGICAL SERVICES IN ENVIRONMENT

One of the most useful services offered by modern biotechnology relates to the management of environment, which is challenged by natural or man made hazards. It is enssential for the survival of man and the health of the planet that the soil, water and the air are not excessively loaded with pollutants. Biotechnology offers very promising means of cleaning the environment and of retarding the rate of further deterioration.

Biodegradation, biomineralization, bioremediation, biodeterioration, biotransformation, bioaccumulation and biosorption are some terms with minor differences but often overlap. All these approaches deploy different species of microbes, depending upon the specific needs of the situation.

Biodegradation is the general term used for all biologically mediated break down of chemical compounds and complete biodegradation leads to mineralization.

14.12 CRIMINAL MISUSE OF BIOTECHNOLOGY

It has been a curse in human history that technologies developed for the benefit of man and his environment have been misused to cause severe damage. A number of new technologies in life sciences have immense benefits. Unfortunately, the same technologies could be used to cause untold harm. Such technologies are called 'dual use' technologies.

Bioterrorism, emerging out of criminal misuse of modrn biotechnology, is one of the most serious threats man faces in the current times. A terrorist or a rougue state can use any of the pathogenic organisms or their toxic products to affect large sections of targeted populations in selected cities, as anthrax bacteria were used in some U.S cities a few years ago.

Biowarfare is the use of biological weapons in a declared war. Many pathogens can be employed in making biological bombs that would serve as 'Weapons for Mass Destruction'. During the II World War, several countries stockpiled bioweapons that contained pathogenic viruses and bacteria. Even though now there are international conventions that prohibit the manufacture and stockpiling of bioweapons, one can never say who or which country resorts to the use of bioweapons.

Bioweapons, whether used by terrorists or warring nations, are easily made as they require only an elementary knowledge of microbiology and human pathology. Bioweapons can be used in any medium like drinking water, food or the air. They are very difficult to detect unlike the conventional weapons and can be used without being noticed till the outbreak of a deadly epidemic after its incubation period.

A number of impressive sophisticated developments in structural biology, systems biology, means to control infectious diseases such as sophisticated vaccines and antibodies, and many others have become dual use technologies, posing a great threat to human health security. A number of western countries have been developing protective measures and plans of action in case of bioterroristic or biowarfare attacks, but the developing countries are grossly unprepared and are even ignorant of the kinds of threats they may face.

14.13 NANOBIOTECHNOLOGY

Microscopic objects are measured in micrometers (μm), the one thousandth part of a millimeter. A nanometer (nm) is one billionth part of a meter. A hydrogen atom is 0.1 nm wide. Technologies that operate at this level are nanotechnologies. Nanotechnology is one of the most rapidly enlarging fields and the horizons are still hazy, though some applications such as washing machines and refrigerators that use silver nanoparticles as antibacterial agents are in the market.

Nanobiology operates at the level of molecules and sizes of less than about 100 nanometers, such as microtubules, microfibres, chromatin, nucleic acids, etc., in such areas as biochemistry, genomics and cell biology. Biologists are excited about the emerging applications of nanobiology that constitute nanobiotechnology, principally involving DNA and believe that this technology will revolutionize several areas of biotechnology, more particularly miniature biodevices and medicine, exploiting the capacity of biological systems for self assembly.

Biological nanofabrication supports material synthesis with fine precision and serves as a model for mimetic approach, enabling the regulation of assembly, control of selective inorganic templating and structuring. The process is safe and has low energy consumption. Built from molecules upwards, nanobiotechnology mimics nature (bottom up) in contrast to conventional industrial processes. Biological systems can be modified for inorganic interactions and can also be conserved.

The following are the principle advantages of using a biomimetic approach in nanofabrication:

1. Intact harvest of desired biological structures that provide a set of building blocks that may be used as templates that can be generated in large scale;
2. *In vitro* material growth that can be regulated by controlling supply of raw materials;
3. Biological structures that can be manipulated at the genetic level; and
4. Exploitation of the ability of the biological structures to self assemble in material synthesis.

Several nanotech based drugs, drug delivery systems, diagnostics and medical devices are in different stages of development and regulatory process. The nanoparticle based Abraxane, a form of taxol that is free from toxic solvents and a nanoengineered synthetic bone material used in the treatment of fractures are some such examples.

14.13.1 Viruses as Tools in Biotechnology

Viruses and bacteria are the principle tools in many nanobiotechnological applications. Viruses range in size from five to 300 nm, although a few are much larger. Strictly speaking viruses are not organisms. They contain only a capsid (protein coat composed of units called capsomers) and a nucleic acid that serves as the genetic material. Viruses depend upon a host to replicate as they are devoid of organelles and other machinery required for independent replication.

Bacteria range in size from 0.1 to 600 ìm, though exceptionally some species are larger. Bacteria replicate independently. The structurally more complex bacteria are more difficult to handle than the simplistic viruses, where it is easier to isolate the capsid and the nucleic acids, which are more frequently used in nanobiotechnology.

14.13.2 Advantages of Using Viruses in Nanofabrication

Viruses are the most suitable candidates for understanding the nanodimensional architecture and assemblies as they have an accumulated expression of 'memory patterns' at the molecular level that serve as a key to material organization. Diversity in genome size, its circular or linear organization, single or duplex state, shapes, geometry, orientation and sizes offer a potential for its application in nanotechnology for nanoarchitecture. The ability of capsomers to self assemble into a pre-destined geometry is an asset in nanofabrication as it eliminates complicated processing.

Viral protein coat contains all the information required for capsid assembly. Capsid proteins can be expressed independently in hosts such as *Escherichia coli*. The purified subunits can be reversibly assembled *in vitro* which is helpful in nanobiotechnology. Mutant and wild type protein units can be mixed at various ratios to generate divergent architectures with nanomaterial properties.

The flexibility of the protein building blocks and the ability to withstand structural alterations introduced by genetic engineering of the interior coat of proteins, and the interface between the subunits without changes in the morphology, enable the designing of quasi-equivalent polymorphic structures of defined shape and size to scaffold, template and pack nanomaterial.

Viral capsids with non-specific packaging compatibility can be used to pack any nucleic acid in a viral capsule from any source and size *in vitro*. This strategy could be used for packaging materials which are non-native. Lee *et al.*, (2002) created liquid crystalline films of gold nanoparticles using engineered M13 bacteriophage. They also engineered a liquid crystal system of viral quantum dots (semiconductor nanostructures) by imparting specific recognition properties through insertion of high affinity inorganic binding peptide sequences in M13 bacteriophage through which nanocrystals like zinc sulphide could be ordered.

Viruses are also used as scaffolds for biopatterning inorganics on organics (Peaboy, 2003). Transformation of linear M13 phage into a ring structure could be used as biological template for formation of metallized magnetic or semiconductor nano-rings.

Nanowires are the most integral components of any nanoelectronic architecture. Viral scaffolds could be used to grow quantum dots and nanowires. The single uninterrupted central channel of TMV which is highly resistant to high pH and temperature, is used to cast cobalt and nickel nanowires (Kenz *et al.*, 2003). Continuous nanowires, 3 nm in diameter and 500nm in length, were laid through the channel. Creation of magnetic and semiconducting nanowires by Flynn *et al.*, (2003) and Mao *et al.*,(2003) are a true projection of a biomimetic approach.

Phage display system is a commercially available platform for studying organic-inorganic material interactions and screening the inorganic compatibility indices of selective affinity peptide sequences and to screen the combinatorial libraries which otherwise are difficult using conventional phage libraries (Smothers *et al.,* 2002).

The biomimetic approach to biomineralization of desired nanomaterials is guided by a set of criteria including a cage like architecture to spatially constrain the shape and size of mineral particles, interior and exterior surfaces with distinct chemical or electro static affinities and accessibility to the interior, are all fulfilled by viruses.

Mechanical strength of viral protein nanocontainers make them ideal for safe transfer as they can withstand mechanical pressure. *Bacillus subtilis* phage f29 could offer mechanoprotection offered by any hard plastic. Several viruses are known to exhibit liquid crystalline phase properties which in future could be utilized to build well ordered 2- or 3-D structures for the next generation optical or electronic and magnetic materials (Lee *et al.,* 2003).

Natural tendency of Tobbaco Mosaic Virus (TMV) particles to associate into nematic liquid crystals can be exploited to prepare silica nanoparticles less than 150 nm in length (Fowler *et al.,* 2001) that can be used to produce a wide variety of semiconductors and metal based nanoparticles

Attachment of fluorescent dyes and gold clusters through maleamide linkers using site specific mutagenesis demonstrates that virus particles can be exploited as nanoblock (Doughlas and Young, 1999). Virus based fabrication of nanofibres is used in biomedical and tissue engineering (Mao *et al.,* 2003).

14.13.3 Bacteria in Nanobiotechnology

The bacterial surface layer (S-layer) is a part of the cell envelope common in bacteria and archaea, composed of identical proteins or glycoproteins. S-layer forms through self-assembly processes and covers the whole cell surface. Components of bacterial S-layer are being used as functional molecules or templates in the ordered arrays of nanoparticles, with applications in nanoelectronics and non-linear optics. S-layer can also be used to form stabilizing and supporting structures for lipid membranes, as an alternative to soft polymer cushions that are now used in industry.

CHAPTER 15

ANTIBIOTICS

15.1 INTRODUCTION

Antimicrobials are a large group of natural or synthetic chemical compounds, used as drugs to inhibit or kill pathogenic bacteria, fungi or other groups of micro-organisms, used in controlling microbial diseases.

Antibiotics are the most well known of antimicrobials and one of very important natural products, that save millions of lives every year and are apparently called as "Magic bullets".

Selman Waksman proposed the term 'antibiotics' in 1945, for that group of metabolic products of living organisms, which in very small amounts inhibit or kill other micro-organisms, more specifically pathogenic bacteria.

Bartolomeo Gasio discovered the first broad-spectrum antibiotic compound in 1893, isolated from the fungus *Penicillium brevicompactum* (*Penicillium glaucum*) growing on spoilt corn, which inhibited *Bacillus anthracis*. This important discovery went unnoticed, till it was rediscovered by Alsberg and Black in 1913, who named the compound as mycophenolic acid. In steps it was found to be antibacterial, antifungal, antiviral, antitumour and antipsoriatic, but was never used as an antibiotic due to its toxicity. Its ester derivative was later approved as a new immunosuppressant for kidney and heart transplantation in 1995 and 1998.

Alexander Fleming is considered as the Father of modern antibiotics, for his discovery of Penicillin in 1928. Since the release of Benzyl penicillin in 1942, every year there have been new antibiotics. The period between 1945 and the mid 1970s is considered as the 'Golden Age' of antibiotics, as the more important antibiotics such as Streptomycin, Chloramphenicol, Neomycin, Oxytetracycline, Procaine, Penicillin, Erythromycin, Tetracycline, Ampicillin, Amikacin and several others were discovered by 1975. However, the subsequently discovered antibiotics and the use of antibiotics as immunosuppressants, are no less significant.

The original use of the term antibiotics was restricted to natural products and did not include synthetic chemicals. Also, the term was applied only to those substances that inhibit bacteria (bacteriostatic) or kill them (bactericidal).

Currently, the term antibiotic is applied generally to all substances that are antimicrobial, irrespective of their origin. Most of the antibiotics that are in clinical use are derived form fungi, the rest from bacteria (bacteriocins) or other groups of plants. Many scientists do not consider those antimicrobials that are derived from the higher plants as antibiotics.

Most antibiotics are peptides or derived from them. They are small molecular weight compounds (less than 2000 Daltons), without enzymic activity. Depending upon the pathogen and the location of the infection, antibiotics are administered orally, or injected or applied topically.

Antibiotics are broadly classified basing on the range of their effectiveness. Some antibiotics are effective against several different kinds of pathogenic bacteria (broad-spectrum antibiotics) while some affect only specific types of bacteria (narrow-spectrum antibiotics). Some antibiotics are effective only against either gram-positive or gram-negative bacteria.

Antibiotics function by interfering with some important metabolic activity of the pathogen, such as cell wall polymerization, membrane integrity, inhibition of DNA or RNA synthesis, and the like, which adversely affects the pathogen. Antibiotics should be carefully chosen, from the point of view of their action, to control specific pathogens. In general, antibiotics are not effective against viruses and most fungi.

Antibiotics are pathogen group specific and so are classified by the pathogen against which they are effective, and/or by the type of infection in which they are useful.

More technical classifications of antibiotics are based on their chemistry. The different chemical classes of antibiotics and their representative examples are given in Box 15.1. The fungal sources of antibiotics are given in Table 15.1

Box 15.1 Chemical Classes of Antibiotics

Aminoglycosides (*Amikacin*, Gentamycin, Neomycin, Streptomycin)

Cephalosporins Generation I (Cephalexin); Generation II (Cephamandole), Generation III (Cefdinir, Cefpodoxin), Generation IV (Cefepime)

Glycopeptides (Vanomycin)

Macrolides (Erythromycin)

Monobactam (Aztreonam)

Penicillins (Ampicilin, Penicillin)

Nocardicins (Nocardicins A and B)

Polypeptides (Bacitracin, Polymixin B)

Quinolones (Gatifaxacin, Oflaxacin)

Tetracyclines (Oxytetracycline, Tetracycline)

Others (Chloramphenicol, Pyrazinamide, Specitnomycin)

Table 15.1 Sources of antibiotics

Source bacteria	Antibiotic
Aerobacillus polymixa	Polymixin
Bacillus brevis	Tyrothrycin
Bacillus licheniformis	Bacitracin
Bacillus polymixa	Polymixin
Nocardia lucida	Ristocetin
Streptomyces ambofaciens	Ravomycin, Spiramycin
Streptomyces antibioticus	Oleandomycin, Actinomycin-D
Streptomyces aureofaciens	Aureomycin, Tetracycline
Streptomyces azureus	Thiostreptin
Streptomyces canus	Amphomycin
Streptomyces chrysomallus	Actionmycin
Streptomyces decaris	Farmycetin
Streptomyces endus	Stendomycin
Streptomyces erythraceus	Erythromycin
Streptomyces fradiae	Neomycin
Streptomyces fragilis	Azaserine
Streptomyces griseochromogenes	Blasticidin
Streptomyces griseoflavus	Griseoflavin
Streptomyces griseofulvus	Horomycin
Streptomyces griseus	Streptomycin, Vanomycin (Ristocetin), Cyclohexamide
Streptomyces hachijoensis	Trichomycin
Streptomyces halstedii	Carbomycin
Streptomyces hygroscopeus	Hygromycin
Streptomyces kanamyceticus	Kannamycin
Streptomyces kitasoensis	Laucomycin
Streptomyces lincolnensis	Linomycin
Streptomyces mediterranei	Rifamycin
Streptomyces nodosus	Amphotericin B, Polymixin
Streptomyces nouresii	Polymixin, Cyclohexamide, Nystatin
Streptomyces orchidaceous	Cycloserine
Streptomyces orientalis	Vanomycin
Streptomyces pimprei	Hamycin
Streptomyces puniceus	Viomycin
Streptomyces rimosus	Terramycin, Paromycin

(Table 15.1 Contd...)

Streptomyces spheroids, *Streptomyces niveus*	Novobiocin (cathomycin)
Streptomyces taxus	Tetracycline (Archromycin)
Streptomyces venezuelae	Chloramphenicol
Streptomyces virginiae *Strepomyces mitakaensis*	Staphylomycin
Source fungus	**Antibiotic**
Aspergillus fumigatus	Fumagillin
Micromonospora purpurea	Gentamycin
Penicillium griseofulvin, *Penicillium nigricans*	Griseofulvin
Penicillium sp.	Hadacidin
Penicilliun notatum, *Penicillium chrysogenum*	Penicillin

Source: Purohit, 1999

15.2 BACTERIOCINS

In addition to the large general group of antibiotics from fungi, there are bacteriocins, the ribosomally synthesized antimicrobial peptides, produced by bacteria (Oscarizi and Pisabarro, 2001). Several groups of bacteriocins are recognized. The Lantibiotics are polycyclic peptides containing uncommon post-translationally modified amino acids like the lanthionine, 2-amino isobutyric acid, and unsaturated amino acids such as dehydroalanine. They are active in low concentration and show substantial specificity for some components of bacterial cell membranes. Actagardin (*Actinoplanes* sp.), Epidermin (*Staphylococcus epidermidis),* Mersacidin (*Bacillus* sp.), Nisin (*Lactococcus lactis)* and Subtilin (*Bacillus subtilis*), are lantibiotics produced by bacteria. Other classes of bacteriocins are Cystibiotics that need at least one disulphide bridge formed by cysteine for activity and include such examples as Mesentericin *Y* (from *Leuconostoc mesenteroides Y*) and Cerecin (from *Bacillus cereus*). Thiolbiotics are thiol activated peptides that have a single –SH residue (*eg*., Lactococcin from *Lactococcus* species) and few groups do not contain cysteine (Lactococcin A).

15.3 SIDE EFFECTS OF ANTIBIOTICS

Antibiotics may have certain side effects such as fever, nausea, diarrhoea and allergic reactions, in certain individuals. The most serious problem is that some antibiotics like Penicillin may cause very severe prophylactic reactions, which are sometimes fatal. Test dosing for allergenicity of an antibiotic in a particular patient is a standard procedure in antibiotic administration. Antibiotics may interact with other drugs taken simultaneously causing complications. Some antibiotics can interfere with the efficacy of birth control pills. Such effects are not common, though studied only for a limited number of antibiotics. Administering them against inappropriate pathogens, such as viruses including those that cause common colds, destroys beneficial gut microflora.

15.4 ACQUIRED RESISTANCE TO ANTIBIOTICS

The usefulness of antibiotics has been at risk now, on account of misuse and abuse over a period of time. Not complying with the full course of antibiotic treatment results in a subset of the pathogen acquiring resistance, with the result that the antibiotic that was very effective would cease to be so against the resistant subset that would spread. A vast majority of people suffering from tuberculosis cannot be treated with the once successful antibiotics, as there is a prevalence of antibiotic resistant strains of tuberculosis bacteria. Similarly, *Staphylococcus aureus* that was very successfully treated with Penicillin earlier cannot now be controlled with either Penicillin or others, though a few strains are still controllable by vancomycin.

15.5 APPLICATIONS OF FAILED ANTIBIOTICS

A number of antibiotics did not find favour for use in controlling pathogens either due to their insufficiency or severe side effects. Several such 'failed antibiotics' have been put to many other novel uses in medicine and agriculture (Umezawa, 1982).

15.5.1 Antibiotics as Immunosuppressive Agents

A serious problem in organ and tissue transplantation surgery is the rejection of the transplant by the recipient's immunological system. Immunosuppressants retard the rejection process, giving the transplant a chance to establish and integrate.

Cyclosporin A and C, originally discovered as narrow spectrum antifungal peptides produced by the mold *Tolypocladium nivenum* (*Tolypocladium inflatum*), with the sexual stage called *Cordyceps sessilis,* were poor antibiotics with serious side effects but were found to be a very efficient immunosuppressants (Borel *et al.,* 1976). This discovery led to their use in heart, liver and kidney transplants.

Now a number of fungi, such as the species of *Acremonium, Aphanocladium, Paecilomyces, Verticillium, Beauveria bassiana, Beauveria brogniartii, Chaunopycnis alba, Fusarium solanii, Isaria felina, Neocomospora vasinifecta, Tolypocladium cylindrocapsum, Tolypocladium geodes, Tolypocladium nubicola*, and *Tolypocladium tundrense,* are known to produce cyclosporin or similar compounds with immunosuppressive potential (Jegorov *et al.,*1990).

FK 506 (Tacrolimus) and Rapamycin (Sirolimus) from *Streptomyces hygroscopieus,*) both narrow spectrum polyketide antifungal agents are 10 times more potent than Cyclosporin as immunosuppressants and less toxic (Vezina *et al.*, 1975; Kino *et al.*, 1987). Several other antibiotics such as Bredinin (*Eupenicillium brefeldanium),* Dunamycin (*Streptomyces diastatochromogenes),* Prodigiosin (*Streptomyces hiroshimensis),* Tetranactin (*Streptomyces aureus),* Myriocin (*Tsaria sinclarii),* Depsidomycin (*Streptomyces lavendulae),* and Hormomycin (*Streptomyces griseofulvus)* are very effective immunosuppressants (Sanglier, 1992). Spergualin from *Bacillus laterosporus* is used in transplant surgery as an immunosuppressant.

Some of the immunosuppressants are useful in treating a number of other ailments such as nephritic syndrome, ulcerative colitis, reheumatic arthritis, paediatric diseases and allergy related ailments.

15.5.2 Antibiotics as Hypocholesterolemic Agents

High blood cholesterol leads to atherosclerosis, which is a causal factor in many kinds of heart diseases and a leading cause of death. About 70 per cent of cholesterol is synthesized in the liver and the rest comes from diet. Many a time a healthy cholesterol level cannot be maintained by diet alone and hypocholesterolemic drugs are needed.

Cholesterol levels are clinically regulated by cholestyramine, a semisynthetic resin that binds to the bile acids. Some natural polymers, collectively called as statins such as compactin (lovastatin), an antifungal product of *Penicillium brevicompactum,* inhibit cholesterol-synthesizing enzymes (Brown *et al.*,1976). Compactin was also found in the broth of *Penicillium citrinum.* Endo *et al.*,1976 and Endo and Monacolin, 1979 discovered a methylated form of compactin known as Lovastatin in the broth of *Monascus ruber* and *Aspergillus terreus.* Another successful derivative is Pravastatin produced by hydrosylation of compactin using Actinomycetes (Peng and Demain, 1998). Natural statins are also produced by species of *Doratomyces, Eupenicillium, Gymnoascus, Hypomyces, Paecilomyces, Penicillium, Phoma, Pleurotus* and *Trichoderma* (Manzoni and Rollini, 2002).

Lovastatin was approved by FDA in 1987 when clinical tests on humans showed 18 to 34 per cent lowering of total blood cholesterol, 19 to 39 per cent decrease in low density lipoprotein cholesterol (bad cholesterol) and an increase in high density lipoprotein cholesterol (good cholesterol) (Demain and Zhang, 2005).

A combination of the synthetic resin and the enzyme inhibitor can reduce cholesterol levels by 50 to 60 per cent. However, since the human system is not accustomed to these synthetic substances, they may affect digestion and the absorption of many nutrients, and can cause abdominal discomfort and other side effects. Hence, natural statins are preferred.

Statins can also prevent stroke, reduce peripheral vascular diseases, thrombosis and inflammation. Statins now are in clinical trials for multiple sclerosis (Brazil, 2002), and in autoimmune disease (Youssef *et al.*, 2002).

15.5.3 Antitumour Activity of Antibiotics

Rapamycin inhibits tumour growth by interfering with angiogenesis (Pray, 2002). It also reverses multidrug resistance to antitumour agents in mammalian cells (Arceci *et al.*, 1992). Anthramycin, Sibromycin, Neothramycin and Tomamycin also have an antitumour activity.

15.5.4 Other Medicinal Applications of Antibiotics

Clindamycin, a semisynthetic derivative of Lincomycin from *Streptomyces lincolnensis,* is an effective antimalarial drug when used with quinine (Lell and Kremser, 2002).

Tetracyclines produced by *Streptomyces aureofaciens* are useful against prion diseases like fatal familial insomnia and Gerstmann-Straussler-Scheinker syndrome in humans (Borman, 2002; Forloni *et al.*, 2002).

Violacein, a toxic antibiotic from *Chromobacterium violaceum* is studied for preventing gastric ulcers when complexed with α-cyclodextrin to reduce toxicity (Duran *et al.*, 2003).

Rapamycin is also useful in cardiology as the antibiotic impregnated stents are less prone to proliferation and arterial restenosis, which usually occur after treatment of coronary disease (Demain and Zhang, 2005).

A topical preparation of FK 506 (Tacrolimus) was shown to be very active against atopic dermatitis. It also reverses multidrug resistance to antitumour agents in mammalian cells (Arceci *et al.*, 1992).

15.5.5 Antibiotics as Enzyme Inhibitors

Desferal produced by *Streptomyces pilosus* (Winkelmann, 1986) has metal binding activity, which led to its use in iron overload diseases (haemochromatosis) and aluminium overload in dialysis patients.

Acarbose, produced by species of *Actinoplanes,* is used as an inhibitor of intestinal α-glucosidase in type I and II diabetes and hyperlipoproteinemia (Truscheit *et al.,* 1981).

Lipostatin produced by *Streptomyces toxytricini,* is a pancreatic lipase inhibitor involved in the digestion of fat (Weibel *et al.,* 1987) and is often used to reduce obesity and diabetes by interfering with gastrointestinal absorption of fat..

Clavulanic acid is a α-lactam with poor antibiotic activity produced by *Streptomyces clavuligenus*. It is an inhibitor of penicillinase and is included with penicillin in combination therapy of penicillin-resistant bacterial infections (Brown, 1986).

15.6 OTHER USES OF ANTIBIOTICS

Antibiotics are used in food preservation. Meat contamination or spoilage is prevented by spraying of antibiotics or by injecting the animals with antibiotics prior to slaughter.

Certain antibiotics are routinely fed to healthy farm animals to make them grow faster. The use of antibiotics as feed supplements is as important as their use in medicine (Purohit, 1999). However, this upsets the balance of their gut flora, besides resulting in resistant pathogens, making antibiotics ineffective when the animals fall sick.

Antibiotics like streptomycin and nystatin are used in tissue culture to control bacterial and fungal contamination. They are used in crop protection and food preservation.

The antibiotics used as feed additives in animal husbandry and their anticipated benefits are given in Table 15.2.

Table 15.2 Antibiotics used as feed additives in animal husbandry

Active ingredient	Animal	Uses
Apramycin	Pigs	Antibacterial, enteritis
Avoparcin	Pigs, Broilers, Cattle	Growth promotion
Bambermycin	Poultry, Cattle, Pigs	Growth promotion
Chlorotetracycline	Calves, Pigs, Poultry	Antibacterial
Levamisole	Cattle, Sheep	Anthelmintic
Lincomycin	Pigs	Antibacterial vs. dysentery
Olaquindox	Pigs	Growth promotion
Monensin	Broilers, Cattle	Anticoccidial, Growth promotion
Narasin	Broilers	Anticoccidial
Salinomycin	Broilers	Anticoccidial
Sulphamethazine	Cattle, Sheep, Pigs	Antibacterial, vs. atropic rhinitis
Tylosin	Pigs	Growth promotion
Virginiamycin	Pigs, Broilers, Turkeys	Growth promotion
Bacitracin	Pigs, Cattle, Sheep, Broilers	Growth promotion

Source: Purohit, 1999

CHAPTER 16

VACCINES

16.1 INTRODUCTION

There are several natural or synthetic therapeutic agents to control a large number of microbial diseases. However, in course of time the pathogens acquire resistance making the therapeutic agents ineffective. Vaccines are a more effective means of controlling infectious diseases.

Even before 200 BCE, there was a practice of inoculating healthy people with a pathogen to prevent the onset of disease. Smallpox was the first disease people tried to prevent by inoculation. The Turks inoculated themselves with fluid taken from mild cases of smallpox. Edward Jenner used infectious fluid from cowpox pustules to inoculate against smallpox, in 1796. It was soon realized that the use of vaccines against microbial pathogens, and more particularly the pathogenic viruses, is a safe and more certain recourse to protect susceptible populations. More importantly, this approach, which involves the human immune system, is both preventive and curative, while other means are only curative.

16.2 VACCINES

A vaccine is a modified pathogenic organism or a low molecular weight protein or a non-protein compound conjugated with the protein, all obtained from the pathogen.

Vaccines are deliberately introduced into the mammalian system in order to impart a 'memory' of the pathogen or its pathogenic component. The memory imparted on the first contact of the vaccine, will enable the system to recognize the pathogenic element on a second contact. Vaccines are antigens as they trigger the production of very specific proteins (antibodies) against themselves, which recognize and bind with the antigen, and form an antigen-antibody complex that neutralizes the harmful effects of the antigens or the organisms that produce them.

Since vaccines employ a part of the biochemical machinery of the organisms themselves, pathogens cannot easily acquire resistance to vaccines, as they do for antibiotics and chemical therapeutic agents.

16.3 IMMUNOTHERAPY

Immunotherapy differs from vaccination in that in the former the protein antibodies, produced in response to an antigen, and isolated from an immunised organism, are used to cure the patient. Immunotherapy becomes essential when there is no time to prepare the patient through vaccination or when the patient is physically and/or physiologically not competent to respond to vaccination.

16.4 COMPOSITION OF VACCINES

Vaccines are saline or buffered saline suspensions of weakened pathogenic organisms or their fractions or the proteins they secrete, that have the potential to cause a disease. A virulent organism cannot be used as a vaccine.

Antigens often need to be coupled with an adjuvant, which holds the antigen and releases it slowly over a longer period of time, in order to enhance the effectiveness of the antigens.

16.5 TYPES OF VACCINES

a) **Inactivated vaccines:** The pathogen is killed using heat or formalin, as for example, typhoid or Salk poliomyelitis vaccines.

b) **Attenuated vaccines:** The pathogen is weakened (attenuated) by aging or altering growth conditions, but is alive as in the case of measles, mumps and rubella vaccines. These vaccines can be administered in low doses and require no adjuvants. They are commonly used to control animal diseases. There is some risk of the concerned virus becoming virulent.

c) **Avirulent organisms:** A non-pathogenic strain of a pathogenic organism is used as a vaccine, as in BCG (*Bacillus Calmette Guerin*) vaccine against *Mycobacterium tuberculosis,* the tuberculosis bacterium.

d) **Toxoids:** The toxin from the pathogen is used as an antigen to produce the vaccine. The severity of the toxicity of the antigen is reduced by treating it with aluminum salts while preparing the 'toxoid', as in the case of diphtheria and tetanus.

   Injecting toxoid in human serves as a vaccine to protect the body from harmful toxin produced by the organism from which the toxoid is made. A toxoid when injected into animals, it serves to generate polyclonal or monoclonal antibodies for use in research.

   Enterotoxin of enteropathogenic *E.coli* consists of two subunits, toxic alpha subunit and the beta subunit that binds the alpha subunit to enteric cells. The purified beta subunit is immunogenic and is an effetive toxoid. In the case of allergy, the allergenic proteins from pollen and other allergenic material are isolated and used to immunize the patient.

e) **Acellular vaccines:** Only the antigenic component of the organism is used instead of the whole organism, as in Haemophilus influenza B vaccine.

f) **Subunit vaccines:** Genetic engineering techniques have now made it possible to use as a vaccine only a part of an organism that is adequate to stimulate the immune response. An appropriate segment of genetic material is isolated from the pathogens and introduced into bacteria or yeasts to transcribe and translate the inserted foreign DNA. The product is used as a vaccine as in the case of Hepatitis B vaccine. The first commercially available subunit vaccine was against Feline Leukemia Virus (FeLV).

g) **DNA vaccines:** Described as the third generation vaccine, DNA vaccines are an offshoot of gene therapy. Selected segments of DNA, when introduced into the patient's system synthesize and deliver proteins that are needed to normalize the defective enzyme system or tag a cell for destruction. Viruses are lipid vehicles used to deliver the DNA into the cells. This recent technology is being tried to produce vaccines against HIV, by a direct injection of plasmid borne DNA.

h) **Multiple vaccines:** While most vaccines contain antigens of a single pathogen, there is a practice of multiple vaccines, which combine antigens of more than one pathogen. For example, diphtheria, tetanus and pertussis are administered together as DTP vaccination.

i) **Antifungal vaccines**: Although avirulent strains of fungi are known to exist, common vaccination approaches which rely on the use of attenuated strains are not applicable in the case of fungi as in the case of viruses or bacteria. In case of deep seated fungal infections, the fungus would have to be totally eliminated by the immune system to prevent the formation of dormant fungal infection that can be reactivated in the future.

   Attenuated forms of deep tissue infecting fungi can be generated by genetically incapacitating growth or the components have been proven useful in *Histoplasma capsulatum* infections.

   Murine monoclonal antibody (mAb) KT4 from *Pichia anomala* killer toxin (PaKT) induces the secretion of anti-idiotype antibodies which have a direct *in vitro* killing activity against PaKT-susceptible *Candida albicans* cells and were able to generate passive protection in unvaccinated animals.

j) **Edible vaccines:** Now transgenic plants are being developed through genetic engineering techniques, where the vaccine is synthesized in the edible part of a food plant (edible vaccines). Transgenic bananas, melons, and tomatoes are choice candidates for carrying edible subunit vaccines, as for example against rabies. Eg: *E.coli* enterotoxin in transgenic potato leaf and TGEV spike peplomer glycoprotein in transgenic potato tuber. The obvious advantage is the ease of transportation and storage of the vaccine bearing material and administration without technical support.

k) **Recombinant vaccines**: A recomobinant vaccine is able to stimulate an immune response distinguishable from that due to a natural infection and hence is also known as a marker vaccine. These vaccines allow vaccination and eradication to proceed simultaneously. These vaccines are under development against animal diseases.

l) **Gene deleted vaccines**: These are engineered by deleting specific gene sequence which code for pathogenecity from the genetic material of a pathogen resulting in, irreversibly attenuated organism. These vaccines are used against pseudorabies in pigs caused by Herpes Virus.

m) **Vectored vaccines**: Genes encoding the protective antigens can be recombinantly inserted into an attenuated viral or bacterial vector that acts as a vaccine delivery system for the antigen. These vaccines are host restricted and overcome immunogenecity associated subunit vaccines. They are free of adverse side effects, highly stable, are suitable for mass vaccination programs and do not need adjuvants.

n) **Auto vaccines**: They are genetically modified cells such as tumour infiltrating lymphocytes implanted with tumour necrosis factor gene for treating cancer.

o) **Microsphere vaccines**: Microencapsulation of vaccines in biocompatible and biodegradable polymers leading to single dose vaccines are known as microsphere vaccines. These are effective in stimulating both humoral and cell mediated immune response and are potential carriers for oral vaccine delivery as the antigens are encapsulated. E.g. Polylactideglycolide (PLG) encapsulated *Pertussis fimbriae* against *Bordatella* induced pertussis and *Salmonella enteritidis* encapsulated with PLG against Salmonellosis.

16.6 DELIVERY OF VACCINES

One of the limitations of vaccine development is the instability of the antigen introduced into the body. To overcome the problem, enteric bacteria that are commensals in the intestine and are capable of replicating are employed as vehicles to deliver the vaccine antigen. With appropriate construction of vaccine using the enteric bacteria, a continuous supply of the antigens is ensured, as they exist as commensals.

An ideal way to develop such vaccines is to genetically engineer and fuse a subunit of toxin to a surface display protein such that toxin can be displayed on the surface of bacteria. *E.coli* and species of *Salmonella* have been used for surface display of several potential vaccine proteins. Immunization of mice with *Salmonella dublin* expressing flagellin with a fused cholera toxin or hepatitis B virus epitope has been quite promising.

The advantage of such vaccine delivery is that the bacteria employed for delivery are part of the intestinal bacterial flora, easy to grow and cloning vectors are readily available. Moreover, no toxins need to be extracted and purified, and no chemical inactivation of the toxin is necessary. In addition, as these bacteria can replicate in the intestine, engineered surface display protein also simultaneously replicate.

16.7 COMMONLY USED VACCINES

Live viral vaccines

Poliomyelitis

Mumps

Measles

Rubella

Yellow fever

Inactivated viral vaccine

Influenza

Varal subunit vaccine

Hepatitis B

Killed bacterial vaccines

Pertussis (whooping cough)

Typhoid fever

Inactivated bacterial toxins (toxoids)

Diphtheria

Tetanus

Source : Glazer and Nikaido, 1995

16.8 MASS VACCINATION PROGRAMMES

Having realized the potential of immunization using vaccines to protect different sections of the world population, the health authorities have implemented mass vaccination programmes. A great success of this approach is the eradication of smallpox. Such programmes have also drastically reduced the number of people afflicted by polio, measles and several other diseases.

16.9 SAFETY OF VACCINE THERAPY

By and large vaccination programmes have proven to be reasonably safe for the human populations. However, at certain times, complications may arise mostly due to an incorrect handling of the vaccines and/or vaccination or due to individual metabolic deficiencies. Such instances have raised controversies regarding the safety of vaccines. In spite of all the adverse campaigns against vaccines, immunization is one of the most effective means of disease prevention, particularly when large segments of human population are under threat. In the case of HIV and epidemic diseases and even cancer, immunization is probably the only hope.

16.10 SOCIO-ECONOMIC VALUE OF VACCINES

Infectious diseases cause 25 per cent of all deaths worldwide. They are also responsible for 45 per cent of deaths in low-income countries and 65 per cent of all child deaths. In addition, several million man-hours are lost due to infectious diseases, causing enormous economic losses. Vaccines can prevent a very significant part of these losses, and so are of a great societal value. Nevertheless, very few pharmaceutical companies involve in research and development of vaccines as this involves heavy financial investment. Pharmaceutical companies are not encouraged into vaccine development, as the financial returns are low because vaccines are more needed in the poor countries, which cannot afford the costs. Vaccine storage, transport and administration are complex and sensitive to several factors, which also restrict the use of vaccines. The current levels of use of vaccines are supported by Governments and charitable international organizations. There is a very urgent need to encourage research, development and a wider deployment of vaccines.

CHAPTER 17

MICROBIAL TOXINS

17.1 INTRODUCTION

Microbial toxins are the products of microbes that act at very low concentration to influence the course of disease development. Different kinds of toxins are produced by different groups of organisms. The toxins can be host specific or non host specific. The host specific toxins are pathotoxins which play a causal role in disease development. Non host specific toxins are generally toxic to plants and are called phytotoxins. They may or may not play a role in disease development. These toxins produced by bacteria fungi and algae can be exploited as a potential source of medicines and have implications in human health .

17.2 BACTERIAL TOXINS

Bacterial toxins produced by bacteria can be exotoxins or endotoxins. Exotoxins are secreted into the medium by the bacteria while endotoxins remain within the cell and are released after their death. Bacterial toxins cause various diseases in animals and humans.

Bacterial toxins affecting the nervous tissue are neurotoxins and those affecting the gut are enterotoxins. A few examples of bacterial toxins are cholera toxin (enterotoxin) and tetanus toxin (neurotoxin).

The toxins which are altered to remove the toxicity are called toxoids. They can be used as antigenic agents to elicit antibody production.

17.3 MYCOTOXINS

A large number of fungi have been demonstrated to produce toxic substances known as mycotoxins, which are organ specific carcinogens causing plant and human diseases. They also threaten the availability and safety of food products. About 18 mycotoxins have been recognized as carcinogens, aflatoxin being the most potent (Chu and Bhatnagar, 2004).

Mycotoxins are produced by fungi as secondary metabolites and they have no apparent significance to the producing organism. They possibly represent one way the fungus resorts to reduce the metabolic precursors that are not required. Many of the mycotoxins cause diseases (mycotoxicoses) and are highly poisonous. The major mycotoxins are discussed here.

Aflatoxins

Aflatoxins are a group of mycotoxins produced by *Aspergillus flavus*, a common soil fungus. Aflatoxin attracted the attention in 1960 when countless turkeys and ducks died in the United Kingdom after the consumption of contaminated feed of groundnut cakes, resulting in liver and kidney damage. Aflatoxins have been demonstrated on fresh beef, ham, bacon, milk, beer, cocoa, raisins, soybean, wheat bread, cheese and apple juice. Aflatoxins are also found on improperly stored nuts, grains, dried fruits and spices. They may form on intact millets and maize in hot humid weather.

Evidence of acute aflatoxicosis in humans has been reported from several parts of the world, mainly Taiwan, Uganda and India. The syndrome is characterized by vomiting, abdominal pain, pulmonary and cerebral edema, convulsions, coma and death.

Aflatoxins are potent liver toxins known to cause cancer in animals and human beings. The carcinogenic effects of aflatoxins have been demonstrated experimentally in a wide variety of domestic and experimental animals and in human beings, who inadvertently consumed contaminated corn, pepper, peanuts, or peanut meal.

Field outbreaks of aflatoxin induced diseases have been observed in turkeys, ducks, chickens, swine, cattle, dogs and trout.

The occurrence of aflatoxins is influenced by certain environmental factors; hence the extent of contamination varies with geographic location, agricultural and agronomic practices.

Aflatoxin M1 and M2 are modified forms of aflatoxin B1 found in bovine milk. (Natori *et.al.*, 1988).

Altoryzine

Altoryzine is a mycotoxin produced by *Aspergillus oryzae* var. *microsporus* on bean sprouts and sprouts used as animal feed. The consumption of the toxin results in muscular paralysis.

Fumonisins

Fumonisins are produced by *Fusarium anthrophilum, Fusarium nupiforme, Fusarium nygamai, Fusarium proliferatum, Fusarium verticillioides,* on corn based foods, sorghum and rice. Fumonisins are hepatotoxins, immunotoxins and carcinogens causing hepatic nodules, adenofibrosis, hepatocellular carcinoma, ductular carcinoma, cholangio carcinoma (Haschek *et al.*, 2001) and tubular nephrosis in animal models. They also cause Equine leuko encephalo malacia and pulmonary edema in rats and horses. In pigs fumonisins cause liver damage and alters cardiac function and in cattle they cause renal injury and hepatic lesions.

F-2 Ze

F-2 Ze is a phytoestrogen produced by *Fusarium graminearum* on wheat. It causes hyper esterogenic effects and reproductive problems such as onset of premature puberty in female animals, especially swine.

Species of *Fusarium* also produce several other mycotoxins such as moniliformin, fusaproliferin and beauvericin (Chu, 2002).

Moniliformin

Moniliformin is produced by *Fusarium verticellioides* which causes cardiomyopathy and Keshan disease in humans.

Nitropropionic acid (NPA)

Nitropropionic acid is produced by *Aspergillus flavus, Aspergillus wentii, Aspergillus oryzae* and *Penicillium atrovenetum.* This mycotoxin causes convulsions, lung congestion and liver damage in test animals.

Production of NPA in sugarcane by *Arthrinium sacchari, Arthrinium saccharicola and Arthrinium phaeospermum* has been found to result in fatal food poisoning in humans (Liu *et al.*, 1988).

Ochratoxins

Ochratoxins are produced by *Aspergillus citricus, Aspergillus fonsecaeus, Aspergillus mellus, Aspergillus ochraceus, Aspergillus ostianus, Penicillium cyclopium, Penicillium variable, Penicillium viridicatum* and *Petromyces alliceus.* They are produced in stored barley, oats, rye, corn and wheat. They are also found in bread, coffee beans, nuts, olives, moldy tobacco, raisin, pepper, cheese, fish, pork, milk powder, wine and fruit juices (Pietri *et al.*, 2001).

Ochratoxin is a potent nephrotoxin causing kidney damage, liver necrosis and enteritis in animal species and tumors in rats (Schlatter *et al.*, 1996; Chu 2002). It interferes with hormonal calcium signaling, leading to altered cell proliferation. In humans ochratoxin induced nephropathy and Balkan endemic nephropathy. Ochratoxins also cause dose dependant immuno modulation and reduce the degree of pneumonia.

Patulin

Patulin is a hepatotoxin produced by *Aspergillus clavatus, Aspergillus gigantea, Aspergillus terreus, Byssochlamys flava, Byssochlamys nivea,*

Penicillium claviforme and *Penicillium expansum* and was found on unpolished rice, apple juice, grapes, cider, pears and bananas.

Penicillium toxins

Species of *Penicillium* produce citreoviridin, citrinin, cyclochlorotine, cyclopiazonic acid, luteoskyrin (LS), penicillic acid, penitrems A-F, roquefortines A-C, rubratoxin A and B, rugulosin (RS), xanthomegnin with hepatotoxic effects,

Citrinin is produced by *Penicillium citrinum* and *Penicillium viridicatum.* It is a nephrotoxin causing kidney damage and was demonstrated from polished rice, moldy bread, wheat, oat and rye.

Cyclopiazonic (CPA) acid, produced by *Aspergillus flavus, Aspergillus versicolor, Penicillium cyclopium, Penicillium crustosum* and *Penicillium griseofulvin* causes hyperplasia, convulsions, liver, spleen, kidney, salivary gland and myocardial damage (Cast, 2003). It is found in corn, cheese and sausges (Lopez-Diaz *et al.,* 2001).

Penicillic acid is produced by *Penicillium puberulum* and *Penicillium cyclopium* and also by *Aspergillus ochraceus.*

The penitrems are a family of complex indole derivates produced by the mold *Penicillium crustosum* which have been shown to be tremorogenic in animals. Penitrems cause Ljesha shakes in humans and grass staggers in sheep.

Rubratoxin a highly toxic hepatotoxin is prodced by *Penicillium rubrum* and *Penicillium purpurogenum.*

Phalloidin and Amanitin

Phalloidin and Amanitin are produced by *Amanita phalloides* commonly known as angel of death. Both are neurotoxins.

Slaframine

Slaframine is produced by *Rhizoctonia leguminicola* on legume forage crops. It causes excessive salivation in ruminants.

Sporidesmines

Sporidesmines, a group of hepatotoxins causing facial eczema are produced by *Pithomyces chartarum* and *Sporidesmium chartarum.*

Sterigmatocystin (ST) and Dihydrosterigmatocystin (DHST)

Sterigmatocystin (ST) and dihydrosterigmatocystin (DHST) are the precursors of aflatoxins which are hepatotoxic and carcinogenic mycotoxins produced by *Aspergillus nidulans, Aspergillus rugulosus, Aspergillus versicolor, Penicillium luteum* and species of *Bipolaris,* and *Chaetomium* on barley, rice, corn, coffee beans and cheese. They are also genotoxins (Chu, 2002).

Tenuazonic acid

Tenuazonic acid is produced by *Phoma sorghina* and *Pyricularia oryzae* and is found to cause a haematological disorder, known as "onyalai" in people living in South of Sahara. It is found in apples, tomato and cereals.

Trichothecenes

Trichothecenes are sesquiterpernoid mycotoxins produced by species of *Acremonium, Cephalosporium, Cylindrocarpon, Dendrodochium, Fusarium, Gliocladium, Myrothecium, Stachybotrys, Trichoderma, Trichothecium* and *Verticimonosporium* on barley, maize, wheat and rye. Trichothecenes are dermatotoxic, cytotoxic and phytotoxic. They are contact toxins and cause severe blistering and necrosis of the target tissue. They are of great concern as food and feed contaminants (Brown *et al.,* 2001). They are suggested to play a role in plant pathogenesis as virulent factors. They are being used to stimulate and suppress immune function and in treatment of tumours.

T2 Toxin (Type A TCTC)

T2 toxin, a highly toxic type A TCTC is produced by *Fusarium tricinctum, Fusarium sporotrichioides, Fusarium poae, Fusarium sulphureum, Fusarium acuminatam, Fusarium sambucinum,* species of *Cephalosporium, Myrothecium* and *Stachybotrys* on cereal grains such as barley, corn, oat, and wheat. TCTC toxins are cytotoxic, causing haemorrhage, edema, skin necrosis, emesis, tachycardia, diarrhoea, anorexia, leukocytosis, leucopaenia, fore stomach ulceration and hepatomegaly in animals.

TCTCs are hazardous to humans. They cause 'Sick building syndrome (SBS) in humans residing in mold-damaged buildings. The symptoms include hair loss, diarrhoea, fatigue, dermatitis, sore throat, headache, lesions in lymph nodes, spleen, thymus and liver(Toumi *et al.,* 1993). The causal organisms are yeasts such as *Saccharomyces atra* and *Saccharomyces chartarum.*

Vomitotoxin (Deoxynivalenol-DON)

DON is a major type of Type B TCTC mycotoxin produced by *Fusarium graminearum, Fusarium culmorum* and *Fusarium crookwellense* on barley, oat, sorghum and safflower seeds. DON produces anorexia and emesis in both humans and animals. It also causes "Hole in the head" disease and Alimentary toxic alekiia.

17.4 PHYTOTOXINS

Phytotoxins are produced by a microorganism which are toxic to plants. (Table 17.1) Certain parasitic fungi and bacteria produce a variety of chemical substances, which are known to reproduce part of the symptoms of plant diseases even in the absence of disease causing organisms. The presence of such a toxin was first visualized in the disease caused by *Corynebacterium diphtheriae*.

Phytotoxins can be vivotoxins or pathotoxins.Vivotoxins are substances produced in the infected host by pathogens. They promote the production of the disease but are not the agents of the disease. Pathotoxins play a causal role in disease production and produce symptoms of the disease.

The toxins interfere with the permeability of plasma membrane, affecting water translocation, which leads to irreversible pathological wilting. The following are important phytotoxins. (Mehrotra and Agarwal, 2003).

Alternaria toxins

Alternaria alternata, Alternaria citri, Alternaria solani and *Alternaria tenuissima* causing Alternaria diseases produce toxic substance in apple, tomato, sorghum, barley, oat and blue berries. The toxins include alternariol, altertoxin I, altenuene, altenuisol, alternaric acid, altenusin and dehydroaltenusin.

Colletotin

Colletotin is produced by *Colletotrichum fuscum* which causes anthracnose of *Digitalis.*

Fusaric acid

Fusaric acid is produced by *Fusarium oxysporum* causing wilts in tomato, pea and banana. It is also produced by *Fusarium heterosporum* and *Fusarium vasinfectum.* Fusaric acid is called "wilt toxin" and is known to play a major role in wilt disease.

Fusarin C (FC)

Fusarin C is a mutagen produced by *Fusarium subglutinans* and *Fusarium graminearum* causing wilts.

Lycomarasmin

Lycomarasmin is also a wilt toxin produced by *Fusarium oxysporum* f. sp. *lycopersici,* in tomato.

Periconin or PC toxin

Periconin is produced by *Periconia circinata,* which causes milo disease of *Sorghum vulgare* var. *subglabrescens.*

Piricularin and Pyriculol

Are produced by *Pyricularia grisea* which is the causal organism of blast disease of rice. Piricularin has been identified as a metabolic product of the fungus, which is responsible for characteristic symptoms of the disease.

Phaseolotoxin

Phaseolotoxin is produced by *Pseudomonas syringae* p.v. *phaseolicola* which causes halo blight of beans and other legumes.

Tentoxin

Tentoxin is produced by *Alternaria alternata*, which causes chlorosis in many plants.

T-toxin

T-toxin is produced by *Helminthosporium heterostrophus*, which causes Southern corn leaf blight.

Tabtoxin (wild fire toxin)

Tabtoxin is produced by *Pseudomonas syringae* p.v. *tabaci*, which causes wild fire disease of tobacco. The toxin is also produced by other pathovars of *Pseudomonas syringae*, causing diseases in soybean, maize, oats and coffee.

Victorin

Victorin is produced by *Helminthosporium victoriae*, which causes seedling blight of oat.

Table 17.1 Phytotoxins produced by various fungi

Phytotoxin	Fungal species
Alternaric acid	*Alternaria solani*
Altenuene	*Alternaria tenuis*
Alternariol monomethyl ether	*Alternaria tenuis*
AK toxin(phytoalternarin)	*Alternaria kikuchiana* *Phyllosticta maydis*

Ceratoulmin	*Ceratocystis ulmi*
Colletotin	*Colletotrichum fuscum*
Diaporthin	*Endothia parasitica*
Fusaric acid	*Fusarium oxysporum* f.sp. *vasinfectum* *Fusarium heterosporum* *Fusarium cubense* *Fusarium moniliforme* *Fusarium oxysporum* *Fusarium lycopersici* *Nectria cinnabarina*
Fusicoccin	*Fusicoccum amygdali*
Fumaric acid	*Rhizopus* species
HC toxin	*Helminthosporium carbonum*
HS toxin	*Helminthosporium sacchari*
HM-T toxin	*Helminthosporium maydis*
Helminthosporal	*Helminthosporium sativum*
Lycomarasmin	*Fusarium oxysporum* f.sp *lycopersici*
Ophiobolin A (cochliobolin)	*Helminthosporium oryzae*
Oxalic acid	*Aspergillus niger*, species of *Sclerotium*
Piricularin, α-Picolinic acid	*Pyricularia grisea*
Pyriculol	*Pyricularia grisea*
Periconin or PC toxin	*Periconia circinata*
PM toxin	*Phyllosticta maydis*
Rhizobitoxin	*Rhizobium japonicum*
Tentoxin	*Alternaria tenuis* *Alternaria alternata*
Tenuazonic acid	*Alternaria longipes* *Alternaria alternata* *Pyricularia grisea*
Victorin	*Helminthosporium victoriae*

Source: Mehrotra and Agarwal, 2003

17.5 ALGAL TOXINS

The bloom producing algae such as Cyanobacteria and dinoflagellates are the major producers of toxins. These toxins are neurotoxins and hepatotoxins which also cause allergic responses in humans. These toxins are discussed in Chapter 10 under water blooms.

CHAPTER 18

NATURAL PRODUCTS FROM ENDOPHYTES

18.1 INTRODUCTION

Endophytic micro-organisms reside in the living tissues of a host plant, in a symbiotic to slightly pathogenic relationship, virtually in every plant.

Fungi are the frequently isolated endophytic microbes. The most common endophytic fungi belong to the group Fungi Imperfecti (Deuteromycetes), a temporary assemblage till the perfect sexual stage of the species was identified.

The endophytes may produce a variety of chemical substances that help the hosts in combating other pathogens and pests. Such compounds are potentially useful in modern medicine, agriculture, and industry. Novel antibiotics, antimycotics, immunosuppressants, and anticancer compounds are only some examples of what has been found after the isolation, culture, purification, and characterization of some selected endophytes in recent times (Strobel and Daisy, 2003). The potential prospects of finding new drugs that may be effective candidates for treating newly developing diseases in humans, plants, and animals, are extensive.

Endophytic microbes are relatively unstudied as a source of bioactive novel natural products for exploitation (Strobel *et al.,* 2005). The endophytes produce toxic compounds only in their respective hosts. Though innumerable species of higher plants are hosts for one or more endophytes, only grasses have been studied more extensively in relation to endophytic biology (Bacon and White, 2000).

18.2 ANTIBIOTICS FROM ENDOPHYTIC BACTERIA

One of the most common bacterial genus associated with plants belong to *Pseudomonas. Pseudomonas viridiflora* in the leaves of grasses produces a phytotoxic ecomycin, a lipopeptide active against human pathogenic yeasts such as *Cryptococcus neoformans* and *Candida albicans* (Miller *et al.,* 1998).

The pseudomycins produced by plant-associated *Pseudomonas* are antifungal peptides active against *Candida albicans, Cryptococcus neoformans* and *Ceratocystis ulmi* (Dutch Elm disease). They are also active against the black Sigatoka disease of bananas caused by *Mycosphaerella fijiensis* (Harrison *et al.*, 1991; Ballio *et al.*, 1994).

An endophytic *Streptomyces* species isolated from *Lolium perenne* (grass) (Guerny and Mantle, 1993), produces the antibiotic munumbicin that showed activity against *Bacillus anthracis* and *Mycobacterium tuberculosis* multidrug resistant strains and against the malarial parasite *Plasmodium falciparum* (Castillo, *et al.*, 2002).

Streptomyces species (NRRL 30566) from *Grevillea pteridifolia* (Proteaceae, a flowering plant family) produces kakadumycin A with a broad spectrum activity particularly against *Bacillus subtilis* (Castillo, *et al.*, 2003).

18.3 ANTIBIOTICS FROM ENDOPHYTIC FUNGI

Cryptosporiopsis quercina, an imperfect stage of *Pezicula cinnamomea,* isolated as an endophyte from *Tripterigium wilfordii* (Strobel *et al.*, 1999) produces peptide antimycotic cryptocandin that contains 3-hydroxy-4-hydroxy methyl proline, which was active against human yeast and fungal pathogens such as *Candida albicans* and *Trichophyton* species. It is currently being considered for use against fungal diseases of the skin and nails in humans.

A strain of *Pyricularia microspora* isolated from *Torreya taxifolia* produces pestaloside, pestalopyrone and hydroxy pestalopyrone (Lee *et al.,* 1995 a,b). These compounds are phytotoxic. Cryptocin, produced by *Cryptosporiopsis quercina,* is active against *Pyricularia oryzae,* the causal organism of one of the world's worst plant diseases, the rice blast and is being considered as natural chemical control agent (Li *et al.*, 2000).

Phomopsichalasin, a cytochalasin type of compound from endophytic *Phomopsis* species was active against *Bacillus gallinarum* and *Staphylococcus aureus.* (Horn *et al.*, 1995), and so have a promise to be useful antibacterials.

An endophytic species of *Fusarium* from *Selaginella pallescens* (Pteridophyte) is potent against *Candida albicans* where as *Colletotrichum gloeosporioides* isolated from *Artemisia mongolica* (flowering plant) was active against the fungus *Helminthosporium sativum* (Zou, *et al.*, 2000).

Though at its infancy, the use of products from endophytic fungi in the inhibition of viruses is a promising area. Cytonic acids A and B isolated from *Cytonaema* (endophytic fungus) function as inhibitor of human cytomegalo virus (hcMV) (Guo, *et al*, 2000).

18.4 ANTICANCER AGENTS FROM ENDOPHYTIC FUNGI

18.4.1 Paclitaxel

Taxol is the trade name of paclitaxel, one of a group of chemical compounds called taxanes. Paclitaxel prevents polymerization of tubulin, causing disruption of mitosis at the metaphase stage. This retards proliferation of cancer cells.

Paclitaxel, a diterpenoid isolated from the Pacific yew (*Taxus brevifolia*) and other species of *Taxus* such as *Taxus baccata, Taxus canadensis and Taxus wallichiana*, is a major but very expensive drug being routinely used in the treatment of ovarian and breast cancer since 1995. Paclitaxel is also effective against kaposis sarcoma, bladder, prostrate, esophageal, head, neck, cervical and endometrial cancer (Kowalski *et al.*, 1997). Paclitaxel inhibits cell division and induces programmed cell death in the cancerous tissue.

Pacific yew is one of the slowest growing trees in the world, and treating a single patient requires cutting down and processing of six 100-year old trees. Total synthesis of paclitaxel being difficult and economically inviable, scientists looked for alternate sources. Initially, actinobacteria were identified, from which paclitaxel-like compounds could be obtained by fermentation. Unfortunately, these compounds from different sources, are not as potent as paclitaxel itself. Semi-synthesis of paclitaxel from other compounds was then explored. 10-Deacetylbaccatin, extracted in relatively large amounts from the trees or cell cultures of the yew and its related species, could be converted into paclitaxel.

It is now understood that paclitaxel and its derivatives are produced by an endophytic fungus *Taxomyces andreanae* in *Taxus brevifolia* (Strobel, *et al.*, 1993). C^{14} labeling studies showed the presence of fungal derived paclitaxel in the medium in which the fungus was cultured (Lee, *et al.*, 1995).

Species of *Pestalotiopsis* are the most common endophytic fungi associated with species of *Taxus*. *Pestalotiospsis microspora* is most frequently isolated from not only species of *Taxus* but also *Torreya taxifolia* (Lee *et al.*, 1996) and the bald cypress (*Taxodium distichum*) (Strobel, 2002). This fungus and *Pestalotiospsis guepini* isolated from *Wollemia nobilis* (Wollemi pine), produce paclitaxel (Strobel *et al.*, 1997). *Maguireothamus speciosus* (Rubiaceae, a flowering plant family) harbours a paclitaxel producing novel fungus *Seimatoantlerium tepuiense*.

Other fungi that produce paclitaxel are species of *Periconia* (Li *et al.*, 1998), *Seimatoantlerium nepalense* (Bashyal *et al.*, 1999), and species of *Tubercularia* from *Taxus mairei*, the Chinese yew (Wang *et al.*, 2000), *Sporomia minima* and a species of *Tricothecium* from *Taxus wallichiana* which also occurs in the Indian Himalayan region (Shreshta *et al.*, 2001) and *Nodulisporium sylviforme* (Jacoby, 2005).

The relationship between the taxane producing fungi and their hosts is symbiotic since the fungi do not cause any apparent disease and the compounds protect the host from other pathogens, while the fungus gets a safe habitat to thrive (Young *et al.*, 1992).

Torreyanic acid, a selectively cytotoxic quinone and a potential anticancer agent was isolated from a strain of *Pestalotiopsis microspora,* a fungus isolated from a strain of *Pestalotiopsis microspora,* a fungus associated with *Torreya taxifolia* (Lee *et al.,* 1996).

Antitumour alkaloids like cytochalasins are commonly produced by endophytic fungi such as *Xylaria, Hypoxylon* and *Chalara.* Three new cytochalasins were reported from a species of *Rhinocladiella,* an endophyte of *Tripterigium wilfordii* (Celastraceae, a flowering plant family) (Wagenaar *et al.,* 2000). However, on account of their cellular toxicity these compounds have not yet been developed into pharmaceuticals.

18.4.2 Camptothecin

Camptothecin, an indole alkaloid originally known from the Chinese plant *Camptotheca acuminata* (Nyssaceae) which also occurs in Tibet, is an anti-cancer drug in clinical use against solid malignancies. Camptothecin inhibits the enzyme Topoisomerase I required for DNA transcription, synthesis, and replication and prevents proliferation of cancer cells.

The discovery of camptothecin in the Indian tree *Nothapodytes nimmoniana* (=*Nothapodytes foetida, Mappia foetida,* Icacinaceae) two decades ago has generated a lot of interest and the excessive use of the tree for its drug containing bark has severely affected the density of natural populations in the Western Ghats of India.

Interestingly, camptothecin and related alkaloids seem to be synthesized by an endophytic fungus *Entrophospora infrequens* symbiotic in *Nothapodytes nimmoniana,* and not by the host species *per se.* Fifty two strains of *Entrophospora infrequens,* isolated from the inner bark of *Nothapodytes nimmoniana,* were tested for the ability to produce the anticancer alkaloid camptothecin (Amna *et al.,* 2006). Bioreactors gave 24 times of camptothecin, in half the time than in shake flask cultures. No camptothecin was detected in zero time cultures or in uninoculated culture broth. Several studies on tissue culture of *Nothapodytes nimmoniana* have failed to detect any appreciable quantities of camptothecin in the cultures, reiterating the need for the endophyte to synthesize camptothecin.

The endophyte *Entrophospora infrequens* and its product camptothecin have also been isolated from *Ophiorrhiza pumila* (Kitajima *et al.,* 2002), *Ophiorrhiza mungos* and *Ophiorrhiza rugosa* (Roja, 2006). However, *Nothapodytes nimmoniana* gave higher yields of camptothecin than some species of *Ophiorrhiza* (Roja, 2006).

Camptothecin is now known from several Indian angiosperms such as *Ophiorrhiza prostrata,* (Beegum et al., 2007), *Ophiorrhiza rugosa* var. *decumbens* (Vineesh *et al.,* 2007) and *Trema orientalis* (Erika et al., 2004).

In addition to its anti-cancer activity, camptothecin from *Entrophospora infrequens* was demonstrated to possess immunomodulatory activity (Pur *et al.,* 2007).

18.5 ANTIDIABETIC AGENTS FROM ENDOPHYTIC FUNGI

A non-peptidal fungal metabolite (L-783, 281), isolated from an endophytic fungus *Pseudomassaria* species (Zhang *et al.,* 1999), has insulin mimicking effect and was not denatured in the digestive tract on per-oral administration (Strobel *et al.,* 2005). This metabolite offers a chance for development of a future antidiabetic drug.

18.6 IMMUNOSUPRESSIVE AGENTS FROM ENDOPHYTIC FUNGI

Fusarium subglutinans isolated from *Taxus wilfordii* produces immunosuppressive diterpene pyrones- subglutinols A and B which are non-toxic. These compounds can be the future immunosuppressive drugs for prevention of allograft rejection in transplant patients, rheumatoid arthritic and insulin dependant diabetic patients.

18.7 ANTIOXIDANTS FROM ENDOPHYTES

Pestacin and isopestacin, found in the culture fluids of *Pestalotiopsis microspora,* an endophyte isolated from *Terminalia morobensis* (Combretaceae, a flowering plant family) (Strobel *et al.*, 2002; Harper *et al.*, 2003), exhibit more potent antioxidant activity than that of trolox, a vitamin E derivative.

18.8 ENDOPHYTES IN BIOCHEMICAL RESEARCH

Pyricularia microspora, a common rainforest endophyic fungus produces ambuic acid and terrein that have been used as models to develop new solid-state nuclear magnetic resonance sensors to assist in the characterization of the molecular stereochemistry of organic molecules.

CHAPTER 19

FERMENTED FOODS AND BEVERAGES

19.1 INTRODUCTION

Oldest of all biotechnological processes is fermentation by microbial systems or their products. Bacteria, yeasts and fungi, the mediators of this chemical process have yielded alcoholic beverages, industrial alcohols and food products that are not only nutritious but are of culinary interest. In this chapter, industrial alcoholic beverages, fermented indigenous beverages and foods unique to the country are considered.

19.2 INDUSTRIAL BEVERAGES

19.2.1 Industrial Fermented Beverages

Micro-organisms are involved in the production of almost all alcoholic beverages. Starch containing grains are sprouted, dried and ground, in a process called malting, which induces the production of starch degrading amylases. Starch is converted by the amylases into glucose and maltose, which are then fermented by microbes into ethanol and carbon dioxide.

Different types of alcoholic beverages are given in Table 19.1

Table 19.1 Types of alcoholic beverages

Beverage	Source species
Beer	Made by fermentation of barley by *Saccharomyces cerevisiae.*
Corn beer	Beer from corn
Wheat beer	Beer from wheat
Rye beer	Beer from rye
Potato beer	Beer from potato
Ale	Beer with higher percent of alcohol
Stout	Brown bear
Shandy	Beer with orange juice
Whisky	Prepared by fermentation of corn, rye and barley by *Saccharomyces cerevisiae.*

American corn whisky	Made from corn
American rye whisky	Made from rye
Canadian whisky	Prepared from corn, rye and barley malt
Grain whisky	Prepared from malted barley with unmalted maize
Irish whisky	Whisky from rye from Ireland
Scotch whisky	A blend of malt whisky and grain whisky made from barley from Scotland
Bourbon	Whisky from maize
Brandy	Distilled from fermented grapes
Cognac	A type of brandy made in the city of Cognac in France.
Gin	Liquor flavoured by *Juniperus* berries
Cider	Made from apples
Betsa, Cachaca	Made from sugar cane
Tequila, Mezcal	Made from *Agave* from Mexico
Rum	Distilled fermented molasses
Vodka	Liquor made from potato starch from Russia
Wine	Made by alcoholic fermentation of grapes by *Sacchoromyces cerevisiae*
Champagne	Sparkling wine made from grapes grown in Champagne distsrict in France
Dry wine	Wine with less of sugar
Fortified wine	Wine mixed with brandy
Red wine	Wine made from red grapes with skin
White wine	Wine from white grapes with skin
Vermouth	White wine flavoured with worm wood (*Artemisia abysynthum*)
Arrack	Crudely distilled from toddy
Sake	Liquor made from fermented rice and hydrolysed by *Aspergillus oryzae*
Schochu (Japanese spirit)	Distilled from potato.
Riceshoch	Rice spirit.

19.2.2 Types of Cocktails

A mixture of choice distilled alcoholic beverages such as whisky, rum, vodka, brandy etc., in precise proportions is called a cocktail.

1. Cocktails with Beer

These are not true cocktails, because beer is not a distilled beverage.

Black velvet: Guinness beer or stout and champagne.

Boiler maker: Cocktail of whisky or vodka added to a glass of beer.

Irish carbomb: Guinness beer or stout and whisky with Irish cream.

Dr. Pepper: Amaretto (Italian liquer (Liquer is a potent alcoholic preparation flavoured, perfumed and sweetened) flavoured with almonds) (1oz) dropped into quarter pint of beer and quarter pint of Coca Cola.

Dogs Bollocks: A mixture of melon vodka, lager beer and lime cordial.

Michelada: A Mexican drink similar to Bloody Mary with Mexican beer instead of vodka.

Mexican Piss: Cocktail of corona mixed with Tequila.

Red Eye: A cocktail of tomato juice and beer.

Skippys: A drink containing equal proportion of beer, vodka and lemonade.

Snake bite: A cocktail of beer, cider and blackcurrent cordial.

Tom Bass: Cocktail of Bass with a shot of Jagermeister.

Yorsh: Beer mixed with vodka.

French connection: Cognac with orange liquer.

2. Cocktails with Brandy and Cognac

Brandy Alexander: 1 shot brandy, 1 shot Dekuyper cre'me, Decacao-white and 1 shot of cream.

Incredible Hulk: 2 oz Cognac, 3 oz hypnotique (a blend of vodka, cognac and tropical fruit juice).

Side Car: 2 oz brandy, 1 oz lemon juice, 1 oz orange peel spirit.

3. Cocktails with Gin

Gin and Tonic: A mixture of gin and carbonated soft dirnk flavoured with quinine.

Hanky Panky: 2 dashes of bitters (Angostura, campari, amerpicon), 1/2 oz Italian vermouth, 1/2 oz dry gin.

Salty Dog: 5 oz grape fruit juice, 1 oz gin with 1/4 tsp salt.

Pink Gin: 1 oz dry gin, splash of cranberry juice.

Singapore Sling: 1 oz gin, 1 oz club soda, 1/2 oz cherry and pomegranate syrup.

4. Cocktails with Rum

Dr. Pepper Slurpee: A mix of Baccardi, vanilla and rum.

Blue Hawaiian: A mix of rum and blue curacao (artificial colour) blended with pineapple juice and coconut cream.

5. Cocktails with Sake

Duncan MacLeod: Cocktail of Coca Cola, scotch whisky and sake.

Saketini: Vodka martini with sake instead of vermouth.

6. Cocktails with Tequila

Margarita: Mix of tequila, triple sec and fresh lime juice.

Tequila Slammer: Equal parts of tequila and champagne or ginger ale.

Tequila Sunrise: two parts tequila, two parts orange juice and two dashes of grenadine syrup.

TNT: Tequila mixed with Tobasco sauce.

7. Cocktails with Vodka

Black Cossack: Chilled Guinness with vodka.

Black Marble: Vodka with black olive and orange peel.

Black Celebration: One part vodka with two parts of coca cola.

Bloody Mary: A mix of tomato juice and vodka seasoned with salt and pepper.

Cape God: Vodka with cranberry juice.

Chi-chi: Pineapple juice and coconut milk with vodka.

Lawnmower: Vodka with wheat grass juice.

Red lotus: Lichee juice and cranberry juice with vodka.

Vodka McGovern: Equal parts of vodka, club soda and orange juice with a squeeze of lime.

8. Cocktails with Whisky

Blacj Jack: 1 oz whisky, 1 oz coffee liquer, 1/2 oz orange liquer and 1/2 oz lemon juice.

Rusty Nail: 6/10 scotch whisky and 4/10 drambuie.

Manhattan: 3/4 oz sweet vermouth, 2 oz bourbon whisky, one maraschino cherry and 1 twist of orange peel.

9. Clocktails with Wine

Martini: Liquor made by mixing vodka and wine

Sherry: A mixture of wine and whisky.

Port: A mixture of brandy and wine.

19.2.3 Fruit Based Alcoholic Beverages

Yeasts are used in preparation of alcoholic beverages from fruits. The quality of the beverage depends upon the variety of fruit, yeast strains, and preservation method.

Apricot wine – *Saccharomyces cerevisiae.*

Date wine – *Bacillus, Gluconobacter, Klebsiella, Candida, Saccharomyces, Torulopsis.*

Honey wine (Apple, Pear and Plum) – *Saccharomyces cerevisiae.*

Jackfruit wine – *Endomycopsis.*

Jambal wine – *Saccharomyces cerevisiae.*

Kiwi fruit wine – *Saccharomyces cerevisiae.*

Litchi wine – *Saccharomyces cerevisiae.*

Mango wine – *Saccharomyces cerevisiae* var. *ellipsoids.*

Plum wine – *Saccharomyces pombe.*

Sparkling plum wine – *Saccharomyces cerevisiae.*

Coconut wine – Species of *Candida* and *Pichia.*

19.3 TRADITIONAL FERMENTED ALCOHOLIC BEVERAGES

These are primitive, untreated traditional liquors prepared by villagers. They are a good source of vitamins, amino acids and calories (Ng, 2004 c).

African Kaffir (Sorghum beer)

Sorghum grains are ground and allowed to ferment by lactic acid bacteria to cause souring and *Sacharomyces cerevisiae* to produce alcohol (Hesseltine, 1979). The drink has sour yoghurt like flavour.

Chinese Lao-chao

Is made from fermenting glutinous rice. *Amylomyces rouxii, Rhizopus chinensis* and *Rhizopus oryzae* ferment the starch in the rice.

Ethiopian Teji

Is a home processed honey wine fermented by *Saccharomyces* species. Honey is mixed with hop stems and fermented in mud pots with daily stirring and the filtrate is collected.

Indian Ruhi

Is prepared from fermentation of boiled rice. Boiled rice is cooked and mixed with an inoculum of *Mucor*, *Rhizopus* and yeasts. The fermented liquid at the top is collected.

Japanese Sake

Is rice wine made from steamed rice and allowed to ferment in an open system, with a starter culture of yeast and lactic acid bacteria. *Saccharomyces sake* ferments the rice slowly reaching 20 per cent alcohol content (Murakami, 1972). *Aspergillus oryzae* hydrolyses starch.

Kenyan Urwaga

Is a sour drink prepared from green banana, maize, pearl millet or sorghum. The green bananas are ripened in soil with banana leaves. The peeled bananas are mixed with grass and the juice is forced out by squeezing. Roasted flour of maize, and sorghum is added to the banana juice and allowed to ferment in an underground pit (Harkishore, 1977). Yeasts and lactic acid bacteria carry out fermentation.

Mexican Tesguino

Is prepared from fermentation of germinated maize or maize stalk. Maize kernels or stems are boiled in water and yeasts such as *Candida guilliermondii* and *Saccharomyces cerevisiae* are added as fermenters (Lappe and Ulloa, 1989).

Nigerian Pito

Is a bitter, sweet sour beverage with a fruity flavor produced by fermenting maize or sorghum. Fungi such as *Aspergillus flavus, Penicillium citrinum, Penicillium funiculosum* and *Rhizopus oryzae,* bacteria such as species of *Lactobacillus, Leuconostoc,* yeasts such as *Candida* and *Saccharomyces* play a role in fermentation (Ekundayo 1977).

Palm Toddy

Exudate from cut ends of vegetative buds, young leaves and inflorecences of coconut and other palms like *Borassus flabellifer* and *Caryota urens* is tapped into earthenware pots and fermented by *Kloeckera javanica, Saccharomyces marxianus, Saccharomyces cerevisiae, Saccharomyces chevielier,* and *Pichia ohmeri.*

Philippine Basi

Is a sugarcane wine prepared by boiling sugar cane juice and fermented in earthern pots by the addition of molds, yeasts and bacteria (Sakai and Caldo, 1984).

Philippine Tapuy

Is a sweet rice wine prepared by cooking rice in a clay pot and mixed with powdered *Endomycopsis*, *Saccharomyces uvarum* and lactic acid bacteria to ferment.

South American Chichi

Is a drink prepared from germinated maize mixed with human saliva. The drink has cider flavor. *Aspergillus* sp, *Mycoderma vini, Penicillium* sp, *Saccharamyces cerevisiae, Saccharomyces pastorianus, Saccharomyces apiculata* and lactic acid bacteria carry out fermentation (Gomez, 1949). Chichi prepared from using human saliva reportedly has a better flavour.

19.4 FERMENTATION OF FRUITS AND VEGETABLES

Olives

Olives are processed either unripe (green) or ripe (black)by allowing fermentation by *Leuconostoc mesenteroides*, *Lactobacillus plantarum* and species of *Aeromonas, Flavobacterium and Klebsiella* and preserved in brine.

Pickles

Pickle making is a process of fermenting various fruits and vegetables by *Lactobacillus plantarum* and *Pediococcus cerevisiae.*

Sauerkraut

Is shredded, salted, cabbage fermented by lactic acid bacteria. The coliform bacteria perform the initial fermentation. The rest of the fermentation is carried out by *Lactobacillus plantarum* and *Leuconostoc mesenteroides*.

19.5 FERMENTED MEAT PRODUCTS AND SEA FOODS

Sausage

Chopped or minced meat with fat and cereals, is seasoned and stuffed into bovine intestinal or synthetic tubes. Fermentation is achieved by starter cultures containing *Candida famata, Penicillium chrysogenum, Penicillium nalgiovense, Pediococcus* and *Micrococcus*.The inclusion of yeasts encourages development of colour and aroma in sausages. The fermented meat is preserved by lactic acid bacteria (Ng, 2005).

Cambodian Phaak (Mamchas)

Is fish paste mixed with glutinous rice and fermented by yeasts.

Japanese Katsiobushi

Is a fermented paste made from fish, which is used to season Japanese dishes. *Aspergillus glaucus* is involved in fermentation process.

Korean Jeotkal

Fish such as herring, sardine, shrimp, cuttle fish, oyster are salted and fermented by species of *Torulopsis* and *Saccharomyces.*

Vietnamese Nuoc-mam

Is a brown liquid obtained by fermenting fresh water fishes in earthernware pots and buried in ground for several months. *Aspergillus oryzae* and yeasts are added for flavour.

19.6 TRADITIONAL FERMENTED FOODS FROM DIFFERENT COUNTRIES

Fermented foods are those that are subjected to the action of micro-organisms like bacteria, moulds and yeasts that cause desirable chemical changes and significant modification of food.

The major chemical change is microbe mediated fermentation which was developed even before the recognition of existence of micro- organisms. These traditional foods were an important part of the diet in many Asian and African countries providing nutritious food. These grew from culture and experience rather than scientific knowledge.

It is very necessary to understand, document and update these foods and harness them to future generation failing which ethnoculture is lost forever. Fermented foods are mainly prepared from tubers, cereals and legumes and are part of diet and unique to each country. Many of such foods are highly nutritious and free of toxins.

19.6.1 Tuber Based Foods

Tuber crops like cassava, yam, potato and sweet-potato are fermented to yield food products.

African Fufu

Is made by soaking peeled and cut cassava roots in water. The softened root is rolled into balls and cooked to form a paste called Fufu. *Saccharomyces cerevisiae* is involved in the fermentation.

Hawain Poi

Is a semi-solid dish made from taro corms served with fish or meat. The fermentation is carried out by *Candida vini* (Padmaja and George, 1999).

Indonesian Tape

Is a staple food of Indonesia made from cassava roots. The roots are peeled, boiled and spread on bamboo trays and covered with banana leaves. Fermentation is carried out by *Endomycopsis fibuliger, Rhizopus oryzae* and *Saccharomyces cerevisiae.*

Nigerian Gari

Is a product of cassava. Peeled roots of cassava are grated and juice is taken out. Fermentation is allowed to proceed normally. Species of *Candida* with *Leuconostoc mesenteroides* take part in fermentation.

West-African Gari

Is made from cassava roots which are fermented, then sun dried and milled into flour and a paste is prepared for consumption. Species of *Candida* are the fermenters.

19.6.2 Cereal Based Foods

Chinese Minchin

Is made from wheat gluten and fermented by species of *Aspergillus, Cladosporium* and *Paecilomyces.*

Chinese Red rice (Anka)

Is fermented rice with various strains of *Monascus purpureus.* It is used to color fish, wine, soybean cheese, pickled vegetables and meat (Su and Wang, 1977). It is also used to treat indigestion and dysentery in China.

Ethiopian Enjera

Enjera is also known wanjera, manjeriya, Kissra habashiya. It is made by grinding teff grass, wheat, barley, sorghum and corn mixture into flour. The flour is mixed with water and a starter culture of various fungi such as species of *Aspergillus, Candida, Hormodendron, Penicillium, Pullullaria, Rhodotorula* and *Torulopsis* for fermentation.

Ghanian Kenkey

Maize kernels are washed and soaked in water before grinding. The ground paste is packed in plantain or maize leaves for 3-4 days for fermentation. The fermented dough is salted and only a portion of the dough is cooked and mixed with the rest of uncooked dough and made into balls and again wrapped in maize or plantain leaves prior to thorough cooking. The fermenting fungi involved in kenkey are species of *Aspergillus, Pencillium, Rhizopus* and *Saccharomyces.* Gram negative bacteria are also involved in fermentation which increases thiamine, riboflavin and protein content (Christian, 1970).

Indian Idli

Indian idli is the equivalent of western sour dough bread. Idli is prepared by bacterial fermentation of rice and black gram batter. The leavening of the batter is achieved by *Leuconostoc mesenteroides.* Yeasts such as *Debaryomyces hansenii, Hansenula anomala, Oidium lactis, Saccharomyces cerevisiae, Torulopsis holmii, Torulopsis pullulans,* and *Trichosporon beigelii* also can be added to the batter for enhancing the texture and aroma of the idli (Soni and Sandhu, 1999).

Indian Dosa

Is a pancake prepared by soaked and ground black gram and rice fermented by *Candida kefyr, Cadida krusei, Kluyveromyces marxianus, Leuconostoc mesenteroides, Oosporidium margaritiferum* and species of *Bacillus* and *Streptococcus* (Soni and Sandhu, 1999).

Nigerian Kamu

Is prepared from fermented millet. The soaked millet is ground with pepper and ginger and allowed to ferment by *Candida krusei, Leuconostoc mesenteroides* and *Saccharomyces*. The filtered liquid is consumed.

Nigerian Ogi

Ground maize or sorghum is fermented by *Aspergillus, Cephalosporium, Cercospora, Fusarium, Oospora, Penicillium, Rhizopus,* and *Rhodotorula.* Bacteria like *Acetobacter, Corynebacterium, Lactobacillus brevis,* and *Lactobacillus plantarum* are also responsible for the flavour and taste of orgi.

Philippine Puto

Is similar to Indian idli except there are no legumes used in Puto. The batter is prepared by grinding soaked rice with sugar and a starter culture of *Leuconostoc mesenteroides* and yeasts such as *Saccharomyces cerevisiae* and *Streptococcus faecalis* is added to ferment puto.

Sri Lankan Hopper

Rice and wheat flour, either alone or mixed with sugar, coconut water, coconut toddy and is allowed to ferment by an inoculum of baker's yeast and acid producing bactera. Following fermentation, coconut milk and salt are added. The batter is poured on hot pan and cooked.

Sudanese Kisra

Is a bread or stiff porridge prepared by fermenting sorghum flour which is mixed with wheat or millet flour and a starter culture is added to aid fermentation. The fungal species *Aspergillus,* yeasts like *Candida intermedia, Fusarium, Penicillum, Rhizopus,* and *Saccharomyces cerevisiae* and bacteria such as *Acetobacter, Erwinia ananas,* and *Klebsiella pneumoniae* are fementers (Mohammed *et al.*, 1991).

19.6.3 Legume Based Foods

Legumes are a major source of diet in Asia. Fermentation of legumes improves digestibility and removes antinutritional factors.

Chinese Chee-fan

Is prepared from soybean curd and fermented by *Aspergillus glauca* and *Mucor,* and eaten like cheese.

Chinese Jiang

Soaked, steamed soybeans are coated with wheat flour and allowed to ferment at high temperature in brine solution. *Aspergillus* species carry out fermentation.

Chinese Meitauza

Is a Chinese soybean cake which can be fried in oil or cooked with other vegetables. It is a product of fermentation by *Actinomucor elegans.*

Chinese Sufu

Sufu is made by fermenting soybean curd with A*ctinomucor* and *Rhizopus.* Fermentation is allowed to proceed in brine and rice wine mixture. Red sufu is prepared by using *Monascus purpureus* on rice, while the white sufu is untreated.

Cheonggukjang

Cheonggukjang stew is an important menu in traditional Korean restaurants. It is also available as a dry powder that is sprinkled on other foods. The word

Cheonggukjang in Chinese means "Quing dyanstry in war". The Quing army used to carry the fermented soybean as military ration for a supply of nutritious food (Lee 2001).

The newly harvested soy beans are cooked, covered with straw and placed on stone floor heated by fire underneath for three days until the mucous string is formed with a pungent odour. It is mixed with roasted soybean powder, salt, ginger, garlic, pounded and stored. Major fermenter is *Bacillus subtilis.*

Kinema

Kinema is a naturally fermented food of north eastern states of India. Kinema is called by various names. viz. *Hawaijar* in Manipur, *Axuina* in Nagaland, *Trombai* in Meghalaya and *Bekanthu* in Mizoram.

Soy beans are washed, soaked overnight in spring water and cooked. The cooked beans are crushed and transferred to bamboo baskets. Crushed beans are mixed with firewood ash and wrapped in fern leaves for 1 -3 days. The desired state of fermentation is determined by ammonia odour and development of rough white long stringy threads. Longer the threads, better is the quality of Kinema. Kinema is fried in oil and mixed with vegetables, spices and made into thick curry and eaten with rice. Fermentation of Kinema is carried out by *Bacillus subtilis, Bacillus licheniformis, Bacillus cereus, Bacillus circulans* and *Bacillus sphaericus.*

Natto

Natto is a popular breakfast and dinner dish in Japan which is eaten with rice and soy sauce, spices, mustard and miso soup. Three types of natto are traditionally prepared in Japan. *Hamma – natto, Yukiwari natto and Itohiki natto.* Special small variety of soybean (Natto bean) are soaked and boiled until tender. Cooked beans are partially air dried in bamboo trays covered with wax paper. These trays are stacked covered with straw mats and allowed to ferment. During the fermentation, bacteria produce long viscous slimy strings which is the characteristic feature of natto.The major fermentors of natto are *Bacillus subtilis and Bacillus natto.*

Natto is priced for its nutritional value, improved digestibility and its therapeutic value in preventing heart attack, stroke, cancer, osteoporosis, obesity, dementia, high blood cholesterol, dryness of skin and shoulder crick (Sarkar 2005)

Oncom (Ontjon)

Is a cake like product prepared from peanut press cake, which is soaked, crumbled and mixed with solid waste of tapioca. It is steamed and pressed into flat cakes and inoculated with molds such as *Rhizopus* or *Neurospora* and covered with banana leaves and allowed to ferment.

Tempe (Kedele)

Is a fermented soybean based food, popular with American vegetarians. The beans are soaked, dehulled, cooked and inoculated and packed in banana leaves and allowed to ferment for 2 days. The fermenting species include *Aspergillus, Bacillus, Micrococcus, Penicillium,* and *Rhizopus* (Yeoh and Merican, 1977). Tempe like food is also prepared by fermenting broad bean and cowpea, by *Rhizopus arrhizus, Rhizopus oligosporus* and *Rhizopus oryzae.*

Thua-Nao

Thua – Nao is a traditional food of Thailand used as a paste for flavouring vegetable dishes. By itself it is a part of the diet. Soy beans are soaked, cooked and put in bamboo baskets covered with banana leaves and allowed to naturally ferment for 3 -4 days. The fermentation carried out by *Bacillus subtilis* results in ammonia like odour and stickiness. The mass is made into a paste by grinding with garlic, onion and red pepper. The paste is cooked in banana leaves (Leejeerajummean *et al.,* 2001).

Tofu

Soy milk proteins clotted to form a curd tofu is fermented by *Actinomucor elegans, Mucor silvaticus, Mucor dispersus and Rhizopus chinensis,*

Ugba

Ugba, a Nigerian food which is prepared by fermentation of African oil bean *Pentaclethra macrophylla.* The seeds are boiled without the seed coat overnight and thoroughly washed after which they are cut into thin slices, salted and fermented for five days by wrapping in wilted banana leaves. Ugba is also used as a flavouring ingredient. The main fermentor of Ugba is *Bacillus subtilis* and species of *Staphylococcus.*

West African Dawa Dawa

Is prepared by locust beans which are soaked, cooked and spread on trays, covered with banana leaves and fermented by lactic acid bacteria.

Yukiwari Natto

Is prepared by mixing soaked, steamed soybean fermented by *Bacillus natto* with salt and rice koji prepared by using *Aspergillus oryzae.*

19.6.4 Soybean Sauces

Soy sauce is prepared by fermentation of ground soybean seeds (steeped in brine) with or without ground wheat and fermented by *Aspergillus oryzae* and *Aspergillus soyae.* Various types of soy sauces are given below.

Indonesian Soy Sauce (Kecap)

Soaked and boiled soybeans are allowed to ferment using species of *Aspergillus* and *Rhizopus*.

Japanese Soy Sauce

The most popular Japanese soy sauces are *Koikuchi, Shiro, Saishikomi, Tamari* and *Usukuchi*. They are prepared by mixing salted, cooked soybeans with roasted and ground wheat and the mixture is inoculated with *Aspergillus oryzae, Aspergillus soyae, Pediococcus soyae, Saccharomyces rouxii and Torulopsis versatile*. Fermentation is allowed to proceed for 3 years, after which it is pasteurized and bottled (Yokotsuka, 1977).

Korean Soy Sauce (Kanjang)

Steamed soybeans are crushed and left to ferment in brine. Fungi such as *Aspergillus, Mucor, Penicillium and Rhizopus*, bacteria like, *Bacillus pumilis, Bacillus subtilis* and *Leuconostoc mesenteroides* and yeasts such as *Rhodotorula flava, Saccharomyces rouxii* and *Torulopsis dattila* carry out fermentation (Lee and Cho, 1971).

Malaysian Soy Sauce

Soybeans are boiled, mixed with wheat flour and spread on bamboo trays to promote fungal growth. It is then transferred to earthenware jars, covered with salt and incubated in sun for 3 months. The extracted sauce is mixed with sugar, monosodium glutamate and benzoic acid. *Aspergillus flavus, Aspergillus niger, Aspergillus oryzae, Aspergillus soyae, Pencillium* and *Rhizopus* are involved in fermentation (Ong, 1977).

19.6.5 Soybean Pastes

Soybean pastes are used in making soups or side dishes. The five major types of pastes are Japanese *Miso*, Indonesian *Tauco*, Korean *Doenjang*, Korean *Meju* and *Kochuzang*. These are prepared by boiling soaked soy seeds, with or without the addition of rice or barley and incubated for several days where *Aspergillus oryzae, Aspergillus soyae, Hansenula, Penicillium* and *Rhizopus* species carry out fermentation. They are put in brine, along with palm sugar, cooked ground and bottled. (Chang *et al.*, 1977; Winarno *et al.*, 1977).

19.6.6 Milk Based Products

Several milk based products are highly nutrient with hypocholesterolemic activity. They are cooling and energy boosters.

Cheese

Cheese requires the formation of curd by fermentation of milk. Inoculation of lactic acid bacteria as a starter culture is prerequisite for fermentation. Cheese is a solid product which is obtained by separating casein protein from whey and allowed to ripen under appropriate conditions. *Streptococcus thermophilus* and *Streptococcus lactis* are employed in cheese production.The varieties of cheese are classified according to the texture, degree of ripening, and the fermenter (bacteria or by molds) (Hansen and Jakobsen, 2004).

Texture based cheeses are hard, semi-hard and soft. Hard cheese includes swiss, cheddar, provolone, Romano, Edam and paramesan; semihard cheese include roquefort, blue, muenster and gouda; soft cheese include lim burger (ripened by bacteria) brie and camembert (ripened by molds). Camembert cheese is ripened with spores of *Penicillium camembertii.* Swiss cheese is made using *Lactobacillus bulgaricus* and *Streptococcus thermophilus.* Roquefort cheese is made by inoculating curd with *Penicillium roquefortii* that gives a characteristic blue veined appearance.

Unripened cheeses include cottage cheese, cream cheese and neuf chattel. Cottage cheese contains the yeast species *Candida lipolytica, Candida spherica, Candida sake, Kluyveromyces marxianus and Yarrowia lipolytica.*

Surface ripened cheeses contain yeasts such as *Candida* species, *Debaryomyces hansenii, Kluyveromyces lactis, Trichosporon* species and *Yarrowia lipolytica* and bacteria such as *Arthobacter, Brevibacterium, Corynebacterium, Micrococcus* and *Staphylococcus.*

White mold cheeses contain the yeast species *Debaryomyces hansenii, Galactomyces geotrichum, Kluyveromyces lactis, Kluyveromyces marxianus, Saccharomyces cerevisiae and Zygosaccharomyces rouxii.* Yeast species on blue mold cheese are *Candida catenulata. Candida famata, Candida lipolytica, Candida spherica, Debaryomyces hansenii, Kluyveromyces lactis, Trichosporon cutaneum* and *Yarrowia lipolytica.*

Feta cheese, from Greece and the Middle East are made from goat or ewe's milk and ripened by *Candida famata, Debaryomyces hansenii, Dekkera anomala, Pichia farnosa, Pichia membranaefaciens, Saccharomyces cerevisiae, Candida sake, Dekkera bruxellensis, Geotrichum candidum, Kluyveromyces lactis, Kluyveromyces marxianus, Yarrowia lipolytica and Torulaspora delbrueckii.* Danish or Scandanavian Feta cheese is also available. Cheese is also prepared from soybean, fermented by *Monascus purpureus.* The flavour of cheese is obtained from *Propionibacterium shermanii.* (Hansen and Jakobsen, 2004).

Buttermilk and Butter

Butter is made by churning the milk cream until the fatty globules of the butter separate. The remaining liquid is butter milk. The typical flavour and aroma of butter and buttermilk are from diaceyls that are metabolic end products of fermentation by lactic acid bacteria like *Lactococcus cremoris*, *Lactobacillus plantarum* and *Lecuconostoc citrovorum*. Buttermilk now is artificially made by inoculation of skimmed milk with lactic acid bacteria.

Yoghurt

Yoghurt is made from low fat milk from which much of the water has been evaporated. The resulting thick milk is inoculated with *Streptococcus thermophilus* for acid production followed by the inoculation of *Lactobacillus bulgaricus* for flavour and aroma, where as Indian curds is a product of fermentation of whole milk.

Sour cream

Sour cream is obtained by inoculating milk cream with *Leuconostoc dextranicum* and *Streptococcus lactis*. The aroma and flavour are obtained by inoculating with lactic acid bacteria as starter culture.

Kefyr (Kefir)

Kefyr is a fermented milk beverage popular in Eastern Europe. Kefyr is prepared using Kefyr grains which are hard grain like bacterial clots containing *Lactobacillus bulgaricus* and *Streptococcus lactis*. Kefyr is mixed with the milk of sheep, goat or cow and fermented in bags made of goat skin (Motaghi *et al.*, 1997). There are fascinating stories about Kefyr (Box 19.1)

Box 19.1 Kefyr

Two versions of a tale are given as an explanation for Kefyr grains and the receipe of Kefyr was a well guarded secret for centuries (Hansen and Jacobsen, 2004). Allah granted a chosen tribe the first Kefyr grains as a symbol of immortality, and hence the grains were called grains of the Prophet. In another version, Kefyr grains were presented by Prophet Mohammed as a sign of immortality together with the receipe of Kefyr, but it was forbidden to give away the grains or the receipe otherwise the grains would lose their healing power. Even though, Kefyr was made for centuries (Wouter *et al.*, 2002), the first production of retail Kefyr started in the USSR in the 1930s (Wyder, 1998).

Kefyr is acidic and mildly alcoholic with a high level of carbon dioxide. It originated from Caucasus in Central Asia. Kefyr grains are white, yellowish hard granules. They are formed by encapsulation of symbiotic microflora in a fibrous matrix of coagulated casein, polysaccharides, fat and lysed cells. The polysaccharides present in the kefyr grains are called kefryran and are different from other known polysaccharides (Wyder, 1998). *Candida kefyr, Kluyveromyces marxianus, Saccharomyces cerevisiae* and *Saccharomyces delbrueckii* are the major yeast species detected in Kefyr (Wouters *et al.*, 2002).

Koumiss

Koumiss originating from Khazakisthan in Russia is similar to Kefyr, but is traditionally made from mare's milk. Koumiss is a type of fermented milk produced by natural yeast microflora, with a low level of alcohol. Thermophilic lactic acid bacteria and *Saccharomyces unisporus* and *Kluyveromyces marxianus* are the major fermentors present in koumiss.

Russian Koumiss

Russian Koumiss is prepared from boiled ass milk, camel milk or mare milk. Lactic acid bacteria, *Saccharomyces lactis* and *Torula koumiss* cause fermentation. It is used in Russia for pulmonary tuberculosis.

Egyptian Kishk

Kishk is a fermented wheat milk beverage. Wheat grains are boiled, air dried and ground with spices and is rolled into balls and dried. Species of *Penicillium, Mucor* and *Lactobacillus* are the main fermentors involved.

Lebanese Labneh

Labneh is prepared from goat or sheep's milk. Lactose fermenting bacteria and yeast cause fermentation.

19.6.7 Other Fermented Milk Products

Amasi/hodzeko from Zimbabwe, *Banik* from Senegal, *Rob* from Sudan, *Nono* from Nigeria and *Ergo* and *Ithu* from Ethiopia, Lactofil from Sweden, *Quark* from Germany, *Viili* from Finland, *Yakult* from Japan, *Leben* from Iraq, *Leben raib* from Egypt and Long milk from Scandinavia are few other traditional fermented milk products (Hansen and Jakobsen, 2004).

CHAPTER 20

MICROBE MEDIATED PRODUCTION OF INDUSTRIAL RAW MATERIAL

20.1 INTRODUCTION

Most of the raw materials for innumerable industries such as food, pharmaceuticals, leather, paper and detergents come from microbial metabolites. The raw material for these industries are organic acids, amino acids, vitamins, sugars, lipids, enzymes, polysaccharides, biopolymers, cellulose, alternative sweeteners, biosurfactants, bioinsecticides, biofungicides and single cell proteins.

20.2 ORGANIC ACIDS

Several organic acids which are of industrial importance are produced by microbial fermentation carried out by bacteria, fungi and yeasts. The production of acetic acid has been known as long as 10,000 years. The Romans and Greeks produced vinegar by leaving wine open to air and used diluted vinegar as a refreshing drink. Acetic acid is formed from oxidation of ethanol by fermentative bacteria.

The most important organic acid is citric acid which is used as a preservating agent in food industry and used in flavouring fruit juices, candy, ice cream and marmalades. In pharmaceutical industry it is used in tablets, ointments, cosmetics, as a preservative for stored blood, in the chemical industry it is used as an anti foam agent and as a softener for textiles. It is employed in the tanning and detergent industry also.

Itaconic acid is used as wetting agent in the manufacture of resin; it also forms co-polymers with its esters and is used in making wall paper and adhesives. Gluconic acid is used in toothpaste, inks, dyes, leather industry and in the manufacture of metals and in electroplating. Sodium gluconate is used as a sequestering agent in detergents while gluconolactone formed in the first step of gluconic acid production is used as baking powder additive. Sorbic acid is used in preservation of food, cosmetics, pharmaceuticals and in animal feed. Propionic acid is used as antifungal additive in bakery products and cheese. Propionic acid and fumaric acid are used in the manufacture of dyes, perfumes and lacquers. The microbial source of different organic acids from bacteria, fungi and yeasts are given in (Table 20.1).

Table 20.1 Microbial source of organic acids

Source species	Organic acid
Acetobacter aceti, *Acetobacter pasteurianus,* *Gluconobacter suboxydans,* *Tetragenococcus halophilius*	Acetic acid
Aspergillus flavus, *Aspergillus oryzae*	Kojic acid
Aspergillus flavus, *Aspergillus niger,* *Aspergillus wentii,* *Aspergillus clavatus,* *Arthobacter paraffineus,* *Candida guilliermondii,* *Citromyces* sp., *Citromyces pfefferianus,* *Corynebacterium* sp., *Mucor pyriformis,* *Penicillium sp.,* *Penicillium luteum,* *Penicillium citrinum,* *Paecilomyces divaricatum,* *Saccharomyces lipolytica,* *Trichoderma viride*	Citric acid
Aspergillus itaconicus, *Aspergillus niger,* *Aspergillus terreus*	Gallic acid
Candida, Rhodotorula sp.,	Itaconic acid
Bacillus sp.,	Acrylic acid
Fusarium moniliforme	Gibberellic acid
Lactobacillus bulgaricus, *Lactobacillus delbruckii,* *Penicillium chrysogenum,* *Rhizopus oryzae*	Lactic acid
Acetobacter suboxydans, *Aspergillus niger*	Gluconic acid

Endomycopsis, *Gluconobacter suboxydans*, *Gonatobotrys* , *Penicillium* sp., *Penicillium luteum*, *Penicillium purpurogenum*, *Pseudomonas*, *Pullullaria*, *Vibrio* sp.,	Gluconic acid
Propionibacterium shermanii	Propionic acid
Candida sp., *Rhizopus nigricans*	Fumaric acid
Candida sp., *Leuconostoc brevis*	Malic acid
Aerobacter sp., *Candida* sp.,	α – Ketoglutaric acid
Serratia marcescens	2- Ketogluconic acid
Acetobacter suboxydans	5 – Ketogluconic acid
Ustilago zeae	Ustilagic acid

Source: Purohit, 1999

20.3 AMINO ACIDS

In nature, microbes do not produce amino acids in excess of their needs because feed back inhibition prevents wasteful production. Commercial production depends on selected mutants and manipulation of metabolic pathways. Amino acids have become major industrial products from micro-organisms. About 66% of the amino acids produced are used in food industry, 31% as feed additives, and 4% in medicine, cosmetic and chemical industry.

Amino acids are employed in medicine as ingredients of infusion in post operative treatment. In the chemical industry, amino acids are used as starting materials for the manufacture of polymers, resin and synthetic leather. Glycine is used as a starting material for the production of herbicide glyphosphate while threonine is used in the production of azthreonam.In the cosmetic industry, serine and uronic acid are used in sunscreen lotions.

In the food industry, amino acids are used to enhance flavours.The commercial synthesis of essential amino acids has become important to compensate for the deficiencies in cereal foods.As plant proteins are deficient in essential amino acids such as L-lysine, L-methionine, L-threonine they are added to food and feed products to compensate for the deficiency. L- alanine is added to fruit juices; L-cysteine improves the quality of bread during baking and added to fruit juices as an anti – oxidant; L-tryptophan with L-histidine is added to milk powder to prevent rancidity; Lysine and aspargine are added to bread.

Phenylalanine and aspartic acid, sold as Aspartame, are commercially important low calorie alternate sweetners used in soft drinks and 'diet' soda. Glutamic acid is used in the production of monosodium glutamate (MSG) used as a flavouring agent in Chinese cooking; in the production of biopolymers; and as a thickener for animal feed. Alanine and aspartic acid are used as flavor enhancers. The industrial source of amino acids are given in Table 20.2.

Table 20.2 Microbial source of amino acids

Source species	Amino acid
Arthobacter sp., *Brevibacterium lactofermentum, Corynebacterium glutamicum, Micrococcus* sp., *Microbacterium* sp., *Proteus* sp.,	Glutamic acid
Brevibacterium subtilis	Isoleucine
Corynebacterium glutamicum	Phenylalanine
Achromobacter obae, Arthrobacter paraffineus , Brevibacterium divaricatum , Candida periculosa, Corynebacterium glutamicum, Corynebacterium hydrocarbolastus, Cryptococcus laurentii, Enterobacter aerogenes, Escherichia coli, Methylomonas methylovora, Saccharomyces lipolytica, Saccharomyces cerevisiae	Lysine
Achromobacter liquidum, Bacillus subtilis, Brevibacterium flavum S- 225,	Tryptophan
Corynebacterium glutamicum, Enterobacter cloacae TA599, *Escherichia coli, Flavobacterium aminogenes Enterobacter* sp.,	Valine

Source: Purohit, 1999; Crueger and Crueger 2005

20.4 VITAMINS

Vitamins are essential nutrients required in small quantities and have an accepted value to health of humans and animals. There is a large requirement for vitamins, other than those derived from plant and animal food sources both in food and pharmaceutical industry.

These vitamins are produced biotechnologically via fermentation by microbes or bioconversion processes. Five types of vitamins are produced commercially, Ascorbic acid (Vitamin C), Riboflavin (B2), Panthothenic acid (Biotine) and Cyanocobalamin (Vitamin B12).

Vitamin B12: also known as cyanocobalamin is exclusively synthesized in nature by microorganisms. Humans cannot synthesize this vitamin and they obtain it through food.

Riboflavin: also called as lactoflavin or Vitamin B2 was first isolated from whey. It is present in milk in a free state and in eggs, liver and heart as flavoprotein containing prosthetic groups FMN (Flavin mononucleotide) or FAD (Flavin adenine dinucleotide). Riboflavin deficiency in humans causes Ariboflavinosis appearing as a type of dermatitis. A number of vitamins are industrially sourced in microbes (Table 20.3).

Table 20.3 Microbial source of vitamins

Source species	Vitamin
Ashbya gossypii, *Bacillus pumilus,* *Bacillus subtilis,* *Candida famata,* *Candida flareri,* *Clostridium acetobutylinum,* *Eremothecium ashbyii,* *Mycobacterium smegmatis,* *Mycocandida riboflavina,* *Pichia miso*	Riboflavin (Vitamin B2)
Aspergillus niger, *Gluconobacter suboxydans*	Vitamin C
Arthrobacter hyalinus, *Bacillus megaterium,* *Butyribacterium rettgeri,* *Klebsiella pneumoniae,* *Methanobacillus omelianski,* *Methanobacterium soehngenii,* *Methanosarcina barkeri,* *Micromonospora* sp.,	Vitamin B_{12} (Cyanacobalamin)

Nocardia gardneri,	
Propionibacterium denitrificans,	
Propionibacterium freudenreichii,	
Propionibacterium shermanii,	
Protominobacter ruber,	
Pseudomonas aureofaciens,	
Pseudomonas denitrificans,	
Pseudomonas ovalis,	
Rhodopseudomonas aureofaciens	Vitamin B_{12} (Cyanocobalamin)
Rhodopseudomonas protamicus,	
Saccharomyces *cerevisiae,*	
Streptomyces olivaceus,	
Saccharomyces cerevisiae	Vitamin D2
Escherichia coli	Pantothenic acid
Aerobacter aerogenes ssp., alvei,	Biotin
Bacillus cereus,	
Candida utilis,	
Escherichia coli,	
Proteus morganii,	
Proteus mirabilis,	
Proteus vulgaris,	
Saccharomyces cirevisiae	
Mucor cercinelloides	Vitamin F
Gluconobacter sp.	Ascorbic acid

Source: Perlman 1979 b.

20.5 SUGARS

Several sugars that are produced using microbes (Table 20.4) have applications in food industry.

D-Ribose is used as a flavour enhancer, in health foods, in pharmaceuticals and cosmetic industry.

Table 20.4 Microbial source of sugars

Source species	Sugars
Gluconobacter suboxydans	Sorbose and 5-ketogluconic acid
Rhodococcus erythropolis	Trehalose
Actinoplanes missouriensis, Bacillus coagulans, Flavobacterium arborescens, Streptomyces rubiginosus	Fructose

Bacillus pumilus	D-ribose
Saccharomyces cerevisiae	Raffinose
Thermoanaerobacter	Cyclodextrin

20.6 LIPIDS

Micro-organisms have a great potential as source of oil. The most expensive oils being produced commercially are those containing polyunsaturated fatty acids (PUFA). Y-linolenic acid (GLA) acts as a precursor of prostaglandins.

PUFAs which are of dietery importance are required for the functioning of brain and vision in infants. Two important PUFAs are dihomogammalinolenic acid (DGLA) and arachidonic acid (ARA) which are normally obtained from fish oil and pig liver. PUFAs are the precursors for the industrial production of eicosanoids (e.g. Prostaglandins, Prostacyclin, Leucotrienes) which are used in the treatment of high blood pressure, asthma, menstrual pain and induction of labour. However, *Mortierella alpina* and *Mortierella elongata* are identified as producers of ARA and *Mortierella alpina* produces DGLA (Sullia and Shantharam, 2005). *Mucor circinelloides, Mortierella isabellena, Mortierella vinacea, Mortierella ramanniana, Mortierella nana* among fungi, *Spirulina major, Nitzchia alba* among algae are identified as potential producers of GLA.

Most of the detergents are man made synthetic materials. They are resistant to biodegradation and are responsible for environmental pollution. Several microbial products particularly glycolipids are used in developing environmental friendly detergents. These can also be used as emulsifiers, wetting agents and solubilizers. *Pseudomonas aeruginosa, Rhodococcus erythropolis* and *Bacillus subtilis* produe a glycolipid rhamnolipid R3 which can be used as a biodetergent. It is already being used as a dispersing agent in bioremediation of oil spills.

20.7 GROWTH HORMONES

Micro-organisms in soil such as bacteria and fungi produce hormones like indole acetic acid, gibberellins and cytokinins (Table 20.5).

Table 20.5 Microbial source of growth hormones

Growth Hormones	Bacteria	Fungi
Indole acetic acid	*Agrobacterium tumefaciens* and *Pseudomonas fluorescens*	*Endophyllum sempervivi, Gymnosporangium juniperi-virginiae, Rhizobium, Rhizopus, Suinus* and *Synchytrium endobioticum.*

Gibberellins	Species of *Arthobacter, Azospirillum, Bacillus, Brevibacterium, Flavobacterium, Nocardia, Pseudomonas,* and *Rhizobium*	*Alternaria, Aspergillus, Fusarium, Gibberella fujikurai Penicillium, Rhizopus, Sphacelloma, Suillus*
Cytokinins	Species of *Azotobacter, Azosporium, Azospirillum Agrobacterium, Arthobacter, Bacillus, Corynebacterium, Pseudomonas* and *Rhizobium.*	
Ethylene	*Species of Aeromonas, Arthobacter, Citrobacter Clostridium, Enterobacter, Erwinia, Escherichia, Klebsiella, Pseudomonas, Serratia* and *Streptomyces.*	Species of *Acremonium, Agaricus, Alternaria, Ascochyta, Aspergillus, Blastomyces, Botrytis, Candida, Fusarium, Mucor, Neurospora, Penicillium, Pythium* and *Rhizoctonia.*

20.8 STEROIDS

Steroid hormones from microbes have a number of therpaeutic applications. Cortisone relieves rheumatoid arthritic pain, alleviates allergic and inflammatory responses and is a contraceptive. Microbial source of steriods and its commercial products are given in Table 20.6 and Table 20.7.

Table 20.6 Microbial source of steroids

Source species	Steroid
Corynebacterium simplex	Cortisol
Curvularia lunata, Streptomyces roseus	Hydrocortisone
Pseudomonas testosteroni	19-nortestosterone
Rhizopus nigricans	*Progesterone*

Source: Purohit, 1999

Table 20.7 List of commercially produced steroids

Products	Source species
Progesterone	*Rhizopus nigricans*
Cortisol	*Curvularia lunata* *Corynebacterium simplex*
Hydroxycortisol	*Streptomyces roseochromogenes* *Streptomyces roseus*
Prednisolone	*Arthobacter simplex*

Dehydrotestolactone	*Cylindrocarpon radicicola*
Androstadiendione	*Mycobacterium fortuitum*
Triendiol	*Streptomyces affinis*
19 - nortestosterone	*Pseudomonas testosteroni*

Source: Sebek and Perlman, 1979

20.9 INDUSTRIAL SOLVENTS

Microbioal fermentation has given us very important industrial solvents like acetone, ethyl alcohol (ethanol) and butyl alcohol (butanol) which are used in pharmaceuticals, cosmetics and textile industry. Acetone and glycerol which are by products of butanol fermentation is used in the manufacture of methyl isobutyl ketone, methacrylates, methyl isobutyl carbinol, isophorone, diacetone alcohol, trinitro glycerine (TNT), as solvents for paints, lacquers, resins, varnishes, nitrocellulose and cellulose acetate.

Butanol is used in the manufacture of dibutylphthalate, butyl acetate, butyl acrylate, glycol ethers, amine resins and oils used in pharmaceutical and perfume industry. Butanol is also used in the plastic industry as a feed stock chemical, as food grade extraction solvent, as solvency enhancer in the formation of nitrocellulose lacquers (Karcher *et al.,* 2005) and in making synthetic rubber. N-butanol is used in brake fluids, resin and lacquers and used as protective coat in automobiles.

Glycerol is used as a solvent for food colour, as a lubricant in toothpaste, skin creams, lotions, shampoos and laxatives. Vinegar is widely used as a souring agent and preservative in number of food and other products (Table 20.8).

Table 20.8 Microbial source of industrial solvents

Source species	Solvent
Acetobacter, Gluconobacter	Acetic acid, vinegar
Clostridium acetobutylicum,	Alcohol and acetone
Saccharomyces cerevisiae	Ethyl alcohol
Clostridium acetobutyricum,	Glycerol
Bacillus subtilis	

Source: Sullia and Shantharam, 2005

20.10 MICROBIAL ENZYMES

Microbes such as bacteria, yeast and fungi excrete into the medium, the enzymes they synthesize. Such enzymes can be harvested in large quantities and used for various purposes in the industry. Many roles have been found for microbial enzymes in solving environmental problems. The main industries, which utilize enzymes, are detergents, textiles, bakery, animal feed, breweries, and dairy industry. The major producers of commercial enzymes are Novo-Nordisk (Denmark), Gist Brocades (Belgium), Rohm (Germany), Miles (United States) and Hansens (Netherlands) (Saxena *et al.*, 2004).

1. Amylases

α – amylases are produced by bacteria like *Bacillus acidocaldarius, Bacillus amyloliquefaciens, Bacillus caldolyticus, Bacillus coagulans, Bacillus cereus, Bacillus circulans, Bacillus licheniformis, Bacillus megaterium, Bacillus polymyxa, Bacillus stearothermophilus, Bacillus subtilis, Geobacillus Thermoleovorans,Thermomonospora curvata* and species of *Arthobacter, Lactobacillus, Micrococcus, Proteus, Pseudomonas, Serratia and Streptomyces.*

α – amylases are also produced by fungi like *Aspergillus oryzae,* species of *Cephalosporium, Humicola, Mucor, Neurospora, Penicillium, Rhizopus* and *Thermomonospora.*

Isoamylases are produced by species of *Agrobacterium, Erwinia, Lactobacillus, Leuconostoc, Nocardia, Pediococcus, Pseudomonas amylodermosa, Serratia and Staphylococcus* are used in various industries as exemplified below.

Applications of amylases

Food industry

Amylases are used as sweeteners in soft drinks; as clarifying agents in fruit juices; in the preparation of pasta and as a bread additive (Pritchard, 1992).

Textile and paper industry

Amylases are used for desizing starch-sized textiles without harming the fibers; in paper sizing to protect the paper against mechanical damage; and to improve the quality of finished paper.

Detergent industry

Purafect OxAm and Duramyl are new detergents containing amylases for use in dish washing machines.

In medicine

á-Amylases have been used to prepare a hybrid membrane in sol-gel processing. They are used as enzyme thermistor for biochemical analysis of cyclodextrins (Kolb *et al.*, 1996).

2. Cellulase

In 1943, scientists in the United States Army Laboratory isolated a mold *Trichoderma reesei* from cellulase based equipment deteriorating in storage. This discovery led to the commercial production of cellulase in food and drink industries. Cellulases are widely employed in textile manufacturing, in removing complexes which interfere with the processing of fruit juices and in digestive aids and as a supplement in animal feed. Main producers are *Penicillium funiculosum, Sclerotium rolfsii, Trichoderma reesei* , fungi such as *Trichoderma longibracteatum* and species of *Cephalosporium.*

3. Chitinases

Chitinases are produced by bacteria like *Aeromonas* and *Alteromonas* species, *Bacillus circulans, Chromobacterium violaceum, Enterobacter agglomerans, Nocardia orientalis, Pseudomonas fluorescens, Serratia* species, *Streptomyces olivaceoviridis, Vibrio harveyi* and by fungi like *Aspergillus fumigatus* and species of *Kitasatosporia, Herbidospora, Micromonospora, Microbispora and Saccharopolyspora.*

Applications of chitinases

Chitinases have numerous applications in agriculture, medicine , biotechnology and waste management. They are used in biocontrol synergistically with *Bacillus thuringiensis*; used in medicine to combat opportunistic invaders of burn wounds (Fenton and Eveleigh 1981); in ulcerative colitis and gastrointestinal inflammatory disorders (Friedman and Skehan 1980); in cancer therapy and bone diseases (Kenji *et al.,* 1989); in treating arteriosclerosis (Boot *et al.,*1999) ; colorectal cancer (Cintin *et al.,* 1999) and as diagnostic marker in Gaucher disease (Hollak *et al.,* 1994); used as an efficient degrader of chitin waste; Chitinase – colloidal gold labeled complexes have been used for immunocytochemical localization of chitin in biological systems; (Manocha and Zhonghua 1997).

4. Glucose isomerase

Glucose isomerase obtained from *Bacillus coagulans, Brevibacterium imperialis, Microbispora rosea, Microcellobiospora flavea, Micromonospora coerula, Nocardia asteroides, Nocardia dassonvillei, Streptomyces albus, Streptomyces olivochromogenes* and *Streptomyces phaeochromogenes* is used for the production of high fructose syrup used as an alternate sweetener.

5. Glucose oxidase

Glucose oxidase obtained from *Penicillium notatum*, is used clinically in glucose assays.

6. Invertase

Invertase obtained by *Saccharomyces cerevisiae,* is used in the preparation of high fructose syrups and candies.

7. Lactase

Lactase is obtained from fungi like *Aspergillus oryzae, Aspergillus niger, Rhizopus oryzae* and yeasts like *Kluyveromyces lactis, Kluyveromyces fragilis, Saccharomyces lactis* and *Torula cremoris.*

Application of lactase:

Lactases are used as digestive enzymes in lactose intolerance; as feed additives to increase the feed value of animal feeds and in dairy industry to break down lactose in milk and in processing whey.

8. Lipases

Lipases are of great importance due to their versatile catalytic behaviour. Purified lipase preparations are needed in industries employing these for biocatalytic production of pharmaceuticals and cosmetics. Lipases catalyze hydrolysis of triglyceride into diglycerides, monoglycerides, glycerol and fatty acids.

Lipases are produced by bacteria like *Acinetobacter calcoaceticus, Bacillus subtilis, Bacillus thermocatenulatus, Bacillus stearothermophilus, Chromobacterium viscosum, Propionibacterium acnes (Corynebacterium acnes), Propionibacterium parvum, Pseudomonas fragi, Pseudomonas pseudoalcaligenes, Pseudomonas putida, Staphylococcus aureus, Staphylococcus hyicus and Staphylococcus carnosus.*

Lipases are also produced by yeasts such as *Candida antarctica, Candida cylindracea, Candida rugosa, Candida paralipolytica, Saccharomycopsis lipolytica* and fungi like *Aspergillus niger, Aspergillus oryzae, Aspergillus terreus, Aspergillus carneus, Botrytis cinerea, Fusarium oxysporum, Geotrichum candidum, Humicola lanuginosa, Mucor javanicus,Neurospora crassa,Penicillium cyclopium, Penicillium simplicissimum, Penicillium camembertii, Penicillium expansum, Penicillium roqueforti,Rhizomucor arhizus, Rhizomucor delmer, Rhizomucor meihei, Rhizopus niveus, Rhizopus oryzae* and *Ustilago maydis* (Godtfredsen, 1990, Saxena *et al.*, 1998) and are used in the following industries:

Applications of lipases

Food industry

Lipases are used in bakery and beverages for improving the flavor and increasing shelf life. They are used in cheese ripening, hydrolysis of milk fat, as flavouring agents in meat, bakery products, fish products and in digestive aids (Saxena *et al,* 2004). Lipases are used to impart desirable flavours in cheese made from goat and ewe milk and such cheeses are called enzyme modified cheese (EMC) (Kilara, 1985). Lipase catalyzed interesterification is used in the food industry to alter the composition and nutritional properties of oils and fats in the production of cocoa butter from palm oil.

Cosmetics and perfume industry

Lipases are surfactants and aroma producers of immense value in cosmetics. Miyamoto *et al.,* (1995) obtained a patent for the use of lipase in the production of glycerine that is used as a smoothening agent in skin lotions. Nippon Oil and Fats (1986) obtained a patent for the preparation of Propylene-glycerol mono fatty acid ester in the presence of lipase, that is used as a pearling agent in cosmetics.

Environmental management

The increasing use of lipases in bioremediation has achieved a greater importance with its successful application in upgrading of waster fat (Salleh *et al.,* 1993). Oil spills from rigging and refining and lipid tinged wastes in lipid processing factories can be treated by lipases.

Leather industry

Lipases are used with proteases in processing animal hides and skins.

Detergent industry

Lipases form constituents of formulations used in cleaning drains clogged with food. They are used for removing oil stains from textiles and as automatic dish washing detergents. The new products are Purafect OxAm and Duramyl.

In medicine

Lipase catalyzes esterification of Naproxen, a nonsteroid anti-inflammatory drug.

Oleochemical industry

Lipases are used in the synthesis of polyunsaturated fatty acids (PUFA) such as eicosa pentanoic acid and docosa hexanoic acid that are beneficial in cardiovascular and inflammatory diseases. Lipases are also used in the synthesis of 'zero-trans' margarines and the synthesis of fatty acids and glycerol.

9. Pectinase

Pectinase is obtained from *Aspergillus oryzae* and *Aspergillus niger.*

Applications of pectinase

Pectinase is used in clarification of fruit juices and wine. Widely used in the fabric industry, for retting of plant fibres such as flax, hemp and jute, biopreparation of cotton fabrics,m enzymatic polishing of jute/cotton blended fabrics, in paper industry to solve the retention problems in mechanical pulp bleaching and for improving the quality of tea (Sathyanarayana *et al.,* 2005)

10. Phytase

Species of *Aspergillus niger, Mucor, Penicillium* and *Rhizopus* species are the most active fungal phytase producers. Phytases are the major enzymes used in the following industry:

Applications of phytase

Feed industry

Phytase is used for the removal of phytic acid in animal feed. The availability of phosphorus can be improved by adding microbial phytase to feeds or by supplementing with phytase rich cereal diets. They are used in the productionof myo-inositol phosphate which on degradation yields myo-inositol that improves the digestibility and nutritive value of feeds.

Chemical industry

Inositol-phosphate derivatives can be used as enzyme inhibitors and as chiral building blocks (Laumen and Ghisalba, 1994).

11. Penicillin acylases

Penicillin acylases are of two types - type I and type II.

Type I acylase producers are fungi like *Aspergillus ochraceus, Cephalosporium* sp., *Epidermophyton floccosum, Fusarium semitectum, Penicillium chrysogenum, Streptomyces lavendulae and Trichophyton mentagrophytes*

Type II acylase producers are species of bacteria like *Aerobacter, Alcaligenes, Corynebacterium, Erwinia, Flavobacterium, Micrococcus,*

Nocardia, Proteus, Pseudomonas, Salmonella, Sarcina and Xanthomonas. They are used in the production of semi-synthetic penicillins

12. Proteases

Three groups of microbial proteases are widely used in the food industry. They are alkaline, acid and neutral proteases.

Sporotrichum organi is the major source of proteases. Other major producers are fungi like *Aspergillus flavus, Aspergillus niger, Aspergillus oryzae, Mucor, Rhizomucor pusillus, Rhizomucor meihei* and *Rhizopus oligosporus.*

Proteases are used in the preparation of oriental foods such as Tempeh and Koji; used as a substitute for papain, rennin and pepsin; for preparing protein hydrolysates which are used in infant food formulations, in therapeutic dietery products and for fortifying fruit juices and soft drinks; used for partial hydrolysis of the wheat protein glutin which is used in bread making and as meat softner. They are also known for their catalytic use in synthetic chemistry (Zaks, 1991). The largest application of proteases is in formulation of household detergents to remove protein based stains.

12.1 Alkaline proteases

Alkaline proteases are produced by bacteria like *Bacillus licheniformis,Bacillus amyloliquefaciens, Bacillus pumilus, Bacillus mojavensis.*

Applications of alkaline proteases

They are used in silk degumming and bioprocessing in photography industry; contact lens cleaning solutions; in the synthesis of dipeptides and selective hydrolysis of peptide esters. Novo Nordisk in Japan has developed a detergent protease called Kannase.

In food and dairy industry

Alkaline proteases are used to produce functional soy protein hydrolysate with good whipping characteristics and added to protein fortified soft drinks as diet beverage and in the production of whey hydrolysate from cheese whey.

In leather industry

Alkaline proteases are used in leather processing especially for dehairing hides.Though traditionally pancreatic proteases are used, the fungal alkaline proteases have become popular as they give a better texture (Malathi and Chakraborthy, 1991)

In medicine

Alkaline proteases are being used for developing products used in treating burns, purulent wounds, carbuncles, furuncles and abscesses.

Other industrial uses

Alkaline proteases are used in peptide synthesis and in the resolution of racemates of amino acids. They are used in the management of wastes from various food processing industries and in the management of waste feather in poultry slaughter houses. They are used as supplement of soap and detergents, for removing stains on textiles and recovering fibres from flax.

12.2 Acid proteases (Rennet/Rennin)

Rennet, one of the acid proteases is a milk coagulating enzyme from the fourth stomach of 3-4 week old calves, in its purified form is called rennin, chymase or chymosin. They are used in cheese making as substitutes for natural rennet. Fungi such as *Endothia parasitica, Mucor meihei, Mucor pusillus var., lindt, Penicillium camembertii, Penicillium roqueforti, Rhizomucor meihei and Rhizomucor pusillus* (Sternberg, 1976) and bacteria such as *Bacillus mesentericus, Bacillus polymyxa, Bacillus subtilis, Lactobacillus bulgaricus, Lactobacillus helveticus, Streptococcus cremoris, Streptococcus lactis and Streptococcus thermophilus* are important sources of milk clotting enzymes. Microbial rennin is preferred as it does not involve calves and vegetarian ethical concerns.

Pepsin, an other important acid protease from species of *Aspergillus* and *Rhizopus* is used to produce soy hydrolysate with good whipping characters.

12.3 Neutral proteases

Neutral proteases from *Bacillus subtilis* and *Aspergillus oryzae* in combination with lipases are used to enhance flavour intensity of cheese without producing bitterness, called enzyme modified cheese (EMC). The fungal proteases include exo and endopeptidases. Endopeptidase of *Aspergillus oryzae* are used in bread making as the proteases improve the dough handling and to increase loaf volume, where as exopeptidases releases amino acid from gluten thereby contributing to the crust and crust flavour.

13. Pullulanase

Pullulanase produced by bacteria like *Bacillus polymyxa, Klebsiella aerogenes* and *Pseudomonas saccharophila* is used in starch processing.

Amylopullulanases (Type II pullulanase) produced by *Bacillus circulans, Bacillus stearothermophilus* and *Geobacillus thermoleovorans* is used in the liquefaction and saccharification of starch for the production of sugar syrups such as DP2 and DP4 syrups. Addition of amylopullulanases also help in improving texture, volume, colour, flavour and shelf life of bread; used as an additive in dish washing and laundry detergents.

14. Streptokinase

Streptokinase is an important enzyme produced by *Streptococcus equisimilis,* used in dissolving blood clots in patients suffering from cardiac thrombosis.

15. Tannase

Tannase is produced by fungi like *Aspergillus flavus* and *Aspergillus niger.*

Applications of tannase

Food and feed industry

Tannase produced by *Aspergillus niger* catalyses oleo-esterification of tea leaf compounds leading to enhancement of natural level of gallic acid and epicatechin and this in turn favours the formation of epitheaflavic acid which is responsible for bright reddish black color with cold water solubility.

Tannase produced by a strain of *Aspergillus flavus* significantly reduces chill haze formation in beer (Massechelin and Batum, 1981). It is used in treating grape juice, to remove phenols and to stabilize the beverage. Tannase is used in the treatment of olive oil waste water. It is also used to lessen the hydrolyzable tannin in poultry feeds.

Chemical industry

Tannase is used in the manufacure of printing inks, dyes and as a constituent of photographic developers. It is used in the production of gallic acid from gallo tannins from *Rhus coriaria* and in the transesterification of gallic acid esters such as propyl gallate, a widely used food preservative (Macrae, 1983).

16. Xylanases

Xylanases are produced by bacteria such as *Bacillus firmus, Bacillus stearothermophilus, Streptomyces lividans, Streptomyces thermoviolaceus, Thermoanaerobacterium saccharolyticum* and *Thermotoga maritime* and fungi and yeasts such as *Aspergillus fumigatus, Chaetomium cellulolyticum, Gliocladium virens, Melanocarpus albomyces, Myceliophthora thermophilum, Schizophyllum commune, Thermoascus aurantiacus, Thermomyces lanuginosus* and *Trichoderma viride, Humicola grisea, Humicola lanuginosa, Saccharomonospora viridis, Myceliophthora albomyces, Myceliophthora pusillus, Neurospora crassa, Sporotrichum dimorphosum, Talaromycess aurantiacus, Talaromyces emersoni, Talaromyces lanuginosus, Talaromyces terrestris, Trichoderma reesei.* Xylanases are employed in the following industries.

Applications of xylanases

Xylanases can be used for tailor designing of drugs and modifying the properties of food; in biobleaching of kraft pulp as a replacement for chlorine; for improving quality of bakery products; as animal feed additive; to improve feed conversion of rye based diet of poultry birds and to decrease the anti - nutritive effect and viscoscity of feed stuff; used in retting of hemp, flax, jute and sisal fibre and as specific tools for elucidating the structure of plant cell walls (Chada *et al.*, 2005)

Bioconversion

Xylan rich liquors are used by xylanolytic microbes to produce a mixture of fermentable sugars. Xylose obtained from the hydrolysis of xylan is used as a substrate for the production of ethanol, xylitol and xylonic acid. Pentose produced by the hydrolysis of xylan is used as a substrate in the production of protein rich biomass (SCP) (Rawat and Johri, 2004).

Food and feed industry

Xylanolytic enzymes are widely used in clarification of juices and wines and to enhance, flavour, volume, texture and color of cereal flours. The main application of xylanases are in saccharification of agricultural, industrial and muncipal wastes for production of sugar syrups, and as food thickeners.

Xylanases are used as a feed supplement or pretreatment of forage crops that improves the nutritional value of silage and poultry diet. They remove antinutritional factors (ANF) such as glucans, arabinoxylan and phytic acid which adversely affect the animal performance.

Xylanases are used in extraction of coffee and plant oil, maceration of cell walls during processing of fruits and vegetables, processing of oil seeds and production of xylo-oligosaccharides.

Textile industry

Xylanases are used for accelerating retting of hemp, flax, jute and bast fiber and in the production of rayon, cellulose esters and cellulose ethers (Saxena *et al.*, 2004).

Paper industry

Xylanases are used for enzymatic pre bleaching of pulp for kraft paper manufacture. They are used in debarking and pulp refining. As industrial process takes place at high temperature and alkalinity, xylanases provide biofriendly and cost effective solution.

They are also used to lower the lignin content of unbleached soft wood and hard wood kraft pulp (Chauvet *et al.*, 1987) and to enhance pulp brightness, pulp fibrillation, water retention and reduction in beating time in virgin pulps. They increase freeness in recycled fibres.

Other uses

The hydrolytic products of xylan can be converted into liquid fuel, SCP and solvents (Kuhad *et al.*, 1997). The increased industrial demand for thermostable enzymes with high pH stability makes xylanases a potential candidate for biotechnological applications.

17. Other microbial enzymes

Many of the microbial enzymes are used as therapeutic agents in treating various ailments *viz.*, Collagenase in treating skin ulcers; trypsin as an anti-inflammatory agent; urease in treating gout and urokinase as thrombolytic agents. Several novel enzymes are obtained by screening bacterial species (Table 20.9).

Table 20.9 Novel microbial enzymes found through screening and their final products

Enzyme	Source species	Product
D-Hydantoinase	*Bacillus* sp. *Pseudomonas putida*	D-Amino acids
D-Decarbamoylase	*Agrobactericum* sp. *Blastobacter* sp.	D-Amino acids
L-Hydantoinase	*Pseudomonas putida,*	L-Amino acids
L-Decarbamoylase	*Alcaligenes xylosoxidans*	L-Amino acids
α-Tyrosinase	*Erwinia herbicola*	L-3, 4-dihydroxy phenyl alanine
Dioxygenase	*Dactylosporangium* sp., *Streptomyces* sp.	Trans-4-hydroxy-L-proline, Cis-3-hydroxy-L-proline

Serine hydroxy-methyl transferase	*Methylobacterium* sp.	L-Serine
Aldehyde reductase	*Sporobolomyces salmonicolor*	Ethyl R-4-chloro-3-hydroxy butanoate
Carbonyl reductase	*Candida magnoliae*	Ethyl S-4-chloro-3-hydroxybutanoate
Nitrile hydratase	*Rhodococcus rhodochrous*	Acrylamide, Nicotinamide
Nitrilase	*Rhodococcus rhodochrous*	Acrylic acid, Nicotinic acid
Nitrilase and dehydrogenase	*Agrobacterium* sp.	5-hydroxypyrazine-2-carboxylic acid
Lipase	*Serratia marcescens*	2S, 3R-3-(4-methoxyphenyl-glycidic acid methyl ester).
O-phthalyl amidase	*Xanthobacter agilis*	Carbacephem
Alkene monooxygenase	*Nocardia corollina*	Chiral epoxide
D-Hydroxylase	*Beauveria bassiana*	R-2-(4- hydroxy phenoxy) propionic acid
Alcohol dehydrogenase	*Rhodococcus erythropolis*	S-p-Chlorophenyl ethanol
	Lactobacillus kefyr	R-p-Chlorophenyl ethanol.
Halohydrin hydrogen halide lyase	*Alcaligenes* sp.,	Chiral 2,3-dichloro 1-propanol
	Pseudomonas sp.,	Chiral 3-chloro- 1, 2-propanediol
Alcohol dehydro-genase reductase	*Candida parapsilosis*	S-1, 2-pentanediol.
Malto-oligosyl treharose synthase	*Arthrobacter* sp.	Malto-oligosyl treharose
Trehalohydrolase	*Sulfolobus acido-caldarius*	Treharose
Carbonyl reductase	*Candida parapsilosis*	D-Pantoyl lactone

Lactonase	*Fusarium oxysporum*	D-pantoic acid
Oxygenase	*Pseudomonas putida*	Theobromine
Adenosyl methionine synthetase	*Saccharomyces sake*	Adenosyl methionine
Adenosyl homocysteine	*Alcaligenes faecalis*	Adenosyl hydrolase
Nucleoside phosphorylase	*Enterobacter aerogenes,*	Adenine arabinoside
	Erwinia carotovora, *Escherichia coli*	Ribavirine, 5 - methyluridine Dideoxyadenine

Source: Ogawa and Shimuzu, 1999.

20.11 MICROBIAL CELLULASE

The strong, tough and pure cellulase synthesized by micro-organisms is a new industrial product with exciting application in food and medicine. *Acetobacter xylinium* is being harnessed as a potential source of cellulase (Bungay *et. al.,*1997). Microbial cellulase has been used in foods in Asian countries. *Nata de coco* – a cellulose product from Phillipines is consumed as dessert.

Application and prospect for commercialization of bacterial cellulase were evaluated in detail. (Yamanaka *et al.,* 1989, White and Brown 1989, Cannon and Anderson 1991). Current applications range from food additives to separation membranes (Shibazaki *et al.,* 1993; Ammon *et al.,* 1995). The food appli cations include low calorie additive, thickener, stabilizer, in ice cream and condiments (Okiyama *et al.,* 1993).

Microbial cellulase is added to strawberry pulp in disk reactors and dried into thin sheets. These sheets can be placed right on toast and muffins which is customer friendly since leaking and spooning is avoided.

Microbial cellulase has wide industrial applications. Cellulase is used as a matrix for attaching enzymes for better diffusion and for co-immobilizing two or more enzymes to carry out coupled or sequential reactions and as a substitute for plastic and special grades of paper. This eliminates the pollutants which emerge from pulping and bleaching of wood cellulase.

Perhaps the most publicized product using cellulase already in use is the acoustic diaphragm for audio speakers developed by Sony Corporation for clarity and higher sound (Yamanaka and Watanabe, 1994).

20.12 BIOPOLYMERS

Microbial polysaccharides are cytoplasmic reserve source of carbon. Though glycogen is retained in the cytoplasm, others like exopolysaccharides (EPS), which are excreted outside are of commercial interest. They are useful as thickening, gelling or suspending agents. Alginates, curdlan, dextran, gellan, pullulan and xanthon are a few of the commercially important EPS produced by bacteria. The following is the list of polysaccharides produced by microbes and their uses.

Alginate

Alginate produced by *Azotobacter vinelandii* and *Pseudomonas* species is similar to alginates obtained from seaweeds. It is used as beads for enzyme immobilization and as a gelling agent (Sullia and Shantharam, 2005).

Curdlan

Curdlan produced by *Alcaligenes faecalis* var. *myxogenes* finds use in the pharmaceutical industry.

Cyclodextrins

Cyclodextrins produced by species of *Thermoanaerobacter* are useful in pharmaceuticals for increasing solubility, for masking bad odour and bitterness of drugs.

Dextran

Dextran is obtained from species of *Acetobacter, Klebsiella* and *Leuconostoc.* The richest source of dextran is *Leuconostoc mesenteroides* and *Leuconostoc dextranicum.* Dextran is used as a protective hydrophilic covering layer on burns as an absorbent in wound dressing and in prevention of thrombosis.

Gellan

Gellan produced by *Pseudomonas elodea, Sclerotium rolfsii* and *Sclerotium glucanicum* is used for stabilizing latex paints, printing inks and for immobilization of thermophilic bacteria as it has high setting temperature.

Pullulan

Pullulan is obtained from *Pullullaria pullulans* which is the imperfect stage of *Aureobasidium pullulans,* forms strong films and fibres similar to cellophane. It is useful in making biodegradable plastic.

Xanthan

Xanthan gum obtained from *Xanthomonas campestris,* has a wide range of application in paints, ceramic glazing, dyeing industry for migration of dyes, textiles printing, abrasives, detergents, deodorants and insecticidal sprays. It is used in food products such as sauces, cakes, ice cream, shakes, syrups and spreading cheese, as a stabilizing, thickening and gelling agent. It is employed in petroleum industry to recover residual oil and as a lubricant for drilling machines used in oil exploration.

20.13 PRODUCTION OF ALTERNATIVE SWEETENERS BY MICROBES

Alternate sweeteners are organic compounds other than sucrose many of which are derived from microbes. á-amylases, glucoamylases, pullulanases from microbes are used for hydrolysis of starch. The conversion of statrch slurry to low DE (dextrose equivalent) maltodextrins is the first step of hydrolysis followed by saccharification to glucose syrup and then finally to fructose syrup.

Maltodextrins are employed as raw material for enzymic saccharification, thickeners, fillers, stabilizers and glues. Maltose syrup is used in confectionary, jams, jellies, ice cream, sauces, and soft drinks; High DE syrup is employed in confectionary, soft drinks, jams and sauces; Glucose syrup is used in soft drinks, caramel, baking, and brewing; High fructose corn syrup is a sweetener derived from corn starch contributing to more than one third of the world's market of sweeteners. Fructose is twice as sweet as sucrose used in soft drinks, jams and yoghurt. (Kennedy *et al.*, 1988).

Erythritol is a non carcinogenic, non caloric, diabetic safe sweetener formed from fermentation by *Aureobasidium, Candia magnoliae* and *Trichosporon* species; Mannitol from *Leuconostoc* sp., and *Candida magnoliae*, Sorbitol from *Zygomonas mobilis* and xylitol from *Candida tropicalis* are sugar alcohols used as low calorie alternate sweeteners used in food and pharmaceutical industry.

Aspartame an ester of L-aspartic acid and L-phenylalanine produced by *Corynebacterium glutamicum* and *Pseudomonas fluorescens* is used as low calorie sweetener in soft drinks.

20.14 BIODEGRADABLE PLASTICS

Polyhydroxy alkanoates (PHA) are intracellular energy reserve compounds in several bacteria. PHAs are easily broken down by microbial activity and are being used in biodegradable plastics called bioplastics. *Alcaligenes eutrophus, Alcaligenes latus, Pseudomonas acidovorans, Pseudomonas oleovarans, Rhodococcus ruber* and species of *Nocardia* are a few bacterial examples producing PHA.

Polyhydroxy butyrate (PHB) is a polyester compound which is produced during nutrient exhaustion phase or during non availability of oxygen. Biopol is the trade name for the bioplastic developed by ICI plastics U.K. Genetically engineered *Escherichia coil* produce hight levels of poly-β-hydroxy butyric acid. Bioplastics can be used for implants such as bone plates, rods and wound dressing in human body (Kovacs *et al.*, 2002, Zadvorny *et al.*, 2004; Sullia and Shantharam, 2005)

20.15 BIOSURFACTANTS

Biosurfactants are those surface active chemicals produced by micro-organisms such as bacteria and yeasts which have defined hydrophilic and hydrophobic groups. The four main types of biosurfactants are glycolipids, phospholipids, lipoproteins and polymers. Micobial biosurfactants and bioemulsifiers are given in Table 20.10.

Chemically synthesized surfactants are often not biodegradable and can be toxic to the environment while the natural surfactants are less toxic. They can be of interest in mobilization of heavy crude oil and residual sludge and remediating oil spills in water and soil. Thus they are likely candidates for replacement of chemically synthesized surfactants in bioremediation with broad range of industrial applications (Jone 1997).

Biosurfactants have found use in oil recovery, oil spill cleanup, textiles, cosmetics, pharmaceuticals and in detergent formulations. They are widely used in agriculture, production of elastomers, plastics, paints, petrochemicals, foods and beverages. The diversity of biosurfactants make them likely candidates for replacement of chemically synthesized surfactants in bioremediation efforts.

Rhamnolipids from *Pseudomonas aeruginosa* and *Pseudomonas testosteroni* degrade hexacane and PCB while sophorolipids and glycolipids from yeasts such as *Rhodococcus aurantiacus* and *Torulopsis bombicola* degrade alkanes and hexadecane. (Kosaric *et al.*, 1987, Rouse *et al.*, 1994).

Table 20.10 Microbial biosurfactants and bioemulsifiers

Microbial species	Biosurfactant/ Bioemulsifier
Bacteria	
Acinetobacter sp.,	Glycerides, emulsan
Bacillus licheniformis	Cyclic lipopeptides
Bacillus pumilis, *Bacillus subtilis*	Cyclic lipopeptides and surfactins
Corynebacterium hydrocarbolastus	Proteolipids
Corynebacterium insidiosum	Phospholipids
Corynebacterium lepus	Corynemycolic acids

Nocardia erythropolis	Neutral lipids
Pseudomonas aeruginosa, Pseudomonas fluorescens	Rhamnolipids
Yeasts	
Candida bogoriensis, Candida bombicola, Candida lipolytica	Sophorolipids
Candida antarctica	Erythritol lipids
Candida petrophilum	Peptidolipids
Candida sp.,	Mannosyl erythritiol lipids
Candida tropicalis	Mannan - fatty acid complex
Phaffia rhodozyma	Carbohydrate lipids
Rhodococcus erythropolis	Trehalose lipids and trehalose mycolates
Torulopsis bombicola	Sophorolipids

Source: Healy *et al.*, 1996, Banat 1993.

20.16 BIOFUELS

Biofuels are derived from organisms to be used as substitutes for fossil fuels such as natural gas, petroleum and diesel in order to conserve the latter. Green butanol, ethanol, hydrogen and methane are a few important alternative fuels obtained from microorganisms.

20.16.1 Green butanol

The acetone butanol ethanol (ABE) fermentation by *Clostridium acetobutylicum, Clostridium beijerincki, Clostridium saccharoperbutylacetonicum and Clostridium tetanomorphum* were utilized first during World War I in which *Clostridium acetobutylicum* was used to produce acetone for the manufacture of smokeless gun powder (Jones and Keis 1995). Recently, there has been a renewed interest in fermentation derived butanol (green butanol) both as a potential fuel and as a chemical food stock. Green butanol shares many benefits of other biofuels like ethanol including reduction of CO_2 emission and high energy content.

20.16.2 Ethanol

Ethanol produced by microbial processes using various carbohydrate sources is a very potential biofuel. In Brazil, ethanol mixed with petroleum fuels is marketed as gasohol for running automobiles. (see Chapter 7).

However, use of bacteria in the production of ethanol is not very productive due to their high temperature tolerances. Only *Zymomonas mobilis* is being used to achieve high productivity of ethanol.

Lignocellulosic materials such as cellulose, hemicellulose and lignin are the most abundant renewable organic resources on earth which can be converted to useful products by bacteria. Lignocellulose is produced by bacteria like *Erwinia carotovora, Erwinia chrysanthemi, Klebsiella oxytoca and Escherichia coli*. The application of metabolic engineering offers the potential to revolutionize the chemical industry. Many bacteria have evolved sophisticated systems of hydrolases and lyases that aid the solubilization of lignocellulose. Genetic engineering of these bacteria for ethanol production is an attractive alternative to the solubilization of lignocellulosic biomass.

20.16.3 Methane

Methane emanates from organic remains degraded by methanogenic bacteria such as *Bacillus methanicus, Methylomonas, Methylococcus, Methylocystis, Methylosinus, Methanosarcina, Methanospirillum, Methanothrix, and Pseudomonas methanica*. These bacterial species are being explored for industrial production of methane which has a great potential for mass use as a biofuel (see Chapter 10).

20.16.4 Hydrogen

The potential of hydrogen is being widely explored throughout the world. In view of the growing demand for alternative and renewable sources of energy, hydrogen is being projected as an eco-friendly fuel. Major photobacteria which produce hydrogen are *Allochromatium vinosum, Ectothiorhodospira mobilis, Ectothiorhodospira vacuolata, Phaeospirillum molischianum, Rhodoblastus acidophila, Rhodoblastus sphaeroids, Rhodopseudomonas rutica, Rhodospirillum rubrum, Rhodovibrio gelatinosus* (Khanna *et al*., 1999).

20.17 INSECTICIDES AND HERBICIDES FROM MICROBES

Microbial insecticides are produced in living hosts or in non-living substrates using microbial processes. Commercial production of biopesticides is restricted to few bacteria, fungi and viruses. Artificially formulated media with carbon and nitrogen source serve as substrate for bacteria and fungi, whereas others need live insect larvae grown in biofermenters as substrate because of their obligatory nature.

There are over 90 species of bacteria and fungi pathogenic to insect pests and serve as new tools in biological control of plant pests. Most popular bacterial insecticides come from *Bacillus popillae* produced on a commercial scale in the form of dust on larvae of Japanese beetles where as *Bacillus thuringiensis* and *Bacillus mortiai* require fermenters.

Insecticidal viruses fall into five major groups, Nucleopolyhedrosis viruses, Cytoplasmic polyhedrosis viruses, Entomopox viruses, Granulosis, inclusion viruses and Non - inclusion viruses. The following viruses, bacteria and fungi are the likely source of future insecticides against the pests indicated.

20.17.1 Viral Insecticides

A. Nucleopolyhedrosis viruses (NPV)

Alfalfa caterpillar
Alfalfa looper
Army worm
Boll worm
Douglas fir tursock moth
Forest tent caterpillars
Gypsy moth
Pine sawfly
Lymantia dispar
Oryza pseudostugata
Spruce sawfly
Tobacoo budworm
Wattle bagworm

B. Granulosis Inclusion Viruses (GIV)

Cabbage worm
Codling moth
Potato tuber worm
Red-banded leaf roller

C. Non-Inclusion Viruses (NIV)

Citrus red mite
Coconut palm rhinoceros beetle

D. Cytoplasmic Polyhedrosis Viruses (CPV)

Pine processionary caterpillar

E. Entomopox Viruses (EPV)

Spruce bud worm

20.17.2 Bacterial Insecticides

Aerobacter aerogenes
Bacillus cereus
Bacillus lentimorbus
Bacillus moritai
Bacillus popillae
Bacillus thuringiensis
Clostridium malacosoma
Coccobacillus acridiorium
Proteus vulgaris
Pseudomonas aeruginosa
Serratia marcescens.

20.17.3 Fungal Insecticides

Acrostalagmus aphidium
Aschersonia aleyrodis
Beauveria bassiana
Beauveria tenella
Coelomomyces opifex
Coelomomyces punctatus
Entomophthora grylli
Entomophthora sphaerosperma
Entomophthora thaxiteriana
Hirsutella thompsonii
Metarrhizium anisopliae
Nomuraea rileyi

20.18 BIOHERBICIDES

Bioherbicides are described as intentionally used, mass produced plant pathogens formulated and applied at high inoculum rates as chemicals to control weeds (Boyetchko and Peng, 2004).

20.18.1 Bacterial herbicide

Rhodovibrio gelatinosus
Rhodopseudomonas palustris
Rhodoblastus acidophila sphaeroides
Paecilomyces eriophytis

20.18.2 Fungal herbicides

The following fungi are used to control growth of various weeds:

1. *Puccinia chondrillina* is used to control *Chondrilla juncea*
2. *Cerosporella riparia* is used to control *Ageratina riparia*
3. *Cercospordium* (= *Cercospora) rodmanii* controls *Eichhornia crassipes* (Water hyacinth).
4. *Colletotrichum gloeosporioides* controls *Aeshynomene virginica* in rice fields.
5. *Phytophthora citrophthora* controls *Morrenia odorata* (Milk weed wine)
6. *Phragmidium violaceum* controls *Rubus* sp. (Black berry).

Eight fungal herbicides have been registered in various countries (Table 20.11).

Table 20.11 Patented fungal herbicides

Product	Source species	Country
Devine	*Phytophthora palmivora*	USA
Collego	*Colletotrichum gloeosporioides*	USA
Stumpout	*Colletotrichum leave*	South Africa
Biochon	*Colletotrichum purpureum*	Netherlands
Chontrol	*Colletotrichum purpureum*	Canda
Smolder	*Alternaria destruens*	USA
Mallet WP	*Colletotrichum gloeosporioides*	Canada

20.18.3 Biofungicides

Biofungicides are antifungal microbial metabolites with desired chemical and biological characteristics. They can be used directly or can be a lead for designing novel synthetic products (Kim and Hwang, 2004).

The following antifungal metabolites are in commercial use as biofungicides:

1. Blasticidin from *Streptomyces griseochromogenes*
2. Kasugamycin from *Streptomyces kasugaensis* and *Streptomyces kasugaspinus*
3. Polyoxins from *Streptomyces cacaoi* var., *asoensis.*
4. Validamycin A from *Streptomyces hygroscopicus* var., *limoneus.*
5. Mildiomycin from *Streptomyces rimofaciens*

20.19 SINGLE CELL PROTEINS

Single Cell Protein (SCP) is a product of unicellular microbes with potential for use as food and feed. Single celled yeasts, bacteria, fungi, and algae (For algae see Chapter 10) are cultured on hydrocarbon wastes (molasses, whey, coffee waste, paper and methanol) and harvested. SCP from fungi are mycoproteins and are discussed under section 20.19.2.

SCPs may be used without further processing, in animal feed formulations or may be used as a processed protein ingredient in human food.

SCPs are rich in nitrogen and amino acid content comparable to animal and plant sources with the exception of methionine, which is lower. They are also rich in carbohydrates, lipids, vitamin B and minerals. The fat content varies among the sources, with algal cells containing highest levels and bacteria the lowest.

20.19.1 Single Cell Proteins for Human Use

For human use, the likely products are SCP concentrates that may be further processed into functional single cell protein products. To obtain functional SCPs, the cells are mechanically disrupted, cell walls are removed by centrifugation, proteins are precipitated and the resulting proteins are extracted in acetate buffers or acetic acid. The SCP thus formed can be used to form textured protein products. When bacteria are used as source of SCP, their endotoxins are removed.

Consumption of SCP products may result in gastrointestinal disturbances particularly following the consumption of algal and bacterial SCP. The high nucleic acid (16 per cent) content of SCP may lead to kidney stones or gout in humans.

Cellulosic cell walls of green algae and the diaminopimelic acid, pimelic acid, and muramic acid of the blue green algae, may not be easily digested by humans. Sources of SCPs from bacteria, yeasts and algae are given below.

Bacterial SCP

Acinetobacter calcoaceticus, Alcaligenes eutrophus, Methylococcus capsulatus, Methylomonas methanica, species of *Cellulomonas, Nocardia, Pseudomonas* and *Rhodopseudomonas.*

Yeasts SCP

Candida albicans, Candida guillermondii, Candida intermedia, Candida lipolytica, Candida pseudotropicalis, Candida utilis, Debaryomyces kloeckeri, Hansenula polymorpha, Kluyveromyces fragilis, Kluyveromyces marxianus var. *lactis, Rhodotorula* species, *Saccharomyces cerevisiae, Saccharmomyces fragilis, Saccharomyces febuliger, Torulopsis candida* and *Pichia.*

20.19.2 Mycoproteins (Fungal SCP)

Mycoproteins are derived from multicellular fungi and yeasts. In addition to proteins, they have abundant nucleic acids, carbohydrates, minerals and vitamins, which are currently utilized as a source of food (Castrillo and Ugalde, 2004). Many yeasts and filamentous fungi have been approved by the US Food and Drug Administration for human consumption. In addition to the nutritional benefits, they reduce total and low density lipoprotein (LDL), serum cholesterol level and post meal glycaemia.

Several speciality products are derived from microbial proteins (Box 20.1). Baker's and Brewer's yeasts are used to compliment the properties of other foods and inactive feed yeasts which are non fermentative, are classified as dried yeasts (category 96.1, official publication AAFCO 2002). *Saccharomyces cerevisiae*, *Kluyveromyces marxianus,* and *Candida utilis* are accepted for use in bakery products and as dough conditioner.

Diamond V and the Chinese Academy of Agriculture report the use of edible selenium enriched yeast and fungi (http:// www. diamonddv.com; CN patent 1121531, 1996). Diamond V produce active yeast cultures for fermenting raw material for animal feed.The sophisticated mycoprotein based food technology will not only facilitate the production of speciality products but also for fast production of food during demanding conditions such as war, famine and epidemics.

Mycoproteins are currently appreciated foods as they have bland taste, light colour which renders them as ideal products for flavoring and colouring. They are favoured food for patients with high blood cholesterol level or diabetics. *Aspergillus niger, Aspergillus oryzae, Fusarium graminearum, Fusarium venenatum, Penicillium pastoris, Penicillium rhodozyma, Penicillium roquerfortii, Penicillium variotii, Rhizopus niveus, Rhizopus oryzae,* species of *Mucor* and *Trichoderma viride* are the source of fungal SCPs.

Box 20.1 Mycoproteins

Biomass production of microbial proteins as bulk food for human and animal consumption is being developed by Rank Hovis McDougall (RHM) in co-operation with imperial chemical industries. Marlow foods is producing mycoproteins under the commercial trade mark QUORN. The product is manipulated to achieve a texture and taste which resemble meat and is used as a meat alternative for vegetarians. QUORN products are the only fungal based products obtained form *Fusarium graminearum* as a source of proteins approved by US food and drug administration.

20.20 FLAVOURS AND AROMAS

The demand for natural flavouring substances have been on the increase. Autolysates and hydrolysates from yeasts and fungi are obtained by hydrolysis of protein fraction and can be used as natural flavours and to improve the physical properties (texture) and nutritional value of food products. They are used to enhance the flavour of cheese, meat, sausages, soups and sauces. These flavour enhancers are at present produced by Diamond V, Provests, Burns Philip, Lallemand Inc, Lesaffre and Beldem (Castrillo and Ugalde, 2004, Agarwal 2004)

Fungal biotransformation of low cost substrates into flavour and aroma compounds has been accepted by United State Food and Drug Administration (FDA) and by the European Flavour Commission, as being natural and falling under the Generally Regarded As Safe (GRAS) category (Taylor and Mottram, 1996). These flavouring agents are optically pure, extra cellular and suitable for commercial exploitation with high yield (Agarwal 2004).

20.20.1 Production of Microbial Flavours

Filamentous fungi are used for industrial production of natural flavours, either *de novo* or by biotransformation. Fungi also play an important role in biotransformation of terpenes, esters, lactones, pyrones, aldehydes and ionones which are the constituents of various flavours.

Basidiomycetes are the most important filamentous fungi which are employed in producing benzoate type of flavours such as vanilla, bitter, almond, grape, rasberry and orange.

Flavours such as sour (*Aspergillus niger*), Pine (*Aspergillus oryzae*), fruity (*Aspergillus terreus*), coconut (*Fusarium poae, Trichoderma viride*), banana (*Ceratocystis fimbriata*), peach (*Fusarium poae*) and tobacco (*Aspergillus niger*) can be obtained from fungi by *de novo* synthesis.

Various chemical compounds in fungi which are employed in flavouring are listed in Table 20.12. Benzoate flavours from basidiomycetes are listed in Table 20.14.

Table 20.12 Fungal flavoring constituents

Constituent	Source species
Limonene	*Aspergillus cellulosa, Corynespora cassiicola, Diplodia gossypina, Penicillium digitatum*
Pinene	*Aspergillus niger, Geotrichum candidum, Geotrichum fragrans, Mucor meihei, Rhizopus delemer*
Esters	*Rhizopus oryzae*
Lactones	*Mucor* species, *Penicillium roqueforti, Sporobolomyces odorus, Trichoderma harzianum, Ustilago zeae*

Pyrones	*Trichoderma harzianum, Trichoderma viride*
Aldehydes	*Aspergillus niger, Ischnoderma benzoicum, Penicillium digitatum, Trichosporon beigelii*
Ionone (tobacco flavor)	*Aspergillus niger*
Citronellol	*Botrytis cinerea*
Geraniol	*Saccharomyces cerevisiae*
Phenyl propenol	*Fusarium moniliforme*
Phenyl ethanol (rose flavour)	*Aspergillus niger*
Octanediol	*Botrytis cinerea*

Source: Agarwal, 2004.

20.20.2 Complex Flavours

Complex flavours are flavours produced by combining and blending individual flavour chemicals together. Dairy flavours such as cheese, milk and bread flavours are contributions of fungal constituents mainly from *Penicillium* species. (Kritofferoson, 1973; Maga, 1974).

Mushroom flavors could be produced by *Aspergillus oryzae, Boletus edulis, Coprinus micaceus, Lentinus edodes, Lepiota rhacodes, Merulins rufus, Morchella crassipes, Poria* species and *Trichoderma nudum* (Sugihara and Humfeld, 1954; Schindler and Schmid, 1982).

20.20.3 Flavour Enhancers

Microbially produced nucleosides and nucleotides have found wide use as flavour enhancers in food industry in soups and sauces and in removing metallic taste of canned foods.

They are also used in therapeutics e.g. 8-azaguanine and 6-mercaptopurine are used in cancer chemotherapy; orotic acid is used to treat liver diseases; cAMP in diabetes and asthma; 5- AMP in rheumatism and adenosine in arteriosclerosis and high blood pressure. The major producers of nucleosides and nucleotides are given in Table 20.13.

Table 20.13 Major producers of nucleosides and nucleotides

Product	Source species
Adenosine	*Bacillus subtilis* P53 -18 *Brevibacterium ammoniagenes*
ATP	*Candida boidinii*

cAMP	*Escherichia coli* *Corynebacterium murisepticum* *Microbacterium* sp., *Brevibacterium liquefaciens*
Inosine	*Brevibacterium ammoniagenes* *Corynebacterium petrophilum* *Microbacterium* sp., *Saccharomyces* sp., *Streptomyces* sp.,
Guanosine	*Bacillus subtilis* *Bacillus pumilus* *Bacillus licheniformis* *Corynebacterium petrophilum* *Corynebacterium guanofaciens* *Streptomyces griseus*
Nucleotides	*Candida utilis* *Saccharomyces cerevisiae* *Streptomyces aureus* *Penicillium citrinum*
5' - IMP	*Bacillus subtilis* *Corynebacterium ammoniagenes* *Corynebacterium glutamicum*

Source: Crueger and Crueger , 2005.

Table 20.14 Benzoate-type flavours produced by Basidiomycetes

	De novo synthesis		Production by biotransformation	
Chemical	**Flavour**	**Species**	**Species**	**Precursor**
Vanilin	Vanilla-like		*Phanerochaete chrysosporium* *Pycnoporus cinnabarinus*	*Vanilic acid, Ferulic acid or vanilic acid*
Benzaldehyde	Bitter almond	*Agaricus bisporus*, *Agaricus subrefecens*, *Armillaria mellea*, *Armillaria ostoyae*,	*Bjerkandera adusta* *Dichomitus squalens*, *Ischnoderma benzoinum* *Polyporus* sp.	Phenylalanine Phenylalanine Phenylalanine Castor oil and lignin
		Bjerkandera sp., *Dichomitus squalens*,	*Polyporus tuberaster*	Phenylalanine or benzoic acid
		Ischnoderma benzoinum,		Asparagine
		Hericium erinaceus,	*Poria cocos*	Phynylalanine
		Hirschioporus pergamenus, *Mycena pura*, *Phlebia radiata*, *Pleurotus sapidus*, *Polyporus* sp., *Polyporus tuberaster*, *Poria subacida*, *Trametes suaveolens*, *Tyromyces sambuceus*, *Sarcodontia setosa*	*Poria xantha* *Tyromyces sambuceus*	Asparagines
Methyl anthranilate	Orange blossom, wood, strawberry	*Poria cocos* *Pycnoporus cinnabarinus*	*Trametes* sp. *Polyporus* sp. *Pycnoporus cinnabarinus*.	Methyl *N*-methyl anthranilate Anthranilic acid

(*Table 20.11 Contd...*)

Anisaldehyde	Vanilla-like	*Armillaria mellea,*	*Bjerkandera adusta,*	Tyrosine
	Anise-like	*Bjerkandera* sp., *Bjerkandera adusta,* *Bjekandera* sp. *BOS55,* *Daedalus juparana,* *Dichomitus squalens,* *Ischnoderma benzoinum,* *Lentinus lepidus,* *Mycoacia* sp., *Pleurotus sapidus,* *Phellinus* sp., *Polyporus benzoinus,* *Trametes suaveolens,*	*Ischnoderma benzoinum,* *Pleurotus cornucopia,* *Pleurotus eryngii,* *Pleurotus floridanus,* *Pleurotus sajor-caju* *Pleuratics ostreatus*	Lignin
Veratraldehyde	Vanilla-like	*Armillaria lutea,* *Armillaria mellea,* *Bjerkandera adusta,* *Bjerkandera* sp., *Dichomitus squalens,* *Hirschioporus pergamenus*	*Pleurotus* sp.,	Lignin and straw
p-Hydroxy- benzaldehyde	Vanilla-like	*Sirodesmium diversum*		Lignin and straw
p-Tolualdehyde	Bitter almond	*Lentinus* sp., *Mycoacia uda,* *Phellinus* sp., *Trametes* sp.	*Pleurotus sp.,*	Lignin and straw
Acetophenone	Orange blossom	*Dichomitus squalens,* *Phellinus contiguus* *Pycnoporus cinnabarinus*	*Pleurotus* sp.,	Castor oil and lignin

p-Methyl-acetophenone	Fruity	*Mycoacia uda*, *Lentinus* sp., *Phellinus* sp., *Trametes* sp.,	*Polyporus* sp.,	Castor oil and lignin
p-Methyl benzyl alcohol	Fruity	*Mycoacia uda*, *Lentinus* sp., *Phellinus* sp., *Trametes* sp.,	*Polyporus* sp.,	Castor oil and lignin
Methyl benzoate	Fruity	*Agaricus subrufescens*, *Mycena pura*, *Phellinus* sp., *Polyporus* sp.	*Polyporus* sp.,	Castor oil and lignin
Ethyl benzoate	Fruity	*Lentinus* sp., *Lenzites betulina*, *Phellinus* sp., *Tyromyces sambuceus*	*Polyporus* sp.	Castor oil
Anisic acid	Anise-like	*Trametes odorata*, *Trametes suaveolens*	*Polyporus* sp.,	Castor oil
Methyl anisate·	Anise-like	*Lentinus* sp., *Mycoacia* sp., *Phellinus* sp., *Trametes odorata*, *Trametes suaveolens*	Polyporus sp.,	Castor oil
Methyl salicilate	Anise-like	*Phellinus ignarius*	*Polyporus* sp.,	Castor oil

Source: Lomascola *et al.*, 1999

20.21 FOOD COLOURANTS

Microorganisms such as bacteria, algae and fungi are known to produce variety of pigments that are a promising source of colorants in food and cosmetic industry. Fungi and algae are widely studied for their potential as a source of food colourants.

Fungi such as *Monascus purpureus* and *Monascus ruber* are used for the commercial production of *Monascus* pigments. *Monascus* pigments have been used for centuries in orient. In China, it is known as *Ang – Kak,* in Japan as *Beni – Koji* and *Aga – Koji*. Chinese red rice, red wine, red soybean, red cheese, pickled vegetables, sauces and meats are some examples of drink and food coloured red by *Monascus* pigments.

A blue colourant extracted from Cyanobacterium, *Spirulina platensis* and red colourant from red algae *Porphyridium and Porphyra* is commercially produced in Japan, Australia and America. It is used as a colourant in chewing gum, soft drink, fermented milk products such as yoghurt.

Major sources of carotenoids are plant and animal tissue. α–carotene (Provitamin A) from plants and animals is converted into Vitamin A in the intestinal mucous membrane of humans and stored in the liver as palmitate esters. A deficiency of Vitamin A causes night blindness and change in the skin texture.

Microbes have been successfully employed in the production of α-carotene, lycopene and zeaxanthin with wide range of uses in food and feed additives. β-carotene is produced by *Blakeslea trispora* (fungus), *Dunaliella salina* (alga); lycopene is produced by S*treptomyces chrestomyceticus* ssp., *rubescens* and *Blakeslea trispora*; Zeaxanthin (xanthophyll) by *Flavobacterium* and *Erwinia* species. Lycopene and xanthophylls are used as food colouring agents in cheese, margarine, meat and animal feed supplements and Vitamin A precursors (Ninent and Renaut, 1979). Source of carotenoids are given in Table 20.15.

Melanoid pigments (Biomelanins) from *Streptococcus avermitilis* have been used in sun tan creams. An undesirable, nephrotoxic mycotoxin Citrinin is formed during the production of pigments from fungi and should be eliminated to get non-toxic food colourants.

Table 20.15 Production of carotenoids by microorganisms

Carotenoid	Source species
α– carotene	*Blakeslea trispora*
Lycopene	*Streptomyces chrestomyceticus* ssp., *rubescens*
	Blakeslea trispora
Zeaxanthin	*Flavobacterium* sp.,

Source: Ninent and Renaut, 1979

CHAPTER 21

FOOD SPOILAGE BY MICROBES

21.1 INTRODUCTION

Most foods are generally an excellent media for microbial growth. Microbial growth occurs not only in natural plant products, such as fruits, vegetables, cereal grains and millets, but also animal products like meat, egg, fish, milk, curds, cheese, buttermilk and butter. Even the canned foods and bakery products are susceptible to microbial growth, due to improper processing procedures resulting in food spoilage.

21.2 MICROBIAL CONTAMINATION OF FOODS

The growth of microorganisms like yeasts, fungi and bacteria on food results in food spoilage and the production of toxins, which when consumed causes various disorders, infections and poisoning. Food spoilage is usually due to the growth of various bacteria and fungi. The spoiled food has an unpleasant odour, bitter or pungent taste and is unsuitable for consumption. The microbial contamination of food occurs through air, water and humans. Insects also contribute to microbial attack. The food contamination depends on temperature, humidity and the quality of food itself. Types of food spoilage and the causative micro-organisms are given in Table 21.1.

Table 21.1 Types of food spoilage and causative organisms

Types of food	Types of spoilage	Causal organism
Meat	Putrefaction, Souring	Species of *Bacillus, Clostridium, Pseudomonas, Lactobacillus, Leuconostoc, Micrococcus* and *Proteus.*
	Sliming	*Leuconostoc* species
	Greening	*Lactobacillus,* and *Streptococcus* species
Sugar	Pink syrup	*Micrococcus roseus*
	Ropy syrup	*Aerobacter* and *Saccharomyces* species
	Molding	*Aspergillus, Penicillium* and *Rhizopus* species

Bread	Red bread	*Bacillus subtilis, Monila* (asexual stage of *Neurospora nistophila*)
	Moldy bread	*Rhizopus stolonifer*
	Ropy bread	*Bacillus subtilis*
Pickles	Soft pickle	*Bacillus* species
	Slimy kraut	*Lactobacillus* species
Fish	Putrefaction	*Chromobacterium, Flavobacterium, Micrococcus* and *Pseudomonas* species
Fruits and vegetables	Soft rot, Mold rot	*Mucor* and *Rhizopus* species *Aspergillus* species

21.2.1 Milk

Milk is an ideal medium for yeast and bacterial growth. Contamination of milk by bacteria and yeast occurs through human handling, from the animals through its udder, improper handling, incomplete heating, unclean utensils and water. The most common type of milk spoilage occurs due to action of *Bacillus, Lactobacillus, Lactococcus, Micrococcus, Proteus* species and yeasts such as *Candida famata, Candida lipolytica* and *Trichosporon cutaneum.* This results in curdling of milk and sourness.

The effects of different micro-organisms on milk are given in Table 21.2.

Table 21.2 Biochemical types of microorganisms in milk

Activity	Causal micro-organism
Acid producers	*Enterobacter aerogenes, Escherichia coli, Lactobacillus* species,*Microbacterium lacticum, Micrococcus luteus, Streptococcus cremoris* and *Streptococcus lactis.*
Gas producers	*Clostridium butyricum* and *Torula cremoris.*
Ropy or stringy fermenters	*Alcaligenes viscolactis, Clostridium perfringens, Enterobacter aerogenes* and *Streptococcus cremoris.*
Proteolytic	*Bacillus subtilis, Bacillus cereus, Pseudomonas* and *Streptococcus* species.
Lipolytic	*Achromobacter lipolyticum, Candida lipolytica, Penicillium* species and *Pseudomonas fluorescens.*
Red milk	*Serratia marcescens*
Blue milk	*Pseudomonas syncyanea*

Source: Sullia and Shantharam, 2005.

21.2.2 Cream

Milk cream is spoiled by *Candida diffluens, Candida famata, Cryptococcus laurentii, Rhodotorula glutinis* and *Rhodotorula rubra*, resulting in sourness or bitterness.

21.2.3 Yoghurt

The most common yeast species responsible for the spoilage of yoghurt are *Candida famata, Candida versatilis, Debaryomyces hansenii, Kluyveromyces lactis, Pichia anomala,* and *Saccharomyces cerevisiae* (Hansen and Jakobsen, 2004).

21.2.4 Buttermilk and Curds

Bacteria like *Lactobacillus, Leuconostoc,* and *Srteptococcus* and fungi like *Penicillium* cause sourness, colour change and bitter taste to butter milk and curds.

21.2.5 Butter

Butter turns rancid due to bacteria like *Achromobacter, Alcaligenes, Klebsiella,* and *Pseudomonas,*fungi like *Alternaria, Aspergillus, Mucor* and *Cladosporium,* and yeasts such as *Candida guilliermondii, Candida kefyr, Rhodotorula glutinis, Rhodotorula rubra* and *Saccharomyces cerevisiae.*

21.2.6 Cheese

Cheese is spoilt by bacteria such as *Brevibacterium, Lactobacillus* and fungi like *Penicillium* which give bitter taste and flavour.

21.2.7 Fish and Meat

Contamination of fish and meat can occur due to unhealthy conditions of the slaughter houses and animals or due to highly contaminated water used. Processed meat products like sausages, ham and minced meat are easily spoiled. The important bacteria which bring about the spoilage are *Alteromonas, Clostridium, Lactobacillus, Leuconostoc, Micrococcus, Moravella, Pseudomonas, Staphylococcus* and *Vibrio.*

21.2.8 Eggs

Microbes cause contamination of eggs, specially the yolk which is more susceptible to spoilage. The bacteria which cause spoilage of eggs are species of *Proteus, Pseudomonas* and *Serratia. Pseudomonas* causes green colouration and *Serratia marcescens* imparts red colouration to the eggs with a foul smell. Fungi like *Cladosporium* and *Penicillium*, also contribute for foul smell and black coloration of egg.

21.2.9 Fruits and Vegetables

The spoilage of fruits and vegetables occurs due to micro-organisms which gain entry from soil in which they were grown, and water in which they were washed. Species of *Bacillus, Erwinia, Flavobacterium, Lactobacillus, Pseudomonas* and *Xanthomonas* among the bacteria cause spoilage. Fungi like *Aspergillus, Botrytis, Penicillium, Rhizopus* and *Torulopsis* are also responsible for spoilage.

Fruit juices can also be contaminated by bacterial species such as *Acetobacter, Leuconostoc* and *Lactobacillus* (Purohit, 1999).

21.2.10 Food Grains

Food grains such as cereals, millets, nuts and oil seeds are highly susceptible to microbial attack. The microbial spoilage occurs due to fungi like *Aspergillus, Fusarium, Penicillium* and *Rhizopus*, carried over from soil during harvest and then grow during storage. Cereal flours are attacked by species of *Bacillus, Flavobacterium* and *Pseudomonas. Micrococcus* imparts a bitter taste to flours.

21.2.11 Bakery Products

Bread, cakes, biscuits, pastries and pizza base are also subjected to microbial spoilage. Microbes normally attack carbohydrates in these products. The cream and icing sugar used in cakes and biscuits are degraded by *Bacillus, Lactobacillus*, and *Sterptococcus* and fungi like *Aspergillus, Mucor, Neurospora, Penicillium* and *Rhizopus* to produce sour smelling and tasting fatty acids, which impart rancidity. Candies are spoilt by species of *Bacillus, Clostridium, Flavobacterium* and *Pseudomonas.*

21.2.12 Canned Foods

Many of the canned food are attacked by microbes due to improper canning and handling procedures resulting in spoilage. *Bacillus stearothermophilus, Clostridium botulinum, Clostridium thermosaccharrolyticum, Staphylococcus aureus*, species of *Lactobacillus, Leuconostoc, Micrococcus, Pseudomonas* and *Streptococcus* cause sliming of canned food.

Jams and jellies are spoilt by species of *Bacillus, Clostridium, Flavobacterium* and *Pseudomonas.*

21.2.13 Refrigerated Foods

Refrigerated foods are spoilt by species of *Lactobacillus* and *Streptococcus.*

21.3 FOOD POISIONING AND INFECTION

Consumption of spoilt food results in food poisoning or infection. Food poisoning occurs due to microbes or their toxins resulting in disorders of the digestive tract, where as food infection results in diseases caused by microorganisms in the contaminated food.

21.3.1 Types of Food Poisoning

21.3.1.1 Staphylococcus poisoning

Staphylococcus poisoning is caused by ingestion of enterotoxins produced by *Staphylococcus aureus*. The symptoms of poisoning include nausea, vomiting, abdominal cramps, diarrhoea and headache. Generally meat is the carrier of *Staphylococci.*

21.3.1.2 Clostridium poisoning

Clostridium perfringens is a major cause of food poisoning. Ingestion of foods like cheese containing toxins produced by the bacterium and their absoption into the cell lining of the gastrointestinal tract initiates vomiting, diarrhoea and abdominal pain but no fever.

21.3.1.3 Ergot poisoning (See Chapter 8)

21.3.1.4 Botulinum poisoning

The anaerobic *Clostridium botulinum* produces highly toxic exotoxins in foods, which are powerful and highly poisonous neurotoxins. There are seven types of such toxins produced. They are desigrated A, B, C, D, E, F and G. The foods frequently contaminated are usually pickled and canned food. The symptoms of poisioning include nausea, vomiting, dizziness, dryness of the mouth and throat, double vision, paralysis, respiratory failure and death (See Chapter 6).

21.3.2 Types of Food Infection

21.3.2.1 Salmonellosis

Salmonellosis is caused by *Salmonella* serotypes. The infection results in abdominal cramps, pain, nausea, fever and diarrhoea. Egg, meat and milk are the main source of infection.

21.3.2.2 Gastroenteritis

Gastroenteritis is caused by *Bacillus aureus,* resulting in diarrhoea and vomiting (See chapter 6). Mycotoxins (see Chapter 17), also cause various types of food posioning.

CHAPTER 22

MICROBES IN BIOGEOCHEMICAL CYCLES

22.1 INTRODUCTION

Cyclic transformation of elements and compounds is essential for sustenance of life on earth. Without such recycling process, the biologically useful elements like carbon, nitrogen, sulphur or oxygen become restricted and affect their availability for biological processes of organisms. Billions of micro-organisms form a major community in the soil, of which the bacteria are major component. In biological cycles the elements are oxidized and reduced by the micro-organisms.

22.2 CARBON CYCLE

Carbon is the most important element for all living beings as it forms the skeleton of all organic substances that constitute their body. It accounts for 50 per cent of dry matter of cells. The ultimate source of this element is carbon dioxide present in air and water.

Photosynthesis and respiration are the two biological processes which help to maintain the balance. Various other biotic and abiotic factors disturb the natural balance of carbon dioxide in the atmosphere. Among the biotic factors, the methanogenic bacteria such as *Methanobacillus*, *Methanobacterium*, *Methanococcus* reduce carbon dioxide to methane which is released into the atmosphere. Methane has the ability to react with the ozone layer of the atmosphere which acts as a screen for UV rays of the sun. The destruction of ozone layer increases the amount of UV rays reaching the earth.

22.3 SULPHUR CYCLE

Sulphur is present in proteins and other compounds. In nature, sulphur occurs in an organic form, as sulphates under aerobic conditions and as sulphides under anaerobic conditions.

Elemental sulphur is not available to plants, and animals. Only sulphur oxidizing bacteria like purple sulphur bacteria and nonphotosynthetic sulphur bacteria can oxidize sulphide to sulphur and use the energy for their autotrophic mode of nutrition. Thiobacilli are the best representatives of this group.

The cyclic transformation of sulphur in nature involves its conversion from sulphate (inorganic) to sulphur containing amino acids and their incorporation into proteins and finally its release into the atmosphere in inorganic form after the death and decay of micro-organisms.

A group of anaerobic bacteria has the ability to reduce sulphate to sulphide (desulfication). *Desulfovibrio desulfuricans* is the typical example which is involved in the process.

22.4 NITROGEN CYCLE

Next to carbon, nitrogen is the most important element found in proteins and nucleic acids. It is the major component of earth's atmosphere. However, it is unavailable to most of the living organisms. The conversion of nitrogen to nitrogenous compounds is called nitrogen fixation. Only microorganisms such as algae, fungi and bacteria can fix atmospheric nitrogen (See Chapter 22, Section 22.7). In higher plants the *Rhizobia* present in the root nodules of legumes can fix atmospheric nitrogen (Subba Rao, 1999).

22.5 PHOSPHORUS CYCLE

In the living organisms, phosphorus occurs in the organic form in the protoplasm. On death of the living organism, the organic phosphorus is converted into inorganic phosphoric acid and finally into insoluble salts of calcium, iron, magnesium etc. Insoluble phosphorus is made soluble by various phosphate solubilizing microbes (See Section 22.8). Nitrifying bacteria, sulphur oxidizing bacteria and arbuscular micorrhizal fungi are known to help plants in better uptake of phosphorus from soil.

22.6 OXIDIZING PROCESSES

22.6.1 Hydrogen Oxidizing Bacteria

Bacteria posses the ability to oxidize molecular hydrogen and use it as energy source for autotrophic growth. Some of the well known hydrogen oxidizing bacteria are *Alcaligenes eutrophus, Alcaligenes ruhlandii, Azospirillum lipoferum, Nocardia opaca, Nocordia autotrophicum, Paracoccus denitrificans, Pseudomonas faecalis, Psuedomonas saccharophyla* and *Xanthobacter autotrophicum.*

22.6.2 Carbon Oxidizing Bacteria

A group of aerobic bacteria grow chaemoautotrophically by their ability to oxidise carbon monoxide to carbon dioxide. The important bacteria with the ability to oxidize carbon are species *Alcaligenes* and *Pseudomonas*.

22.6.3 Iron Oxidizing Bacteria

A group of bacteria such as *Ferrobacillus ferro-oxidans (Thiobacillus ferro-oxidans)* oxidize ferrous oxide into insoluble ferric oxide and the oxidization energy is utilized for carbon fixation. *Ferruginea* sp and *Leptothrix ochracea* in water can oxidize ferrous oxide which is deposited on their wall.

22.6.4 Sulphur Oxidizing Bacteria

Oxidation of elemental sulphur and various reduced sulphur compounds like sulphide and thiosulphate takes place in soil and water and is mediated by a variety of bacteria. The best known among them are *Chromatium, Chlorobium, Thiobacillus novellas, Thiobacillus thio-oxidans, Thioparus, Thiomargarita namibiensis* and species of *Beggiatoa, Sulfolobus, Thermoproteus* and *Thiothrix.*

Thiobacillus ferro-oxidans can oxidize elemental sulphur to sulphate and this property is commercially utilized for removal of sulphur from iron to make extraction of metallic iron easy.

22.6.5 Nitrifying Bacteria

Nitrification, a process of fixing nitrogen, is carried out by nitrifying bacteria occurring in soil and water. The two-step process of nitrification is carried out by two different kinds of bacteria. The first step involving oxidation of ammonia to nitrous oxide is carried out by *Nitrosomonas*. The second step involving oxidation of nitrous acid to nitric oxide is carried out by *Nitrococcus.*

22.7 NITROGEN FIXATION

22.7.1 Nitrogen Fixation by Bacteria

The free living bacteria having the ability to fix molecular nitrogen can be obligate, aerobic, facultative aerobic and anaerobic. Bacteria of the family Azotobacteraceae constitute the majority of heterotrophic free living nitrogen fixing bacteria.

22.7.2 Nitrogen Fixation by Actinorhizal Plants. (*Frankia* induced nodulation)

Actinorhizal palnts are those where nodules are formed by Actinomycetes tentatively classified in the genus *Frankia*. There are 24 genera belonging to 8 angiosperm families which possess actinorhizal root nodules.

Rubus (Rubiaceae), *Eleagnus* (Eleagnaceae), *Casuarina* (Casuarinaceae) are a few of actinorhizal plants. It has been demonstrated that *Casuarina* nodules are capable of fixing nitrogen and they possess nitrogenase enzyme similar to legume nodule nitrogenase.

22.7.3 Nitrogen Fixation by Free Living Cyanobacteria

Blue green algae constitute an important group of micro-organisms capable of fixing atmospheric nitrogen.

The following are the nitrogen fixing blue-green algae: *Anabaena, Anabaenopsis, Aphanothece, Calothrix, Chlorogloea, Cylindrospermum, Dermocarpa, Gloeocapsa, Haplosiphon, Lyngbya, Mastigocladus, Nostoc, Nostocopsis, Oscillatoria, Plectonema, Pleurocapsa, Rivularia, Schizothrix, Scytonema, Synechococcus, Tolypothrix, Trichodesmium, Westiella, Westiellopsis* and *Xenococcus* species.

The nitrogen fixing blue-green algae possess nitrogenase which is localized in the heterocysts of the blue greens. Heterocysts which are considered as the sites of nitrogen fixation are impermeable to oxygen which is liberated during algal photosynthesis. Nitrogenase activity is sensitive to oxygen and is degraded by it and hetetocysts offer a protective mechanism to nitrogenase.

Recently it has been demonstrated that non heterycystous Cyanobacteria are also N_2 fixers. In such forms, a compartmentalization of photosynthesis and N_2 fixation is suggested.

Heterocystous cyanobacteria as nitrogen fixers

Species of *Anabaena, Anabaenopsis, Aulosira fertilissima, Calothrix, Chlorogloea, Cylindrospermum, Fischerella, Haplosiphon, Mastigocladus, Nostoc, Scytonema ,Stigonema* , *Tolypothrix* and, *Westiellopsis* are a few examples of heterocystous cyanobacteria which fix nitrogen.

Non heterocystous cyanobacteria as nitrogen fixers

Aphanothece, Dermocarpa, Gloeocapsa, Synechococcus and *Trichodesmium* are the non-heterocystouncyanobacteria which are active N_2 fixers

22.7.4 Nitrogen Fixation by Symbiotic Cyanobacteria

22.7.4.1 Nitrogen fixation in lichens

The algal and fungal association to form lichens is an instance of symbiosis where species of *Nostoc, Calothrix* and other blue green algae fix nitrogen. In such symbiotic association the algal partner in either a true alga, known as "Phycobiont" which forms 85% of lichens and Cyanobacteria known as "cyanobiont" making up the rest of 15%. Both coexist in lichens. The phycobiont is situated in the thallus of the lichen where as cyanobiont exists in special nitrogen fixing structures known as cephalobia.

Nitrogen fixed by the cyanobacteria in lichens is generally 2-3 times more than in free living state. The nitrogen fixing lichens are species of *Collema, Leptogium, Lichinia, Lobaria, Massalongia, Nephroma, Pannaria, Parmeliella, Placopsis, Peltigera, Placynthium, Polychidum, Stereocaulon,* and *Sticta.*

22.7.4.2 Nitrogen Fixation in Pteridophytes and Gymnosperms

An excellent example of algal association with higher plants is the occurrence of endophytic Cyanobacteria, like species of *Anabaena, Calothrix and Nostoc* in the fronds of *Azolla.*

Anabeana azollae associated with the water fern *Azolla* is present in the pores on the leaves. This endophyte fixes atmospheric nitrogen. Application of *Azolla* as an organic manure in rice cultivation is due to high level of nitrogen present in the leaves of *Azolla.*

The most important cycads with coralloid roots are species of *Cycas, Encephalartos, Zamia, Ceratozamia, Macrozamia* and *Stangeria.* The blue green alga *Nostoc cycadeae* in *Cycas, Macrozamia* and *Encephalartos* have been shown to fix atmospheric nitrogen in free state as well as in association with their host plants.

Species of *Podocarpus* possess nodules on the root system. There is enough evidence to show that these nodules have the capacity to fix nitrogen but not as vigorously as legume root nodules.

22.7.5 Nitrogen Fixing Fungi

Fungi also play an important role in fixing free nitrogen in soil. The fungi that are responsible for nitrogen fixation are *Alternaria tenuis, Cladosporium herbarum, Mactosporium commune, Phoma,* and *Rhodotorula.*

22.7.6 Nitrogen Fixation in Legumes

Almost all species of the legumes produce tubercular structures on their roots called the root nodules. Species of the bacterial genus *Rhizobium* are aerobic, and are present in the soil. They attach to the root hairs of legume plants and invade the developing roots of the legumes. An excessive multiplication of the bacteria in the cortical cells of the roots trigger development of nodules, which are the sites of a very efficient process of nitrogen fixation.

22.8 MICROBE MEDIATED CHANGES IN MINERAL STATUS

Many metallic elements and their ionic forms undergo transformation in nature mediated by micro-organisms. Such transformation involves oxidation—reduction reactions resulting in valency changes. Bacteria and fungi are involved in solubilizing, fixing and precipitating certain minerals. The following are the list of actinomycetes, bacteria, fungi, yeasts and cyanobacteria involved in changing the status of mineral such as manganese, copper, iron and phosphorus.

Manganese solubilizing micro-organisms

Aerobacter sp., *Azotobacter chroococcum, Cephalosporium* sp., *Chromobacterium* sp., *Cladosporium* sp., *Cryptococcus* sp., *Curvularia* sp., *Corynebacterium* sp., *Flavobacterium* sp., *Helminthosporium* sp., *Leptothrix* sp., *Proteus* sp., *Pseudomonas fluorescens, Pseudomonas trifolii.*

Copper fixing micro-organisms

Clostridium lentoputrescens, Desulfovibrio desulfuricans, Proteus vulgaris.

Iron precipitating micro-organisms

Colnothrix sp., *Crenothrix* sp., *Ferribacterium* sp., *Ferrobacillus ferro-oxidans, Naumanniela* sp., *Ochrobium* sp., *Siderobacter* sp., *Siderocapsa* sp., *Siderococcus* sp., *Sideromonas* sp., *Sideronema* sp., *Siderophaera* sp., *Sphaerotilus* sp., *Toxothrix* sp.,

Gram positive bacteria known to solubilize mineral phosphate

Bacillus brevis, Bacillus cereus, Bacillus cereus var., *albolactis, Bacillus circulans, Bacillus coagulans, Bacillus lichenformis, Bacillus megaterium, Bacillus megaterium var., phosphaticum, Bacillus mesentericum, Bacillus mycoides, Bacillus polymyxa, Bacillus pumilis, Bacillus pulvifaciens, Bacillus sphaericus, Bacillus subtilis, Clostridium* species.

Gram negative bacteria known to solubilize mineral phosphate

Acetobacter diazotropicus, Achromobacter sp., *Aerobacter aerogenes, Agrobacterium radiobacter, Alcaligenes* sp., *Bradyrhizobium* sp.,

Brevibacterium sp., *Burkholderia cepacia, Citrobacter freundii, Enterobacter aerogenes, Enterobacter agglomerans, Enterobacter asburiae, Enterobacter cloacae, Escherichia freundii, Escherichia intermedia, Erwinia herbicola, Flavobacterium* sp., *Gluconobacter diazotrophicus, Micrococcus* sp., *Mycobacterium* sp., *Nitrosomonas* sp., *Nitrobacter* sp., *Pseudomonas calcis, Pseudomonas cepacia, Pseudomonas fluorescens, Pseudomonas putida, Pseudomonas rathonia, Pseudomonas striata, Pseudomonas syringae, Serratia marcescens, Serratia phosphaticum, Thiobacillus ferroxidans, Thiobacillus thio-oxidans, Rahnella aquatilis, Rhizobium meliloti, Xanthomonas* sp.,

Actinomycetes known to solubilize mineral phosphate

Actinomyces coelicolour, Streptomyces sp.,

Fungi known to solubilize mineral phosphate

Achrothecium sp., *Aspergillus aculeatus, Aspergillus awamori, Aspergillus carbonum, Aspergillus flavus, Aspergillus foetidus, Aspergillus fumigatus, Aspergillus japonicus, Aspergillus nidulans, Aspergillus nidulans* var., *acristatus, Aspergillus niger, Aspergillus rugulosus, Aspergillus terreus, Aspergillus wentii, Cephalosporium* sp., *Chaetomium globosum, Cladosporium herbarum, Cunninghamella* sp., *Curvularia lunata, Fusarium oxysporum, Helminthosporium* sp., *Morteirella* sp., *Micromonospora* sp., *Mucor* sp., *Myrothecium roridum, Oideodendron* sp., *Paecilomyces lilacinus, Paecilomyces fusisporous, Penicillium aurantiogriseum, Penicillium bilaji, Penicillium digitatum, Penicillium funiculosum, Penicillium lilacinium, Penicillium oxalicum, Penicillium pinophilum, Penicillium rubrum, Penicillium rugulosum, Penicillium simplicissimum, Penicillium variable, Phoma* sp., *Populospora mytilina, Pythium* sp., *Rhizoctonia solani, Rhizopus* sp., *Sclerotium rolfsii, Trichoderma harzianum, Trichoderma viride.*

Yeasts known to solubilize mineral phosphate

Candida sp., *Geotrichum* sp., *Humicola inslens, Humicola lanuginosa, Schwanniomyces occidentalis, Torula thermophila.*

Cyanobacteria known to solubilize mineral phosphate

Anabaena sp., *Calothrix braunii, Haplosiphon fontinalis, Nostoc* sp., *Scytonema, Scytonema cincinnatum, Tolypothrix tenuis, Tolypothrix ceylonica, Westiellopsis prolifica.*

Source: Tripura *et al.*, 2005

CHAPTER 23

BIODEGRADATION BY MICROBES

23.1 INTRODUCTION

Biodegradation, biomineralization, bioremediation are terms with minor differences. Biodegradation is the general term used for all biologically mediated break down of chemical compounds of detergents, transformer fluids, pestcides, oils, hydrocarbons, etc. Biodeterioration refers to the breakdown of economically useful plant compounds or/and chemicals produced by metabolic pathways of specific groups of microbes and chemicals formed through biogeochemical process but often the term is used to refer to biological degradation of normally resistant substances such as metals, plastics, drugs, sculpture, glass and equipments. Bioremediation refers to the use of biological systems to degrade toxic compounds in the environment.

Biodegradation is carried out by biological systems such as bacteria, fungi, yeast and algae which use organic (natural or synthetic) substances as source of nutrients and energy. By itself the term biodegradable does not apply to any particular extent of degradation. It is the transformation involving several reactions with insignificant effect. Biodegradability of a compound results in mineralization. Mineralization is complete degradation to the end products.

Biodegradation is not always desirable. Sometimes degradation may convert an innocuous compound into a toxic one or may alter a toxic compound to generate a highly toxic product.

Though fungi are the major micro-organisms in biodegradation of various substances exemplified below, yeasts such as species of *Candida* and *Saccharomyces*, bacteria like *Cellulomonas* and *Pseudomonas* are also active biodegraders.

Illustrations of representative forms involved in biodegradation are given in Fig. 23.1 to provide an overview of their appearance.

23.2 MAJOR ENVIRONMENTAL POLLUTANTS

Major environmental pollutants are the products of large scale manufacturing, processing and handling of chemicals in various industries and that has led to serious surface and subsurface soil and water contamination. A wide variety of hazardous and toxic hydrocarbons such as polychlorinated biphenyls (PCB's), trichloroethylene (TCE), polyaromatic hydrocarbons (PAHs) and pesticides are designated as xenobiotics because of their relative recalcitrance to biodegradation and these compounds (Box 23.1) have the potential to bioaccumulate in living systems (Heitkamp *et al.,* 1988, Love and Eilers, 1982; Kanaly and Harayama, 2000). Nitroaromatic pesticides including organochlorine (DDT, α-hexachlorocyclohexane, endosulfan), organophosphates (chlorophyriphos, parathion and methyl parathion), nitro phenols (dinoseb, dinitrocresol) and their partially hydrolyzed products can be leached into the ground water (Bumpus and Aust, 1987; Lenke and Knackmuss, 1996, Raushel, 2002.)

On the other hand, nitrophenols, nitrobenzenes and nitrotoluenes have been primarily used as chemical intermediates in the production of dyes (azo and sulphur dyes), photographic developers, wood preservatives, hydraulics, fire retardants, pharmaceuticals, plasticizers, solvents etc. and are often released to surface water in water streams (Heiss *et al.,* 2003).

Explosives such as picric acid, trinitrophenol, trinitrotoluene (TNT), RDX and HMX are less biodegradable and often persist for extended periods in soil or ground water (Won *et al.,* 1976; Rosenblatt *et al.,* 1991). Several of these have been categorized as either priority pollutants or toxic and/or hazardous chemicals by U.S. Environmental Protection Agency (USEPA). Another group of problematic pollutants are heavy metals and radionuclides (Daly, 2000).

Morever, as many of these chemical compounds have been introduced relatively recently, micro-organisms have not yet had enough time to evolve metabolic pathways for their degradation. Hence, endeavors are being made to genetically modify microbes into more efficient degraders using modern tools of molecular biology and metabolic engineering. In addition, the untapped genetic diversity can also provide us with novel as well as difficult to synthesize compounds (Osburne *et al.,* 2000)

23.3 BIODEGRADATION OF NATURAL SUBSTANCES

Soil organic matter comprises residues of plants and animals at all stages of decomposition mediated by soil micro-oganisms. These residues include complex carbohydrates, sugars, starch, cellulose, hemicellulose, pectin, gum, proteins, waxes. resins, lignin, tannin, hydrocarbons, amino acids, etc.

Cellulose

The main ground material for all plants as well as plant products like paper, cloth, and fibre is cellulose. Micro-organisms such as bacteria and fungi can degrade cellulose.

Among the cellulolytic aerobic bacteria the species of *Archangium, Cellulomonas, Cytophaga, Eubacteria, Mycobacteria, Pseudomonas,* and *Sporocytophaga* decompose cellulose.

Under anaerobic conditions, *Clostridium cellobioparum* and *Clostridium thermocellum* are the active cellulose decomposers. The cellulose in the rumen of animals is degraded by *Bacteroides succinogenes, Butyrivibrio fibrisolvens, Cellulomonas* and *Ruminococcus flavegaciens.*

The fungi which bring about degradation of cellulose are species of *Aspergillus, Botrytis, Chaetomium, Fusarium, Myrothecium, Rhizoctonia* and *Trichoderma.*

Hemicellulose

Among the hemicelluloses, xylans are the most important. They are present in straw, wood and bast fibres. Hemicelluloses are degraded by bacteria like *Bacillus* sp., under aerobic conditions. Under anaerobic conditions species of *Clostridium* and *Vibrio* are dominant.

Pectin

Pectin is present in the middle lamella of plant cells. Pectin is degraded by bacteria like *Bacillus macerans, Bacillus polymyxa, Clostridium pectinovorum, Erwinia caratovora* and fungi like *Aspergillus niger, Botrytis cinerea* and *Fusarium oxysporum* by using the enzyme pectinase. Pectinases from microbes have been used in retting of fibres like jute, flax and hemp and to clarify fruit juices.

Chitin

Chitin is a cell wall component of higher fungi and the exoskeleton of arthopods and planktonic animals. Many soil and aquatic micro-organisms cause degradation of chitin. Bacteria like *Bacillus, Flavobacterium, Micromonospora, Nocardia, Pseudomonas, Streptomyces* and fungi like *Aspergillus* and *Mortierella* degrade chitin.

Lignin

Lignin is a major constituent of woody tissues imparting mechanical strength. It is highly resistant to microbial attack and is not easily degraded.

Among the wood rotting fungi, *Ganoderma applanatum, Polystictus versicolor, Stercum hirsutum,* and species of *Lenzites, Phanerochaete, Polyporus, Poria, Trametes,* are active lignin degraders. Among the bacteria, *Arthobacter, Flavobacterium, Micrococcus, Pseudomonas, Streptomyces and Xanthomonas* species degrade lignin. (Krik *et al.*, 1976; Lamar *et al.*, 1993)

Proteins

The remains of organisms contain proteins that are degraded by proteolytic enzymes produced by micro-organisms. Proteins in the soil are converted into ammonia by ammonifying bacteria and ammonium salt is taken up by plants as a source of nitrogen. Ammonia is also oxidized to nitrate by nitrifying bacteria (nitrification).

Apart from the transient decomposition of organic matter, a dark coloured soil organic matter called humus is also an intergral part of the organic matter complex in soil. It is a lignoprotein complex containing 45% lignin, 35% amino acid, 11% carbohydrates, 4% cellulose, 7% hemicellulose, 3% waxes and resins.

Bacterial and algal populations in the soil contribute to the nutritive value of humus. Some fungi such as *Aspergillus* and *Penicillium* contribute to the synthesis of humic substances. Humic substances improve seed germination, root growth, mineral uptake and enzyme action.

Actinomycetes, bacteria and fungi are the major decomposers of organic matter in the soil (Table 23.1)

Table 23.1 Genera of micro-organisms capable of degrading different components of organic matter

Natural substance	Fungi	Bacteria	Actinomycetes
Cellulose	*Alternaria*	*Achromobacter*	*Micromonospora*
	Aspergillus	*Angiococcus*	*Nocardia*
	Chaetomium	*Bacillus*	*Streptosporangium*
	Coprinus	*Cellulomonas*	*Streptomyces*
	Fomes	*Clostridium*	*Sporocytophaga*
	Fusarium	*Cytophaga*	
	Myrothecium	*Polyangium*	
	Penicillium	*Pseudomonas*	
	Polyporus	*Vibrio.*	
	Rhizopus		
	Trametes		
	Trichoderma		
	Trichothecium		
	Verticillium		
	Zygorynchus		
Hemicellulose	*Alternaria*	*Achromobacter*	*Streptomyces*
	Aspergillus	*Bacillus*	
	Chaetomium	*Cytophaga*	
	Coriolus	*Lactobacillus*	
	Fomes	*Pseudomonas*	
	Fusarium		*Sporocytophaga*
	Helminthosporium	*Vibrio*	
	Penicillium		
	Polyporus		
	Rhizopus		
	Trichothecium		
	Zygorynchus		

Natural substance	Fungi	Bacteria	Actinomycetes
Lignin	*Arthobotrys* *Cephalosporium* *Clavaria* *Clitocybe* *Collybia* *Humicola* *Hypholoma* *Lepiota* *Mycena* *Pholiota*	*Flavobacterium* *Pseudomonas*	
Starch	*Aspergillus* *Fomes* *Fusarium* *Polyporus* *Rhizopus*	*Achromobacter* *Bacillus* *Chromobacterium* *Clostridium* *Cytophaga*	*Micromonospora* *Nocardia* *Streptomyces*
Pectin	*Fusarium* *Verticillium*	*Bacillus* *Clostridium* *Pseudomonas*	
Chitin	*Absidia* *Aspergillus* *Fusarium* *Gliocladium* *Mortierella* *Mucor* *Penicillium* *Thamnidium* *Trichoderma*	*Achromobacter* *Bacillus* *Beneckea* *Chromobacterium* *Cytophaga* *Flavobacterium* *Micrococcus* *Pseudomonas*	*Micromonospora* *Nocardia* *Streptomyces*
Proteins and nucleic acids		*Bacillus* *Clostridium* *Micrococcus* *Pseudomonas* *Serratia*	
Cutin	*Mortierella* *Penicillium* *Rhodotorula*	*Bacillus*	*Streptomyces*
Tannin	*Aspergilllus* *Penicillium*		
Humic acid	*Aspergillus* *Penicillium* *Polystictus*		
Fulvic acid	*Poria*		

Source: Subba Rao, 1999

23.4 BIODEGRADATION OF HYDROCARBONS

Hydrocarbons are abundant where coal and petroleum are stored, processed and burnt and are toxic with biological effects. Occupational exposure to polycyclic aromatic hydrocarbons in aluminium and steel industries has been linked to lung and bladder cancer (Mastrangelo *et al.*, 1996) and hydrocarbon fuels result in neurotoxicity.

Yeasts and filamentous fungi can transform hydrocarbons. They are also involved in recycling of naturally occurring hydrocarbons in the environment as well as in the biodeterioration of liquid fuels (Lindley, 1992). These are now being exploited for the bioremediation of soil contaminated with hydrocarbon wastes, and in oil polluted seawater.

The degradation of hydrocarbons by filamentous fungi like *Hormoconis resinae* and yeasts are used to decompose petroleum residues in estuaries. The fungal and yeast genera that biotransform hydrocarbons are given in Table 23.2.

Table 23.2 Degradation of hydrocarbons by fungi and yeasts

Compound	Species
1. **Biphenyl**	*Candida elegans* *Debaryomyces vanrijae* *Trichosporon mucoides*
2. **Monocyclic aromatic hycrocarbons**	*Cladophialophora* sp. *Cladosporium sphaerospermum*
a. Toluene	*Exophiala* sp. *Pseudeurotium zonatum* *Scedosporium apiospermum*
b. Ethyl benzene	*Phanerochaete chrysosporium*
c. Methyl benzene	*Fusarium moniliforme*
d. Styrene	*Cladophialophora* *Exophiala jeanselmei*
e. Xylene	*Sporotrichum sulfurescens*
3. **Polycyclic aromatic hydrocarbons**	
a. Naphthalene	*Aspergillus niger* *Candida elegans*
b. Acenaphthene	*Candida elegans* *Coriolopsis gallica*
c. Flourene	*Candida elegans* *Phanerochaete chrysosporium*
d. Anthracene	*Aspergillus niger* *Candida elegans* *Rhizopus arrhizus* *Ulocladium chartarum*

e. Phenanthrene	*Candida elegans* *Syncephalastrum racemosum*
f. Fluoranthene	*Aspergillus cylindrospora* *Aspergillus niger* *Candida elegans* *Cryptococcus albidus* *Phanerochaete chrysosporium* *Phanerochaete ostreatus*
g. Pyrene	*Aspergillus niger* *Candida elegans* *Penicillium janthinellum*
h. Lanthracene	*Candida elegans* *Phanerochete ostreatus* *Phanerochaete sajor-caju*
i. Chrysene	*Amanita muscaria* *Boletus edulis* *Candida elegans*

Source: Sutherland, 2004

Hydrocarbons are also degraded by purple non-sulphur bacteria. *Phaeospirillum fulvum, Rhodocyclus purpureus* and *Blastochloris sulfoviridis* degrades benzoate and *Blastochloris sulfoviridis* degrades toluene, *Pseudomonas citronellolis* and *Brevibacterium erythrogenes* degrades cycloalkanes.

23.5 BIODEGRADATION OF EXPLOSIVES

Explosive compounds are a very significant environmental contaminants (Faull, *et al.*, 2004). Their production, testing and disposal has led to contamination of water and soil leading to toxicity to plants, animals and humans.

Various remediation strategies have been adopted so far. Physical and chemical remediations are successful only to a certain extent.

Biological remediation involves biostimulation of existing indigenous microflora and bioaugmentation of the explosive contaminated environment. Though cost effective, it has limited success due to microbial sensitivity to certain toxic explosives.

This approach was used in United States for cleaning military site (USAEC 1999). But the approach was criticized due to high cost of setup and maintenance, and the interaction of fungi with explosive might result in toxic metabolites that may bind to soil organic matter and not undergo mineralization.

Explosive contaminated areas can be cleaned by treating the soil with soil piles and windrows. Soil piles and windrows are created by mixing soil with wood chips, corn cobs and other ligno-cellulosic materials and adding fungal inoculum containing *Phanerochaete chrysosporium, Stropharia rugosoannulata* and *Trametes versicolor.*

Fungi have the capability to biodegrade explosives using existing co-metabolic pathways leading to mineralization. *Phanerochaete chrysosporium* is the best known TNT degrader followed by other fungi and yeasts. (Fernando *et al* 1990; Esteve-Nunez *et al.*, 2001). Like wise, biodegradation of RDX and HMX is achieved by *Cladosporium resinae, Cunninghamella echinulata* var. *elegans* and *Cyathus pallides.* (Table 23.3). Bioreactors are used to degrade TNT and RDX where in *Phanaerochaete chrysosporium* is attached to rotating discs of the bioreactor (Sublette *et al.,*1992).

Table 23.3 Species of fungi capable of TNT degradation

Species	Habitat
Agaricus aestivalis	Litter
Agaricus bisporus	Litter
Agaricus praecox	Litter
Alternaria sp.	Ubiquitous
Amanita muscaria	Ectomycorrhizal
Amorphotheca resinae	Ubiquitous
Aspergillus terreus	Ubiquitous
Bjerkandera adusta	Wood
Candida odora	Wood
Candida dusenii	Wood
Candida nebularis	Wood
Candida stercorus	Wood
Candida versicolor	Wood
Cunninghamella echinulata	Ubiquitous
Fusarium fomentarius	Wood
Fusarium solanii	Ubiquitous
Geotrichum separium	Wood
Heterobasidion annosum	Wood
Hypholoma fasciculare	Wood
Irpex lacteus	Wood
Nematoloma frowardii	Wood
Neurospora crassa	Ubiquitous
Penicillium chrysogenum	Ubiquitous
Phanerochaete chrysosporium	Wood
Phanerochaete sordida	Wood
Phlebia brevispora	Wood

Phlebia ochracheofulva	Wood
Phlebia radiata	Wood
Pisolitus frequentens	Ubiquitous
Pisolitus tinctorius	Ectomycorrhizal
Rhizopus nigricans	Ubiquitous
Saccharomyces commune	Wood
Saccharomyces granulatus	Ectomycorrhizal
Saccharomyces variegatus	Ectomycorrhizal
Trametes suaveolens	Wood
Trametes versicolor	Wood

Source: Faull *et al.*, 2004.

23.6 BIOREMEDIATION OF RADIOACTIVE WASTES

Since microbes used for bioremediation need to survive radiotoxicity of effluents containing highly active radionuclides such as 235U, 99Tc and 241Pu, radiation resistant organisms need to be used. The radiation resistant organism discovered is the non pathogenic soil bacteria. *Deinococcus radiodurans* is able to reduce Cr, Tc and UV. *Deinococcus geothermalis* is another extremely radiation resistant thermophilic bacterium that is used for bioremediation of radioactive wastes (Hassan *et al.,* 2003, Maruthi Mohan *et al.*, 2005).

23.7 BIODEGRADATION OF PESTICIDES AND HERBICIDES

Pesticides are chemicals used to control insects and nematodes that affect crop plants. Herbicides are chemical substances, which are used to destroy weeds. Continuous application of synthetic pesticides and herbicides, particularly those containing heavy metals, results in an accumulation of these metals and other residues in the soil and water, as these are not completely degraded. Pesticide residues enter different cycles in the environment, including the food chain causing harm to birds and animals that feed on pesticide treated plants and humans who consume such animals and crop produce. In addition to finding pesticides and herbicides from biological sources, which can be completely degraded in nature, finding ways and means of degrading pesticide residues has been a major conern of research for decades. A successful attempt has been made to degrade 2,4-D and 2,4,5-T from water by using immobilized *Rhodopseudomonas sphaeroides* on magnetites. Now a number of bacteria and fungi are identified to perfom this task (Table 23.4).

Table 23.4 Pesticides/pollutant degrading microbes

Species	Pesticide
Pseudomonas aeruginosa *Pseudomonas stutzeri*	Parathion (insecticide)
Arthobacter species	Isopropyl-N-phenyl carbamate (IPC) (herbicide)
Pseudomonas species	Chloropropham (herbicide)
Arthobacter chlorophenolicus	Pentachlorophenol (fungicide)
Flavobacterium *Rhodococcus*	
Arthobacter *Corynebacterium* *Flavobacterium* *Pseudomonas*	2,4-D (fungicide)
Gram negative cocci *Pseudomonas cepacia*	Pyrazon (herbicide)
Phanerochaete chrysosporium *Agrobacterium* *Alcaligenes* *Arthobacter*	2,4,5-T (herbicide)
Micrococcus	TCA and Dalapon
Phanerochaete chrysosporium	Dicamba, Chlorodane, Lindane
Aerobacter	DDT
Pseudomonas	4-nitrophenol
Aspergillus *Fusarium* *Penicillium* *Rhizopus* *Trichoderma*	Atrazine and simazine
Aspergillus nidulans *Achromobacter* *Agrobacterium*	Linuron
Flavobacter	Chloropropham
Methanogenic bacteria	Perchloroethane
Pseudomonads	Mono- and dichlorobenzene
Acinobacter and *Alcaligenes*	Pentachlorobenzene
Sulphur bacteria	Tetrachloroethane

Source: Subba Rao, 1999.

23.8 BIODEGRADATION OF AZO DYES BY FUNGI

Use of dyes is a multibillion dollar industry. Of all the dyes, azo dyes are the most widely used colourants. They are available in almost every color across the spectrum and are color fast, with an ability to bind to natural and synthetic fabrics. Unfortunately they are resistant to biodegradation. (Rose 1981; Bumpus, 2004).

Fungi have the ability to degrade azodyes. The white rot fungi are the major degraders of azo dyes (Table 23.5).

Table 23.5 Biodegradation of azo dyes by white rot fungi

Fungal species	Dye
Bjerkandera B0855	Amaranth, Remazol black, Remazol brilliant orange, Tropaeolin-O
Fusarium trogii	Astrazon red
Ganoderma sp.,	Orange II
Myrothecium sp.,	Orange II
Phanerochaete chrysosporium	Congo red, orange II, azure B, tropaeolin O
Pleurotus ostreatus	Congo red, Methyl orange
Pleurotus sajor-caju	Amaranth, Orange.G
Pycnoporus sanguineus	Orange.G, Amaranth
Trametes hirsuta	Remazol black B
Trametes versicolor	Acid violet, Acid orange, Acid black

Source: Bumpus, 2004.

The major pollutant of azo dye industry is its waste water. Thus, textile industry is a major contributor to industrial waste water due to discharging of synthetic dyes. Over 70,000 tonnes of synthetic dyes which are produced annually world wide pollute the environment via air, water and soil as they are non-biodegradable. They are resistant to biodegradation as these dyes are designed to be resistant to sunlight, washing and microbial action to be successful as a textile dye.

The goal of waste water treatment is to reduce the level of toxicity, odour and colour of water. *Myrothecium verrucaria* is very effective in clearing effluents from textile dyeing factories. Decolorization and remediation of waste water from azo dye industry is achieved by using immobilized white rot fungi such as *Bjerkandera adusta*, *Phanerochaete chysosporium* and *Trametes versicolor*, in continuous-flow packed bioreactors (Ollikka *et al*., 1993, Nigam *et al*., 1995, SchilePhake *et al*., 2004). Decolourization of dyes is also achieved by intact cultures or the enzymes of fungi.(Table 23.6).

Decolorization and biodegradation of Azo dyes is also achieved by bacteria like *Bacillus subtilis, Pseudomonas cepacia, Pseudomonas stutzeri, Klebsiella pneumoniae*, fungi like *Streptomyces chromofuscus, Geotrichum candidum* and algae like *Chlorella pyrenoidosa, Chlorella vulgaris and Oscillatoria tenuis* .

Table 23.6 Decolorization of dyes by intact cultures or the enzymes of fungi

Fungal species	Dye
Aspergillus foetidus	Drimarene red, Drimarene blue
Aspergillus sojae B-10	Sudan II, Congo red
Curvularia inaequalis	Chicago sky blue
Neurospora crassa	Vermelho reanil P813

23.9 BIODEGRADATION OF METALS IN WASTE WATER THROUGH BIOSORPTION

Biosorption is a term that describes the process by which biomass of micro-organisms remove metals from water and the sorbed metal ions are recovered for reuse or safer disposal. Suitable biotrap for biosorption is important. Fungi and yeast can serve as biotrap for heavy metals as chitin in fungal cell walls are well known metal ion adsorbers.

Fungi, algae and yeasts have received much attention as large amount of biomass is generated as byproducts of industrial alcohol fermentation and antibiotics production. One such system is *Aspergullus niger* which is shown to remove zinc, magnetite and metal sulfides from waste water of citric acid production. *Fusarium flocciferum* (dried powder) can adsorb cadmium and nickel. Similarly *Aspergillus carbonarius* is able to adsorb copper and chromium.

Yeasts similarly are effective in metal adsorption. Immobilized *Saccharomyces cerevisiae* has the capability to adsorb copper, cadmium, nickel, zinc and chromium. *Kluyveromyces marxianus* was found to take up uramium. (Crusberg *et al.*, 2004). It is proposed that biosorption, utilizing biomass can be a replacement technology for the existing metal removal process.

Diverse range of bacteria are able to use oxidized metals such as Fe (III), Cr (VI) or Mn (IV) as terminal electron acceptors. Most of them are anaerobes with a few being facultative anaerobes. A strain of *Shewanella (=Alteromonas) putrefaciens* reduced Fe(III), Mn(IV), Ur (VI) thus providing a potential alternative to conventional chemical technologies. However, Fe and Mn are not significantly toxic. Other toxic metals are potential targets in waste treatments and are biotechnologically precipitated by bacteria. Molybdenum (VI) was reduced to Molybdenum blue by *Enterobacter cloacae* which also reduced Cr (VI) to Cr(III) under similar conditions.

Metal reduction process may be useful as pretreatment for removing the metals by biosorption. But the most promising potential application of biological metal reduction is uranium precipitation in waste treatment and in concentrating uranium from low grade sources (Lovley 2003).

The hydroxides and carbonates of uranium have low solubility and readily form precipitates. Bacterial uranium reduction has also been combined with chemical extraction to produce a potential process for soil bioremediation. (Philips *et al.,* 1995).

23.10 REMOVAL OF METAL IONS IN INDUSTRIAL WASTE WATER

Although metal ions are essential to living organism, excessive intake is toxic as they can be transported along the food chain that enters the human body. They also cause disturbance to biogeochemical cycles which in turn elevate their concentration level in the ecosystem. Metal ions are released not only from natural sources but also from human activities such as mining, agriculture, industrial processes and other urbanization related activities which contribute to mine wastes, pesticides, fertilizers, preservatives and sewage sludge.

Removal of metal ions from industrial effluents is achieved by metal resistant strains of micro-organisms. Sulphate reducing anaerobic heterotrophs such as *Desulfovibrio* and *Desulfotomaculum* are being utilized in laboratory scale packedbed reactors using a synthetic gas with a substantial proportion of hydrogen to remove sulphates from industrial waste water (Uhrie *et al.*, 1996, Christensen *et al.*, 1996). Among them bacteria have the greatest tolerance to high salinity as they can adsorb and precipitate metal ions from solution. Copper, nickel, zinc are the major metals ions whose toxicity can be hazardous to members of the food chain of both aquatic and terrestrial ecosystem. Toxic mercury containing wastes are being discharged from industry posing as a health hazard. Many micro-organisms such as species of *Bacillus, Staphylococcus, Pseudomonas syringae, Pseudomonas putida, Pseudomonas chromatophila* and *Shigella* are resistant to toxic ionic forms of mercury and are being employed to remove mercury from the aquatic environment. List of biosorbent micro-organisms are given in Table 23.7.

Tracer studies have long been used to collect data on retention time and flow patterns in waste water treatment plants. Currently, the tracer studies use lithium salts or fluorescent dyes which is highly expensive. As an alternative, Interbio has developed a biological method where concentrated slurry of *Bacillus globigii* (MicroTrace) is used (Rumbaugh *et al.,* 1997).

Table 23.7 Biosorbent potential of micro-organisms

Source species	Biosorbate
Bacteria	
Streptomyces clavuligens	Metal ions
Bacillus cereus	Metal ions

Streptomyces viridosporus	Dyes
Streptomyces levi	Metal ions
Pseudomonas putida	Metal ions (Cu)
Enterobacter	Metal ions (Ni)
Fungi	
Aspergillus niger	Metal ions
Cyathus bulleri	Metal ions
Ganoderma lucidum	Metal ions
Rhizopus oryzae	Metal ions
Fusarium oxysporum	Metal ions (Cu)
Algae	
Ascophyllum nodosum	Metal ions
Fucus serratus	Metal ions
Laminaria digitata	Metal ions
Sargassum fluitans	Metal ions
Sargassum natans	Metal ions

Source: Gallagher *et al.*, 1997

23.11 MICROBES IN SEWAGE AND WASTE WATER TREATMENT

The amount of waste from industries and agriculture is discharged into water bodies. Increase in population, urbanization, fertilizers and pestcides application has contributed to the volume of sewage and waste water.

The challenges of waste water treatment is to remove compounds with high biochemical oxygen demand and pathogens causing communicable water borne diseases.

The following three steps are involved in waste water treatment:

1. Primary treatment to remove large objects and sedimentation of solid particles.
2. Secondary treatment which is aided by microbes and their degradative capabilities (microbial oxidation) and transfer to an anaerobic digestion for degradation by different population of micro-organisms.
3. Tertiary treatment to render the treated water potable by removing phosphates and nitrates followed by chlorination.

Gram negative bacteria such as *Zoogloea, Pseudomonas, Alcaligenes, Achromobacter, Flavibacterium* and gram positive bacteria like *Corynebacterium* along with fungal genera such as *Subbaromyces, Fusarium, Geotrichum* and *Trichosporon* carry out the secondary aerobic treatment. The autotrophic nitrifying bacteria oxidize ammonia to nitrite (*Nitrosomonas*) and nitrite to nitrate (*Nitrobacter*).

In the aerobic secondary treatment, the degradation and fermentative reaction is carried out by acid and methane forming bacteria such as *Methanobacterium, Methanobacillus, Methanococcus* and *Methanosarcina*.

Box 23.1 Biodegradation of Xenobiotics

Synthetic novel organic compounds are called xenobiotics (xenos means foreign in Greek). The evergrowing list of xenobiotics released into the environment include aliphatic and aromatic compounds, phthalate esters, dibenzofurans, polycyclic aromatic hydrocarbons, etc. These are called as priority pollutants by the US Environmental Agency. These enter the atmosphere through fertilizers, pestcides, industrial effuents and PCB formulations that are found in hydraulic fluids, plastcisers, lubricants, adhesives, dielectric fluids in capacitors and transformers.

23.12 BIOTREATMENT OF PHENOLIC WASTES

Phenolic compounds when discharged into waste water have both direct and indirect effects on humans and environment. These wastes impart a carbolic odour to river water and are strong skin irritants. Drinking such water can cause severe pain, vomiting and capillary damage (Brett and Middle brooks1986). Phenolic wastes are discharged from various industries like textile, pharmaceutical, coal carbonization and petrochemical. *Pseudomonas putida, Nocardia rubra* and *Nocardia convoluta* are being employed in biodegradation of phenols (Nordland *et al.,* 1983).

23.13 BIOCONVERSION OF DISTILLERY WASTE

After the distillation of alcoholic beverages, all the ingredients of the fermented mash remain in the waste called stillage, spent wash, slop or vinasse. The composition of this waste are sugar cane, sugar beet molasses, corn, wheat, barley, potato, rice, fruits, etc. with high content of organic substance, high biological oxygen demand (BOD) and chemical oxygen demand (COD). Though there are many methods to convert these organic wastes, biotechnological process for the production of commercially valuable products is the most interesting. Fungi and yeasts are the most suitable for bioprocessing the distillery wastes. Species of *Candida, Saccharomyces, Hansenula* and *Kluyveromyces* among yeasts, species of *Aspergillus, Gongronella, Penicillium, Trichoderma* are the major fungal genera in bioprocessing. In addition, other products such as ethanol, enzymes, polysaccharides, pigments and growth hormones can also be obtained during stillage bioconversion (Friedrich, 2004).

23.14 BIODEGRADATION AS A TOOL IN MINING

Microbes play an important role in leaching of metal ions during mining opera tions. During the mining activity, there is a large amount of metal ions remaining in the ore. The dumping of untreated ore in the mining site results in leaching of metal ions into groundwater causing serious pollution. However, complete extraction of remaining metal ions from the ore is impractical. In order to minimize the metal ion contamination and also to recover remaining ions from the ore, microbial leaching is a very attractive option. Sulphate oxidizing bacteria such as *Thiobacillus ferro-oxidans*, sulphide oxidizing bacteria such as *Leptospirillum ferro – oxidans, Sulfolobus, Thiobacillus acidophilus, Thiobacillus organoparus* have a role to play in facilitating mineral leaching.

23.15 MICROBIAL DESULFURIZATION OF COAL

Coal is the major source of energy for the production of electricity. During the combustion of coal, high concentration of sulphur in coal leads to the production of SO_2 which when mixed with water forms corrosive sulphuric acid forming acid rain. Environmental damage by acid rain is well documented. Although there are physical and chemical methods to reduce the emission of sulphur during coal combustion they are relatively expensive.

Microbial technology has been proven to be a good option for coal desulphurization where sulphur is removed from organic and inorganic component of coal or by oxidation of sulphur to sulphate. The use of microorganism desulphurization is simple, energy saving and economical.

Bacteria such as *Corynebacterium, Thiobacillus thio-oxidans, Leptospirillum ferro-xidans, Sulfulobus* sp., *Pseudomonas, Xanthomonas maltophila, Moraxella* sp., *and Rhodococcus* have been used in desulphurization of coal.

23.16 BIODEGRADATION OF COMMERCIAL PRODUCTS

Biodegradation generally refers to biodeterioration of commercially useful materials like paper, wood, textiles, leather, metals, paints and glass leading to economical loss. No material is absolutely resistant to microbial deterioration. Even fuel, concrete, computer chips, glass lenses, paintings and several others are attacked by microbes. Biodeterioration of some commercially important materials is discussed below

Wood

Decay of wood by fungi results in rotting of wood and appearance of spots and holes in the wood. Basidiomycetes cause white rot and brown rot of wood where as Ascomycetes cause soft rot. Bacteria like *Sphaerotilus mutans,* yeasts like *Torula* and *Rhodotorula,* and fungi like *Trichoderma, Fusarium, Penicillium* and *Mucor,* cause slime of wood. Wood degrading white rot fungi include *Aspergillus nidulans, Cereporiopsis subvermispora, Coriolus pruinosum, Cyathus bulleri, Phanerochaete chrysosporium, Phlebia tremellosus, Pleurotus ostreatus, Podospora anserina and Xylaria polymorpha.*

Paper

Micro-organisms affect paper in different stages of the manufacturing process. Wood, the raw material of paper is attacked by micro-organisms. The pulping process of raw material involves removal of lignin and hemicellulose. The microbial growth during this process results in slimy layer that may affect the quality of the paper. The glazing material such as glue and casein may also be subjected to microbial degradation.

Microbial attack on paper leads to weakening of the paper, discoloration, stains and spots on paper and even holes. Micro-organisms that attack paper include both bacteria and fungi.

Among the fungi *Alternaria, Cladosporium, Trichoderma, Penicillium* and *Rhizopus* and among the bacteria *Cellulomonas,Cytophaga* and *Pseudomonas* cause biodegradation of paper.

Textiles

Plant fibres like cotton, flax, jute, hemp and animal skin are all susceptible to microbial deterioration. Materials produced from these fibres like, textiles, wool, fur, silk, ropes and cordage are also prone to microbial attack. The sizing materials used in textiles are also a target of microbial attack. The organisms that attack textiles include fungi like *Myrothecium verrucaria, Chaetomium globosum,* species of *Aspergillus, Rhizoctonia,Cladosporium, Trichoderma, Humicola, Botrytis* and Spirochaetes.

Leather goods

Under high moisture and low temperature and during rainy season, leather goods are attacked by species of *Aspergillus* and *Penicillium* resulting in moldy growth on leather. It results in chaffing of leather and discoloration.

Paints

The organic compounds of paints and the painted surfaces are prone to attack by micro-organisms especially fungi. As a result, paints are discolored and spots appear on painted surface. Species of *Alternaria, Aspergillus,*

Cladosporium Penicillium, Phoma glomerata, and *Pullullaria pullulans,* are the main fungi that attack paints.

Metals

Among metals, iron is highly susceptible to microbial deterioration under anaerobic conditions. It results in corrosion of iron. Among bacteria, *Desulfovibrio* and methanogenic *Archaebacteria* are largely responsible for corrosion of iron.

Glass

Even a hard material like glass is damaged by microbes. Optical instruments, microscopic lenses, camera, binocular lenses and slide projectors are affected by the growth of *Aspergillus* and *Penicillium.* The glass surface is etched due to fungal attack resulting in loss of transparency and clarity.

Concrete

The cement used in making concrete is damaged by *Thiobacilli.* These bacteria oxidise sulphides present in the pyrite to produce sulphuric acid that dissolves concrete. *Thiobacillus thio-oxidans* causes extensive damage to concrete roads.

23.17 MICROBES IN COMPOSTING

Composting is accelerated decomposition of organic materials of biological origin such as plant wastes and animal excreta by micro-organisms and invertebrates to form a stabilized organic product. As the aerobic decomposition is exothermic, heat is released raising the temperature of the composting mass. Microbial decomposition of polysaccharides and proteins raises the temperature further in the mesophilic stage. A further increase in the thermophilic stage kills mesophiles and thermophiles take over at 45° - 55 °C. Above this temperature thermophilic Actinomycetes dominate which attack hemicellulose and cellulose. At temperatures higher than 60° C, spore forming micro-organisms remain viable and others vanish and the temperature of the composting mass decreases. Mesophilic fungi such as basidiomycetes and bacteria such as Actinomycetes reinvade to totally degrade hemicellulose, cellulose and lignin in this cooling stage. It is followed by humification and nitrification that occur in the maturation stage.

Thus several metabolic stages that represent different microbial populations can be seen during composting. Compost should be cured, in addition to being low in pathogenic microbes to make it viable for utilization posing no harm to crop and public health (Table 23.8).

Table 23.8 Biological composition of microbial genera of a composting heap

Decomposers	Bacteria	Fungi	Actinomycetes
Mesophiles	*Bacillus* *Pseudomonas* *Enterobacter* *Clostridium*	*Rhizopus* *Coprinus* *Polyporus* *Pleurotus* *Penicillium* *Aspergillus* *Geotrichum* *Trichoderma* *Cephalosporium*	*Nocardia*
Thermophiles	*Lactobacillus*	*Thermoascus* *Chaetomium* *Sporotrichum*	*Streptomyces* *Thermoactinomyces* *Thermomonospora* *Humicola*

Source: Chu, L.M and Wong, P.K. 2006.

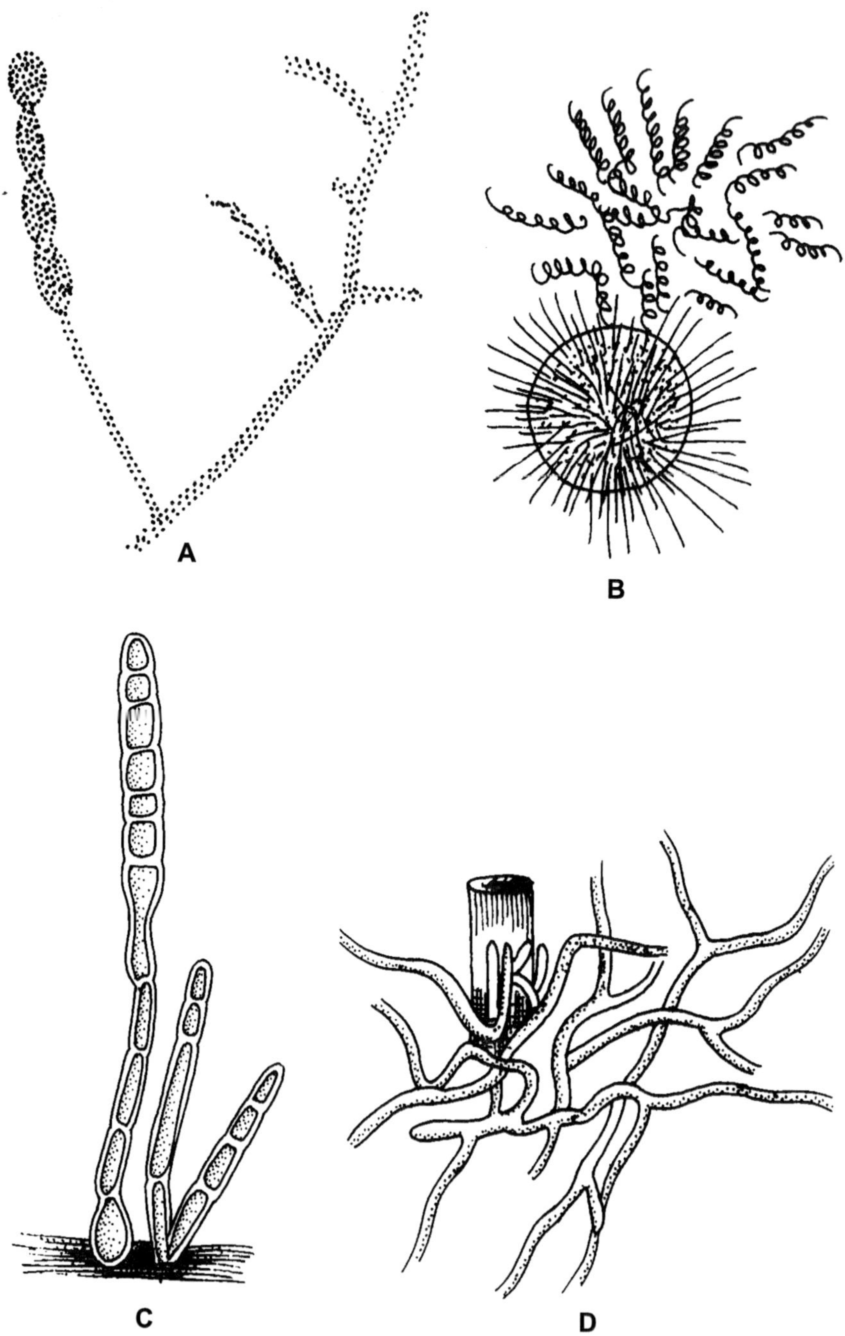

Fig 23.1 **Fungal species in biodegradation. A** : *Alternaria tenuis*; **B** : *Chaetomium aterrimum*; **C** : *Helminthosporium teres*; **D** : *Mortierella rostafinskii*.
Fig. 23.1. Redrawn from Bessey, E.A. (1950)

CHAPTER 24

LECTINS IN MICROBES

24.1 INTRODUCTION

Lectins are sugar binding proteins (or glycoproteins) of non-immune origin with the ability to agglutinate cells and/or precipitate glyconjugates. Lectins can recognize the specific sugars on cell surfaces resulting in cell agglutination. Lectin induced agglutination is inhibited by a specific sugar and this distinguishes lectins from mammalian antigens. Lectins are present in almost every organism but are readily detectable in different parts of a large number of species of plants.

Lectins, due to their specificity to cell surfaces, can be used to identify cell types, can be employed in the determination of human and animal blood groups and to immobilize human sperms through agglutination, or detect the onset of cancer. Some lectins induce cell division (mitogenic lectins).

Lectins are predominant in the flowering plant species but were also detected in all groups of non-flowering plants and microbes.

24.2 MICROBIAL LECTINS

Microbial lectins were first reported by Guyot in 1908 in *Escherichia coli*. More recently numerous micro-organisms have been shown to possess agglutinating properties. Proteins responsible for these sugar specific agglutinating activities have been isolated from various microbial strains. In most cases, surface bound agglutinins have been investigated.

The main reason for the intense interest in microbial lectins is that they may play a role in mediating adherence to surfaces colonized by micro-organisms and are considered to be determinants of virulence in infection of both plants and animals. Thus the microbial lectins serve as molecules of recognition in cell-cell interaction, for example, bacteria-epithelial or phagocytic cell interactions. They may also play a role as recognition determinants during development.

One of the most interesting aspects of microbial lectins are their location in the cell. They appear on the cell on bacterial fimbriae or pili and convert the organism into a particle with multiple binding sites capable of haemagglutination.

Lack of haemagglutination by whole organisms, in contrast to agglutination by microbial extracts may not necessarily indicate lack of lectins. Lectins have been identified in *Mycoplasma*, viruses, bacteria, fungi, yeasts lichens and algae (Table 24.1)

24.2.1 *Mycoplasmal* Lectins

Most Mycoplasmas adhere to and colonize the epithelial lining of respiratory and urinogenital tracts causing serious infections. Several strains of *Mycoplasma* also agglutinate erythrocytes and bind to eurkaryotic cell surfaces. *Mycoplasma pneumoniae, Mycoplasma gallisepticum* and *Mycoplasma genitalium* possess special organelles at the tip of the cells functioning as attachment organelles containing sialic acid moieties arranged on host membrane sialoglycoproteins or sialoglycolipids, named as adhesins (lectins). So far only *Mycoplasma pneumoniae* lectin (PI) has been partially characterized (Razin 1986). Attempt to isolate and fully characterize lectins from *Mycoplasma* have not been successful.

24.2.2 Viral Lectins

The first viral lectin was isolated from Sendai virus. Human influenza A virus agglutinates fowl erythrocytes while measles virus agglutinates most of the primate erythrocytes except humans. Both non enveloped and enveloped viruses display haemagglutinating ability. The viral attachment proteins responsible for, haemagglutination are a component of the capsid in non-enveloped virus while it is the component of the viral envelope in the enveloped viruses. The viral component involved in both haemagglutination and infection is a protein referred as virus attachment protein (VAP).

The use of viral lectins has opened up several new areas of research. First, by using the cytopathic effect of viral infections, cells resistant to infection can be selected. Second, many of these viruses can be used to deliver substances to target cells; third, VAPs are suitable for cloning and expression. Genetic engineering techniques are used to design new viral lectins (Markwell 1986) as high affinity probes and new generation tools to explore the surface of eukaryotic cells.

24.2.3 Bacterial Lectins

Lectins in bacteria are the principle mediators of adherence, colonization and establishment of bacterial communities in causing infections.

Bacterial surface lectins or lectin like substances are involved in the initiation of infection by mediating bacterial adherence to epithelial cells.

The ability of *E. coli* to agglutinate erythrocytes was first observed by Guyot in 1908. Later it was shown that *E.coli* with mannose specific lectins can agglutinate leukocytes, sperms, pollen and fungal spores also (Rosenthal 1943).

Many bacteria possess the ability to bind to and agglutinate erythrocytes that are inhibited by sugars, suggesting the presence of lectins on the bacterial surface (Ofek *et al.,* 1977).

Lectins can induce bacterial cell agglutination and inhibit their growth and are useful as an effective tool in the identification of pathogens and control infections caused by them. In addition, lectins can pre-empt the binding sites on bacterial cell surfaces preventing the binding of the pathogen to human skin, gut and other surfaces and membranes, which is an important event in pathogenesis.

Bacterial infection can be regarded as a lectin mediated metastasis of bacteria into special target organs. The carbohydrate receptor for the lectin can be on the bacterial surface or on the surface of the metastatic tumour cells and are found to recognize specific sugars. Sugar specific bacterial lectins can be successfully used to block and prevent metastasis and also prevent infection (Dietz *et al.,* 1988)

Non- immune phagocytosis mediated by bacterial surface lectins of *E.coli* may be of clinical importance in non-immune hosts. (Sharon, 1977)

Pseudomonas lectins (PA-I) is known to stimulate growth of T-lymphocytes from human peripheral blood to lymphoblasts (Sharabi and Garber 1979). The blastogenic response with lectins is one of the tests used for measuring the effectiveness of T-cells to participate in cell mediated immune response. This test is performed on cancer patients who exhibit a diminished response when compared to healthy ones (Avichezer and Garber 1983).

Purified lectins of *Pseudomonas aeruginosa* is used as vaccine for active and passive immunization (Garber and Sudakevitz 1982). It also employed in studying chromosomal structures and in diagnosis of disease states; stimulation of growth and physiological functions of other eurkaryotic cells and contributes to a better understanding of structural and functional relationship (Garber 1986).

Lectins of nitrogen fixing bacteria, *Rhizobium japonicum, Klebsiella aerogenes, Enterobacter agglomerans* and *Rhizobium leguminosarum* are responsible for the interaction with biological surface of the hosts for fixing nitrogen (Dazzo, 1986) and to maintain symbiotic relationship.

24.2.4 Fungal Lectins

Lectins are present in many fungi, mainly in the fruiting bodies of higher fungi. They are also detected in the mycelia of lower fungi and cellular slime molds (Hertz and Chet 1986). Several of them were galactose specific and N-acetyl-galactosamine specific lectins. Fructose and lactose binding lectins were detected in *Aleuria aurantia* and *Laccaria amethystina.*

Knowledge of fungal lectin function is still in its infancy. Many possible roles have been assigned to fungal lectins. They may be involved in fungal – nematode, fungal-fungal interactions, fungal-algal interaction and fungal-plant interactions as recognition proteins. They specifically play a role in pathogenecity which is species specific. In mycoparasitic relationships, lectins might be involved in linkage between two different fungi.

Lectins from *Grifola frondosa* are reported to by cytotoxic against Hela cells and that of *Tricholoma mongoli* against mouse monocyte-macrophage PU5 cells indicating the antitumour activity. Lectin from the fruiting body of *Volvoriella volvacea* is reported to be immunomodulatory in activity.

24.2.5 Algal Lectins

The presence of lectins in algae was first demonstrated by Boyd *et al.*, in 1966 in marine algae from their cell sap with haemagglutinating activity. Among the algae that were screened, *Ptilota plumosa* is the only record of algal lectin agglutinating B erythrocytes of humans. Major genera of red algae agglutinated animal erythrocytes and leukaemic cells.

Algal lectins have been demonstrated to function in gamete recognition and fertilization as in *Chlamydomonas* (Weise and Shoemaker 1970) and *Fucus serratus* (Bolwell *et al.,* 1979). Algal lectins also play a role in algal ecology, colony formation and assembly of phycobilisomes (Maki and Mitchell 1986).

Table 24.1 Source of lectins from micro-organisms

	Bacteria
1.	*Actinomyces naeslundii*
2.	*Actinomyces viscosus*
3.	*Aeromonas hydrophila*
4.	*Aeromonas liquefaciens*
5.	*Salmonella arizonae* sp.
6.	*Bacteroides asacharolyticus*
7.	*Bacteroides fragilis*
8.	*Bacteroides gingivalis*
9.	*Bacteroides melaninogenicus*

10.	*Bordetella bronchiseptica*
11.	*Bordetella pertusis*
12.	*Citrobacter freundii*
13.	*Citrobacter lerupensis*
14.	*Clostridium botulinum*
15.	*Corynebacterium diphtheriae*
16.	*Corynebacterium parvum*
17.	*Corynebacterium renale*
18.	*Eikenella corrodens*
19.	*Enterobacter aerogenes*
20.	*Enterobacter agglomerans*
21.	*Enterobacter amnigenus*
22.	*Enterobacter cloacae*
23.	*Enterobacter gergoviae*
24.	*Enterobacter intermedium*
25.	*Enterobacter sakazakii*
26.	*Erwinia caratovora*
27.	*Escherichia coli*
28.	*Fusobacterium nucleatum*
29.	*Haemophilus influenzae*
30.	*Haemophilus parainfluenzae*
31.	*Klebsiella aerogenes*
32.	*Klebsiella pneumoniae*
33.	*Leptotrichia buccalis*
34.	*Moraxella bovis*
35.	*Morganella morganii*
36.	*Myxococcus xanthus*
37.	*Neisseria gonorrhoeae*
38.	*Neisseria meningitidis*
39.	*Pasteurella multocida*
40.	*Proteus mirabilis*
41.	*Proteus myxofaciens*
42.	*Proteus vulgaris*
43.	*Providencia alcalifaciens*
44.	*Providencia rettgeri*
45.	*Providencia stuarti*
46.	*Pseudomonas aeruginosa*
47.	*Pseudomonas echinoides*
48.	*Pseudomonas multivorans*
49.	*Salmonella entertidis*
50.	*Salmonella typhii*
51.	*Salmonella typhimurium*
52.	*Serratia liquefaciens*
53.	*Serratia marcescens*

54. *Serratia marinorubra*
55. *Serratia plymuthica*
56. *Shigella flexneri*
57. *Staphylococcus saprophyticus*
58. *Streptococcus mutans*
59. *Streptococcus pneumoniae*
60. *Streptococcus salivarius*
61. *Streptococcus sanguis*
62. *Vibrio anguillarum*
63. *Vibrio cholerae*
64. *Vibrio ordalii*
65. *Vibrio parahaemolyticus*
66. *Yersinia enterolitica*
67. *Yersinia frederiksenii*
68. *Yersinia intermedia*
69. *Yersinia pseudotuberculosis*
70. *Yersinia ruckeri*

Viruses

Paramyxoviruses
Sendai virus

Myxoviruses
Influenza A virus
Influenza B virus
Influenza C virus

Papoviruses
Japanese encephalitis virus
Polyoma virus

Picornaviruses
Aphtho virus
Entero virus
Rhinovirus
Cardio virus

Non-enveloped viruses (Hemagglutins)
Adeno virus
Parvo virus

Reoviruses
Reo virus
Rota virus

Enveloped viruses
Bunya virus
Corona virus
Paramyxo virus
Morbilli virus
Pneumo virus
Pox virus
Retroviruses
Murine leukemia virus
Rhabdo virus
Toga virus
Alpha virus
Flavi virus
Rubi virus

Fungi
Agaricus bisporus
Agaricus campestris
Aleuria aurantia
Arthrobotrys oligospora
Aspergillus niger
Clitocybe nebularis
Conidiobulus lamprauges
Flammulina velutipes
Fomes fomentarius
Laccaria amethystina (LAL)
Laccaria amethystina (LAF)
Marasmius oreades
Meria coniospora
Sclerotium rolfsii
Rhizoctonia solani

GLOSSARY OF MEDICAL TERMS

Abortifacient	:	An agent which causes abortion.
Abscess	:	A localized collection of pus anywhere in the body, surrounded and walled off by damaged and inflamed tissues.
Acne	:	A skin disorder in which the sebaceous glands become inflamed.
Adenocarcinoma	:	Nodular benign lesions on the breast.
Adenofibrosis	:	A benign nodular breast lesion consisting of hyperplastic distorted lobules of acinar tissue.
Adenoma	:	A non-malignant tumour of glandular tissue.
Adenomatous polyp	:	Non-malignant pedunculated tumour of the glandular tissue.
Adjuvant	:	A substance included in a prescription to aid the action of other drugs.
Aflatoxicosis	:	Poisoning caused by consumption of substances or foods contaminated with aflatoxin.
Albuminuria	:	Presence of serum albumin, serum globulin or other serum proteins in the urine.
Alopecia	:	Absence of hair from areas where it normally grows.
Ambylopia	:	Defective vision approaching blindness.
Amenorrhoea	:	The absence or stopping of the menstrual cycle.
Amnesic shell fish poisoning	:	Shell fish poisoning leading to amnesia
Amnesia	:	Loss of memory.

Anaphylactic : An agent which causes allergy.

Angina pectoris : Severe but temporary attack of cardiac pain which may radiate to the arms; results from myocardial ischaemia.

Angiogenesis : The process of vascularisation of a tissue during the development of new capillary blood vessel.

Angular chelitis : Inflammation of the lip.

Anodyne : Any treatment or drug that soothes and caeses pain.

Anthelmintic : Any drug or chemical agent used to destroy parasitic worm.

Antibacterial : An agent which destroys bacteria.

Anticoagulant : An agent that prevents the clotting of blood.

Antifungal : Agent which inhibits the action of fungi.

Antimicrobial : An agent which destroys microbes.

Antimycobacterial : An agent which destroys fungi and bacteria.

Antinociceptive : A substance that reduces sensitivity to painful stimuli.

Antioxidant : Any substance which delays the process of oxidation.

Antiperiodic : Preventing the regular recurrence of a disease.

Antiprotozoal : An agent which prevents or destroys protozoans.

Antiseptic : A chemical that destroys or inhibit the growth of disease-causing bacteria and other micro-organisms.

Antispasmodic : Preventing or relieving spasms.

Antitubercular : Agent used to prevent or cure tuberculosis.

Antiviral : An agent which prevents or destroys virus.

Aphrodisiac : An agent that stimulates sexual excitement.

Aplastic crisis : Incapable of forming a new tissue.

Apoptotic : One which causes programmed cell death.

Arthralgia : Pain in the joint.

Arthritis	:	Inflammation of one or more joints, characterized by swelling, warmth, redness, of the overlying skin, pain and restriction of motion.
Astringent	:	An agent which contracts organic tissue thus lessening secretion.
Asymptomatic	:	Absence of symptoms.
Athelete's foot	:	Ringworm of the foot caused by *Tinea pedia/ Tinea pedis.*
Atherosclerosis	:	A disease of the arteries in which fatty plaques develop on their inner walls, with eventual obstruction of blood flow.
Atony	:	Weakness.
Atopic dermatitis	:	Inflammation of the skin.
Atrophy	:	Wasting, emaciation, diminution in size and function.
Autoimmune disease (Autoimmunization)	:	Sensitization of a person to a substance elaborated in his own body.
Autolysate	:	An end product of autolysis.
Autolysis	:	Breakdown of all or part of a cell or tissue by self-produced enzymes.
Bacteraemia	:	The presence of bacteria in the blood.
Bacteriocide	:	Any agent which destroys bacteria.
Bacteriostat	:	Agent which arrests bacterial growth.
Balanitis	:	Inflammation of the glans penis and prepuce (foreskin).
Balkan nephropathy	:	Chronic intestinal nephropathy.
Bovine spongiform encephalopathy	:	A degenerative disorder of the brain characterized by tiny holes which give the brain a spongy appearance.
Blisters	:	A swelling containing watery fluid and sometimes also blood (blood blister) or pus, within or just beneath the skin.
Bradycardia	:	Slow rate of heart contraction resulting in slow pulse rate.
Bright's disease	:	Inflammatory affection of the kidney.

Bronchial carcinoma	:	Malignant growth of bronchi.
Bronchiolitis	:	Inflammation of the bronchioles.
Bronchodilator	:	An agent that causes widening of the air passages by relaxing bronchial smooth muscle.
Bronchopneumonia	:	Type of pneumonia localised often in the bronchioles and surrounding alveoli.
Bronchorrhoea	:	An excessive discharge of mucus from the bronchial mucous membrane.
Bulimia	:	Excessive appetite.
Burkitt's lymphoma:	:	An aggressive type of B-cell Non-Hodgkin lymphoma that occurs most often in children and young adults.
Calculi/Calculus	:	An abnormal concretion, usually composed of mineral salts, occurring within the body.
Callosites	:	A piece of skin that has thickened due to repeated friction.
Cancerous ulcers	:	An ulcer which transforms itself into a cancerous form.
Capillary bronchitis	:	Inflammation of the tiny thin walled vessel of the bronchi.
Carbuncles	:	An acute inflammation caused by *Staphylococcus* involving several hair follicles and surrounding subcutaneous tissue.
Carcinogenic	:	Any substance that when exposed to living tissue may initiate cancer.
Carcinoma	:	A malignant new growth made up of epithelial cells tending to infiltrate surrounding tissues to give rise to metastatis.
Cardiomyopathy	:	Disease of the heart muscle (the myocardium).
Carminative	:	An agent which stimulates or improves the digestive system.
Cataract	:	Opacity in the lens of eye, resulting in blurred vision.
Catarrh	:	Inflammation of mucous membrane with constant flow of mucus.
Catatonia	:	A syndrome of psychomotor disturbance in schizophrenics.

Cathartic	:	A drug used to stimulate or increase the frequency of the bowel evacuation or to encourage a softer or bulkier stool.
Catheters	:	A hollow tube of variable length and bore usually having one fluted end and a tip of varying size and shape.
Cephalalgia/Cephalgia	:	Headache.
Cervical adenitis	:	Inflammation of gland or a group of glands in the neck region.
Chancre	:	The primary syphilitic ulcer associated with swelling of local lymph glands; it is painless, indurated and highly infectious.
Chicken pox	:	A highly infectious disease caused by a Herpes virus that is transmitted by air borne droplets.
Chilblain	:	Severe itching, burning sensation in reaction to cold in the feet.
Cholagogue	:	An agent that stimulates gall bladder contraction to promote bile flow.
Cholangio carcinoma	:	Malignancy of biliary duct.
Choleretic	:	An agent that stimulates the secretion of bile.
Cholinergic	:	Liberation of acetyl choline by parasympathetic nerves.
Chorea	:	The ceaseless occurrence of rapid, jerky involuntary movements.
Chorioretinitis	:	Inflammation of the choroids and retina.
Cirrhosis	:	Interstitial inflammation of an organ, particularly the liver.
CNS depressant	:	An agent which depresses the central nervous system.
Coccidia	:	A parasitic protozoa causing alimentary canal disease in live stock.
Colic	:	Severe pain resulting from periodic spasm in abdominal region.
Colitis	:	Inflammation of the colon.
Collyrium	:	An eye salve or eye wash.

Colonic adenocarcinoma : Malignant tumour of the lymphoid tissue of the colon.

Colonorrhoea : Rarely used term for diarrhoea.

Condyloma/Condylomata : Dry warts on the reproductive organs or on the perianal region.

Congenital : Existing from birth or before.

Conjunctivitis : Inflammation of the conjunctiva (the delicate membrane lining the eyelids covering the eye ball).

Convulsions : Involuntary contractions of muscles resulting from abnormal cerebral stimulation with or without loss of consciousness.

Corn : An area of hard thickened skin on or between the toes.

Cosmeceuticals : Are topically applied and contain ingredients that influence the function of the skin.

Creutzfeldt-Jacob disease : It is a rare type of degenerative dementia.

Cutaneous melanoma ; A tumour arising from the pigment- producing cells of the cutaneous layer.

Cystitis : Inflammation of the urinary bladder.

Cytotoxic : A drug that damages or destroy cells and is used to treat various types of cancer.

Debility : Lack or loss of strength; weakness.

Delirium : An acute disorder of the mental process accompanying organic brain diseases.

Dementia : Irreversible organic deterioration of mental faculties.

Demulcent : A mucilagenous substance which allays irritation and smoothens inflammation, especially of mucous membrane.

Dental caries : Decay and crumbling of the substance of a tooth resulting in a cavity.

Dentifrice : A powder or paste for cleaning the teeth.

Deobstruent : Medicine which removes obstruction.

Depurative : An agent which removes impurities.

Dermatitis : Inflammation of the skin.

Dermatoxin	:	An exotoxin formed by bacteria, and characterized by its ability to cause pathologic changes in skin.
Detoxicant	:	Any agent which neutralizes effect of toxins.
Dhobi's itch	:	A type of allergic dermatitis, caused by wearing clothes, laundry marked with ink used in India by native laundry men.
Diabetes mellitus	:	A disorder of carbohydrate metabolism in which sugars in the body are not oxidized to produce energy due to lack of pancreatic hormone insulin.
Dialysis	:	A process of removing waste products from and replacing essential constituents in blood.
Diaphoretic	:	An agent that promotes profuse perspiration.
Diarrhoetic shell fish poisoning	:	Shell fish poisoning resulting in diarrhoea.
Diphtheria	:	An acute infectious disease caused by *Corynebacterium diphtheriae* and its toxin, affecting the membranes of the nose, throat, or larynx causing fever, pain and respiratory obstruction.
Diuretic	:	A agent which increases the amount of urine.
Dropsy	:	Abnormal accumulation of serous fluids in cellular tissues or in body cavity.
Ductular carcinoma	:	Abnormal accumulation of serous fluids in the lining of a breast duct.
Dysmenorrhoea	:	Absence of menstruation.
Dyspareunia	:	Painful or difficult coitus.
Dyspepsia	:	Impairment of the power or function of digestion; usually applied to epigastric discomfort after meals.
Dysphagia	:	Difficulty in swallowing.
Dyspnoea	:	Difficult or laboured breathing.
Dysuria	:	Painful or difficult urination.
Eczema	:	A superficial inflammatory process involving primarily the epidermis, marked early by redness, itching, weeping, oozing and crusting,

and later by scaling, lichenification and often pigmentation.

Edema/Oedema : Abnormal infiltration of tissues with fluids.

Ehrlichiosis : Is an infection of white blood cells that affects mammals.

Elephantiasis : A chronic filarial disease caused due to infection with *Brugia malayia* or *Wachereria bancrofti* marked by inflammation of the lymphatics and hypertrophy of the skin, affecting the legs and genitals.

Emetic : An agent that cause vomiting.

Emmenogogue : An agent that promotes menstruation.

Emollient : An agent which softens and soothens mucous membrane.

Emphysema : A pathological condition of the lungs marked by an abnormal increase in the size of the air spaces, resulting in labored breathing and an increased susceptibility to infection.

Encephalopathy : Disease of the brain.

Endocarditis : Inflammation of the inner lining of the heart.

Endophthalmitis : Internal infection of the eye as a result of perforated injury.

Enema : A quantity of fluid infused into the rectum through a tube passed through the anus.

Enteritis : Inflammation of the intestine.

Enterorrhagia : Bleeding within the intestinal tract.

Enuresis : Incontinence of urine (bed wetting).

Epilepsy : Abnormal electrical discharge that disturbs cerebration and usually results in fits and loss of consciousness.

Epistasis : Bleeding from the nose which may be caused by physical injury or may be associated with fever, high blood pressure or blood disorder.

Epithelioma : A tumour of epithelium, a covering of internal and external surfaces of the body: a former term for carcinoma but now also used to describe benign tumours.

Erectile dysfunction	:	Abnormal functioning of erectile tissue.
Erysipelas	:	An infection of the skin and underlying tissues with the bacterium *Streptococcus pyogenes*.
Erythema	:	Normal flushing of the skin caused by dilation of blood capillaries.
Estrogenic	:	One of a group of steroid hormones (including estriol, estrone and estradiol) that control female sexual development and secondary sexual characteristics.
Expectorant	:	A drug that enhances the secretion of sputum by the air passages so that it is easier to cough up.
Familial insomnia	:	Sleeplessness affecting several members of the same family.
Febrifuge	:	A treatment or a drug that reduces or prevents fever.
Febrile attacks	:	Feverishness.
Febrile pharyngitis	:	Fever followed by inflammation of the pharynx.
Fibrinolysis/Fibrinolytic	:	The dissolution of fibrin in blood clots.
Fibroids	:	A benign tumour of fibrous and muscular tissues, one or more of which may develop in the muscular wall of the womb.
Fibroma	:	Mass composed of fibrous or connective tissue. Also called a fibroid tumour or fibroid.
Fibromyalgia syndrome	:	Pain in the muscles, ligaments and tendons.
Fibrosarcoma	:	A malignant tumour derived from fibroblastic cells.
Fibrosis	:	Formation of excessive fibrous tissue in a organ.
Fissure	:	A split or cleft.
Fistula	:	An abnormal communication between two hollow organs or between a hollow organ and the exterior.
Five day fever	:	Same as Trench fever.
Flatulence	:	Expulsion of gas or air from the stomach through the mouth by belching or from the bowel by passing flatus.

Flatus	:	Gas in the stomach or intestine.
Follicular lymphoma	:	A benign tumour of the follicle.
Folliculitis	:	Infection of hair follicles by *Staphylococcus.*
Freckles	:	A brown spot of a skin commonly found on the arms and face of young people with fair complexion.
Fungistatic	:	A chemical which inhibits the growth of fungi.
Fungitoxic	:	A chemical toxic to fungi.
Furuncle	:	An acute inflammatory condition surrounding a hair follicle.
Galactagogue	:	An agent inducing or increasing the flow of milk.
Ganglion	:	Group of nerve-cells from which nerve fibres radiate.
Gangrene	:	Death and decay of part of the body due to deficiency or cessation of blood supply.
Gastroenteritis	:	Inflammation of the mucous coat of the stomach and intestines due to bacterial infection.
Gastritis	:	Inflammation of the mucous membrane lining of the stomach.
Gastropathy	:	Any disease of the stomach.
Genotoxic	:	A poisonous substance which harms an organism by affecting its DNA.
Germicide	:	Destructive of micro organisms.
Gerstmann-Straussler -Scheinker syndrome	:	Extremely rare neurological brain disorder. Symptoms include lack of muscle coordination, unsteadiness leading to dementia.
Gingivitis	:	Inflammation of the gum.
Gingivostomatitis	:	Inflammation of the gum and mouth.
Glomerullo nephritis	:	Non-suppurative inflammation of the glomerulus of the kidneys.
Glossitis	:	Inflammation of the tongue.
Goitre	:	Enlargement of the thyroid gland causing a swelling in the front part of the neck.

Gonorrhea/Gonorrhoea	:	An infectious disease of venereal origin caused by bacterium *Neisseria gonorrhoeae* that affects the genital mucous membranes of either sex.
Gout	:	A form of metabolic disorder in which sodium biurate is deposited in the cartilages of the joints, ears and elsewhere.
Grass staggers	:	A condition commonly seen in cows, resulting in decreased milk production, weight loss and increased nervousness.
Gravel	:	A sandy deposit, which if present in the bladder may be passed with the urine.
Haemangioendo-thelioma	:	A tumour of endothelial cells lining the inner surface of blood vessels.
Haematemesis	:	Vomiting of blood.
Haematinic	:	Any substance which is required for the production of the red blood cells and its constituents.
Haematogenous	:	Originating in blood.
Haematuria	:	The presence of blood in the urine.
Haemolytic	:	Destruction of red blood cells.
Haemoptysis	:	Spitting of blood.
Haemorrhoidal diathesis	:	Condition where in tissue of the haemorrhoids reacts to stimuli.
Haernorrhoids	:	Enlarged veins in the wall of/around the anus usually a consequence of prolonged constipation or, occasionally diarrhoea.
Haemostasis	:	Arrest of bleeding / stagnation of blood within its vessel.
Hairy cell leukaemia	:	Chronic, lymphocytic leukaemia where lymphocytes have hairy appearance.
Hay fever	:	A form of allergic rhinitis.
Helminthiasis	:	Condition resulting from infestation with worms.
Hemicrania	:	A headache affecting only one side of the head as in migraine.

Hemiplegia	:	Paralysis of one side of the body due to a cerebro vascular accident on the opposite side.
Hepatic fever	:	Increase in body temperature due to liver disorder.
Hepatic nodules	:	Presence of nodules in the liver.
Hepatitis	:	Inflammation of the liver due to virus infection or such disease as amoebic dysentry and lupus.
Hepatocellular carcinoma	:	Malignant growth of liver cells.
Hepatomegaly	:	Enlargement of the liver to such an extent that it can be felt below the rib margin.
Hepatopathy	:	Relating to diseases of liver.
Hepatoprotective	:	An agent or drug which protects thc liver.
Hepatosplenomegaly	:	Enlargement of liver and spleen.
Hernia	:	The protrusion of an organ or tissue out of the body cavity in which it normally lies.
Herpes	:	Any inflammatory skin disease caused by Herpes virus.
Hodgkin's disease	:	A malignant lymphoma causing progressive enlargement of lymph nodes and involvement of reticuloendothelial tissues including bone marrow.
Hydrolysate	:	A product of hydrolysis.
Hydrophobia	:	Fear of water.
Hyperacidity	:	Excessive acidity.
Hyperalbuminuria	:	Excessive albumin.
Hyperbilirubinemia	:	Excessive bilirubin in the blood, found in physiological jaundice in new born.
Hypercholesterolemia/ Hypercholesterolemic	:	Excessive reduction of the enzyme choline esterase.
Hyperdipsia	:	Intense thirst.
Hyperglycemia	:	Excess of glucose in the blood stream.
Hyperlipoproteinaemia	:	Higher level of lipid and proteins.
Hyperplasia	:	Excessive formation of cells.
Hypertriglyceridemia	:	Excess of triglycerides in the body.

Hypertrophy	:	Increase in the size of tissue or structures, independent of natural growth.
Hypocholesterolemic	:	Deficiency in production of the enzyme choline esterase.
Hypoglycaemia	:	Reduced glucose level in blood stream.
Hypokalaemic/ Hypokalaemia	:	Abnormally low potassium level of the blood.
Hypolipaemia	:	Low level of fat in the body.
Hypomagnesaemia	:	Same as Grass staggers.
Hypothermia	:	Greatly decreased body temperature.
Hypothyroidism	:	Underaction of the parathyroid glands with decrease in serum calcium levels.
Hysteria	:	Neurosis usually arising from mental conflict and repression.
Immunoenhancer	:	An agent that ncreases the immune reaction.
Immunosuppressant	:	That which prevents the occurrence of an immune reaction.
Impetigo	:	*Streptococcal* or *Staphylococcal* skin infection marked by vesicles or bullae.
Indolent	:	Describing a disease process that fails to heal or has persisted.
Induration	:	Hardening of tissue.
Infantile diarrhoea	:	Diarrhoea of a child or baby less than one year old.
Infantile paralysis	:	An epidemic viral infection which attacks the motor neurons of the interior horns in brain stem and spinal cord in infants.
Infarction	:	It is a rapid development of myocardial necrosis caused by a critical imbalance between the oxygen supply and demand of the myocardium
Intermittent fever	:	A fever that rises, subsides, then returns again.
Ischaemia	:	Deficient blood supply.
Jakob-Creutzfeldt disease	:	Also called Madcow disease. (See Bovine Spongiform Encephalopathy).

Jaundice : A condition characterised by a raised bilirubin level in the blood (hyper bilirubinaemia).

Kaposi's sarcoma : A type of vascular cancer.

Keloid tumours An over growth of scar tissue, which may produce a contraction deformity.

Keshan's disease : Is a type of heart disease called cardiomyopathy that affects young women and children in a selenium deficient region of China.

Keratitis : Inflammation of the cornea of the eye.

Keratoconjunctivitis : Inflammation of the cornea and conjuntiva.

Lacrimation : An outflow of tears.

Laryngitis : Inflammation of the larynx and vocal cords due to bacteria or viruses or irritation due to gases and chemicals.

Laxative : One which promotes bowel evacuation by increasing faecal volume.

Lesch - Nyhan syndrome : A rare inherited disorder caused by deficiency of Hypoxanthine - guanine phosphoribosyl transferase causing build of uric acid leading to severe gout.

Leucorrhoea : An abnormal mucous discharge from the vagina.

Leukaemia/leukemia : A malignant proliferation of the leucopoietic cells usually producing an abnormal increase in leukocytes in the blood with immature cells among them.

Leukocytosis : An increase in the number of white blood cells.

Leukoencephalomalacia: : Degenerative brain disorder seen in horses and pigs.

Leukoplakia : White thickened patch occuring on mucous membranes of lips, mouth and genitalia due to syphilis or pre-cancerous changes.

Liniment : A medicinal preparation that is rubbed onto the skin or applied on a surgical dressing.

Lipolytic : Relating to or causing lipolysis (Hydrolysis of lipids).

Lochia	:	The normal uterine discharge of blood, tissue, and mucous from the vagina after childbirth.
Leukopenia	:	A condition in which the number of leucocytes (White blood cells) is reduced.
Lumbago	:	Incapacitating pain low down in the back; a symptom of fibrositis of lumbar muscles, spondylitis and prolapsed intervertebral disc.
Lupus	:	Any of several chronic skin diseases.
Lymphodentitis	:	Inflammation of a lymph node.
Lymphocyte	:	A type of white blood cell.
Lymphoblastic leukaemia	:	Malignant lymphoma in which single or multiple tumours arise from lymphoblasts in lymph nodes.
Lymphoid leukaemia	:	A malignant proliferation of leucocytes in the lymph.
Lymphoma	:	Cancer of the lymphocytes.
Mad Cow disease	:	See Bovine spongiform encephalopathy.
Mastoditis	:	Inflammation of the air cells of the bone behind the ears.
Melancholy	:	Term in psychiatry to mean severe forms of depression.
Meningitis	:	Inflammation of the meninges due to infection by viruses or by bacteria responsible for pneumonia, syphilis or tuberculosis.
Menorrhagia	:	Painful menstruation.
Metastatic/Metastasis	:	Spreading of disease from one part of the body to another.
Metrorrhagia	:	Bleeding from the womb when menstruation is not due.
Micturition	:	The act of passing urine.
Migraine	:	A recurrent throbbing headache that characteristically affects one side of the head.
Mink encelphalopathy	:	A prion spongiform encephalopathy
Mitogenic	:	Any substance which causes mitosis. (division of the cells)
Multiple sclerosis	:	Abnormal hardening or fibrosis of a tissue.

Monoclonal antibodies	:	Identical antibodies.
Mumps	:	A common viral infection mainly affecting children between the age of five and fifteen.
Mutagenic	:	An agent which can increase the rate of mutation.
Myalgia	:	Pain in the muscles.
Mycosis	:	Disease caused by any fungus.
Mydriatic	:	Drug which causes dilation of the pupil.
Myeloid	:	Relating to the granulocyte precursor cells in the bone marrow.
Myeloma	:	Cancer of plasma cells.
Myelosuppression	:	Suppression of myclocytes resulting in myeloma.
Myocardium	:	Middle layer of the heart wall.
Myocarditis	:	Inflammation of the myocardium.
Myopia	:	Short sightedness.
Nanotechnology	:	Is a field of applied science at the nano scale to design, synthesize and characterize materials and devices.
Nasopharyngeal carcinoma	:	A malignant tumour arising in the epithelial lining of the nasopharynx.
Neonatal	:	Pertaining to first four weeks after birth.
Neoplasm	:	Any new growth; refers to pathological process in tumour formation.
Nephritis	:	Inflammation of the kidney.
Nephropathy	:	Diseases of the kidney.
Nephrotic syndrome	:	A chronic condition of unknown cause characterized by massive oedema and heavy proteinuria, resulting in reduction in blood plasma albumen.
Nettlerash	:	Weals of the skin.
Neuralgia	:	Paroxysmal pain extending along the course of one or more nerves.
Neutraceuticals	:	Natural substances found in food and have medicinal properties to treat or prevent diseases.

Neutropenia	:	Shortage of neutrophils
Nocardiosis	:	A disease in humans caused by *Nocardia asteroides*, characterised by pulmonary lesions that may be subclinical or chronic and may spread to other organs of the body, especially the brain.
Nocturnal emissions	:	Emission during the night.
Odontalgia	:	Toothache.
Odontopathy	:	Any diseases of the teeth.
Onyalai	:	A form of thrombocytopenic purpura due to a nutritional disorder occuring in blacks in Africa.
Onychomycosis	:	A fungal infection of the nails.
Opthalmic/Opthalmia	:	Pertaining to the eye
Opthalmopathy	:	Relating to disease of the eye.
Orchitis	:	Inflammation of the testis.
Osteoarthritis	:	Degenerative arthritis.
Osteomalacia	:	A disease occurring from a deficiency of vitamin D or calcium and characterized by softening of bones accompanied with pain and weakness.
Osteoporosis	:	Loss of bone density due to excessive absorption of calcium and phosphorus from the bone due to loss of protein matrix of bone.
Otalgia	:	Pain in the ear.
Otitis	:	Inflammation of the ear.
Otopathy	:	Diseased condition of the ear.
Otorrhoea	:	Discharge from the external auditory meatus.
Palpitation	:	Rapid forceful beating of the heart of which the patient is conscious.
Papillary carcinoma	:	Malignant growth of a papilla.
Papilloma	:	A tumour arising from a non glandular epithelial surface.
Paralytic shell fish poisoning	:	Shell fish poisoning leading to paralysis.

Parkinsonism : A disorder of middle-aged and elderly people characterized by tremor, rigidity, and a poverty of spontaneous movements.

Papule : A small circumscribed elevation of the skin.

Parturition : The process of giving birth to a child.

Pericarditis : Inflammation of the outer serous covering the heart.

Periodontal : Treatment of tissues associated with teeth and gum diseases.

Peritonitis : Is an inflammation of the peritoneum- the membrane that lines the wall of the abdomen, covering the abdominal organs.

Pharyngitis : Inflammation of the part of the throat behind the soft palate.

Phlebitis : Inflammation of a vein

Phthisis : An old term for pulmonary tuberculosis.

Pleurisy : Inflammation of the pleura (serous membrane investing the lungs and lining the walls of the thoracic cavity).

Pneumonitis : Inflammation of lung tissue.

Polyp : A pedunculated tumour arising from any mucous surface eg. cervical, uterine, nasal etc., usually benign but may become malignant.

Polyuria : The production of large volumes of urine, which is dilute and of a pale colour.

Post natal : After delivery.

Prophylactic : Pertaining to the prevention or the development of diseases.

Proteolytic : An enzyme that promotes proteolysis (the splitting of proteins by hydrolysis of the peptide bonds with formation of smaller polypeptides).

Protothecosis : Infection caused by *Prototheca*, varying from cutaneous lesions to systemic invasion, occuring as an opportunistic infection.

Pruritus	:	Itching, caused by local irritation of the skin or sometimes nervous disorders.
Psoriasis	:	A chronic skin disease in which itchy scaly red patches form on the elbows, forearms, knees, legs, scalp and other parts of the body.
Pulmonary oedema	:	Oedema of the lungs.
Pulmonary tuberculosis	:	Any infectious disease caused by the bacillus *Mycobacterium tuberculosis* and characterised by the formation of nodular lesions (tubercles) in the tissues.
Purgative	:	A drug causing evacuation of fluid faeces.
Purpura	:	A disorder characterized by spontaneous extravasation of blood from the capillaries into the skin, manifested by red spots or oozing.
Purulent	:	Forming, consisting of or containing pus.
Pustules	:	A small pus containing blister on the skin.
Putrifaction	:	The process of decomposing or disintegration.
Pyorrhoea	:	A flow of pus, usually referring to that from teeth sockets.
Pyuria	:	Pus in the urine.
Q fever	:	Febrile disease caused by *Coxiella burnetti.*
Relapsing fever	:	Fever after apparent recovery from bout of fever.
Renal calculi	:	Kidney stones.
Resolvent	:	Promoting resolution or the dissipation of a pathological growth.
Restenosis	:	Reccurrence of abnormal narrowing in a blood vessel or other tubular organ.
Rheumatism	:	Disease of connective tissue associated with pain, stiffness, swelling of muscles and joints.
Rheumatoid arthritis	:	An autoimmune disease causing chronic joint inflammation.
Rhinitis	:	Inflammation of the mucous membrane of the nose.

Rickettsial pox : An infectious disease spread by a mite and characterised by a chicken pox like rash.

Rocky mountain fever : A disease which results in abrupt symptoms such as head ache, fever, chills, muscle pain, inflammation of blood vessels and blockage by blood clots. First reported from rocky mountain state.

Rubefacient : An agent that causes reddening or warming of the skin; often used as counter irritants for relief from muscular pain

Saccharification : Process of converting a substance to sugar.

Sarcoma : Malignant growth of mesodermal tissue (Connective tissuc, muscle, bone).

Scabies : A parasitic contagious skin disease caused by itch mite.

Sciatica : Pain in the line of distribution of the sciatic nerve in the leg.

Scirrhus/Cirrhus tumour : A cancerous tumour which is grey or blue in colour and emits a creaking sound when incised.

Scrapie : A viral disease affecting sheep and goats resulting in itching.

Scrofula : Tuberculosis of lymph nodes, usually those in the neck, causing the formation of abscesses.

Scrub typhus : Same as Tsutsugamushi fever.

Scurvy : A disease caused by lack of vitamin C (ascorbic acid); clinical features include fatigue and haemorrhage.

Seborrheic dermatitis : Greasy condition of the scalp, face, sternal region due to over activity of sebaceous glands leading to various conditions like seborrheic dermatitis.

Sepsis : Infected with pus producing organisms

Septicaemia : The persistence and multiplication of living bacteria in the blood stream.

Serotype : A type which tests for recognizable antigens on the surface.

Shingels (Herpes)	:	Vesicular eruption due to Herpes viral infection.
Sick building syndromes:	:	An illness affecting workers in office buildings, characterised by skin irritations, headache and respiratory problems.
Silicosis	:	Also known as Grinder's diseases or Potter's rot is a form of occupational lung disease caused by inhalation of silica dust.
Sinews	:	A ligament or tendon.
Sinusitis	:	Inflammation of a sinus.
Spasms	:	Convulsive, involuntary muscular contraction.
Spermatorrhoea	:	Involuntary discharge of semen without orgasm.
Splenomegaly	:	Enlargement of the spleen.
Stomachic	:	Agent which increases the appetite.
Stomatitis	:	Inflammation of the mouth
Strangury	:	Slow and painful urine flow.
Stupor	:	A state of marked impairment without complete loss of conciousness.
Stye	:	An abscess in the follicle of an eyelash.
Styptic	:	An astringent applied to stop bleeding.
Subacute Sclerosing Panencephalitis (SSPE)	:	A neurodegenerative disease caused by persistant 'slow virus infection' with a mutated measles virus.
Sudorific	:	A drug that causes an increase in sweating.
Suppuration	:	The formation of pus.
Swimmer's itch	:	Itchy rash on the skin also known as Cercarial dermatitis.
Syphilis	:	A chronic venereal disease caused by *Treponema pallidum* resulting in the formation of lesions throughout the body.
T-cell	:	A type of white blood cell which plays a central role in immunity. T means thymus, the organ in which T cell finally develops.
Tachycardia	:	Excessively rapid action of the heart.

Taenifuge : An agent that expels tapeworms.

Teeth caries : Cavity in the teeth.

Teratogenic : Any substances, agent or process that induces the formation of developmental abnormalities in a foetus.

Tetanus : An acute infectious disease, affecting the nervous system, caused by the bacterium *Clostridium tetani*, in which muscles contract and stiffen.

Thrombocytopenic purpura: Life threatening disorder in which blood does not clot like it should, due to platelet count.

Thrombosis : Intravascular formation of a blood clot.

Thrush : Acute pseudomembranous candidiasis

Thymus : A gland lying behind the breast bone.

Tonsillitis : Inflammation of the tonsils.

Toxaemia : A generalized poisoning of the body due to bacteria or damaged tissue.

Tracheotomy : A surgical procedure performed on the neck to open a direct airway through an incision in the trachea.

Tremor : A rhythamical alternating movement that may affect any part of the body.

Trench fever : Is a blood borne infection.

Tridosha : An imbalance of the three elements - Vata, Pitta and Kafa in Ayurvedic system of medicine.

Tsutsugamuschi fever (Scrub fever) : Mite born infection with fever, head ache and raised rashes, swollen glands and dark coloured ulcers.

Tubercular glands : Small rounded swellings or nodules not necessarily caused by tuberculosis.

Tubular nephrosis : Degenerative non-inflammatory change in the tubules of the kidney.

Tumours : A new growth of tissue due to uncontrolled and progressive multiplication of cells.

Tympanitis : Abdominal distension due to accumulation of gas in the intestine.

Typhus fever	:	Disease caused by lice borne bacteria with macular eruption on the upper trunk spreading all over the body, usually not on the face, palms or soles; rash starts as maculopapular becoming hemorrhagic.
Ulcerative colitis	:	Inflammatory and ulcerative condition of the colon.
Urethritis	:	Inflammation of the urethra.
Urethrorrhagia	:	Flow of blood from the urethra.
Urethrorrhoea	:	Discharge from the urethra.
Urrorrhoea	:	Urinary infection.
Urticaria	:	A vascular reaction of the upper epidermis marked by red elevated patches with severe itching.
Vaginitis	:	Inflammation of the vagina characterized by intensely irritating discharge due to ciliated protozoans inhabiting the bowel.
Varicose veins	:	Dilated veins, the valves of which become incompetent so that the blood flow may be reversed; most commonly found in the lower limbs.
Vascular tumour	:	Benign growth in the circulatory system.
Vasoconstrictor	:	An agent which causes narrowing of the lumen of blood vessels.
Venereal	:	Pertaining to or caused by sexual intercourse.
Venerupin poisoning	:	A form of shell fish poisoning that may cause death.
Ventilation pneumonitis	:	Stiffness in the lung along with inflammation of lung tissue
Vermifuge	:	Any drug or chemical agent used to expel worms from the intestine.
Verrucous	:	Low grade variant of squamous cell carcinoma.
Vertigo	:	A sensation of rotation or movement of one's self or one's surroundings in any plane sometimes used erroneously to mean any form of dizziness.

Vesicular rickettsiosis	:	Same as Rickettsial pox.
Vitiligo	:	Leucoderma.
Vulnerary/Vulnerant	:	Drugs used for healing wounds.
Vulvar pruritus	:	Itching in the vulva.
Warts	:	A small (often hard) benign growth in the skin.
Wens	:	A sebaceous cyst.
Whitlow	:	A purulent infection involving the pulp of the distal phalange of the finger.
Whooping cough	:	An acute contagious disease, primarily affecting children, due to infection of the mucous membrane lining the air pasages by the bacterium *Haemophilus pertussis*.
Yaws	:	Disease characterized by yellowish or reddish tumours.

REFERENCES

Acheson, N.H. 2007. Fundamentals of Molecular Virology. John Wiley and Sons, Inc.

Agarwal, R. 2004. Flavours and Aromas. In: Fungal Biotechnology in Agricultural, Food and Environmental Applications. Arora, D.K. (Ed.). Vol. 21, Pp. 281-286. Marcel Dekker, Inc., New York.

Agrios, G.N. 1997. Plant Pathology. 4th Edition. Academic Press, London, UK.

Akhtar, M., Attridge, M.C., Blanchette, R.A., Myers, G.C., Wall, M.B., Sykes, M.S., Koning, Jr. J.W., Burgess, R.R., Wegner, T.H. and Kirk, T.K. 1992. The White rot fungus *Ceriporiopsis subvermispora* saves electrical energy and improves strength properties during biochemical pulping of wood. In: Biotechnology in the Pulp and Paper industry. Kuwahara, M. and Shimada, M. (Eds) Proceedings of the 5th International conference on Biotechnology in the pulp and paper industry, Tokyo, Uni publishers, pp. 3-8.

*Akhtar, M., Attridge, M.C., Meyers, G.C., Kirk, T.K and Blanchette, R.A. 1992. *TAPPI J.*, 75: 105-108.

Akin, D.E. 1993. In: Genetics, Biochemistry, Ecology of lignocellulose degradation Shimada, K., Ohmiya, Y., Kobayashi, S., Hoshino, K., Sakka and Karita, S. (Eds.) UNI Tokyo. pp 95-103.

Al Hasan, R.H., Sorkhoh, N.A., Al Bader, D. and Radwan, S.S. 1994. *Appl. Microbiol. Biotechnol.*, 41:615-619.

Alexander, M. 1994. Biodegradation and Bioremediation. Academic Press, San Diego, California.

*Alsberg, C.L. and Black, O.F. 1913. Contributions to the study of maize deterioration. Biochemical and toxicological investigations of *Penicillium puberulum* and *Penicillium stoloniferum*. *Bur. Plant Ind., Bull.*, No.270, Govt. Printing off., Washington, DC.

Ambasta, S.P. 1986. The Useful Plants of India. Publications and Information Directorate, CSIR, New Delhi, India.

Amma, T., Puri, S.C., Verma, V., Sharma, J.P., Khajuria, R., Musarrat, J., Spiteller, M. and Oazi, G.N. 2006. Bioreactor studies on the endophytic fungus *Entrophospora infrequens* for the production of an anticancer alkaloid camptothecin. *Can. J. Microbiol.,* 52:189-196.

Ammon, H., Ege, W., Oppermann, M., Gopel, W. and Eisele, S. 1995. Improvement in the long term stability of an amperometric gluocose sensor system by introducing a cellulose membrane of bacterial origin. *Anal. Chem.*, 67:466-471.

*Anderson, J.H. 1985. *Ann. Allergy.*, 54:390.

Anonymous. 1994. IUCN Red List Categories. IUCN, Switzerland.

Anonymous. 1999. IUCN Red List Criteria Review Provisional Report: Draft of the Proposed Changes and Recommendations. *Species*, 31-32:43-57.

Aquaron, R., Delange, F., Marchal, P., Longnone, V. and Ninane, L. 2002. Bioavailability of seaweed iodine in human beings. *Cell Mol. Biol., 48*(5): 563-9.

Arceci, R.J., Stieglitz, K. and Bierer, B.E. 1992. Immunosuppressants FK506 and rapamycin function as reversal agents of the multidrug resistance phenotype. *Blood.* 80:1528-1536.

Arora, D.K 2004. Handbook of Fungal Biotechnology. 2nd Edition. Vol. 20, Marcel Dekker, Inc., New York.

Arora,D.K.(Ed.) 2004. Fungal Biotechnology in Agricultural, Food and Environmental Applications. Vol, 21., Marcel Dekker, Inc., New York.

Arunasri, K., Sasikala, C.H., Ramana, C.V., Suling, J. and Imhoff, J.F. 2005. *Int. J. Syst. Evol. Microbiol.*, 55: 673-679.

Ashoka, C., Manjunath, K. and Sullia, S.B. 2000. Biological treatement of combined textile dye effluent. In: Ecology of Fungi. Bhat, D.J. and Raghykumar, S. (Eds.) pp. 63-67.

Ashoka, C., Geetha, M.S. and Sullia, S.B. 2002. Bioleaching of composite textile dye effluent using bacterial consortia. *Asian Jr. Microbiol. Biotech. Env. Sci.*, 4: 65-76.

Atlas, R.M. and Unterman. 1999. Bioremediation: Industrial Microbiology and Biotechnology. 2nd Edition. 666-681. A5M Press, Washington.

Atmaca, M., Tezcan, E., Kuloglu, M., Ustundag, S. and Kirtas, O. 2005. The effect of extract of *Ginkgo biloba* addition to olanzapine on therapeutic effect and antioxidant enzyme levels in patients with schizophrenia. *Psychiatry Clin. Neurosci.,* 59 (6):652-656.

Avalos, J. and Olmedo, E.C. 2004. Fungal Carotenoid Production. In: Handbook of Fungal Biotechnology 2nd Edition. Arora, D.K. (Ed.). Vol. 20. Pp. 367-375. Marcel Dekker, Inc., New York.

Avichezer, D. and Garber, G.N. 1983. *Israel J. Med. Sci.*, 19:680.

Bacon, C. and White, J.F. (Eds.). 2000. Microbial Endophytes. Marcel Dekker, New York.

Bagai, R. and Madamwar, D. 1997. *Appl. Biochem. Biotechnol.*, 62:213-218.

Bagratee, J.S. and Moodley, J. 1990. Synthetic *Laminaria* tent for cervical ripening. *S. Afr. Med. J.,* 78 (12):738.

Balakrishna, P. 1999. Biological conservation and impact assessment. *Curr. Sci.,* 76: 129-131.

Ballio, A., Bossa, F. and Digiogio, P. 1994. Structure of the pseudomycins, new lipodepsipeptides produced by *Pseudomonas syringae. FEBS Lett.,* 355:96-100.

Banat, I.M. 1993. The isolation of the thermophilic biosurfactant producing *Bacillus* sp. *Biotech. Lett.*, 15: 591-594.

Banerjee, A. and Banerjee, N. 2006. Fundamentals of Microbiology and Immunology. New Central Book Agency (P) Ltd., New Delhi, India.

Barr, D.P. and Aust, S.D. 1994. Mecanisms of the white rot fungi used to degrade pollutants . *Env. Sci. Techonol.*, 28:79A-87A.

Bashyal, B., Li, J.Y., Strobel, G.A. and Hess, W.M. 1999. *Seimatoantlerium nepalense,* an endophytic taxol producing coelomycete from Himalayan yew (*Taxus wallichiana*). *Mycotaxon.* 72:33-42.

*Becker, L.E. and Skipworth, G.B. 1975. *JAMA.* 231:1162.

Beddome, R.H. 1970. The Ferns of Southern India. Today and Tomorrow's Printers and Publishers. New Delhi, India.

Bedzan, L., Biliekiewicz, A. and Turczynski, J. 2005. Preliminary assessment of *Ginkgo biloba* (Ginkofar) in patients with dementia. *Psychiatr Pol.,* 39 (3):559-566.

Begum, A.S., Martin, K.P., Zhang, C., Nishitha, J.K., Ligimol, Slater, A. and Madhusoodanan, P.U. 2007. Organogenesis from leaf and internode explants of *Ophiorrhiza prostrata*, an anticancer drug (camptothecin) producing plant. *Electronic. J. Biotech.*, 10 (5).

*Behrens, P.W., Sicotte, V.J. and Delente, J. 1994. *J. Appl. Phycol.,* 6:113-121.

*Belay, A., Ota, Y., Niyakawa, K. and Shimamatsu, H. 1993. *J. Appl. Phycol.,* 5:235-241.

*Benedict, H. 1999. Arabinogalactans: New immune boosting fiber. *Natural Pharmacy.* 3:12.

Benitez, T. and Codon, A.C. 2004. Ethanol -Tolerance and Production by Yeast. In: Handbook of Fungal Biotechnology 2nd Edition. Arora, D.K. (Ed.). Vol. 20. Marcel Dekker, Inc., New York.

Bennett, J.W. and Faison, B.D. 1997. Use of Fungi in Biodegradation. In: Environmental Microbiology, ASM Press, Washington.

Bentley, R. 1997. Microbial secondary metabolites play important roles in medicine; prospects for discovery of new drugs. *Persp Biol. Med.*, 40 : 364-394

Bessey, E.A. 1950. Morphology and Taxonomy of Fungi. Blakiston Company, Philadelphia.

Bhatia, K.N. 1984. A Treatise on Algae. 4th Edition. R. Chand and Co. Publishers. New Delhi, India.

Bhatnagar, S.P. and Moitra, A. 1996. Gymnosperms. New Age International Pvt. Ltd., Publishers, New Delhi, India.

Blanchette, R.A., Akhtar, M. and Attridge, M.C. 1992. Using Simons stain to evaluate fibre characteristics of biomechanical pulps. *Tappi J.*, 75: 121-124.

Blumenthal, M. and King, P. 1995. MaHuang: Ancient Herb, Modern Medicine. Regulatory Dilemma. *Herbal Gram.*, 34:22.

Blunden, G., Rogers, D.J. and Farnham, F. 1975. *Lloydia.* 38: 162.

Boelsma, E., Lamers, R.J., Hendriks, H.F., Van Nesselrooij, J.H. and Roza, L. 2004. Evidence of the regulatory effect of *Ginkgo biloba* extract on skin blood flow and study of its effects on urinary metabolites in healthy humans. *Planta Med.,* 70 (11): 1052-1057.

Boericke, W. 1991. Materia Medica With Repertory. Pocket Manual of Homoeopathy Materia Medica. Jain Pub. Pvt. Ltd., New Delhi, India.

Bog-Hansen, T. C. and Freed, D. L.J. (Eds.) 1987. Lectins, Biology, Biochemistry, Clinical Biochemistry. Vol 6. Proceedings of the 9th Lectin meeting Cambridge, Sigma Chemical Company, USA.

*Bolwell, G.P., Callow, J.A., Callow, M.E. and Evans, L.V. 1979. *J. Cell Sci.*, 36:19.

*Boot, R.G., van A. Chterberg, T.A., van Aken, B.E., Renkema, G.H., Jacobs, M.J., Aerts, J.M. and de Vries, C.J. 1999. *Arterioscler. Thromb. Vasc. Biol.*, 19: 687-694

* Borel, J.F., Feurer, C., Gabler, H.U. and Stahelin, H. 1976. Biological effects of cyclosporin A: a new antilymphocytic agent. *Agents and Actions*. 6:468-475.

Borgotta, L. and Barad, D. 2004. Prolonged retention of *Laminaria* fragments: an unusual complication of *Laminaria* usage. *Obstet Gynecol.* ,78: 988.

* Borman, S. 2002. Potential treatment for prion diseases. *Chem. Eng News.*, 80 (33):42.

Boruch, T., *et al.,* 1984. Extracts of plants and their cosmetic application, Part V. Extracts from *Equisetum arvense. Chem. Abstracts.,* 100. q.f. The Review of Natural Products. Marderosian, A.D. and Beutler, J.A. (Eds.) 2002 3rd Edition. Missouri, USA.

Boyd, W.C., Almodovar, L.R. and Boyd, l.g. 1966. *Transfusion*. 6:82.

Boyetchko, S.M. and Peng, G. 2004. Challenges and Strategies for Development of Mycoherbicides. In: Fungal Biotechnology in Agriculultral, Food and Environmental Applications. Arora, D.K. (Ed.) Vol. 21. Marcel Dekker Inc., New York.

*Brasch, J. and Jacobsen, P. 1991. *Flechten und ihre Allergene. Allergologie*, 14: 99.

Brazil, M. 2002. A new use for statins. *Nature Revs/Drug Dis.,* 1:934.

Breinbauer, R., Vetler, I.R. and Waldmann, H. 2002. From protein domains to drug candidates-natural products as guiding principles in the design and synthesis of compound libraries. *Angewchem. Int Ed. Engl.,* 41:2879-2890.

Brett, B. and Middle brooks, E.J. 1986. Pollution control for the chemical industry. Lewis Publishers, Inc., New York.

*Brown, A.G. 1986. Clavulanic acid, a novel β-lactamase inhibitor-a case study in drug discovery and development. *Drug Design Devel.,* 1:1-21.

*Brown, A.G., Smale, T.C., King, T.J., Hasenkamp, R. and Thompson, R.H. 1976. Crystal and molecular structure of compactin, a new antifungal metabolite from *Penicillium brevicompactum. J. Chem. Soc. Perkin., Trans I.,* 1165-1170.

Brown, D.W., McCormick, S.P., Alexander, N.J., Proctor, R.H. and Desjardins, A.E. 2001. A genetic and biochemical approach to study trichothecene diversity in *Fusarium sporotrichioides* and *Fusarium graminearum. Fungal Genet Biol.,* 32: 121-133.

Bruce, N.C., Wilmot, C.J., Jordan, K.N., Trebilcock, A.E., Stevens, G.L.D. and Lowe, C.R. 1990. Microbial degradation of the morphine alkaloids; identification of morphinone as an intermediate in the metabolism of morphine by *Pseudomonas putida* M10. *Arch. Microbiol.*, 154: 465-470.

Bumpus, J.A. 2004. Biodegradation of Azo Dyes by Fungi. In: Fungal Biotechnology in Agricultural, Food and Environmental Applications. Arora. D.K. (Ed.). Vol.21. Marcel Dekker Inc., New York.

Bumpus, J.A. and Aust, S.D. 1987. *Appl. Environ. Microbiol.*, 53:2001-2008.

*Bumpus, J. A. and Tatarko, M. 1994. *Curr. Micobiol.*, 28: 155-190.

Bungay, H.R., Serafica, G. and Mormino, R. 1997. Environmental Implications of Microbial Cellulase. In: Global environmental Biotechnology. Wise, D.L. (Ed.). Elsevier Scince, Netherlands.

Cannon, R.E. and Anderson, S.M. 1991. Biogenesis of bacterial cellulose. *Crit. Rev. Microbiol.*, 17:435-447.

Cardarelli, M., Serino, G., Campanella, L., Ercole, P., De Cicco-nardone, F., Alesiani, O. and Rossiello, F. 1997. Anti-mitotic effects of usnic acid on different biological systems. *Cell. Mol. Life Sci.,* 53: 667-672.

Carlos, A.G. and Carlos, M.L. 2004. Candidiasis. In: Handbook of Fungal Biotechnology. 2nd Edn. Arora, D.K. (Ed.). Vol. 20. pp. 503-509. Marcel Dekker, Inc., New York.

*Carmeli, S., Paik, S., Moore, R.E., Patterson, G.M.L. and Yoshida, W.Y. 1993. *Tetrahedron Lett.,* 34:6681-6684.

*Carmichael, W.W and Gorham,P.R. 1977. Factors influencing the toxicity and animal susceptibility of *Anabaena flos-aquae* (cyanophyta) blooms. *J. Phycol.*, 13: 97-100.

Carroll, G. C. and Wicklow, D. T. 1992. The Fungal Community. Its Organization and role in the Ecosystem. 2nd Edition. Marcel Dekker, Inc. New York.

CAST. 2003. Mycotoxins: risks in plant, animal and human systems. (Richard, J.L., Payne, G.A. (Eds.) Pp. 199. Council for Agricultural Science and Technology Task Force Report No 139, Ames, Iowa.

Castillo, U.F., Strobel, G.A. and Ford, E.J. 2002. Munumbicins, wide-spectrum antibiotics produced by *Streptomyces* NRRL 30562, endophytic on *Kennedia nigriscens. Microbiology.* 148:2675-2685.

Castillo, U. F., Harper, J.K. and Strobel, G.A. 2003. Kakadumycins, novel antibiotics from *Streptomyces* NRRL 30566, an endophyte of *Grevillea pteridifolia. FEMS Lett.,* 224:183-190.

Castrillo, J.I. and Ugalde, V. 2004. Mycoprotein and Related Microbial Protein Products. In: Fungal Biotechnology in Agriculatural, Food, and Environmental Applications. Arora, D.K. (Ed.). Vol. 21, Pp. 247-254. Marcel Dekker, Inc., New York.

Chadha, B.S., Sonia, K.G., Badhan, A.K., Saini, H.S. and Bhat, M.K. 2005. Heterogenecity of xylanases and related enzymes. In: Microbial Diversity. Current Perspectives and Potential Applications. Satyanarayana, T. and Johri, B. N. (Eds.). I.K International Pvt. Ltd., New Delhi.

Chamberlain, C.J. 1986. Gymnosperms-Structure and Evolution. Indian Edition. CBS Publishers and Distributors, New Delhi, India.

Chandran, S. and Pandey. A. Role of glucoamylases in Biotechnology. 2005. In: Microbial Diversity: Current Perspectives and Potential Applications. Satyanarayana, T and Johri, B.N. (Eds.) I. K. International Pvt. Ltd. New Delhi.

Chang, C.H., Lee, S.R., Lee, K.H., Mheen, T.I., Kwon, T.W. and Park, K.I. 1977. Fermented Soybean foods. Symposium on Indigenous Fermented Foods, Bangkok, Thailand.

Chapman, V.J. and Chapman, D.J. 1980. Sea Weeds and Their Uses. 3rd Edition. Chapman and Hall, London.

Chauhan, K.P.S. 1993. Research options and priorities for biodiversity conservation: forests, wetlands and deserts. Frame,B., Victor,J. and Joshi, Y. (Eds.) TERI, New Delhi, India.

*Chauvet, J.M., Comtat, J. and Nue, P. 1987. Assistance in bleaching of never dried pulp by the use of xylanase, consequences on pulp properties. Proc. *4th Int Symp Wood Pulp Chem.,* Paris. 4: 27-30.

Chevallier, A. 1996. Encyclopedia of Medicinal Plants. Pp.224.DK Publishing. New York .

Chopra, R.N. and Kumra, P.K. 1988. Biology of Bryophytes. New Age International Publishers, New Delhi, India.

Chopra, R.N., Nayar, S.L. and Chopra, I.C. 1956. Glossary of Indian Medicinal Plants. Publications and Information Directorate, CSIR.

*Chorney, W., Sculley, N.J., Crepi, H. and Katz, J.J. 1960. *Biochem. Biophys. Acta.,* 37:280.

Chow, T., Browne, V., Heileson, H.L., Wallace, D., Anholm, J. and Green, S.M. 2005. *Ginkgo biloba* and acetazolamide prophylaxis for acute mountain sickness: a randomized, placebo-controlled trial. *Arch. Intern. Med.,*165(3):296-301.

Chow, V.T.K., Inglis, T.J.J. and Song, K.P. 2006. Microbes: Friends or Foes? Chapter 15. In: Microbial Biotechnology Principles and Applications. Kun, L.Y. (Ed.) World Scientific Publishing Co Ltd., Singapore.

*Christian, W.F.K. 1970. Lactic acid bacteria in fermenting maize dough. *Ghana J. Sci.,* 10: 22-28,

Chu, F.S. 2002. Mycotoxins. In: Food Borne Diseases, 2nd Edition. Cliver, D.O. and Richmann, H.P. (Eds.) Pp. 271-303.: Academic Press, Inc., New York.

Chu, F.S. and Bhatnagar, D. 2004. Mycotoxins. In : Fungal Biotechnology in Agricultural ,Food and Environmental Applications. Arora,D.K.(Ed.) Vol,21., Marcel Dekker, Inc., New York.

Chu, L.M. and Wong, P.K. 2006. Muncipal and Industrial Solid Waste Treatment. Chaper 18. In: Microbial Biotechnology, Principles and Applications. Kun, L.Y. (Ed.) Scientific Publishing Co. Ltd., Singapore.

Cintin, C., Johasen, J.S., Christensen, I.J., Price, P.A., Sorensen, S. and Nielsen, H.J. 1999. *J. Cancer.,* 79: 1494-1499.

*Cocchietto, M., Skert, N., Nimis, P.L. and Sava, G. 2002. A review on usnic acid, an interesting natural compound. *Naturwissenchaften.* 89: 137-146.

Cohen, Z. 1986. 'Products from Microalgae'. In: CRC Handbook of Microalgal Mass culture. Richmond, A. (Ed.) pp. 421-454. CRC Press. Florida, USA .

Cook, A.A. 1975. Diseases of Tropical and Subtropical Fruits and Nuts. Hafner Press.

Correche, E.R., Carrasco, M., Escudero, M.E., Velaquez, L., De Guzman, A.M.S., Giannini, F., Enriz, R.D., Jauregui, E.A., Cenal, J.P. and Giordano, O.S. 1998. Study of the cytotoxic and antimicrobial activities of usnic acid and derivatives. *Fitoterpia.* 69: 493-501.

Cragg, G.M., Boyd, M.R., Cardellina, J.H., Newman, D.J., Snader, K.M. and McCloud, T.G. 1994. Ethnobotany and Drug Discovery: The Experience of the US National Cancer Institute. In: Ethnobotany and The Search for New Drugs . Chadwick, D.J. and Marsh, J. (Eds.) CIBA Foundation Symposium, Wiley and Sons, UK.

Crueger, W. and Crueger, A. 2005. Biotechnology. A Text Book of industrial Microbiology. 2nd edition. Panima Publishing Corporation, New Delhi, India.

Crusberg,T.C., Mark, S.S. and Dilorio, A. 2004. Biomineralization of Heavy Metals. In : Fungal Biotechnology in Agricultural, Food and Environmental Applications. Arora,D.K.(Ed.) Vol,21. Pp. 409-415, Marcel Dekker, Inc., New York.

Curran, B. and Bugeja, V. 2005. The Biotechnological Exploitation of Heterologous Protein Production in Fungi. In: Fungi, Biology and Applications. Kavanagh, K. (Ed.) John Wiley and Sons Ltd., U.K.

*Dainippon Ink and Chemicals Inc. 1980. Production of highly purified alcoholophilic phycocyanin. Japanese Patent 80, 77, 890.

Daly, M.J. 2000. *Curr. Opin. Biotechnol.,* 11: 280-285.

Davis, M.W., Glaser, J.A., Evans, J.W. and Lamar, R.T. 1993. Field evaluation of lignin degrading fungus *Phanerochaete sordida* to treat creosote contaminated soil. *Environ. Sci. Technol.,* 27: 2572-76.

*Davis, P. US Patent 1969. Fermentation of 9-(β-d-arabinofuranosyl) adenine. 1159290, 26JUL69.

Dazzo, F.B., Kijne, J.W., Haahtela, K. and Korhonen, T.K. 1986. Fimbriae, Lectins and Agglutinins of Nitrogen Fixing Bacteria. In: Microbial Lectins and Agglutinins. Properties and Biological Activity. Mirelman, D. (Ed.) John Wiley and Sons, New york.

*De, P.K. 1939. The role of blue green algae in nitrogen fixation in rice fields. *Proc. Res. Soc. Land.,* 127:121-193.

Demain, A.A. and Zhang, L. 2005. Natural Products and Drug Discovery. In : Natural Products: Drug Discovery and Therapeutic Medicine. Zhang, L. and Demain, A.L. (Eds.). Humana Press Inc., Totowa, New Jersey.

Der Marderosian, A.H. and Liberti, L.E. 1988. Natural Product Medicine. G.F. Stickley Co., Philadelphia.

Desikachary, T.V. 1959. Cyanophyta: Monograph. Indian Council of Agricultural Research, New Delhi, India.

Di Domenico, B. 1999. Novel antifungal drugs. *Curr. Opin. Microbiol.,* 2:509-515.

Dietz, E., Bohmer, G., Solter, J., Uhlenbruck, G., Beuth, J., Ko, H.L., Lutticken, R. and Pulverer, G. 1988. Adhesion of bacteria and tumours: Inhibition by lectin blocking. 27-33. In: Lectins, Biology, Biochemistry, Clinical Biochemistry. Bog-Hansen, T. C. and Freed, D. L.J. (Eds.) 1987. Vol 6. Proceedings of the 9th Lectin meeting Cambridge, Sigma Chemical Company, USA.

Dixon, P.S. 1973. Biology of the Rhodophyta. Oliver and Boyd, Publishers. Edinburgh, UK.

*Doughlas, T. and Young, M. 1999. *Adv. Mater.*, 11: 679-681.

Duran, N., Justo, G.Z., Melo, P.S. 2003. Evaluation of the antiulcerogenic activity of violacein and its modulation by the inclusion complexation with β-cyclodextrin. *Can. J. Physiol. Pharmacol.,* 81:387-396.

*Ekundayo, J.A. 1977. Nigerian Pito. Symposium on Indigenous Fermented Foods, Bangkok, Thailand.

Elias, C.J., Coggins, C., Alvarez, F., Fraser, I.S., Lacarra, M., Lahteenmaki, P., Massai, R., Mishell, D.R.Jr., Phillips, D.M. and Salvatierra, A.M. 1997. Colposcopic evaluation of a vaginal gel formulation of ε carrageenan. *Contraception.* 56 (6):387-389.

Elsabagh, S., Hartley, D.E., Ali, O., Williamson, E.M. and File, S.E. 2005. Differential congnitive effects of *Ginkgo biloba* after acute and chronic treatment in healthy young volunteers. *Psychopharmacology* (Berl). 179 (2):437-446.

*Endo, A. and Monacolin, K. 1979. A new hypocholesterolemic agent produced by a *Monascus* species. *J. Antiobiot.,* 32:852-854.

*Endo, A., Kuroda, M. and Tsujita, Y. 1976. ML-236B and ML-236C, new inhibitors of cholesterogenesis produced by *Penicillium citrinin. J. Antibiot.,* 29:1346-1348.

Erika, S., Fernandes, E. C. M. and Rakotondramasy, H.M. 2004. http:// www. erikastyger.com / Biodiversity files / Erika % 20 Styger % 20 Mycorrhiza % 20 Diversity % 20 Madagascar % 20Poster Pdf.

*Eriksson, K.E. and Vallander, L. 1982. *Svensk Papperstidning.,* 6: R33-R38.

Eriksson, K.E., Ander, P., Henningsson, B., Nilsson, T. and Goodell, B. 1976. Method for producing cellulose pulp. US patent No. 3962033.

Esteve-Nunez, A., Caballero, A. and Ramos, J.C. 2001. Biological degradation of 2,4,6-trinitrotoluene. *Micro Mol. Biol. Revs.,* 65: 335-352.

Faull, J.L., Baker, S.C., Wilkinson, S. and Nicklin, S. 2004. Fungal Degradation of Explosives. In: Fungal Biotechnology in Agricultural, Food and Environmental Applications. Arora, D.K. (Ed.). Vol. 21, Pp. 471-478. Marcel Dekker, Inc., New York.

Fenton, D.M. and Eveleigh, D.E. 1981. *J. Gen. Microbiol.,* 126: 151-154.

Fernando, T., Bumpus, J.A. and Aust, S.D. 1990. Biodegradtion of TNT (2, 4, 6-trinitrotoluene) by *Phanaerochaete chrysosporium. Appl. Environ. Microbial.,* 56: 1666-1671.

Flieger, M., Votruba, J., Kren, V., Pazoutova, S., Rylko, V., Sajid, P. and Rehacek, Z. 1988. Physiological control and process kinetics of clavinet production by *Claviceps purpurea* 129 and 129/35. *Appl. Microbiol. Biotechnol.,* 29: 181-185.

Flieger,M., Mehta,P. and Mehta, A. 2004. Biotechnological Potential of Ergot Alkaloids. In: Fungal Biotechnology in Agricultural, Food and Environmental Applications. Arora, D.K. (Ed.). Vol. 21, Marcel Dekker, Inc., New York.

Flynn, C.E., Mao, C., Hayhurst, A., Williams, J.L., Georgiou, G., Iverson, B. and Belcher, A.M. 2003. *J. Mat. Chem.,* 13: 2412-2421.

Forloni, G., Lussich, S., Awan, T. 2002. Tetracyclines affect Prion activity. *Proc. Natl. Acad. Sci.,* USA. 99:10,849-10,854.

*Fowler, C.E., Wayne, S., Stubbs, G. and Mann, S. 2001. *Adv. Mater.,* 13: 1266-1269.

*Friedman, S.J. and Skehan, P. 1980. *Proc. Natl. Acad. Sci., U.S.A.* 77: 1172-1176.

Friedrich, J. 2004. Bioconversion of Distillery Waste. In: Fungal Biotechnology in Agricultural, Food and Environmental Applications. Arora, D.K. (Ed.). Vol. 21, Pp. 431-440, Marcel Dekker, Inc., New York.

*Friedrich, V.H. and Weiedermeyer, H. 1976. (Quantitative determination of the tannin-precursors and the tannins in *Ephedra helvetica*). *Planta Medica.* 303 (3):223.

Fritsch, F.E. 1935. Structure and Reproduction of the Algae Vol. I. Cambridge University Press, Cambridge, New York.

Fritsch, F.E. 1935. Structure and Reproduction of the Algae Vol. II. Cambridge University Press, Cambridge, New York.

Fujimura, T., Tsukahara, K., Moriwaki, S., Kitahara, T., Sano, T. and Takema, Y. 2002. Treatment of human skin with an extract of *Fucus vesiculosus* changes its thickness and mechanical and developmental properties. *J. Cosmet. Sci.,* 53 (1): 1-9.

Gallagher, K.A., Healy, M.G. and Allen, S.J. 1997. Biosorption of synthetic dye and metal ions from aqueous effluents using fungal biomass. In: Global Environmental Biotechnology. D.L. Wise. (Ed.) Elsevier Science, Netherlands.

Garber, G. N. 1986. Lectins of *Pseudomonas aeruginosa*: Properties, Biological effects and Applications. In: Microbial Lectins and Agglutinins. Mirleman, D. (Ed.). John Wiley and Sons, Inc.,

Garber, N.G. and Sudakevitz, D. 1982. In : Advances in Pathology. Levy, E. (Ed.) Vol I, Pergamon Press, London.

German, B., Schiffrin, E.J., Reniero, R., Mollet, B., Pfeifer, A. and Nesser, J.R. 1999. The Development of Functional Foods: Lessons from the gut. In: Trends in Microbiology. Meran Owen. (Ed.) Vol. 17, Reference Edition, Elsevier Science, Netherlands.

*Gerniglia, C.E., Gibson, D.T. and Van Baalen, C. 1980. *J. Gen. Microbiol.,* 116:495-500.

*Gerwick, W.H., Jiang, Z.D., Agarwal, S.K. and Farmer, B.T. 1992. *Tetrahedron.* 48:2313-2324.

*Gerwick, W.H., Nagle, P.J., Nagle, D.G., Hamel, E., Blokhin, A. and Slate, D.L. 1994. *J. Org. Chem.,* 59:1243-1245.

*Ghione, M., Parrello, D. and Grasso, L. 1988. Usnic acid revisited: its activity on oral flora. *Chemioterapia.* 7: 302-305.

Ghosh, A., Paul, D., Sharma, N.K., Pandey, J., DhanPrakash, Singh, R., Goyal, A., Kaur, H and Jain, R.K. 2005. Microbial Diversity: Potential Applications in Bioremediation. In: Microbial Diversity: Current Perspectives and Potential Applications. Satyanarayana, T and Johri, B.N. (Eds.) I. K. International Pvt. Ltd. New Delhi.

Glazer, A.N. and Nikaido,H. 1995. Microbial Biotechnology. W.H. Freeman And Company, New York.

*Glazer, A.N. and Stryer, L. 1984. *Trends Biochem. Sci.,* 9:423-429.

Godtfredsen, S.E. 1990. Microbial lipases. In: Microbial Enzymes and Biotechnology. Fogarty, W.M. and Kelly, C.T. (Eds.). Pp. 255-273. Elsevier Applied Sciences. Netherlands.

*Gomez, P.J. 1949. La Chicha, Su fabricaciony algunas sugarencias tecnicas adicionables a las disposiciones legales en actual vigenica. *Notas Agronomicas, Estacion Agric. Exp. Palmira.,* 2: 20-42.

*Goyal, S.K. 1982. Blue green algae and rice cultivation. *Proc. Nat. Symp on. Biol. Nit. Fixation.,* Ind. Agric. Res. Instit., New Delhi. Pp. 346-376.

Guerny, K.A. and Mantle, P.G. 1993. Biosynthesis of 1-N-methyl albonoursin by an endophytic *Streptomyces* species. *J. Nat Prod.,* 56:1194-1199.

Guo, B., Dia, J., Ng, S. 2000. Cytonic acids A and B: novel tridepside inhibitors of hCMV protease from the endophytic fungus *Cytonaema* species. *J. Nat. Prod.,* 63:602-604.

Gutierrez, S., Cardoza, R.E., Casqueiro, J. and Martin, J.F. 2004. Commercial Production and Biosynthesis of Fungal Antibiotics: An Overview. In : Handbook of Fungal Biotechnology. 2nd Edition. Arora, D.K. (Ed.) Vol 20, Pp 401-409. Marcel Dekker, Inc., New York.

Hansen, T.K. and Jakobsen, M. 2004. Yeast in the Diary Industry. In: Fungal Biotechnology in Agricultural, Food and Environmental Applications. Arora, D.K. (Ed.) Vol. 21, Pp. 269-276. Marcel Dekker, Inc., New York.

*Harkishore, K.M. 1977. Kenyan sugarcane wine-muratina. Symposium on Indigenous Fermented Foods, Bangkok, Thailand.

Harper, J.K., Ford, E.J., Strobel, G.A. 2003. Pestacin: a 1, 3-dihydro isobenzofuran from *Pestalotiopsis microspora,* possessing antioxidant and antimycotic activities. *Tetrahedron.* 59:2471-2476.

Harrison, L., Teplow, D., Rinaldi, M. and Strobel, G.A. 1991. Pseudomycins, a family of novel peptides from *Pseudomonas syringae,* possessing broad spectrum antifungal activity. *J. Gen. Microbiol.,* 137:2857-2865.

Hartwell, J.L. 1982. Plants Used Against Cancer. A Survey. Quarterman Publications, Inc., Massachusetts, USA.

Haschek, W.M., Gumprecht, L.A., Smith, G., Tumbleson, M.E. and Constable, P.D. 2001. Fumonisin toxicosis in Swine: an overview of porcine pulmonary edema and current perspectives. *Environ. Health Perspect., 109 (Suppl. 2):* 251-257.

Hassan, B., Venkateshwaran, A., Fredrickson, J.K. and Daly, M.J. 2003. *Appl. Environ. Microbiol.,* 69: 4575-4582.

Hayashi, M., Kim, Y.P., Takamatsu, S., Preeprame, S., Komiya, T., Masuma, R., Tanaka, H., Komiyama, K. and Omura, S. 1996. Chlovalicin, a new cytocidal antibiotic produced by *Sporothrix* species. FO-4649. I. Taxonomy, fermentation, isolation and biological activities. *J. Antibiot.,* 49: 4631-634.

*He, M., Zhang, X.M. and Yuan, H.Q. 2005. Clinical study on treatment of pulmonary intestinal fibrosis with *Ginkgo* extract. *Zhongguo. Zhong, Xi Yi Jie. He Za.Zhi.,* 25(3): 222-234.

Healy, M.G., Devine, C.M. and Murphy, R. 1996. Microbial Production of Biosurfactants. Reprinted from Resources, Conservation and Recycling. Vol 18 (1-4).

Heiss, G., Trachtmann, N., Abe, Y., Takeo, M. and Knackmuss, H.J. 2003. *Appl. Environ. Microbiol.,* 69: 2748 -2754.

Heitkamp, M.A., Franklin, W and Creniglia, C.E. 1988. *Appl. Environ. Microbiol.,* 54: 2549-2555.

Horta, M.D. and Chet, I. 1986. Fungal Lectins and Agglutinins. In : Microbial Lectins and Agglutinins. Mirleman, D. (Ed.). John Wiley and Sons, Inc.,

*Hesseltine, C.W. 1979. Some important fermented foods of mid-Asia, the Middle East, and Africa. *J. Am. Oil. Chem. Soc.,* 56: 367-374.

*Hesseltine, C.W. and Shibasaki, K. 1961. Miso. III. Pure culture fermentation with *Sacchoromyces rouxii. Appl. Microbiol.,* 9: 515-518.

Hestbjerg, H.P., Willumsen, A., Christensen, M., Anderson, O., Jacobsen, C.S. 2003. *Environ. Toxicol. Chem.,* 22: 692 -698.

Hirahashi, T., Matsumoto, M., Hazeki, K., Saeki, Y., Ui, M. and Seya, T. 2002. Activation of the human innate immune system by *Spirulina*: augmentation of interferon production and NK cytotoxicity by oral administration of hot water extract of *Spirulina platensis. Int. Immunopharmacol.,* 2(4):423-434.

*Hirono, I. 1987. Naturally Occurring Carcinogens of Plant Origin: Toxicology, Pathology and Biochemistry. Elsvier Science Publishers, Amsterdam.

Hirschberg, J. and Chamoritz, D. 1994. In: The Molecular Biology of Cyanobacteria. Bryant, D.A. (Ed.) Pp. 559-579. Kluwer Academic Publishers. Netherlands.

Ho, M.H., Wang, D.M., Hsieh, H.J., Liu, H .C., Hsein, T.Y., Lai, J. Y. and Hou, L.T. 2005. Preparation and characterization of RGD immobilized chitosan scaffolds. *Biomaterials*. 26: 3197-3206.

Hollak, C.E., van Weely, S., van Oers, M. and Aerts, J. 1994. *J.Clin. Invest.*, 93: 1288-1292.

Holt, P.J., Bruce, N.C. and Lowe, C.R. 1996. Bioluminescent enzyme assays for heroin and heroin metabolites. *Anal. Chem.*, 68: 1877-1882.

Horn, W.S., Simmonds, M.S.J., Schwartz, R.E. and Blaney, W.M. 1995. Phomopsichalasin, a novel antimicrobial agent from an endophytic *Phomopsis* sp. *Tetrahedron.* 14:3969-3978.

Horsh, S. and Walther, C. 2004. *Ginkgo biloba* special extract EGB 761 in the treatment of peripheral arterial occlusive disease (PAOD)-a review based on randomized, controlled studies. *Int. J. Clin. Pharmacol. Ther.,* 42(2):63-72.

Howard, R.L., Abotsi, E., Jansen van Rensberg, E.L. and Howard, S. 2003. *African. J. Biotechnol.*, 2:602-619.

http://www.bpfoods.com

http://www.diamondsdv.com

http://www.fda.gov.

http://www.mushroom98.co.kr/body-02-199811-14.html.

http:// www.tigr.org/tdb/mdb complete.html

Husneck, S. 1999. The significance of Lichens and their metabolites. *Naturwissenschaften.* 88: 559-570.

Husneck, S. and Yoshimura, Y. 1996. Identification of Lichen Substances. Springer-Verlag, Berlin, Heidelberg, New York.

Hutchens, A.R. 1991. Indian Herbology of North America. Shambhala Publications, Inc. Boston, Massachusetts, USA.

Imhoff, J.F. 1995. In: Anoxygenic Photosynthetic Bacteria. Blankenship, R.E., Madigan, M.T. and Bauer, C.E. (Eds.) Kluwer Academic Publishers, The Netherlands.

Ingolfsdottir, K. 2002. Molecules of interest: usnic acid. *Phytochemistry.* 61: 729-736.

Ingolsdottir, K., Chung, G.A.C., Gissurarson, S.R., Skulason, V.G. and Vilhelmsdottir, M. 1998. *In vitro* antimycobacterial activity of lichen metabolites. *Eur. J. Pharm. Sci.,* 6: 141-144.

Inouye, S., Abe, S. and Yamaguchi, H. 2004. Fungal Terpenoid, Antibiotics and Enzyme Inhibitors. In: Handbook of Fungal Biotechnology. 2nd Edition.Arora, D.K. (Ed.) Vol.20, Pp 379-394. Marcel Dekker, Inc., New York.

Ishida, Y., Nakamura, F., Kanzato, H., Sawada, D., Yamamoto, N., Kagata, H., Oh-Ida, M., Takeuchi, H. and Fujiwara, S. 2005. Effect of milk fermented with *Lactobacillus acidophilus* strain L-92 on symptoms of Japanese Cedar pollen allergy: a randomized placebo-controlled trial. *Biosci. Biotechnol. Biochem.,* 69 (9):1652-60.

Issing, W., Klein, P. and Weiser, M. 2005. The homeopathic preparation vertigoheel versus *Ginkgo biloba* in the treatment of vertigo in an elderly population: a double-blinded, randomized, controlled clinical trial. *J Altern. Complement Med.,* 11(1):155-160.

*Jacoby, M. 2005. *Chemical and Engineering News*. 83: 25.

Jain, P.C. and Jain, P.K. 2005. Diversity and Significance of Keratinophilic Fungi. In: Microbial Diversity: Current Perspectives and Potential Applications, Satyanarayana, T and Johri, B.N. (Eds.) I. K. International Pvt. Ltd. New Delhi.

Jain, S.K. 1991, Dictionary of Indian Folk medicine and Ethnobotany. Deep Publications, New Delhi, India.

Jain, S.K. 1991. *Ethnobotany.* Deep Publications, New Delhi, India.

Jansson, H.B. and Llorca, L.V. 2004. Control of nematodes by Fungi. In: Fungal Biotechnology in Agricultural, Food and Environmental Applications. Arora, D.K. (Ed.) Vol. 21, Marcel Dekker, Inc., New York.

Jegorov, A., Matha, V. and Weiser, J. 1990. Production of cyclosporins by entamopathogenic fungi. *Microbiol. Lett.,* 45:65-70.

Jone, W.R. 1997. Biosurfactants, bioavailability and bioremediation. In: Global Environmental Biotechnology. Wise, D.L. (Ed.) Elsevier Science, Netherlands.

Jones, D.T. and Keis, S. 1995. *FEMS Microbiol. Rev.*, 17: 223-232.

Jowers, C., Shih, R., James, J., Deloughery, T.G. and Holden, W.E. 2005. Effects of *Ginkgo biloba* on exhaled nasal nitric oxide during normobaric hypoxia in humans. *High Alt. Med. Biol.,* 5(4):445-449.

Jung, G. 1991. Lantibiotics-ribosomally synthesized biologically active polypeptides containing sulfide bridges and α,β-didehydroamino acids. *Angew Chem. Int. Ed. Engl.,* 30:1051-1068.

Kameswara Rao, C. 2000. Database of Medicinal Plants. Karnataka State Council of Science and Technology, Bangalore, India.

Kameswara Rao, C., Geetha, B.L. and Geetha, S. 2003. Red List of Threatened Vascular Plant Species in India. ENVIS, BSI, Ministry of Environment and Forests, Calcutta, Governoment of India.

Kamiku, T.M., Hayashi, N., Nishio and Nagai, S. 1978. Utilization of non sugar sources for Vitamin B_{12} production. *Appl. Environ. Microbiol.*, 35: 971-973.

Kammerer, V.T., Motzkat, K., Daniel, D., Gertz, H.J., Koller, M., Lorenz, W., Pilartz, H., Zimmer, B., Habs, M., Vonden Driesch, V. and Rychlik, R. 2005. The situation of patients with dementia may be rectified by *Ginkgo biloba. MMW Fortschr. Med.,* 47(42):51.

Kamphues, J. 1990. Refusal of breeding bulls to eat hay contaminated with horse-tail (*Equisetum palustre). Tierarztl Parx.,* 18(4):349.

*Kanaly, R.A. and Harayama, S. 2000. *J. Bacteriol.,* 182: 2059-2067.

Kapoor, R.K., Sharma, K.K., Kuhar, S and Kuhad, R.C. 2005. Diversity of Lignin Degrading Micro-organisms, Ligninolytic Enzymes and Their Biotechnological Applications. In: Microbial Diversity: Current Perspectives and Potential Applications. Satyanarayana, T. and Johri, B.N. (Eds.) I. K. International Pvt. Ltd. New Delhi.

Karcher, P.M., Ezeji, T.C., Qureshi, N. and Blaschek, H.P. 2005. Microbial Production of Butanol: Product Recovery by Extraction. In: Microbial Diversity: Current Perspectives and Potential Applications. Satyanarayana, T. and Johri, B.N. (Eds.) I.K. International Pvt. Ltd., New Delhi.

Karna, R.R., Uma, L., Subramanian, G. and Mohan, P.M. 1999. *World J. Microbiol. and Biotechnol.,* 15:729-732.

Kasting, J.F. 2001. The Rise of Atmospheric Oxygen. *Science*. 293:819-820.

Kenji, N., Nobuyuki, M., Moriaki, H. and Junichi, N. 1989. Jap. Kokai., Tokyo, Japan.

*Kennedy, J.F., Cabalda, V.M. and White, C.A. 1988. Enzyme starch utilization and genetic engineering. *Tr. Biotechnol.,* 6: 184-189.

*Kenz, M., Bittner, M., Boes, F., Wege, C., Jeske, H., Maib, E. and Ken, K. 2003. *Nano Lett.*, 3:1079-1082.

Keys, J.D. 1976. Chinese Hebs. Their Botany, Chemistry and Pharmacodynamics. Charles, E. (Ed.). Tuttle Company, Inc.

Khachatourians, G.G. 2004. Fungi in Food Technology: An Overview. In: Fungal Biotechnology in Agricultural, Food and Environmental Applications. Arora, D.K. (Ed.) Vol. 21, Pp. 217-221. Marcel Dekker, Inc., New York.

Khanna, P., Raj Kumar, B. and Jothi Kumar, N. 1999. *Int. J. Hydrogen energy*, 25: 63-67.

Khoshoo, T.N. 1996. India needs a national biodiversity conservation board. *Curr. Sci.*, 71:506-513.

Kidd, P.M. 2000. The use of mushroom glucans and proteoglycans in cancer treatment. *Altern. Med. Rev.*, 5: 4-27.

*Kilara, A. 1985. Enzyme modified lipid food ingredients. *Process Biochem.*, 20:30-45.

Kim, B.S. and Hwang, B.K. 2004. Biofungicides. In: Fungal Biotechnology in Agricultural, Food and Environmental Applications. Arora. D.K. (Ed.) vol. 21. Marcel Dekker Inc., New York.

Kim, S.H., Lee, E.K., Chang, J.W., Min, W.K., Chi, H.S, and Kim, S.B. 2005. Effects of *Ginkgo biloba* on haemostatic factors and inflammation in chronic peritoneal dialysis patients. *Phytother. Res.*, 19 (6):546-548.

Kino, T., Hatanaka, H., Hashimoto, M. 1987. FK-506, a novel immunosuppressant isolated from *Streptomyces.* 1. Fermentation, isolation and physico-chemical and biological characteristics. *J. Antibiot.,* 40:1249-1255.

Kirk, T.K., Connors, W.J. and Zeikus, J.G. 1976. Requirement of growth substrate during lignin degradation by two wood rotting fungi. *Appl. Environ. Microbiol.,* 32: 192-194.

Kirk, T.K., Burgess, R.R. and Koning, Jr. J.W. 1990. Use of Fungi in Pulping Wood: An Overview of Biopulping Research. In: Frontiers in Industrial Mycology. Leatham, G. (Ed.) Proceedings of Industrial Mycology Symposium.

Kirtikar, K.R. and Basu, B.D. 1935. Indian Medicinal plants. Second Edition. M/S Periodical Experts, New Delhi.

Kitajima, M., Ohida, S., Amagato,K., Ayama, H., Saito, K., Seki, H. and Aimi, N. 2002. Camptothecin related alkaloids from hairy roots of *Ophiorrhiza pumila. Tetrahedron.* 58:9169-9178.

Kobayashi, H., Matsunaga, K. and Oguchi, Y. 1995. Antimetastatic effects of PSK (Krestin), a protein-bound polysaccharide obtained from basidiomycetes an overview. *Cancer Epidermiol. Biomarkers Prev.,* 4:275-281

*Kolb, M., Zentgrat, B., Mattison, B., Arvindson, P. and Danielsson, B. 1996. *Thermochim Acta.,* 277: 1-6.

Kosaric, N., Gray, N.C.C. and Cairns, W.L. 1987. Introduction: Biotechnology and surfactant industry. In: Biosurfactants and Biotechnology. Kosaric, N., Cairns, W.L. and Gray, C.C. (Eds.) Marcel Dekker, New York. Pp. 1-19.

Kovacs, K., Fodder, B., Kovacs, A., Casanadi, G., Marati, G., Balogh, J., Arvani, S. and Rakhely, C. 2002. *Int. J. Hydrogen energy*, 27: 1463-1469.

Kowalski, R.J., Giannakakou, P. and Hamel, E. 1997. Activities of the microtubule-stabilizing agents epothilones A and B with purified tubulin in cells resistant to Paclitaxel (Taxol). *J. Biol. Chem.,* 272:2534-2451.

* Kristofferson, T. 1973. The biogenesis of cheese flavours. *J. Agirc. Food Chem.,* 21: 573-575.

*Kronick, M.N. 1986. *J. Immunol. Methods.,* 92:1-13.

Kudolo, G.B., Delaney, D. and Blodgett, J. 2005. Short-term oral ingestion of *Ginkgo biloba* extract (EGb 761) reduces malondialdehyde levels in washed platelets of type-2 diabetic subjects. *Diabetes Res. Clin. Pract.,* 68(1):29-38.

Kuhad, R.C., Singh, A. and Eriksson, K.E.L. 1997. Micro-organisms and enzymes involved in the degradation of plant fibre cell walls. In: Biotechnology in The Pulp and Paper Industry. Erikson, K.E.L. (Ed.).Pp. 45-127. Springer-Verlag, Berlin.

Kulik, M. 1995. *Eur. J. Pl. Pathol.,* 1: 585-599.

*Kulys, J., Vidziunaite, R. and Schneider, P. 2003. *Enz. Microb. Technol.,*32: 455-463.

*Kumar, H. 1994. *Fresenius Environ. Bull.,* 3:195.

Kumar, H.D. 1990. Introductary Phycology. Affiliated East-West Press Pvt. Ltd., New Delhi.

*Kumazawa, S. and Mitsui, A. 1994. *Biotechnol. Bioeng.,* 44:854-858.

Kun, L.Y. 2006. Micro-organisms and Production of Alternative Energy. In: Microbial Biotechnology Principles and Applications. (Ed.) Kun, L.Y. World Scientific Publishing Co. Ltd., Singapore.

Kurane, 1997. Environmentally Friendly Products and Processes for the 21st Century. In : Global Environment Biotechnology. Wise, D. L. (Ed.) Elsevier Science.

Kurup, V.K. 2004. Fungal Allergy. In: Handbook of Fungal Biotechnology. 2nd Edition. Arora, D.K. (Ed.) Vol.20, Pp. 515-522. Marcel Dekker, Inc, New York.

*Kushner, D.I. and Brevil, C. 1977. *Arch. Microbiol.,* 112:219-223.

*Kyriakidis, N., Waight, E.S., Day, J.B. and Mantle, P.G. *Appl. Environ. Microbiol.,* (in press).

Lamar, R.T., Evans, J.W. and Glaser, J. 1993. Solid phase treatment of pentachorophenol contaminated soil using lignin degrading fungi. *Environ. Sci. Technol.,* 27: 2566-2571.

Lampe, K.F. 1985. AMA Handbook of Poisonous and Injurious Plants. Chicago Review Press, Chicago.

*Lappe, P. and Ulloa, M. 1989. Estudios Etrnicos, Microbianos. Y. Quimicos del Tesquino Tarahumara. Mixico: *Universidad National Autonomo de Mexico.*

Larch, C.A. 1996. Encyclopedia of Medicinal Plants. DK Publishing Company. New York.

Laumen, K. and Ghisalba, O. 1994. Preparative scale chaemoenzymatic synthesis of optically pure D-myoinositol 1-phosphate. *Biosci. Biotech. Biochem.,* 58: 2046-2049.

Lazcano, A. and Miller, S.L. 1994. How long did it take for life to begin and evolve to cyanobacteria? *Journal of Molecular Evolution.* 39: 546-554.

Le-Tien, C., Millette, M., Mateescu, M.A. and Lacroix, M. 2004. Modified alginate and chitosan for lactic acid bacteria immobilization. *Biotechnol. Appl. Biochem.,* 39: 189-198.

LeBlanc, F. and Rao, D.N. 1975. Effects of Air Pollution on Lichens and Bryophytes. In : Responses of Plants to Air Pollution. Mudd, J.B. and Kozlowski, T.T. (Eds.) Pp.237-272. Acadamic Press, New York.

LeCoulant, P. and Lopes, G. 1956. Wood dermatitis, Experimental research on the allergy due to mosses. *J. Med. Bordeaux,* 133: 245.

Ledin, M. and Pedersen, K 1996. The environmental impact of mine wastes: roles of micro-organisms and their significance in the treatment of mine wastes. *Earth Sci. Rev.,* 41: 67-108.

Lee, C.H. 2001. Fermentation Technology in Korea. Korea University Press, Seoul.

Lee, K.H. 2004. Current developments in the discovery and design of a new drug candidates from plant natural product leads. *J. Nat. Prod.,* 67:273-283.

Lee, J., Lobkovsky, E., Pliam, N.B., Strobel, G.A. and Clardy, J. 1995a. Subglutinols A and B:immunosuppressive compounds from the endophytic fungus *Fusarium subglutinans. J. Org. Chem.,* 60:7076-7077.

Lee, J.C., Yang, X., Schwartz, M., Strobel, G.A. and Clardy, J. 1995b. The relationship between an endangered North American tree and an endophytic fungus. *Chem. and Biol.,* 2:721-727.

Lee, J.C., Strobel, G.A., Lobkovsky, E. and Clardy, J.C. 1996. Torreyanic acid; a selectively cytotoxic quinone dimer from the endophytic fungus *Pestalotiopsis microspora. J. Org. Chem.,* 61:3232-3233.

Lee, S.W., Mao, C., Flynn, C.E. and Belcher, A.M. 2002. *Science*. 296: 892-895.

*Lee, S.W., Lee, S.K. and Belcher, A. 2003. *Adv. Mater.*, 15: 689-692.

Lee, W.A., Gu, L., Kikszal, A.R., Chu, N., Leung, K. and Nelson, P.H. 1990. Bioavailability improvement of mycophenolic acid through amino ester derevitalization. *Pharmaceut Res.,* 7:161-166.

*Lee, W.J. and Cho, D.H. 1971. Microbiological studies of Korean native soy sauce fermentation. A study on the microflora changes during Korean native soy sauce fermentation. *J. Korean Agric. Chem. soc.,* 14: 137-148.

Leejeerajummean, A., Duckham, S.C., Owens, J.D. and Ames, J.M. 2001. *J. Sci. Food Agric.*, 81: 525-529.

Leite, J. 2003. *J.Braz. Chem. Soc.,* 14: 297-303.

Lell, B. and Kremser, P.G. 2002. Clindamycin as a antimalarial drug: review of clinical trials. *Antimicrob Agents Chemother.,* 46:2315-2320.

Lenke, H. and Knackmuss, H. 1996. *Appl. Environ. Microbiol.,* 62: 784-790.

Leung, A.Y. 1980. Encyclopedia of Common Natural Ingredients Used in Food, Drugs and Cosmetics. John Wiley and Sons, Inc., New York.

Levy, Y., Zaltsberg, H., Ben-Amotz, A., Kanter, Y. and Aviram, M. 2000. Dietary supplementation of a natural isomer mixture of beta-carotene inhibits oxidation of LDL derived from patients with diabetes mellitus. *Ann. Nutr. Metab.,* 44 (2) : 54-60.

Li, J.Y., Strobel, G.A., Sidhu, R.S., Hess, W.M. and Ford, E. 1996. Endophytic taxol producing fungi from Bald Cypress *Taxodium distichum. Microbiology.* 142 : 2223-2226.

Li, J.Y., Strobel, G.A., Sidhu, R.S., Hess, W.M. and Ford, E. 1998. The induction of taxol production in the endophytic fungus *Periconia sp.* from *Torreya grandifolia. J. Ind. Microbiol.,* 20:259-264.

Li, J.Y., Strobel, G.A., Harper, J.K., Lobkovsky, E. and Clardy, J.C. 2000. Cryptocin a potent tetramic acid antimycotic from the endophytic fungus *Cryptosporiopsis quercina. Org. Lett.,* 2: 767-770.

Li, Y., Xiang, H., Lill, J., Zhou, M. and Tan, H. 2003. *Extremophiles.* 7:401-407.

Lincoln, R.J., Boxshall, G. A. and Clark, P.F. 1982. A Dictionary of Ecology, Evolution and Systematics. Cambridge University Press, Cambridge.

Lindley, N.D. 1992. Hydrocarbon Degrading Yeasts and Filamentous Fungi of Biotechnological Importance. In: Arora, D.K., Elander, R.P., Mukerjii, K.G. (Eds.). Handbook of Applied Mycology. Vol. 4. 1st Edition. Pp. 905-929.

Lister, D.L., Kanungo, G., Rathbone, D.A. and Bruce, N.C. 1999. Transformation of codeine to important semisynthetic opiate derivatives by *Pseudomonas putida* M10. *FEMS Microbiol. Lett.,* 181: 137-144.

Liu, X., Wei, J., Tan, F., Zhou, S., Wurthwein, G. and Rohdewald, P. 2004. Antidiabetic effect of Pycnogenol French maritime pine bark extract in patients with diabetes type II. *Life Sci.,* 75(21):2505-13.

Liu, X.J., Luo, X.Y. and Hu, W.J. 1988. *Arthrinium* sp. and The Etiology of Deteriorated Sugarcane Poisoning. In: Mycotoxins and Phycotoxins. Natori, S., Hashimoto, K., Ueno, Y. (Eds.). 88 Pp. 109-118. Elsevier. Amsterdam.

Lomascolo, A., Stentelaire, C., Asther, M. and Lesage-Meessen, L. 1999. Basidiomycetes as New Biotechnological Tools to Generate Natural Aromatic Flavours for the Food Industry. In: Trends in Biotechnology. Owen, M. (Ed.). Elsevier Science, London.

Long, D.E., Smidmansky, E.D., Archer, A.J. and Strobel, G.A. 1998. *In vivo* addition of telomeric repeats to foreign DNA generates chromosomal DNAs in the taxol-producing fungus *Pestalotiopsis microspora. Fungal Genetics Biol.,* 24:335-344.

Long, M.T., Hailes, A.M., Kirby, G.W. and Bruce, N.C. 1995. Transformation of the morphine alkaloids by *Pseudomonas putida* M10. *Appl. Environ. Microbiol.,* 61: 3645-3649.

Lopez-Diaz, T.M., Santos, J.A., Garcia-Lopez, M.L. and Otero, A. 2001. Surface mycoflora of a Spanish fermented meat sausage and toxigenicity of *Penicillium* isolates. *Int. J. Food. Microbiol.,* 68: 69-74.

Lopez-Macia, A., Jimenez, J.C., Royo, M., Giralt, E. and Albericio, F. 2001. Synthesis and structure determination of kahalalide F[12]. *Am. Chem. Soc.,* 123 (46): 11398-11401.

Love, O.T. (Jr.). and Eilers, R.G. 1982. *J. Am. Water works Assoc.*, 74: 413-425.

Lovely, D.R. 2003. *Nature Rev. Microbiol.*, 1: 35-44

Lozovskaia, M.E. 2005. Effectiveness of using the biologically active additive to food from *Laminaria* in adolescents during complex treatment of the pulmonary tuberculosis. *Vopr Pitan.,* 74 (1): 40-3.

Lu, J. and He, H. 2005. Clinical observation of *Ginkgo biloba* extract injection in treating early diabetic nephropathy. *Clin. J. Integr. Med.,* 11 (3): 226-228.

*Ludianskii, E.A. 1992. The extension of the use of physical methods of treatment to patients with disseminated sclerosis. *Vopr Kurortol Fizioter Lech Fiz Kult.,* 3:34.

Luo, C.Y., Tang, J.Y. and Wang, Y.P. 2004. Homoharringtonine: a new treatment option for myeloid leukemia. *Hematology.* 9 (4): 259-270.

Macaskie, L.E. 1991. The application of biotechnology to the treatment of wastes produced by the nuclear fuel cycle-biodegradation and bioaccumulation as a means of treating radionuclide containing streams. *Crit. Rev. Biotechnol.*, 11: 41-112.

Macaskie, L.E., Thomas, R.A.P. and Lloyd, J.R. 2001. Application of Micro-organisms to The Decontamination of Heavy Metal Bearing Wastes. In: Industrial and Environmental Biotechnology. Ahmed, N., Qureshi, F. M. and Khan, O.Y. (Eds.) Horizon Scientific Press, Wymondham, U.K.

Mackaskie, L.E. 1997. *Res. Microbiol.*, 148: 528- 530.

Mackaskie, L.E., Jeong, B.C. and Tolley, M.R. 1994. *FEMS Microbiol., Rev.*, 14: 351-368.

Mackaskie, L.E., Lioyd, J.R., Thomas, R.A.P. and Tolley, M.R. 1996. *Nuclear Energy.* 35: 257-271.

*Macrae, A.R. 1983. Lipase catalyzed inter esterification of oils and fats. *J. Oil Chem. Soc.,* 60: 291-294.

*Maga, J.A. 1974. Bread flavour. *CRV Rev. Food. Technol.*, 5: 55-141.

Mahalingam, N. (Ed.). 1998. Kisan World. Vol. 25, No. 8. Dosh Publications, Chennai, India.

Maki, J.A. and Mitchell, R. 1986. The Function of Lectins in Interactions Among Marine Bacteria, Invertebrates and Algae. In: Microbial Lectins and Agglutinins. Mirleman, D. (Ed.). John Wiley and Sons, Inc.,

Malathi, S. and Chakraborthy, R. 1991. Production of alkaline protease by a new *Aspergillus flavus* isolate under solid-substrate fermentation conditions for use as a depletion agent. *Appl. Environ. Microbiol.*, 57: 712-716.

Manfredi, J.J. and Horowitz, S.B. 1984. Taxol: an antimitotic agent with a new mechanism of action. *Pharmacol. Ther.*, 25:83-125.

Manocha, M.S. and Zhonghua, Z. 1997. *Mycologia*. 89: 185 -194.

Mansi, E.M.T., Bryce, C.F.A., Demain, A.L. and Allman, A.R. (Eds.) 2007. Fermentation Microbiology and Biotechnology Second edition CRC Press, Taylor and Francis Group, London.

Manzoni, M. and Rollini, M. 2002. Biosynthesis and biotechnological production of statins by filamentous fungi and application of these cholesterol lowering drugs. *Appl. Microbiol. Biotechnol.*, 58:555-564.

Manzoni, M., Bergomi, S., Rollini, M. and Cavazzoni, V. 1999. Production of statins by filamentous fungi. *Biotechnol Lett.*, 21:253-257.

Manzoni, M., Rollini, M., Bergomi, S. and Cavazzoni, V. 1998. Production and purification of statins from *Aspergillus terreus* strains. *Biotechnol Tech.*, 12 : 529-532.

Mao, C., Flynn, C.E., Hayhurst, A., Sweeney, R., Qi, J., Georgiou, G., Iverson, B. and Belcher, A. 2003. *Proc. Natl. Acad. Sci.*, 100: 6946-6951.

Mao, T.K., Van dewater, J. and Gershwin, M.E. 2005. Effects of *Spirulina*-based dietary supplement on cytokine production from allergic rhinitis patients. *J. Med. Food.*, 8 (1):27-30.

Marderosian, A.D. and Beutler, J.A. 2002. The Review of Natural Products. 3rd Edition. Facts and Comparisons, Missouri, USA.

Margesin, R. and Schinner, F. 2001. *Extremophiles*. 5:73-83.

Markwell, M.A.K. 1986. Viruses as Haemagglutinins and Lectins. In : Microbial Lectins and Agglutinins. Mirleman, D. (Ed.). John Wiley and Sons, Inc.,

Maruthi Mohan, P., Rashmi, K., Haritha, A. and Venkateswaran, G. 2005. Bioremediation of Radioactive Waste from Nuclear Power Reactor Effluents: Present Status and Future Perspective. In: Microbial Diversity: Current Perspectives and Potential Applications. Satyanarayana, T and Johri, B.N. (Eds.) I. K. International Pvt. Ltd. New Delhi.

*Massechelin, C.A. and Batum, M.S. 1981. Enzyme degradation and participation of ester linked beer polyphenols in chill haze formation. *Proc. Congr. Eur. Brewers Convention*: 359-370.

*Mastranglelo, G., Fadda, E. and Marzia, V. 1996. Polycyclic Aromatic Hydrocarbons and Cancer. In: Fungal Biotechnology, Agriculture, Food and Envrionmental Applications. Arora, D.K. (Ed.) Vol. 21. Marcel Dekker Inc., New York.

Mathew, B., Sankaranarayanan, R., Nair, P.P., Varghese, C., Somanathan, T., Amma, B.P. and Nair, M.K. 1995. Evaluation and chemoprevention of oral cancer with *Spirulina fusiformis. Nutr. Cancer.*, 24:197.

Matsui, M.S., Mcizzuddin, N., Arad, S. and Marenus, K. 2003. Sulphated polysaccharides from red microalgae have anti-inflammatory properties *in vitro* and *in vivo. Appl. Biochem. Biotechnol.*, 104(1): 13-22.

*Matsumoto, K. 1979. The Mysterious Reishi Mushroom.: Wood Bridge Press Publishing, Santa Barbara, US.

McGinnis, G.D., Boraziani, A., Pope, D.F., Strobel, D.A. and McFarland, L.K. 1991. On site treatment of creosote and PCP sludge's and contaminated soil. EPA 1600/2-91/019, Washington, DC.

Mehrotra, R.S. and Agarwal, A. 2003. Plant Pathology. 2nd Edition. Tata McGraw-Hill Publishing Company Limited, New Delhi, India.

Mehta *et al.*, 2004. In: Economic Microbiology. Secondary Products of Metabolism, Rose, A.H. (Ed.). Academic, London. Pp. 389-420.

Merchant, R.E., Carmack, C.A. and Wise, C.M. 2000. Nutritional supplementation with *Chlorella pyrenoidosa* for patients with fibromyalgia syndrome: a pilot study. *Phytother. Res.*, 14(3): 167-173.

Merchant, R.E. and Andre, C.A. 2001. A review of recent clinical trials of the nutritional supplement of *Chlorella pyrenoidosa* in the treatment of fibromyalgia, hypertension and ulcerative colitis. *Alfern Ther Health Med.*, 7 (3): 79-91.

Merchant, R.E., Andre, C.A. and Sica, D.A. 2002. Nutritional supplementation with *Chlorella pyrenoidosa* for mild to moderate hypertension. *J. Med. Food.*, 5 (3): 141-152.

Messner, K. and Srebotnik, E. 1994. *FEMS Microbiol. Rev.*, 13: 351-364.

Messner, K., Mase, K.S. and Techt, G. 1992. Fungal Pretreatment of Wood Chips for Chemical Pulping. In: Biotechnology in the Pulp and Paper Industry. Euwahara, M. and Shimada, M. (Eds.) Unipublishers, Tokyo.

*Messner, K., Schiefermellier, M., Srebotni, K.E. and Techt, G. 1993. Biosulphite Pulping Current State of Research. In: Proc. FEMS Symposium on Lignin Biodegradation and Transformation. Duarte, J.C., Ferreira, M.C. and Ander, P. (Eds.) Lisboa Forbitech. Edition.

Miller, R.V., Miller, C.M., Garton-Kinney, D. 1998. Ecomycins, unique antimycotics from *Pseudomonas viridiflava. J. Appl. Microbiol.*, 84:937-944.

*Minussi, R.C., Pastore, G.M. and Duran, N. 1998. Brazilian Meeting on the Chemistry of Food and Beverages. Sao Carlos-Brazil. Abstr. 10.

Mirelman, D. (Ed.) 1986. Microbial Lectins and Agglutinins. Properties and Biological Activity. John Wiley and Sons, New york.

Mittermeier, R.A., Myers, N., Gil, P.R. and Mittermeier, C.G. 1999. Hotspots: Earth's Biologically Richest and Most Endangered Ecoregions.. CEMEX Conservation International, Washington DC.

Miyamoto, A., Shigeta, A., Tanaka, Y., Oomura, H., Masui, K., Katada, M., Asahi, M., Komori, T. and Suzuki, T. 1995. Glycerine mixture for cosmetics, perfumes, medicines, lubricants and plastic additives. *Eur. Pat.* EP 658629.

Mohammed, S.I., Steenson, L.R. and Kirleis, A.W. 1991. Isolation and characterization of micro-organisms associated with the traditional sorghum fermentation for production of Sudanese Kispa. *Appl. Environ. Microbiol.*, 57: 2529-2533.

*Mohan, S., Chinnaswamy, P. and Krishnamoorthy, A.S. 2005. A study on the effect of *Ganoderma lucidum* .Lloyd in Indian HIV Carriers. *International Journal of Medicinal mushrooms.*

Morissey, S.R. 2003. Maximizing returns. *Chem. Eng. News.*, 81 (37): 17-20.

Morley, G.F., Sayer, J.A., Wilkinson, S.C., Gharieb, M.M. and Gadd, G.M. 1996. Fungal Sequestration, Mobilization and Transformation of Metals and Metalloids. In: Fungi and Environmental Change. (Eds.) Frankland, J.C., Magan, N. and Gadd, G.M. Cambridge Uni. Press. Pp. 235-256.

Motaghi, M., Mazaheri, M., Moazami, N., Farkhondeh, A., Fooladi, M.H. and Goltapeh, E.M. 1997. Kefir production in Iran. *World J. Microbiol. Biotechnol.*, 13: 579-581.

*Murakami, H. 1972. Some Problems in Sake Brewing. In: Fermentation Technology. Proc. 4th International Fermentation Symposium. Terui, G. (Ed.). Pp. 639-643. Kyoto, Japan.

Murphy, R.A. and Horgan, K.A. 2005. Antibiotics, Enzymes and Chemical Commodities from Fungi. In: Fungi, Biology and Applications. Kavanagh, K. (Ed.) John Wiley and Sons Ltd., U.K.

Myers, N. 1988. Threatened biotas: hot spots in tropical forests. *Environmentalist*, 8: 187-208.

Myers, N., Mittermeler, R.A., Mittermeler, C.G., da Fonseca, G.A.B. and Kent, J. 2000. Biodiversity hot spots for conservation priorities. *Nature*, 403: 853-858.

Nadkarni, K.M. and Nadkarni, A.K. 1991. Indian Materia Medica. Vol. 1 and Vol. 2. Popular Prakashan Pvt. Ltd., Bombay.

*Nakamura, T., Akiyama, Y. and Kawagishi, H. 2005. A study of pharmacological components from submerged culture mycelia of *Phellinus linteus* (Berk. Et Curt.) Teng. *International Journal of Medicinal mushrooms*.

Navratil, M.P., Gemeiner, E., Sturdik, Z., Domeny, D., Smogrovicova, Z. and Antalova. 2000. Fermented beverages produced by yeast cells entrapped in ionotropic hydrogels of polysaccharide nature. *Minerva Biotec.*, 12: 337-344.

Neuman, I., Nahum, H. and Ben-Amotz, A. 1999. Prevention of exercise-induced asthma by a natural isomer mixture. *Ann. Allergy Asthma Immunol.*, 82 (6): 549-553.

Newman, D.J. and Cragg, G.M. 2005. The Discovery of Anticancer Drugs from Natural Sources. In: Natural Products: Drug Discovery and Therapeutic Medicine. Zhang, L. and Demain, A.L. (Eds.), Humana Press Inc., Totowa, New Jersey.

*Newton, B.W. 1972. *Laminaria* tent: relic of the past or modern medical device? *Am. J. Obslet. Gynecol.*, 113: 442.

Ng, T.B. 1998. A review of research on the protein bound polysaccharide (polysaccharopeptide, PSP) from the mushroom *Coriolus versicolor* (Basidiomycetes: Polyporaceae). *Gen. Pharmacol.*, 30:1-4.

Ng, T.B. 2004a. Antitumour and Immunomodulatory compounds from Fungi. In: Handbook of Fungal Biotechnology. 2nd Edition. Arora, D.K. (Ed.) Vol. 20, Pp. 469-474. Marcel Dekker, Inc. New York.

Ng, T.B. 2004b. Bioactive Fungal Polyaccharides and Polysaccharopeptides. In: Handbook of Fungal Biotechnology. 2nd Edition. Arora, D.K. (Ed.). Vol. 20, Pp. 331-336. Marcel Dekker, Inc. New York.

Ng, T.B. 2004c. Fungi and Fermented Food. In: Fungal Biotechnology, in Agricultural, Food and Environmental Applications. Arora, D.K. (Ed.) Vol. 21, Pp. 223-244. Marcel Dekker, Inc., New York.

Niamke, J.N. and Wang, N.S. 2004. Cellulose Degradation by Fungi. In: Fungal Biotechnology, in Agricultural, Food and Environmental Applications. Arora, D.K. (Ed.) Vol. 21, Marcel Dekker, Inc., New York.

Nigam, P., Banat, I.M., McMullan, G., Dalel Singh and Merchant, R. 1995. Microbial degradation of textile effluents containing azo, diazo and reactive dyes by aerobic and anaerobic bacterial and fungal cultures. 36th *Annu. Conf. AMI.*, Hisar, India.

Ninent, L. and Renaut, J. 1979. Carotenoids. In: Microbial Technology. Vol I. 2nd Edition. Peppler, H.J. and Periman, D (Eds.) Academic Press, London. Pp 529-544.

*Nippon-Oil and Fats. 1986. Production of propylene glycol mono fatty acid ester. Japan, Patent number. JP-J61149099.

Noha, M., Mesbah. and Wiegel, J. 2005. Halophilic thermophytes: A novel group of Extremophiles. In : Microbial Diversity. Current Perspectives and Potential Applications. Satyanarayana, T. and Johri, B. N. (Eds.). I.K International Pvt. Ltd., New Delhi.

Nordland, I..O., Powlowski, J., Hastrom, A. and Shirgler, V. 1983. Conservation of regulatory and structural genes from multicomponent phenol catabolizing bacteria that utilize a meta cleavage pathway. *J.G. Microbiol.*, 139: 2695-2703.

O'Connor, E.M. and Shand, R.F. 2002. *Indus. Microbial. Biotechnol.*, 28:23-31.

Ofek, L., Mirleman, D. and Sharon, N. 1977. *Nature.* 265:623-625.

*Ofodile, L.N., Uma, N.V., Kokubun, T., Ogundipe, O.T. and Simmonds, M.S.J. 2005. Antimicrobial colossolactones from a Nigerian Polypore *Ganoderma colossum* (Fr.) C.F. Baker. *International Journal of Medicinal mushrooms.*

Ogawa, J. and Shimizu, S. 1999. Microbial Enzymes: New Industrial Applications from Traditional Screening Methods. In: Trends in Biotechnology. Owen, M. (Ed.) Elsevier Science, London.

Ogunseitan, J.F. 2001. Microbial Divrsity. Blackwell Publishing Company, London.

Okiyama, A., Motoki, M. and Yamanaka, S. 1993. Bacterial Cellulose III. Development of a new form of cellulose. *Food Hydrocolloids.* 6:493-501.

Okuyama, E., Umeyama, K., Yamazaki, M., Kinoshita, Y. and Yamamoto, Y. 1995. Usnic acid and diffractic acid as analgesic and antipyretic components of *Usnea diffracta*. *Planta Med.*, 61 (2): 113-115.

Ollika, P., Alhonmaki, K. Leppanen, V.M., Glumoff, T., Raigola, T. and Suominen, H. 1993. Decolorisation of azo triphenyl methane heterocyclic and polymeric dyes by lignin peroxidase, isoenzymes from *Phanerochaete crysosporium*. *Appl. Environ. Microbiol.,* 59: 4010-4016.

*Ong, K.C. 1977. Malaysian Soy Sauce: Kicap. Symposium on Indigenous Fermented Foods, Bangkok, Thailand.

*Osagualekhor, D.O. and Okhuoya, J.A. 2005. Sociocultural and Ethnomycological uses of mushrooms among the Esan people of Nigeria. *International Journal of Medicinal mushrooms.*

Osburne, M.S., Grossman, T.H., August, P.R. and MacNeil, I.A. 2000. *ASM News.* 66: 411-417.

Oscarizi, J.C. and Pisabarro, A.G. 2001. Classification and mode of action of membrane active bactericins produced by gram-positive bacteria. *Int. Microbiol.*, 4: 13-19.

Padmaja, G. and George, M. 1999. Oriental fermented foods. In: Biotechnology: Food Fermentation. Microbiology. Biochemistry and Fermentation Technology. Joshi, V.K. and Pandey, A. (Eds.). Vol. 2, Pp. 523-581 Educational Publishers and Distributors. India.

Palma-Carlos, A.G. and Palma-Carlos, M.L. 2004. Candidiasis. In: Handbook of Fungal biotechnology . 2nd Edition. Arora, D.K.(Ed.) Vol. 20, Pp 493-501. Marcel Dekker, Inc., New York.

Panlasigui, L.N., Baello, O.Q., Dimatangal, J.M. and Dumelod, B.D. 2003. Blood cholesterol and lipid-lowering effects of carrageenan on human volunteers. *Asia Pac. J. Clin. Nutr.*, 12 (2): 209-214.

Pant, D.D. 1973. Cycas and the Cycadales. 2nd Edition. Central Book Depot. Allahabad, India.

Parkinson, J and Gordon, R. 1999. Beyond Micromachining: The Potential of Diatoms In: Trends in Microbiology. Meran Owen, M.(Ed.) Reference Edition Vol.17.

Parihar, N.S. 1965. An Introduction to Embryophyta. Vol. I Bryophyta. Central Book Depot. Allahabad, India.

Parihar, N.S. 1977. The Biology and Morphology of the Pteridophytes. Central Book Depot. Allahabad, India.

Patra, A., Tripati, S., Bhunia, S. and Mukherjee, B. 1993. Traditional Medicine Proceedings of International Seminar, Oxford IBH Publishing Company, Pvt, Ltd. New Delhi, India.

*Pattenden, G. and Thom, S.M. 1993. *J. Chem. Soc.*, 1:1629-1636.

*Patterson, G.M.L. 1996. *J. Sci. Ind. Res.*, 55:669-684.

Paxton, J.D. 1981. Phytoalexins-a working redefinition. *Phytopathol.*, 101:106-109.

Peaboy, D.S. 2003. *J. Nanotech.*, 1: 5.

Pelczar, M.J. Jr., Chan, E.C.S. and Krieg, N.R. 1993. Microbiology. 5th Edition. Tata McGraw-Hill Publishing Company limited. New Delhi, India.

Peng, Y and Demain, A.L. 1998. A new hydroxylase system in *Actinomadura* species cells converting compactin to pravastatin. *J. Indust. Microbiol. Biotechnol.*, 20:373-375

*Penny, R.H.C., O'Sullivan, B.M., Mantle, P.G. and Shaw, B.I. 1979. *Vet. Rec.*, 105, 392.

Pentacost, A. 1984. Key to Algal genera. The Richmond Publishing Co. Ltd., England.

Penttila, M., Limon, C. and Nevalainen, H. 2004. Molecular Biology of *Trichoderma* and Biotechnological Applications. In: Handbook of Fungal Biotechnology. 2nd Edition. Arora, D.K. (Ed.). Vol. 20. Pp. 413-422. Marcel Dekker, Inc. New York.

* Periasamy, K. 2005. Novel anti-bacterial compounds obtained from some edible mushrooms. *International Journal of medicinal mushrooms.*

Perlman, D. 1978. Vitamins. In: Economic Microbiology: Primary Products of Metabolism. Vol 2. Rose, A.H. (Ed.). Academic press, London.

Perlman, D. 1979a. Microbial production of antibiotics. In : Microbial Technology. Peppler, H.J. and Perlman, D. (Eds.) 2nd Edition, Vol. I, Academic Presss, New York.

Perlman, D. 1979b. Microbial process for riboflavin production. In : Microbial Technology. Peppler, H.J. and Perlman, D. (Eds.) 2nd Edition, Vol. I, Academic Presss, New York.

Perry, N.B., Benn, M.H., Brennan, N.J., Burgess, E.J., Ellis, G., Galloway, D.J., Lorimer, S.D. and Tangeny, R.S. 1999. Antimicribial, antiviral and cytotoxic activity of New Zealand lichens. *Lichenologist.* 31: 627-636.

Phillips, E.J. P., Landa, E.R. and Lovley, D.R. 1995. Remediation of uranium contaminated soils with bicarbonate extraction and microbial U (VI) reduction. *J. Indus. Microbiol.*, 14: 203-207.

Pietri, A., Bertuzzi, T., Pallaroni, L. and Piva, G. 2001. Occurrence of Ochratoxin A in Italian wines. *Food Addit Contam.*, 18: 647-654.

Pinner, R., Teutsch, S., Simonsen, L., Klug, L., Graber, J., Clarke, M. and Berkelman, R. 1996. Trends in infectious diseases mortality in the United States. *J. Am. Med. Assoc.*, 275: 189-193.

*Plotnikoff, G.A. 2005. *Poria cocos* (Schwein.) F.A. Wolf In: Japanese Traditional Herbal Medicine: Insights from Kampo Case Studies and Implications for Contemporary Research. *International Journal of Medicinal Mushrooms.*

*Powell, H.A., Kerby, N.W. and Rowell, P. 1990. *J. Chromotagra.*, 502: 201-207.

*Prabharan, D., Sumathi, M. and Subramanian, G. 1994. *Curr. Microbiol.*, 28: 315-320.

*Pray, L. 2002. Strange bedfellows in transplant drug therapy. *Scientist.* 16 (7): 36.

Pritchard, P.A. 1992. Studies on bread improving mechanisms of fungal α-amylase. *J. Biol. Edu.*, 26: 12-18.

Pur, S.C.,Amma, T., Khajuria, A., Gupta, A., Arora, R., Spiteller, M. and Oazi, G.N. 2007. Immuno modulatory activity of an extract of the novel fungal endophyte *Entrophospora infrequens* isolated from *Nothapodytes foetida* (Wight) Sleumer. *Acta Microbiologica et Immunologica Hungarica.* 54:237-262.

Purohit, S.S. 1997. Microbiology Fundamentals and Applications. 6th Edition. Agro Botanica Publishing Company, New Delhi, India.

*Quang, D.N., Hashimoto, T., Radulovic, N., Stadler, M. and Asakawa, Y. 2005. Antimicrobial Azaphilones from the Xylariaceous inedible mushroom. *International Journal of Medicinal mushrooms.*

Querol, A., Fernandez-Espinar, M.T. and Barrio, E. 2004. Genetic Variability of Yeast in Wine Fermentation. In: Fungal Biotechnology in Agricultural, Food and Environmental Applications. Arora, D.K. (Ed.) Vol. 21, Pp. 257-264. Marcel Dekker, Inc. New York.

Radehaus, P.M. and Schmidt, S.K. 1992. Characterization of a novel *Pseudomonas* sp. that mineralizes high concentration of pentachlorophenol. *Appl. Environ. Microbiol.*, 58: 2879-2885.

Rademaker, M. 2000. Lichens. In: Dermatologic Botany. (Eds.) Avalos, J. and Maibach, H.I. CRC Press, Washington DC.

Rafanelli, S., Bacchilega, R. and Stanganelli. 1995. Contact dermatitis from usnic acid in vaginal ovules (letter). *Contact Derm.,* 33: 271-272.

*Rahman, H., Rahman, S., Rahman, A. and Rahman, K. 2005. Dissemination of case studies on ganotherapy. *International Journal of Medicinal Mushrooms.*

Rai, R.D. 2004. Production of Edible Fungi. In: Fungal Biotechnology in Agricultural, Food and Environmental Applications. Arora, D.K. (Ed.). Vol. 21, Pp. 233-245. Marcel Dekker, Inc., New York

Rao, D.N. 1982. Responses of Bryophytes to Air Pollution. In: Bryophyte Ecology. Smith, A.J.E. (Ed.)., Pp. 445-471.Chapman and Hall, London and New York.

Ratner, P.R., Hampel, F.C., Amar, N.J., Van Bavel, J.H., Mohar, D., Marple, B.F., Roland, P.S., Wall, G.M., Brubaker, M.J., Drake, M., Turner, D. and Silver, L.H. 2005. Safety and efficacy of dopatadine hydrochloride nasal spray for the treatment of seasonal allergic rhinitis to mountain cedar. *Ann. Allergy Asthma Immunol.,* 95 (5): 474-479.

Raushel, F.M. 2002. *Curr. Opin. Microbiol.,* 5: 288-295.

Rawat,S. and Johri,B.N. 2004. Xylanases of Theermophilic Molds and Their Application Potential. In: Hand Book of Fungal Biotechnology. 2nd Edition. Arora, D.K.(Ed.) Vol.20, Pp 299-310. Marcel Dekker, Inc., New York.

Ray, P. and Liewen, M.B. 2004. Antifungal Food Additives. In: Fungal Biotechnology in Agricultural, Food and Environmental Applications. Arora, D.K. (Ed.). Vol. 21, Pp. 291-296.Marcel-Dekker, Inc., New York.

Razin, S. 1986. Mycoplasmal Adhesions and Lectins. In : Microbial Lectins and Agglutinins. Mirleman, D. (Ed.). John Wiley and Sons, Inc.,

Reddy, D.R. 1992. Application of Microbial Lipases in the Synthesis of Chiral Synthons. In: Industrial Biotechnology. Malik, V.S., Sridhar, P. (Eds.) Pp. 387-404. Oxford and IBH Publishing Co., New Delhi.

Resnick, S.M. and Chapman, P.J. 1994. Physiological properties and substrate specificity of a pentachlorophenol-degrading *Pseudomonas* species. *Biodegrad.*, 5: 47-54.

Richmond, A. 1990. In: Algal Biotechnology.. Round, F.E. and Chapman, D.J. (Eds.) Pp. 269-330 Biopress Ltd., Bristol, U.K.

Roja, G. 2006. Comparative studies on the camptothecin content from *Nothapodytes foetida* and *Ophiorrhiza* species. *Natural Product Research.* 20:85-88.

*Rose, A.H. 1981. Microbial Biodeterioration. Vol. 6, Economic Microbiology. Academic Press, London, U.K.

Rosenblatt, D.H., Burrows, E.P., Mitchell, W.R. and Parmer, D.L. 1991. In: The Handbook of Environmental Chemistry. Hutzinger, O. (Ed.). Springer-Verlag, Berlin-Heidelberg.

Rosenthal, L. 1943. J. Bacter., 45:545. In: Microbial Lectins and Agglutinins. Properties and Biological Activity. Mirelman, D. (Ed.) John Wiley and Sons, New york.

Rosser, A. 1997. Report on CITES, the tenth conference of the parties. *Species*, 29: 25-28.

Round, F.E. 1973. The Biology of the Algae. 2nd Edn. Edward Arnold Publishers Limited.

Rouse, J.D., Sabatini, D.A., Sulflita, J.M. and Harwell, J.H. 1994. Influence of surfactants on microbial degradation of organic compounds. *Crit. Rev. Environ. Sci. Technol.*, 24: 325-370.

Rumbaugh, E.R., John, A., Christiansen, Thomas, N. and Tremblay. 1997. The Use of Novel Microbial Organism to Perform Tracer Studies in Waste Water Treatment. In: Global Environmental Biotechnology. Wise, D.L. (Ed.) Elsevier Science, Netherlands.

*Sahu, J., Das, M.K. and Adhikary, S.P. 1992. *Trop. Agric.*, 69:362-364.

*Sakai, H. and Caldo, G.A. 1984. *Saccharomyces cerevisiae* in Philippine Basi (Sugarcane wine). *Philip Agric.*, 67: 183-189.

Salleh, A.B., Musani, R., Basri, M., Ampon, K., Yunus, W.M.Z. and Razak, C.N.A. 1993. Extra and intra-cellular lipases from thermophilic *Rhizopus oryzae* and factors affecting their production. *Can. J. Microbiol.*, 39: 978-981.

*Samson, R. and LeDuy, A. 1983. Influence of mechanical and thermochemical pretreatments on anaerobic digestion of *Spirulina maxima* algal biomass. *Biotechnol. Lett.*, 5: 651-682.

Samuel, R., Mani, U.V., Iyer, U.M. and Nayak, U.S. 2002. Hypocholesterolemic effect of *Spirulina* in patients with hyperlipidemic nephrotic syndrome. *J. Med. Food.*, 5 (2):91-96.

Samy and Radhaune. 1995. Role of lignin peroxidase and manganese peroxidase from *Phanerchaete chrysosporeum* in the decolourisation of olive mill waste water. *Appl. Environ. Microbiol.*, 61:1098-1103.

Sancholle, M., Larulle, F., Losel, D.M. and Muchembled, J. 2004. Biotechnological Potential of Fungal Lipids. In : Handbook of Fungal Biotechnology. 2nd Edition. Arora, D.K.(Ed). Vol 20, Pp 341-347. Marcel Dekker,Inc., New york.

Sanglier, J.J. 1992. Immunosuppressants Based on Microbial Leads. In: New Drugs from Natural Sources. Coombes, J.D. (Ed.) Pp. 63-89. IBC Technical Services, London.

Sarkar, P.K. 2005. Microbiology of Traditional Alkaline Fermented Foods In. Microbial Diversity: Current Perspectives and Potential Applications. Satyanarayana, T. and Johri, B. N. (Eds.). I.K International Pvt. Ltd., New Delhi.

Sarma, Y.S.R.K., Chandarana, R.B., Ramavat,B.K. and Tarwadi, S.J. 1984. Utilization of seaweed for liquid fertliser energy and poultry feed. In: Proceedings seminar on sun, sea and shore, Pp 90-91, Bhavnagar, India.

Satyanarayana, T. and Johri, B. N. (Eds.). 2005. Microbial Diversity. Current Perspectives and Potential Applications. I.K International Pvt. Ltd., New Delhi.

Sawada, T., Nakamura, Y., Kobayashi, F., Mohammed., Watanabe, T., Scheibner, K., Hfrichter, M., Herre, A., Michels, J. and Fritsche, W. 1997. *Appl. Microbiol. Biotechnol.*, 47: 452 -457.

Saxena, R.K., Davidson, W.S. and Gupta, R. 1998. Purification of microbial lipases. In: Fermentation Biotechnology – Industrial perspectives. Proc. Symp. Biotech. Industry.A challenge for 2005 A.D. Chand, S. and Jain, S.C. (Eds.) Dept Biotech., Government of India.

Saxena, R.K., Malhotra, B. and Batra, A. 2004. Commercial Importance of Some Fungal Enzymes. In: Handbook of Fungal Biotechnology. 2nd Edition. Arora, D.K. (Ed.) Vol. 20. Pp. 287-295. Marcel Dekker, Inc. New York.

Schaeffer, D.J. and Krylov, V.S. 2000. Anti-HIV activity of extracts and compounds from algae and cyanobacteria. *Ectoxicol Environ. Saf.*, 45 (3): 208-227.

Schippers, A., Von Rege, H. and Sand, W. 1996. Impact of microbial diversity and sulphur chemistry on safeguarding sulphidic mine waste. *Miner. Eng.*, 9: 1069-1079.

Schlatter, C., Struderrohr, J. and Rasonyi, R. 1996. Carcinogenicity and kinetic aspects of ochratoxin A. *Food Addit. Contam.*, (Suppl.) 13: 435-445.

Schliephake, K., Baker, W. L. and Lonergon, G. T. 2004. Decolourisation of Industrial Wastes and Degradation of Dye Water. In : Fungal Biotechnology, in Agricultural, Food and Environmental Applications. Arora, D. K. (Ed.). Vol, 21, Marcel Dekker Inc., New York.

Schneider, L.S., Dekosky, S.T., Farlow, M.R., Tariot, P.N., Hoerr, R. and Kieser, M. 2005. A randomized, double-blind, placebo controlled trial of two doses of *Ginkgo biloba* extract in dementia of the Alzheimer's type. *Curr. Alzheimer Res.*, 2 (5): 494-496.

*Schindler, J. and Schnid, R.D. 1982. Fragrance and aroma chemicals. Microbial synthesis and enzymatic transformation –a review. *Process. Biochem.*, 17:2-8.

Schorling, S.R., Kortinga, H.C., Froscyb, M. and Muhlschegel, F.A. 2000. The role of *Candida dubliniensis* in oral candidiasis. *Gut. Rev. Microbial.*, 26: 59-68.

Scrippa, P., Scambia, G., Masciul, V., 1999. Battaglia, F., Foti, E., Lopez, R., Villa, P., Malecore, M. and Mancuso, S. 1999. A zinc sulphate and usnic acid preparation used as post-surgical adjuvant therapy in genital lesion by human papilloma virus. *Minerva Ginecol.*, 51: 255-260.

Sebek, O.K. and Perlman 1979. Microbial transformation of steroids and sterols. In: Microbial technology. Peppler, H.J. and Perlman, D. (Eds.) Vol. I Academic Press, New York.

Seifert, P. and Berterm, C. 1995. Usnic acid. Natural preservation for lichens. *Seifen. Ole. Fertte. Wachse.*, 121: 480-485.

Serizawa, N. and Matsuoka, T. 1991. A two component type cytochrome P-450 monooxygenase system in a prokaryote that catalyzes hydroxylation of MC-236B to Pravastatin, a tissue selective inhibitor of 3-hydroxy-3'-methyl glutaryl coenzyme A reductase. *Biochem. Biophys. Acta.*, 1084: 35-40.

Shahbazi, A.M., Mims, M.R., Shirley, V., Ibrahim, S.A. and Morris, A. 2005. Lactic acid production from cheese whey by immobilized bacteria. *Appl. Biochem. Biotechnol.*, 122: 529-540.

Shahi, S.K. and Patra, M. 2003. Microbially synthesized bioactive nanoparticles and their formulation active against human pathogenic fungi. *Rev. Adv. Mater. Sci.*, 501-509.

Sharabi, Y. and Garber, G.N. 1979. *FEMS Lett.*, 5:273.

Sharabi, Y. and Garber, G.N. 1980. *J. Protozool.*, 27:80.

Sharon, N. 1977. Sci. Am., 236(6): 108-119.

Shibazaki, H., Kuga, S., Fumihiko, O. and Usuda, M. 1993. Bacterial cellulase. *J. Appl. Polym. Sci.*, 50: 965-969.

Shirahashi, H., Morimoto, T., Nagatsu, A., Murakami, N., Tatta, K., Sakakibasa, J., Tokuda, H. and Nishino, H. 1996. *Chem. Pharm. Bull.*, (Tokyo). 44: 1404-1406.

Shreshta, K., Strobel, G.A., Prakash, S. and Gewali, M. 2001. Evidence for paclitaxel from three new endophytic fungi of Himalayan yew of Nepal. *Planta Medica.* 67: 374-376.

Simpanya, M.F. 2000. In: Biology of Dermatophytes and Other Keratinophilic Fungi. Kushwaha, R.K.S. and Guaro, J. (Eds.) *Reuista Iberoamericana de Mycologia*, Spain, pp 1-12.

Singh, A. and Ward, O.P. 2005. Relevance of Microbial Diversity to Environmental Biotechnology. In: Microbial Diversity: Current Perspectives and Potential Applications, Satyanarayana, T. and Johri, B.N. (Eds.) I. K. International Pvt. Ltd. New Delhi.

Smith, G. M.. 1955. Cryptogamic Botany, Vol I and II, Mc Graw Hill Book Company, Inc., London.

*Smith, G.D., Ewart, G.D. and Tucker, W. 1992. *Int. Hydrogen Energy.*, 17: 695-698.

*Smith, C.D., Mooberry, S.L., Zang, X. and Helt, A.M. 1994. *Cancer Lett.*, 79: 213-219.

Smothers, F.J., Heinkoff, S. and Carter, P. 2002. *Science*, 298: 621-622.

Soni, S.K. and Sandhu, D.K. 1999. Fermented Cereal Products. In: Biotechnology: Food Fermentation, Microbiology, Biochemistry and Fermentation Technology. Joshi, V.K.and Pandey, A. (Eds.) Vol. 2, Pp. 896-949. Educational Publishers and Distributors. India.

Sodowski, M. 1998. First Ever IUCN Red List of Threatened Plants. *Species.* 30:22-25.

Spiller, G.A. and Dewell, A. 2003. Safety of an astaxanthin-rich *Haematococcus pluvialis* algal extract: a randomized clinical trial. *J. Med. Food.*, 6(1): 51-6.

*Spoerke, D.G. 1980. Herbal Medications.: Woodbridge Press. Santa Barbara .US.

Sporne, K.R. 1965. The Morphology of Gymnosperms. The Structure and Evolution of Primitive Seed-Plants. Hutchinson and Co. Ltd., London.

Sporne, K.R. 1974. The Morphology of Pteridophytes. (Indian edition). B. 9. publications, Madras, India.

Srivastava, J. P. 1962. An Introduction to Fungi. 2nd Edn. Central Book Depot, Allahabad, India.

*Stamets, P. and Plotnikoff, G.A. 2005. Anticancer medicinal mushrooms can provide significant vitamin D2 (Ergocalciferol). *International Journal of Medicinal mushrooms.*

*Stein, J.R. and Borden, C.A. 1984. Causative and beneficial algae in human disease conditions: a review, *Phycologia.* 23: 485-501.

Sternberg, M. 1976. Microbial rennets. *Adv. Appl. Microbiol.*, 20: 135-137.

Stierle, A. Strobel, G. and Stierle, D. 1993. Taxol and taxane production by *Taxomyces andreanae*, an endophytic fungus of Pacific Yew. *Science.* 260: 214-216.

Stradler, W.M., Kugel, T., Shapiro, C., Sosman, J., Clark, J. and Vogelzana, N.J. 1999. Multi-institutional study of the angiogenisis inhibitor. TNP-47 in metastatic renal carcinoma. *J. Clin. Oncol.*, 17: 2542.

Stratmann, K., Burgoyne, D.L., Moore, R.E. and Patterson, G.M.L. 1994. *J. Org. Chem.*, 59: 3779-3784.

Strobel, G.A. 2002. Microbial gifts from rain forests. *Can. J. Plant Path.*, 24: 14-20.

Strobel, G. A. and Daisy, B. 2003. Bioprospecting for Microbial Endophytes and Their Natural Products. *Microbiol. Mol. Boil. Rev.*, 67(4): 491-502.

Strobel, G.A., Stierle, A., Stierle, D. and Hess, W.M. 1993. *Taxomyces andreanae* a proposed new taxon for a bulbilliferous hyphomycete associated with Pacific Yew. *Mycotaxon.* 47: 71-78.

Strobel, G.A., Hess, W.M. and Li, J.Y.1997. *Pestalotiopsis guepinii*, a taxol producing endophyte of the Wollemi Pine, *Wollemia nobilis. Aust. J. Bot.*, 45: 1073-1082.

Strobel, G.A., Miller, R.V., Miller, C., Condron, M., Teplow, D.B. and Hess, W.M. 1999. Cryptocandin, a potent antimycotic from the endophytic fungus *Cryptosporiopsis quercina. Microbiology*. 145: 1919-1926.

Strobel, G.A., Ford, E. and Worapong, J. 2002. Isopoestacin, an isobenzofuranone from *Pestalotiopsis microspora*, possessing antifungal and antioxidant activities. *Phytochemistry*. 60: 179-183.

Strobel, G. A., Daisy, B. and Castillo, U. 2005. Novel Natural Products from Rainforest Endophytes. In: Natural Products: Drug Discovery and Therapeutic Medicine. Zhang, L. and Demain, A.L.(Eds.). Humana Press Inc., Totowa, New Jersey.

Strohl, W. 1997. Industrial Antibiotics: Today and The Future. In; Biotechnology of Antibiotics. Strohl, W. (Ed.). 1-47. 2nd Edition. Marcel Dekker, New York.

Su, Y.C. and Wang, W. 1977. Chinese red rice-Anka. Symposium on Indigenous Fermented Foods, Bangkok, Thailand.

Subba Rao. N.S. 1999. Soil Microbiology. Oxford and IBH Publishing Co. Pvt. Ltd., New Delhi, India.

Subba Reddy, P. and Retnakaran, A. 2003. Biological Control of Forest Pests; A Biotechnological Perspective. In: Forest Products Biotechnology. Bruce, A. and Bruce, A. Palfreyman, J.W. (Eds.) Taylor and Francis Publishers, London.

Sublette, K.L., Ganapathy, E.V. and Schwartz, S. 1992. Degradation of ammunitions wastes by *Phanaerochaetae chrysosporium. Appl. Biochem. Biotechnol.*, 34-35: 709-723.

Sudakevitz, D. and Garber, G.N. 1982. *Microbios.*, 34:159.

*Sudan, B.J.L. 1985. Seborrhoeic dermatitis induced by nicotine of horsetails (*Equisetum arvense* L.). *Contact Dermatitis*. 13(3): 201.

*Sugihara, T.F. and Humfeld, H. 1954. Submerged culture of various species of mushroom. *Appl. Microbiol.*, 2:170-172.

Sullia, S.B. and Shantharam, S. 2005. General Microbiology 2nd Edition. Oxford and IBH Publishing Co. Pvt. Ltd., New Delhi, India.

*Sumba, J.D. 2005. GACOCA Formulation of East African wild mushrooms show promise in combating Kaposi's sarcoma and HIV/AIDS. *International Journal of Medicinal Mushrooms.*

Sutherland, J.B. 2004. Degradation of Hydrocarbons by Yeasts and Filamentous Fungi. In: Fungal Biotecnology in Agricultural Food and Environmental Applications. Arora, D.K. (Ed.). Vol. 21, Pp. 443-451. Marcel Dekker, Inc., New York.

Tanaka, T., Nose, M., Endo, A., Fujii, T. and Taniguchi, M. 2003. *J.Biosci. Bioeng.*, 96: 541-546.

Taylor, A.J. and Mottram, D.S. (Eds.). 1996. In: Flavour Science. Recent Developments. Proc. Eighth Weurman Flavour Research Symposium, The Royal Society of Chemistry Pp. 371-376. Cambridge, UK.

Thayer, A.M. 2000. Busting down a block buster drug. *Chem. Eng. News.*, 78 (45): 20-21.

Thompson, M.F., Rachakonda Sarojini, R.and Nagabhushanam,R (Eds.) 1991. Bioactive Compounds from Marine Organisms With Emphasis on The Indian Ocean. An Indo-United States Symposium. Oxford & IBH Publishing Co. Pvt. Ltd., New Delhi, India.

Tiwari, P., Adholeya, A. and Prakash,A. 2004. Commercialization of Arbuscular Mycorrhizal Biofertilizer. In : Fungal Biotecnology in Agricultural Food and Environmental Applications. Arora, D.K. (Ed.). Vol. 21, Marcel Dekker, Inc., New York.

*Tomikel, J. 1973. Edible Wild Plants of Pennsylvania and New York. Althegheny Press. Pittsburgh.

Torsdottir, I., Alpsten, M., Holm, G., Sandberg, A.S. and Tolli, J. 1991. A small dose of soluble alginate fiber affects post prandial glycemia and gastric emptying in humans with diabetes. *J. Nutr.*, 121 (6): 795-799.

Tortora, G.J., Funke, B.R. and Case, C.L. 1998. Microbiology : An Introduction, 6th Edition. Addison Wesley Longman, Inc. California, USA.

Toumi, T., Reijula, K., Johnsson, T., Hemminki, K., Hintikka, E.l., Lindroos, O., Kalso, S., Koukila-Kahkola, P., Mussalo-Rauhamaa, H. and Haahtela, T. 1993. Quantitative analysis of aflatoxin albumin adducts. *Carcinogenesis.* 14: 1203-1208.

Trimurtulu, G., Ohtani, I., Patterson, G.M.L., Moore, R.E., Corbett, T.H., Vareriote, F.A. and Demchick, L. 1994. *J. Am. Chem. Soc.*, 116: 4729-4737.

Tripura, C.B., Sashidhar, B. and Podile, A.R. 2005. Transgenic Mineral Phosphate Solubilizing Bacteria for Improved Agricultural Productivity. In: Microbial Diversity: Current Perspectives and Potential Applications. Satyanarayana, T and Johri, B.N. (Eds.). I. K. International Pvt. Ltd. New Delhi.

Trivedi, P.C.. 2001. Algal Biotechnology. Pointer Publishers. Jaipur, India.

Trivedi, P.C. 2002. Ethnobotany. Aavishkar Publishers, Distributors, Jaipur, India.

Truscheit, E., Frommer, W., Junge, B., Muller, L., Schmidt, D.D. and Wingender, W. 1981. Chemistry and biochemistry of microbial α-glucosidase inhibitors. *Angew. Chem. Intl. Ed. Engl.*, 20: 744-761.

*Tsukagoshi, S., Hashimoto, Y., Fujii, G., Kobayashi, H., Nomoto, K., Orita, K. and Krestin (PSK). 1984. *Cancer Treat Rev.*, 11: 131-155.

UCK-Hanchun, Simonov, N., Chen, Y. and Britz, M.L. 1996. Continuous Pollution Monitoring using *Photobacterium phosphoreum*. In: Global Environmental Biotechnology. Wise, D.L. (Ed.) Elsevier Science, Netherlands.

*Umezawa, H. 1972. Enzyme Inhibitors of Microbial Origin. University Park, Baltimore, USA.

*Umezawa, H. 1982. Low-molecular-weight inhibitors of microbial origin. *Ann. Rev. Microbiol.,* 36: 75-99.

Urizar, N.L., Liverman, A.B., Dodds, D.T., *et al.* 2002. A natural product that lowers cholesterol as an antagonist ligand for FXR. *Science.* 296: 1703-1706.

Vartio, K.O. 1973. Antibiotics in Lichens. In: The Lichens. Ahmadjian, V.and Hale, M.E. (Eds.) Pp. 547-561. Academic Press, New York.

Vasdev, K., Kuhad, R.C. and Saxena, R.K. 1995. *Curr. Microbiol.*, 30: 1-4.

Vasishta, P. C. 1971. Pteridophyta. Vascular Cryptogams. S.Chand and Co. Publishers, India.

Vasishta, P.C. 1996. Gymnosperms. Vol V, S. Chand and Co. Ltd., New Delhi, India.

Vasishta, P.C., Sinha, A.K. and Kumar, A. 2005. Gymnosperms. S. Chand and Co. Ltd., New Delhi, India.

Vega, A. and Bell, E.A. 1967. α-Amino-α-methyl aminopropionic acid from seeds of *Cycas circinalis*. *Phytochemistry*, 6: 759.

Ventosa, A., Nieto, J.J. and Oren, A. 1998. *Microbiol. Mol. Boil. Rev.*, 62:504-514.

*Vepritskii, A.A., Gromov, B.V. and Kononova, S.K. 1992. *Algologia*. 2: 16-19.

Verdine, G. 1996. The Combinatorial chemistry of nature. *Nature*. 384 (Supp): 11-13.

*Vezina, D., Kudelski, A. and Sehgal, S.N. Rapamycin (AY22, 989), A new anti fungal antibiotic. 1. Taxonomy of the producing streptomycete and isolation of the active principle. *J. Antibiot*., 28: 721-726.

Vijayakumar, C.S., Viswanthan, S., Kannapa-Reddy, M., Parvathavarthini, S., Kundu, S.B. and Sukumar, E. 2000. Anti-inflammatory activity of (+) usnic acid. *Fitoterapia*. 71: 564-566.

Vineesh, V.R., Fijesh, P.V., Jelly Lpuis, C., Jaimsha, V.K. and Padikkala, J. 2007. In vitro production of camptothecin (an anticancer drug) through albino plants of *Ophiorrhiza rugosa*.var.*decumbens*. *Current Science*. 92:1216-1217.

Vining, L.C. and Taber, W.A. 1979. Ergot Alkaloids. In: Economic Microbiology. Secondary Products of Metabolism, Rose, A.H. (Ed.).Vol. E. Academic, London. Pp. 389-420.

Viriani, R. 1992. Eutrophication, marine biotoxins, human health. *Sci. Tot. Environ. Suppl*., 631.

Viswanath, C.R. (Ed.) 1961. Handbook of Agriculture. 5th Edition. Indian Council of Agricultural Research, New Delhi, India.

Wagenaar, M., Corwin, J., Strobel, G.A. and Clardy, J. 2000. Three new chytochalasins produced by an endophytic fungus in the genus *Rhinocladiella*. *J. Nat. Prod*., 63: 1692-1695.

Walker, G.M. and White, N.A. 2005. Introduction to Fungal Physiology. In: Fungi, Biology and Applications. Kavanagh, K. (Ed.) John Wiley and Sons Ltd., U.K.

Wall, M.E. and Wani, M.C. 1995. Camptothecin and taxol: discovery to clinic. *Cancer Res*., 55: 753-760.

Wang, J., Li, G., Lu, H., Zheng, Z., Huang, Y. and Su, W. 2000. Taxol from *Tubercularia* sp. strain TF5, an endophytic fungus of *Taxus mairei*. *FEMS Microbiol Lett*., 193: 249-253.

Watanabe, I. 1973. Isolation of pentachlorophenol decomposing bacteria from soil. *Soil Sci. Plant Nutrit*., 19: 109-116.

Watson, J.H.P., Ellwood, D.C. and Duggleby, C.J. 1996. A chemostat with magnetic feed back for the growth of sulphate reducing bacteria and its application to the removal and recovery of heavy metals from solution. *Min. Eng.*, 9: 937-983.

Webster. 1970. Introduction to Fungi. Cambridge University Press, Cambridge.

*Weibel, E.K., Hadvary, P., Hochuli, E., Kupfer, E. and Lengsfeld, H. 1987. Lipostatin, an inhibitor of pancreatic lipase, produced by *Streptomyces toxytricini. J. Antibiot.*, 40: 1081-1085.

Weise, L. and Shoemaker, D. W. 1970. *Biol. Bull.*, 138:88.

Weiss, R.E. 1988. Herbal Medicine. Gothenburg, Sweden.

White, D. and Gand Brown, R.M. 1989. Prospects for the Commercialization of the Biosynthesis of Microbial Cellulose. In: Cellulose and Wood Chemistry and Technology. Schuerech, C. (Ed.). Wiley, New York. Pp573.

Wilson, B.J., Wilson, C.H. and Hayer, A.W. *Nature* (London), 220: 77.

Wilson, E.O. 1988. The Current State of Biological Diversity. In Biodiversity Wilson, E. O. (Ed.) National Academy, Washington, DC.

Winarno, F.G., Muchtadi, D., Laksani, B.S., Rahman, A., Swastomo, W., Zainuddin, D. and Santaso, S.N. 1977. Indonesian Tauco. Symposium on Indigenous Fermented Foods, Bangkok, Thailand.

Winkelmann, G. 1986. Iron Complex Products (Siderophores). In: Biotechnology. Rehm, H.J. and Reed, G. (Eds.) Vol. 4, 215-243. Weinheim Publishers.

Wolf, K. 1996. Non conventional Yeasts in Biotechnology. Springer-Verlag, Berlin.

Won, W.D., Di Salvo, L.H. and Ng, J. 1976. *Appl, Environ. Microbiol.*, 31:575 -580.

Wouter, J.T.M., Ayad, E.H.E., Hugenholtz, J. and Smit, G. 2002. Microbes from raw milk for fermented dairy products. *Int Dairy J.*, 12: 91-109.

Wu, J., Zhang, M., Ding, D., Tan, T. and Yan, B. 1995. Effect of *Cladonia alpestris* on *Trichomonas in vitro. Chinese J. Parasit. Dis.*, 13: 126-129.

Wyder, M. 1998. Identification and characterization of the yeast flora in kefyr and smear ripened cheese contribution of selected yeasts to cheese ripening. Ph.D Dissertation Swiss Federal Institute of Technology. Zurich, Switzerland.

Xavier, L.J.C. and Boyetchko, S.H. 2004. Arbuscular Mycorhizal Fungi in Plant Disease Control. In: Fungal Biotechnology in Agricultural, Food and Environmental Applications. Arora, D.K. (Ed.). Vol, 21. Marcel Dekker Inc., New york.

Xiaoquiang, W.V., Vennison, J., Huirong, L., Ben-dov, E., Zaritsky, A. and Boussiba, S. 1997. *Appl. Environ. Microbiol.,* 63: 4971-4975.

Yamamoto, Y., Miura, Y., Kinoshita, Y., Higuchi, M., Yamada, Y., Murakami, A., Ohigashi, H. and Koshimizu, K. 1995. Screening of tissue culture and thalli of lichens and some of their active constituents for inhibitions of tumor promoter-induced Epstein-Barr virus activation. *Chem. Pharm. Bull.,* 43 (8): 1388-90.

Yamanaka, S., Watanabe, K. and Kitamura, N. 1989. The structure and mechanical properties of sheets prepared from bacterial cellulase. *J. Mat. Sci.,* 24: 3141-3145.

Yamanaka, S. and Watanabe, K. 1994. Applications of Bacterial Cellulase. Chapter 11. In: Cellulosic Polymers. Gilbert, R. (Ed.) . Hanser, Cincinati, OH.

*Yeoh, Q.L. and Merican, Z. 1977. Malaysian Tempeh. Symposium on Indigenous Fermented Foods, Bangkok, Thailand.

Yokotsuka, T. 1977. Japanese Shoyu. Symposium on Indigenous Fermented Foods, Bangkok, Thailand.

Young, D.H., Michelotti, E.J., Sivendell, C.S. and Krauss, N.E. 1992. Antifungal properties of taxol and various analogues. *Experientia.* 48: 882-885.

Youssef, S., Stuve, O. and Patarroyo, J.C. 2002. The HMG-CoA reductase inhibitor, atorvastatin, promotes a Th2 bias and reverses paralysis in central nervous system autoimmune disease. *Nature.* 420(7)

Yu, J. and Ehrlich, K.C. 2002. Toxins of Filamentous Fungi. In: Fungal Allergy and Pathogenicity. Breitenbach, M., Cremeri, R.and Lehrer, S.B. (Eds.). 81 Pp. 167-206.. *Chem Immunol.,* Basel, Switzerland

*Z'Brun, A. 1995. Schweiz Rundsch. *Med. Prax.,* 84(1): 1-6.

Zadvorny, O.A., Zorin, N.A., Gogotov, I.N. and Gorlenko, V.M. 2004. *Biochemistry*, 69: 164-169.

Zaks, A. 1991. Enzymes in Organic Solvents. In: Biocatalysts for Industry Dordick, J.S. (Ed.). Pp. 161-180.Plenum Press. New York.

*Zeng, X., Liu, M., Yang, Y., Li, Y. and Asplund, K. 2005. *Ginkgo biloba* for acute ischaemic stroke. *Cochrane Database Syst. Rev.,* 19(4.

Zhang, B., Salituro, G., Szalkowski, D. 1999. Discovery of small molecule insulin mimetic with antidiabetic activity in mice. *Science.* 284: 924-981.

Zhang, L. and Demain, A.L. (Eds.). 2005. Natural Products: Drug Discovery and Therapeutic Medicine. Humana Press, Inc., Totowa.New Jersey.

Zhou, W., Chai, H., Lin, P.H., Lumsden, A.B., Yao, Q. and Chen, C. 2004. Clinical use and molecular mechanisms of action of extract of *Ginkgo biloba* leaves in cardiovascular diseases. *Cardiovasc. Drug. Rev.,* 22(4): 309-319.

Zou, W.X., Meng, J.C., Lu, H 2000. Metabolites of *Collectotrichum gloeosporioides,* an endophytic fungus in *Artemisia mongolica. J. Nat. Prod.,* 63: 529-1530.

* Original not seen

INDEX
Scientific Names of Organisms

A

B

C

F

G

H

I

J

K

L

M

N

O

P

S

T

U

V

W

X

Y

Z

GENERAL INDEX

C

D

E

F

O

P

R

S

T

V

W